Springer Texts in Business and Economics

Springer Texts in Business and Economics (STBE) delivers high-quality instructional content for undergraduates and graduates in all areas of Business/Management Science and Economics. The series is comprised of self-contained books with a broad and comprehensive coverage that are suitable for class as well as for individual self-study. All texts are authored by established experts in their fields and offer a solid methodological background, often accompanied by problems and exercises.

Gerhard Larcher

The Art of Quantitative Finance Vol.1

Trading, Derivatives and Basic Concepts

Springer

Gerhard Larcher
Institute for Financial Mathematics
and Applied Number Theory
Johannes Kepler University of Linz
Linz, Austria

ISSN 2192-4333 ISSN 2192-4341 (electronic)
Springer Texts in Business and Economics
ISBN 978-3-031-23872-7 ISBN 978-3-031-23873-4 (eBook)
https://doi.org/10.1007/978-3-031-23873-4

This Springer imprint is published by the registered company Springer Nature Switzerland AG
The registered company address is: Gewerbestrasse 11, 6330 Cham, Switzerland

Foreword

"Quantitative finance" is a broad term!

Yet regardless of how we define the term in these books, one thing is certain: The in-depth study of the fascinating theoretical concepts of quantitative finance and their translation into practice, into ways to generate, analyze and execute intelligent trading strategies, or calculate complex financial products, is a highly challenging and absorbing activity that can quickly become a passion, or even a (positive!) addiction.

When applying judicious and highly (!) creative mathematical thinking to sophisticated alternative trading strategies and refining the requisite methods, we sometimes need to embrace a playful, even "artistic" approach. Practical problems and the theories arising from them often come together in perfect and compact complexes of great beauty (you will agree with me later when we discuss the principles of the Black-Scholes theory or Markowitz's portfolio selection theory, the interdependencies between price developments of stock indices and their implicit volatilities, or the solution to the optimal-hedging problem in Volume III Section 3.11).

You will hardly find another field where theory and practical application interpenetrate so intensely as in the field of quantitative finance. And yet the immense impact that sophisticated mathematical methods and algorithms can have on the measurement of risks and financial products, on the analysis of trading strategies, or on the design of optimal investment portfolios is still frequently underestimated and not sufficiently considered in decision-making processes.

Perhaps one reason (among several) for this diagnosis is the following: There is undoubtedly a great deal of excellent literature on modern financial mathematics; however, most of it is aimed at readers with a (very strong) background in mathematics. Often this literature deals with practical applications only superficially, or limits discussions to highly specific details of such applications.

Conversely, there is also excellent literature that focuses primarily on practical application aspects. Yet this type of literature often comes up short on mathematical relevance and quality. Mathematical methods are sometimes omitted or it is simply assumed that the reader is familiar with them.

Hence, there seems to be a lack of good literature that offers both: a readily understandable introduction to the fundamentally relevant methods of financial mathematics *and* their immediate translation into practical use.

And vice versa: a lack of literature that addresses problems arising directly from real-world practice and how we can use financial mathematical methods to solve them.

Both in my work at the university and in my practical work in the financial industry, especially in fund management, I have repeatedly met users who were eager to obtain more profound but easily understandable information about quantitative methods and gain more confidence in using these methods and assessing their relevance and, on the other hand, financial mathematicians (very often highly interested and talented students) who were eager to get competent information about the realities of the financial markets and about what they can *really* achieve when putting the acquired financial mathematical knowledge to use in the real world.

The goal I am attempting to achieve with these books is to give readers exactly that: books that are easy to read, at the same time captivating and, ideally, highly rewarding for both users interested in financial mathematics and mathematicians interested in the financial markets and their dynamics.

In a nutshell: The aim of these books is to give readers a sound and universally accessible introduction to the fundamental concepts and techniques of quantitative finance, i.e.—and this is my definition of the term—to the fundamental concepts of financial mathematics and financial engineering.

I am convinced that despite its introductory character, these books will also allow professional traders and financial mathematicians to discover new and surprising details.

A key element for achieving the stated goals are the case studies presented and discussed throughout the books, included in a more condensed manner in Volume III Chapter 3. They all deal with problems—sometimes presented in a simplified form—that have arisen in actual projects I have handled and where the solutions—also sometimes presented in a simplified form—are developed based on the methods formulated in these books. All of these case studies have to do with the valuation of financial products, the development of financial software, and the design and analysis of derivative trading strategies.

In addition, the idea is that, by reading and working with the books, readers gain the requisite knowledge and skills for concrete action when trading in financial products, and derivatives in particular.

Given my many years of practical experience, I am convinced that competent, disciplined work with derivatives can lead to significant and sustainable investment gains. I also want to show readers that trading in derivatives is not just a gamble, but actually requires a considerable portion of skill.

Apart from basic mathematical techniques at university-entrance level and basic general knowledge of financial markets and financial products, no further knowledge is required of the reader. By reading the books and exploring the techniques that we present, readers will learn to understand analyses and valuations of complex financial products and trading strategies and will be able to execute them on their own, at least to the extent that this is possible using basic models. In addition, after reading the books, they will be able to work their way through other in-depth literature on financial mathematical techniques and more advanced modeling and valuation methods.

The books are therefore aimed primarily at the following audiences: practitioners in the banking, insurance, or asset management sectors as well as advanced asset advisors with an interest in quantitative methods; expert appraisers in the field of derivative financial products and derivative trading strategies; lawyers dealing with legal proceedings in the field of financial markets and asset management; students of financial mathematics as introductory reading; students of economics and finance; and interested private and institutional investors who wish to learn financial mathematical techniques.

The books are purposely conceived as readily understandable introductory textbooks, allowing maximum learning efficiency even for non-mathematicians. With this in mind, we have generally taken a more heuristic and intuitive approach in discussing the various concepts rather than focusing on mathematical precision in every detail.

Strengthening intuition is something I care deeply about. The multitude of charts and illustrations throughout the books are meant to contribute to that.

Furthermore, as the books focus on providing an *introduction* to basic techniques of financial engineering, it mostly works with basic mathematical models, referring readers interested in more subtle financial models to other sources. The books are therefore less suited for advanced financial mathematicians or researchers in financial mathematics.

The aim of the books is not to strive for completeness, but to impart a basic technical understanding of financial products, their analysis and use, and the role of mathematics in the financial markets.

The focus of these books is on the use of derivative financial products and trading strategies as well as on the use of exact quantitative, mathematical methods as they relate to finance. For the algorithmic and numerical range, I strongly advocate and use the Monte Carlo simulation method. Readers looking for technical chart analyses or fundamental analyses will do so in vain.

At the risk of repeating myself: Despite the occasionally highly complex contents, one of my main objectives is to provide you with books that are easy to follow and comprehend. In writing these books, I sometimes chose—at times perhaps unnecessarily—to provide extra explanations, to present things very simply, even simplistically, omitting details that would distract more than they would be helpful for understanding, rather than sacrificing comprehensibility by being overly concise, elegant, elaborate, complicated, or painstakingly exact in my argumentation.

I will have reached my goal if you feel inspired and intrigued by what you read in these books, following the logical train of thought that I have endeavored to give it.

The contents of the books are based, on the one hand, on a series of lectures and seminars (in the fields of financial mathematics, stochastic financial mathematics, mathematical models in economics, mathematical simulation, and financial market statistics) given by the author as Head of the Institute of Financial Mathematics and Applied Number Theory at the University of Linz, Austria, as well as a large number of seminars and lectures given by the author in various capacities for practitioners and investors, and, on the other hand, on the author's many years of work as a court-appointed expert in the field of derivative products and trading strategies and, above all, on the author's many years of work as a developer and active user of derivative trading strategies in his own asset management company.

The books explore a series of case studies, directly illustrating the application of the developed theory in practical use.

The text is complemented by occasional excursions into a variety of other—especially mathematical—topics, mostly in the fields of probability theory, game theory and applied number theory, but also from the author's practical work. All in all, I strive for comprehensibility of the content even for mathematically less experienced readers.

For these books, my team at the Linz School of Quantitative Finance (LSQF) has created **comprehensive and flexible open source financial software**, which is regularly being further refined and expanded, and which readers can use online at http://www.lsqf.org/. Using this software, readers can execute their own simulations and tests of various trading or hedging strategies, in addition to basic valuations and analyses.

Wherever applicable, the text will point to the software and explain its use. Furthermore, readers will also find up-to-date information, additions, and new developments at this URL, and we strongly encourage readers to use the software and information offered on the site.

Finally, just a couple of **technical details**:

For shorter passages of the books, I had co-authors from my team at the Linz School of Quantitative Finance, and some passages were written completely by one of these colleagues. Wherever that is the case, I explicitly point this out, of course.

Some sections in these books are shown with a light gray background. This is the case when (sometimes more advanced) mathematical derivations are made. Readers less interested in mathematical details can skip these parts, as they are not necessary for understanding the sections that follow.

Occasional excursions into other specialty fields or of a historic or anecdotal nature are shown with a green background.

In all of the following, the "euro = EUR" has been defined as the "home currency." Readers in regions with a different home currency should always replace EUR by their home currency.

I do not want to overwhelm you with an abundance of literature (whether references to books and articles or links to URLs) as it would be impossible to follow up on all of them. You will therefore find—perhaps surprisingly—few references to further literature. My intent is to maintain a clear focus by limiting information strictly to what is relevant in each instance. Most of the cited literature will provide ample further references anyway.

Acknowledgements

The extensive and demanding work on these books has been greatly supported by the following organizations, to which I would like to express my gratitude:

- Johannes Kepler University Linz
- FWF Austrian Science Fund
- Provincial Government of Upper Austria
- LSQF Linz School of Quantitative Finance

In addition to all those who worked directly on creating the books, with great dedication and motivation, above all my assistant Melanie Traxler, who diligently transferred the entire text including a large number of pictures into LaTeX, the translator, Nina Sattler-Hovdar, who with exceptional expertise, diligence and perseverance carried out the extensive and difficult task of translating this monograph from German to English, my editor, Rocio Torragrosa from Springer Verlag, for accompanying the publication of the book in such a pleasant and efficient way, as well as my colleagues from the Linz School of Quantitative Finance, Lucia Del Chicca, Louisa Hofmann, Alexander Brunhuemer, Bernhard Heinzelreiter, Lukas Larcher, and Lukas Woegerer, I want to especially thank the people closest to me who accompanied me at a very personal level throughout the process of writing these books.

Preface to Volume I

We have divided the whole text into three parts:

- Volume I: Trading, Derivatives and Basic Concepts
- Volume II: Volatilities, Stochastic Analysis and Valuation Tools
- Volume III: Risk, Optimal Portfolios and Case Studies

In this first part, we introduce to the properties of basic financial products and financial derivatives, to basic trading strategies, and to the basic techniques of modelling in finance and of derivative pricing. The main goal will be the derivation of the famous Black-Scholes formula and its application to the analysis of trading strategies.

Linz, Austria   Gerhard Larcher

About This Book

This book was translated by Nina Sattler-Hovdar, based on the highly successful German book “Quantitative Finance. Strategien, Investments, Analysen” by Springer Gabler (2020)

Contents

Basic Products and Interest Calculations 1

Abstract

We give an introduction into basic financial products—like bonds, stocks, and stock indices—and into their basic properties. Specifically, we will study the S&P500 stock index in more detail. Further, we provide the basics of discrete and continuous compounding and we give an overview of the most important interest rate and swap rate benchmarks.

Keywords

Bonds · Interest compounding · Interest rates · Swap rates · Stocks · Stock indices · S&P500-index

We assume that you, as a reader of this book, have an interest in the financial markets, in financial products and their characteristics, and in how they are used and traded. Most of you will therefore not need the following introduction to the main characteristics of basic financial products. For the sake of completeness, however, I open this book by providing an overview of the definitions of terms that we will use in everything that follows. In addition, it will be necessary to briefly discuss some fundamental facts concerning interest rates and interest calculations.

In general, we distinguish between risk-free and risk-bearing financial products. Yet drawing a clear line between the two is not as easy as it may seem at first.

When asked about risk-free financial products, most people will probably mention:

- variable-rate savings accounts (demand deposit accounts)
- fixed-rate savings accounts (term deposit accounts)
- government bonds
- cash in the home currency

G. Larcher, *The Art of Quantitative Finance Vol.1*, Springer Texts in Business and Economics, https://doi.org/10.1007/978-3-031-23873-4_1

- fixed-interest loans
- floating-rate loans
- maybe corporate bonds
- ...

Standard examples of risk-bearing financial products include:

- stocks
- junk bonds
- gold
- cash in foreign currency
- convertible bonds
- futures
- options
- swaps
- credit derivatives
- ...

The last five of the above products are examples of derivatives. These are products that *derive their value* from other financial products (more detailed explanations will follow later). For this reason, we do not categorize them as basic financial products.

As regards risk, to what extent should the financial products listed in the first group actually be considered risk-free? Doesn't each of them carry risk? Take the variable-rate savings account: the future development of the applicable interest rate is uncertain. The safety of a fixed-interest savings account hinges on the continued operation and repayment capacity of the bank in question. Government bonds have long ago ceased to be a safe haven, and they are subject to price fluctuations that become relevant if you want to sell them prior to maturity. Cash can be stolen and is subject to depreciation through inflation.

This means that, strictly speaking, no financial product is actually risk-free. For purposes of this book, we therefore define "risk-free" in the context of a financial product in the following—purely technical—sense (although we are well aware that even that definition won't always allow a clear-cut classification):

We classify a financial product as risk-free if it is set up such that the holder of that product can expect clearly defined cash flows in the home currency for a specific period of time.

As noted in the introduction, for the sake of simplicity, we have defined the euro (EUR) as our "home currency". Readers in regions with a different home currency should always replace EUR by their actual home currency.

The above definition of risk-free is independent of whether cash flows actually occur as expected or are prevented, for example, by adverse circumstances (insolvency of the bond issuer, robbery, bankruptcy of the bank, etc.).

Note that the above-mentioned "specific period" is not necessarily a fixed period quantified in advance; it can also be a period given in qualitative terms. For instance, a savings account that accrues fixed interest until the issuing bank applies the next interest rate adjustment (where we assume that the bank informs account holders before the rate is adjusted, giving them the opportunity to take measures before the change takes effect) is a risk-free financial product in this sense. Similarly, based on this definition, every bond is a risk-free financial product, since under ideal circumstances, it provides (as explained in the next section) fixed coupon payments until maturity as well as repayment of a fixed nominal amount at specific predefined times. In this regard, any potential price changes within that period of time are irrelevant.

1.1 Basic Properties of Bonds

The key *defining parameters* of a bond are:

The **issuer** of the bond (e.g. a government, a municipality, a company, etc.)
the **term (life of a bond)** (e.g. 1 March 2015 to 1 March 2030, etc.)
The **nominal value** (also called face or par value) (e.g. EUR 100 million, etc.)
the **coupon** (e.g. 3% per annum, etc.)
the **coupon pay dates** (e.g. on the 1st of March of every year from 2016 to 2030, etc.)
the **currency** in which the bond is denominated (e.g. EUR, USD, etc.)
the **denomination (smallest tradable unit)** (e.g. 1000; 50,000, etc.)

a *variable* parameter is

the **price** of the bond (e.g. 103.50, etc.)

By purchasing a unit of that bond at the prevailing price, an investor grants the issuer a loan equal to the purchased unit's nominal value for the specified period until the bond's maturity. In return, the purchaser receives a coupon payment from the issuer on each coupon pay date in the amount of the per annum interest rate that has been defined as the coupon for that bond, as well as repayment of the bond unit's nominal value at maturity.

However, that description is very crude and too imprecise. The following explanation is more precise, and we will also provide an illustrative example:

By purchasing a unit of the bond at the **initial issue price** (expressed as a percentage of the unit's nominal value), an investor grants the issuer a loan equal to the purchase price for the period of time until the bond's maturity.

In return, the purchaser of that unit receives a coupon payment from the issuer on each coupon pay date—for as long as the purchaser holds that bond—in the amount of the per annum interest rate that has been defined as the coupon for that bond, as well as repayment of the bond unit's nominal value at maturity.

The total lot size (number of units) of the bond is given as follows:

Total lot size = nominal value / denomination

Throughout the term until maturity, the bond can be traded between investors and bond holders through a bond exchange or a bank. In such cases, the purchase price is determined by the prevailing price on that exchange (expressed as a percentage of the bond unit's nominal value) plus the respective pro rata coupon.

1.2 Example of a Bond

We want to illustrate this description with a bond traded on the Stuttgart Stock Exchange. The bonds that are currently traded on this stock exchange (12 May 2016 in our example) can be found on:

www.boerse-stuttgart.de
→ Securities & Markets → Bonds → Bond Finder

On this site you can search for bonds traded on the Stuttgart Stock Exchange according to various criteria.

To illustrate this with an example, let us have a look at the Volkswagen bond shown on

https://www.boerse-stuttgart.de/de-de/produkte/anleihen/stuttgart/a1zutm-volkswagen-intl-finance-nv-eo-medium-term-notes-201530

together with all the relevant data and price movements since the bond was issued.

VOLKSWAGEN INTL FINANCE N.V. EO-MEDIUM-TERM NOTES 2015.
WKN A1ZUTM | ISIN XS1167667283

WKN A1ZUTM
ISIN XS1167667283
Securities Type Corporate Bond
Subtype Technology

Issuer Volkswagen International Finance N.V.
S&P Rating BBB+

Trading Hours 8 am to 6 pm

Interest rate 1.625%
Interest accruing as from 16 January 2015
Next interest payment 16 January

Issue volume 1000 million
Smallest tradable unit 1000.00
Maturity 16 January 2030

Bond price data
Exchange Stuttgart

	Bid	Ask
Quote	108.57	108.95

Quantity 500,000 nom. 1,000,000 nom.
Quotation time 22 September 2021 1:14:30 pm

Daily high/low (Bid) 108.61/108.54
Change from previous day (Bid) -0.10/-0.10%

Last/Yield 108.61/0.48
Quotation time 22 September 2021 1:14:30 pm

Daily volume (nominal) 70,000
Previous day's price (21 September) 108.71
Annual high/low 95.74 (9 May) 83.99 (21 Jan)
52-week high/low 111.08 (4 Aug) 104.38 (25 Sept)

Accrued interest on the nominal amount 1.108%
Currency/Quotation Euro/Percent

Bond callable by issuer No
Bond is subordinated No

The development of the considered VW bond is given in Fig. 1.1:

Let us briefly explain the bond's main data, as they aren't all self-evident:

WKN is short for **Wertpapierkennnummer** and is the bond's German securities identification number.

ISIN is short for **International Securities Identification Number**; it is a 12-digit combination of letters and numbers and represents a security's unique international identification.

The **interest rate** is the bond's **coupon**, expressed as a percentage per annum (p. a.).

Interest accruing as from tells us that the life of the bond (its term) began on the specified date.

Next interest payment: For bonds in the euro area, the common practice is annual coupon payments, and an annual coupon payment can be assumed if, as in this example, no other information is provided. In the US region, coupon payments are customarily made on a semi-annual basis. So, with a coupon of, for example, 3% p.a., the issuer would make semi-annual payments of 1.5%.

The **issue volume** is the bond's **nominal value** (also called principal, face value, or par value).

The **smallest tradable unit** is the bond's **denomination**.

Maturity means "the end of the bond's **term**".

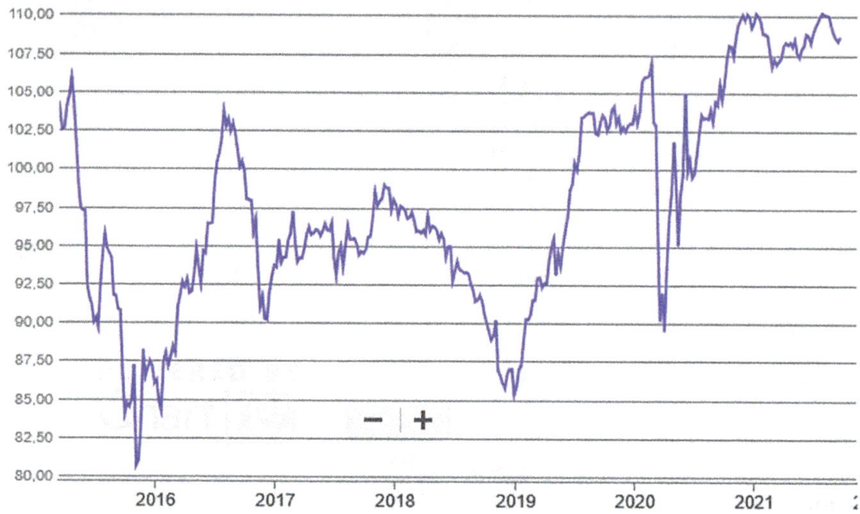

Fig. 1.1 Price history of VW bond (source: Stuttgart Stock Exchange)

Quote and **quantity** for **Bid** and **Ask** tell us that, at the **quotation time**, there was a demand ("Bid") for the purchase of the bond in the nominal amount of EUR 500,000 (i.e. of 500 units) at the price of 108.57 and that, at the same "quotation time", there was an offer ("Ask") for selling the bond in the nominal amount of EUR 1,000,000 (i.e. of 1000 units) at the price of 108.95.

Last along with **price determination time** tells us that this bond was last traded at the stated price determination time at a **price** of 108.61.

The **yield** is an aspect that we will discuss in detail in one of the following sections. Put simply, the yield tells you at which interest rate a fixed-rate savings account would have to bear interest in order for it to generate the same cash flows as the bond does.

The **accrued interest on the nominal amount** is the pro rata coupon amount that has accrued since the last coupon payment.

In our example, the last coupon payment was made on 16 January 2021. The present data were recorded on 22 September 2021, i.e. 249 days later. The pro rata coupon that the bond holder is entitled to as a result of having held the bond since the last coupon payment is

$$\text{pro rata coupon} = \frac{249}{365} \times \text{coupon} = \frac{249}{365} \times 1.625\% = 1.108\%.$$

If the bond is traded between investors before the bond's maturity, the **actual purchase price** (expressed as a percentage of the bond unit's nominal value) is given by the **current price of the bond plus the current pro-rata coupon**.

This means that the last trade of this bond took place at the actual price of $108.61 + 1.108 = 109.718$. Hence, the purchaser paid EUR 1097.18 for one unit of the bond.

An occasionally used terminology refers to the price as the **clean price** and the actual purchase price as the **dirty price**.

The **bond is subordinated** No: Whether or not a bond is subordinated is relevant if the issuer becomes (partially) insolvent. In that case the holders of non-subordinated bonds will be the first to receive payments from the issuer's remaining funds and only then—if funds are still available—will holders of subordinated bonds get any payouts.

The **chart** shows the price history of the bond—i.e. the development of the prices at which the bond was traded—from its issue date on 16 January 2015. Note that what you see in the chart is the development of the price only (i.e. of the "clean price", not the "dirty price").

The characteristics of price developments are discussed in more detail in Sect. 1.4.

1.3 Issuance of a Bond and Initial Issue Price

Another important parameter, which, by the way, is not listed in the statistics provided by the Stuttgart Stock Exchange, is the bond's **initial issue price**. To explain what that means, we need to outline the typical process of a bond issue.

A company wishes to take out a euro loan in the form of a bond for specific purposes and for a specific period of time. Let us assume that the desired loan amount (the nominal amount) is around one billion euros. The desired term of the loan could be 15 years, for example.

The company retains a bank to prepare the bond issue. The bank then proposes a suitable bond offering (e.g. an annual coupon of 4% and a denomination of EUR 1000 per unit) and announces the issue for a certain date. Various parties interested in the product can already express their interest at this point. Based on their feedback, the bank sets an **initial issue price** in an amount that is as high as possible—in the interest of the issuer—but at the same time still attractive enough to ensure that all units of the bond can realistically be expected to be sold at the time of issue. This initial issue price (which is also expressed as a percentage of the nominal value) can be higher than 100 in the case of high demand or a very attractive bond offering—or lower than 100 in the case of low demand or rather unfavourable characteristics.

For our example, let's assume that the initial issue price is 98. Then, on the issue day, one unit of the bond is sold to investors at 98% of the bond's nominal value, i.e. at EUR 980 in our example. The actual loan amount granted to the company is therefore EUR 980 million. This has no bearing, however, on the amount of the annual coupon payments nor on the amount of the principal that the issuer has to repay at maturity. These are always based on the nominal value. This means that the annual coupon payment per unit of the bond is EUR 40 and the principal (nominal amount) to be repaid at maturity is EUR 1 billion (not EUR 980 million).

In the example given in the previous section, the initial issue price, according to the chart, is likely to have been approximately 104.

1.4 Chart of the Bond and Factors Influencing Price Developments

In bond trading, the price of a bond changes over time as a result of the dynamics of supply and demand.

Typically, under normal circumstances, the price of a bond is close to 100. This is particularly true if the coupon amount is roughly equal to the prevailing interest rate level of risk-free investments and if it seems reasonably certain that the issuer will be able to meet all payment obligations for the remaining term of the bond.

As a general rule, it can be said that: **As a bond nears maturity**—assuming the issuer's solvency—**its price tends to move close to 100**. This is because, for a bond nearing maturity, its buyer can soon expect payment of both the bond's nominal amount and the last coupon. The value of this bond a few days before maturity is therefore 100 + the pro rata coupon. Remember, however, that the buyer has to pay the pro rata coupon in addition to the bond's market price anyway (as part of its "dirty price"). Hence, the bond price is almost exactly 100.

As above-mentioned, the price of a bond changes over time as a result of the dynamics of supply and demand. **Yet what are the factors that affect supply and demand, positively or negatively, over the course of the bond's term?**

Changes in the issuer's credit quality or rating: A downgrade in credit quality naturally raises doubts as to the issuer's future ability to make (re-)payments and thus reduces the bond's attractiveness. As the demand for the bond will decrease, its price will fall. With falling prices, the attractiveness of the bond will increase again and become of interest for purchase especially by more speculative investors. For more on credit rating, please refer to Volume III Chapter 1.

Changes in the general interest rate environment: When interest rates for risk-free euro-denominated investment assets increase, the attractiveness of the bond, for which the amount of the disbursements (coupons) until maturity has been fixed in advance, will decrease. As the demand for the bond will decrease, its price will fall. Conversely, falling interest rates for risk-free investment assets increase the attractiveness of the bond; as a result, demand will increase and with that, its price.

Fluctuations in bond prices are relevant only to investors who do not intend to hold the bond until maturity but want to profit from changes in the bond's price by actively trading the bond. Price fluctuations in bonds, assuming the issuer's solvency, are, as we know, only of a temporary nature. Nearing maturity, the price converges towards 100.

The minimum limit for the price of a bond is 0, and it is indeed possible for the price to fall as low as 0 (e.g. for subordinated bonds of a definitively insolvent issuer).

However, it is possible for bonds from insolvent issuers to have a positive—if only a very low—market price and to continue being traded. In such cases, investors hope that the issuer is only temporarily insolvent or that at least partial payments by the issuer are still possible.

See, for example, this chart, dated 28 September 2021, of a bond issued by the Republic of Zambia, with maturity on 20 September 2025, denomination of USD 200,000 per unit, and a coupon of 5.375%. Its Standard & Poor's rating at the time was *D* (for default = nonpayment). Nevertheless, this bond was still offered for trading and had the following quotation and price chart (since 2013); compare Fig. 1.2:

A buyer of 5 units of this bond (total nominal amount: USD 1,000,000) therefore pays a maximum of USD 744,300 for them, hoping to receive at least a couple

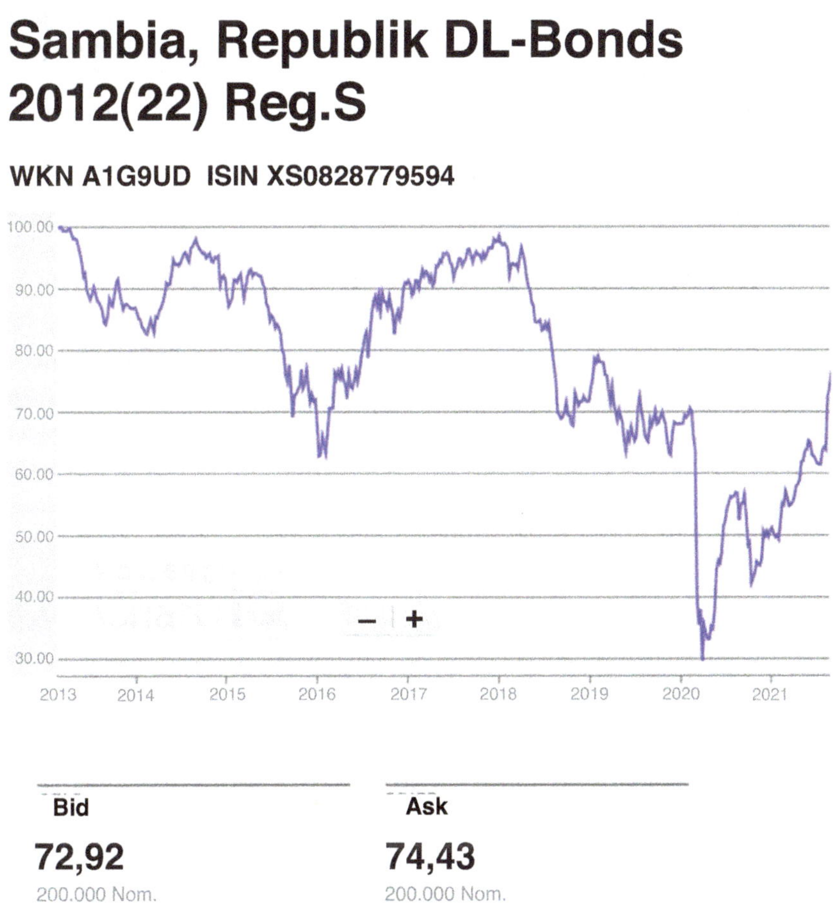

Fig. 1.2 Price history of Republic of Zambia bond (source: Stuttgart Stock Exchange)

of coupon payments of USD 53,750 each until the bond matures in 2025, plus repayment of (at least part of) the nominal amount of USD 1,000,000 at maturity. Or the bond is purchased with the hope that the Republic of Zambia will be able to overcome the apparent difficulties, causing the price of the bond to rise sharply again and thus creating an opportunity for making a profitable sale.

Until just a few years ago, the following natural upper limit for the price of a bond would have been considered binding.

The maximum payments that a bond holder can expect from the bond are capped at the nominal value, i.e. 100%, and the sum total of all coupons. Let us denote the coupon amount as C; assume annual coupon payments and a remaining term to maturity (rounded up to years) of T years. The maximum expected payment amount would then be $100 + T \cdot C$. The price, i.e. the value of the bond, cannot be higher than the highest possible amount of future payments, i.e. $100 + T \cdot C$ at the very most.

In the earlier example of the VW bond with a coupon of 1.625 and a remaining time to maturity of approximately 8 years and 4 months, we would thus arrive at an upper limit for the price of $100 + 9 \times 1.625 = 114.625$.

Why the subjunctive, you may ask—why do we say "we would arrive at such and such limit"? Well, until a few years ago, "negative interest rates" were a purely abstract concept; it was inconceivable that negative interest rates could ever gain a foothold in the reality of the financial market. Yet today—at the time of writing this text—and contrary to all expectations, we, at least the financial institutions of the EUR and CHF regions, and the financial analysts and financial engineers are in fact faced with and have to deal with negative interest rates.

With negative interest rates, future payments have a higher value than payments in the same amount that are made now. Expected future payments of 114.625 can therefore have a present value now that is higher than 114.625. Consequently, in the current interest rate environment, the price of the above VW bond could in principle be higher than 114.625. We will address the questions around discounting cash flows in more detail in later sections of this book.

Despite this objection, the described value of $100 + T \cdot C$ can be used as an approximate benchmark for an upper limit of a bond price (at least as long as the negative interest rates are hovering just slightly below zero).

Before we address the important concept of a bond's "yield", we first need to provide a couple of basic facts regarding "interest compounding" and " types of interest rates". This is what we will do in the following three sections.

1.5 Discrete and Continuous Interest Compounding

In the following, "R" is defined as a given fixed interest rate. " $R(t)$" denotes a variable interest rate with the value "$R(t)$" at any arbitrary point in time "t".

In all of the following, and in the financial mathematics literature more generally, "R" and "$R(t)$" **always** denote an **interest rate per year** (per annum, p.a.), irrespective of how and in what periodic frequency interest is actually calculated.

Unless otherwise noted, we will always state **interest rates R as a percentage**. However, concrete calculations with interest rates will practically always require the value $\frac{R}{100}$.

If we have a **capital K** on which interest is **compounded annually** at the **interest rate R**, within 1 year, that capital K will have a capital

$$K \cdot \left(1 + \frac{R}{100}\right).$$

This means that an initial **capital K** on which interest is compounded annually at an **interest rate R** over a period of **n** years will have a capital

$$K \cdot \left(1 + \frac{R}{100}\right)^n$$

over the course of n years.

Here and below, unless otherwise specifically noted, **time periods always refer to years**!

For clarity, from now on, we will use the **abbreviation** $\mathbf{r} = \frac{\mathbf{R}}{\mathbf{100}}$ **or** $\mathbf{r(t)} = \frac{\mathbf{R(t)}}{\mathbf{100}}$.

Note: *This is a frequent source of error occurring in concrete calculations or in more complex programming tasks! Often, in most cases inadvertently, the interest rate R is used instead of r (or vice versa). Concrete calculations, however, almost always require use of $r = \frac{R}{100}$. In any case, it is important to check whether the interest rate in question is in fact given in the form R or in the form r.*

If, in the last formula, the value n is not an integer (e.g. $n = 1.5$ for one and a half years, or $n = \frac{1}{3}$ for a third of a year), then the same formula applies:

> an initial **capital K** on which interest is **compounded annually** at an **interest rate R for a period of n years** will, over n years, have a capital
>
> $$K \cdot (1 + r)^n$$

where $\mathbf{r} = \frac{\mathbf{R}}{\mathbf{100}}$.

In some cases, for non-integer n, this representation does not correspond entirely to the interest computation practices that banks apply. However, the discrepancies are marginal and will therefore not be further elaborated on in the context of this representation. From the perspective of financial mathematics, the convention we use is the most expedient approach.

In addition to annual compounding, interest can also be compounded in other periodic frequencies, such as semi-annually, monthly, or daily.

Note: Interest rate R is always stated per annum, irrespective of the periodicity in which interest is actually compounded!

And remember, time periods are always expressed in years! Consequently, we quantify:

- 1 month as $\frac{1}{12}$,
- 1 week as $\frac{1}{52}$,
- and 1 day as $\frac{1}{365}$, or $\frac{1}{255}$, depending on whether each calendar day of a year or only the trading days of a year are relevant.

(Note that slightly different conventions may have been agreed on in concrete applications or product specifications.)

When interest is compounded semi-annually, an initial **capital K** on which interest accrues at an **interest rate R for a period of n years** will, over n years, have a capital

$$K \cdot \left(1 + \frac{r}{2}\right)^{2n}$$

where $\mathbf{r} = \frac{\mathbf{R}}{\mathbf{100}}$. So, for this calculation, the per annum interest rate is halved and the number of compounding periods is doubled.

When interest is compounded monthly, an initial **capital K** on which interest accrues at an **interest rate R for a period of n years** will, over n years, have a capital

$$K \cdot \left(1 + \frac{r}{12}\right)^{12n}$$

where $\mathbf{r} = \frac{\mathbf{R}}{\mathbf{100}}$. So, for this calculation, the per annum interest rate is divided by 12 and the number of compounding periods is multiplied by 12.

When interest is compounded weekly, an initial **capital K** on which interest accrues at an **interest rate R for a period of n years** will, over n years, have a capital

$$K \cdot \left(1 + \frac{r}{52}\right)^{52n}$$

where $\mathbf{r} = \frac{\mathbf{R}}{\mathbf{100}}$.

When interest is compounded daily (on a calendar-day basis), an initial **capital K** on which interest accrues at an **interest rate R for a period of n years** will, over n years, have a capital

$$K \cdot \left(1 + \frac{r}{365}\right)^{365n}$$

where $\mathbf{r} = \frac{\mathbf{R}}{\mathbf{100}}$.

General calculation:

Let H be a positive number, **where H expresses the frequency in which interest is compounded within a 1-year period** (e.g. $H = 12$ for monthly compounding, $H = 365$ for daily compounding). Then an initial **capital K** on which interest accrues in frequency H at an **interest rate R for a period of n years** will, over n years, have a capital

$$\boldsymbol{K \cdot \left(1 + \frac{r}{H}\right)^{Hn}}$$

where $\mathbf{r} = \frac{\mathbf{R}}{\mathbf{100}}$.

If we let H get bigger and bigger, i.e. go to infinity, we speak of **continuous compounding**. In this case, interest is compounded in such short intervals as to be applied practically continuously.

Therefore:

When interest is compounded continuously, an initial **capital K** on which interest accrues at an **interest rate R for a period of n years** will, over n years, have a capital

$$\boldsymbol{\lim_{H\to\infty} K \cdot \left(1 + \frac{r}{H}\right)^{Hn} = K \cdot \left(\lim_{H\to\infty} \left(1 + \frac{r}{H}\right)^{H}\right)^{n}}$$

where $\mathbf{r} = \frac{\mathbf{R}}{\mathbf{100}}$.

It is known from mathematical analysis that $\lim\limits_{H\to\infty} \left(1 + \frac{r}{H}\right)^{H} = e^{r}$, where e denotes Euler's number $e = 2.718\ldots$ (see Fig. 1.3).

See what happens when you set $r = 1$:

$$\boldsymbol{\lim_{H\to\infty} \left(1 + \frac{1}{H}\right)^{H} = e^{1} = e = 2.718\ldots}$$

This fact is illustrated in Fig. 1.3.

The red dots show the value $\boldsymbol{\left(1 + \frac{1}{H}\right)^{H}}$ for each H from 1 to 100 (horizontal axis). These values increasingly approximate the value $\boldsymbol{e = 2.718\ldots}$ (blue line).

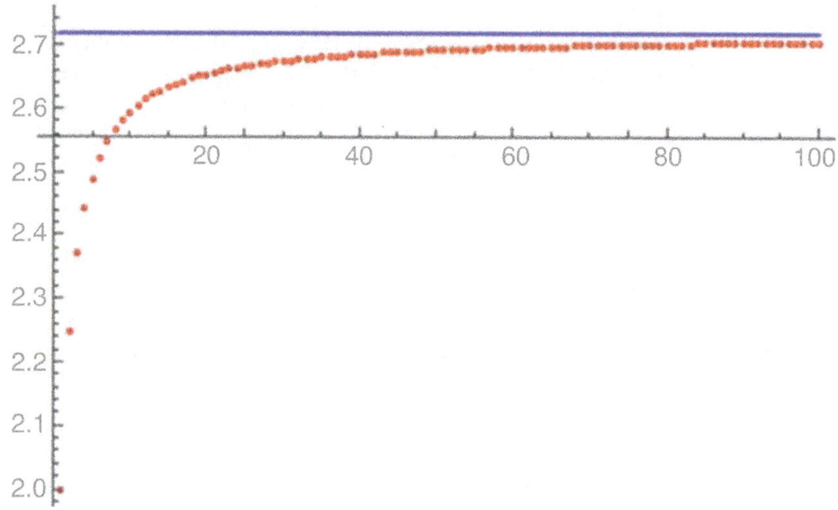

Fig. 1.3 Convergence of $\left(1+\frac{1}{H}\right)^H$ towards e

Conclusion:

When interest is compounded continuously, an initial **capital K on which interest accrues at an interest rate R for a period of n years** will, over n years, have a capital

$$K \cdot e^{r \cdot n}$$

The author vividly recalls his training as a market maker for the derivative market on the Vienna Stock Exchange in 2001. The final exam was preceded by a 6-week seminar at the Vienna Stock Exchange Academy on various financial market topics. During one of the lectures, the trainer addressed the subject of continuous compounding and presented the formula for calculating it. However, he then concluded his presentation by saying, regretfully, "This is the formula for continuous compounding, where, as I have said, e is Euler's number, but unfortunately, as much as I would want to, I cannot tell you why it figures here in this formula . . . ".

Continuous compounding is used almost ubiquitously in financial mathematics, as it is the most advantageous for purposes of mathematical analyses. We will see further below that this is not a limitation, since any other type of interest compounding can be represented through the continuous compounding method, by adjusting the interest rate accordingly. Moreover, the difference in the result when

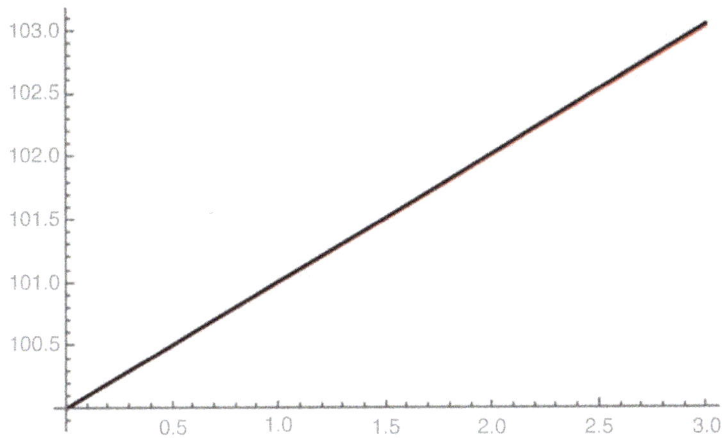

Fig. 1.4 Development of EUR 100 at 1% over a 3-year period using continuous (black) and annual (red) compounding

using different types of compounding is usually negligibly small if the interest rates are not too high and the time periods are not too long.

When a capital K earns interest over a certain period n at a fixed interest rate r, the end capital will be higher the more frequently interest is compounded. So, continuous compounding yields a higher final capital amount than daily interest compounding does, while daily compounding in turn yields a higher final capital than monthly interest compounding, etc. We will use an example to graphically illustrate this fact and then proceed to provide mathematical proof.

Figure 1.4 shows the development of an initial capital of EUR 100 at an interest rate of 1% over a 3-year period. The black line represents the development using continuous compounding. The red line—slightly below—plots the development using annual compounding. The graph also has a green line for monthly compounding, a blue line for weekly compounding, and a pink line for daily compounding (between the black and red lines, respectively). However, these lines are not visible as they are positioned too close to the black line.

Figure 1.5 shows the development of an initial capital of EUR 100 at an interest rate of 10% over a period of 20 years. Again, the black line represents the development using continuous compounding. The red line represents the development using annual compounding. The green line represents the development using monthly compounding. The graph also has a blue line for weekly compounding and a pink line for daily compounding. However, these lines are not visible as they are positioned too close to the black line. Note that these observations also apply if the interest rate r is negative. This may seem counter-intuitive at first, yet the following graph (Fig. 1.6) corroborates this statement, and the below proof of Theorem 1.1 also confirms this fact.

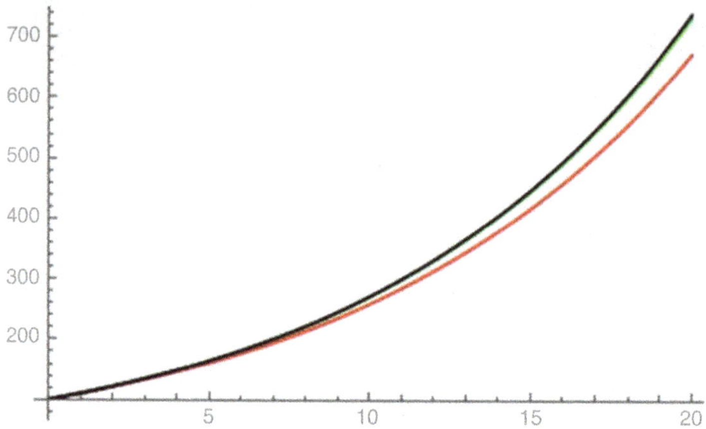

Fig. 1.5 Development of EUR 100 at 10% over a 20-year period using continuous (black), monthly (green), and annual (red) compounding

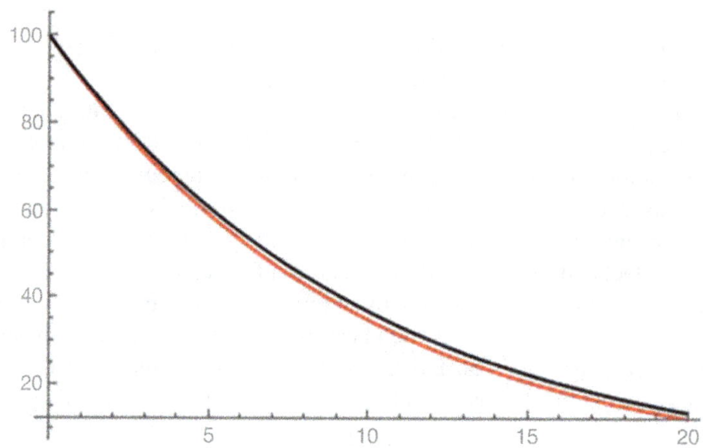

Fig. 1.6 Development of EUR 100 at -10% over 20 years

Figure 1.6 illustrates the development of an initial capital of EUR 100 at an interest rate of **minus** 10% over a period of 20 years: "black" represents continuous compounding, and "red" represents annual compounding.

We will now prove the fact stated above that when a capital K bears interest over a certain period n at a fixed interest rate r, the end capital will be higher and the more frequently interest is compounded.

(continued)

Thus, we want to demonstrate the following:

Theorem 1.1 *For all positive real numbers K and n, all real numbers r, and all real numbers H and L where $L > H > \max(0, -r)$, the following is true:*

$$K \cdot \left(1 + \frac{r}{H}\right)^{Hn} < K \cdot \left(1 + \frac{r}{L}\right)^{Ln}$$

(Note that for $H < \max(0, -r)$, the theorem would not even make any sense.)

Proof It suffices to demonstrate that under the above conditions, $\left(1 + \frac{r}{H}\right)^H < \left(1 + \frac{r}{L}\right)^L$ always holds. Looking at the function $f(x) := \left(1 + \frac{r}{x}\right)^x$, we have to demonstrate that $f(x)$ is monotonically increasing in x for $x > \max(0, -r)$.

Since $f(x) = e^{x \cdot \log(1+\frac{r}{x})}$ and since the exponential function e^y is monotonically increasing in y, this is equivalent with $x \cdot \log\left(1 + \frac{r}{x}\right)$ being monotonically increasing. If we differentiate with respect to x, we get the condition $\log\left(1 + \frac{r}{x}\right) + \frac{1}{1+\frac{r}{x}} - 1 \geq 0$.

We now set $z = 1 + \frac{r}{x}$ and keep in mind that (due to the conditions for x) this z is always greater than 0. This gives us the condition $g(z) := \log z - 1 + \frac{1}{z} \geq 0$ for $z > 0$. Thus, $g(1) = 0$, and $g'(z) = \frac{1}{z} - \frac{1}{z^2}$ is less than 0 for $z < 1$ and greater than 0 for $z > 1$. The result follows from this. □

It has already been pointed out above that in most financial mathematical applications of a more complex nature, limiting the interest computation method to continuous compounding is in fact not a limitation, as any other type of interest compounding can be represented through the continuous compounding method, by adjusting the interest rate accordingly.

An initial **capital K** on which **interest is compounded in an annual frequency H** at an **interest rate R$_H$ for a period of n years** will, over n years, have a capital

$$K \cdot \left(1 + \frac{r_H}{H}\right)^{Hn}$$

When interest is compounded continuously, an initial **capital K** on which interest accrues at an **interest rate R for a period of n years** will, over n years, have a capital

$$K \cdot e^{r \cdot n}$$

If we now want both compounding methods to produce the same capital developments, we equate the two formulas and get

$$\boldsymbol{K \cdot \left(1+\frac{r_H}{H}\right)^{Hn} = K \cdot e^{r \cdot n}}$$

Dividing by K and extracting the n-th root gives us

$$\boldsymbol{\left(1+\frac{r_H}{H}\right)^{H} = e^{r}}$$

from which r_H can easily be expressed through r (at a given constant interest rate r) and r through r_H (at a given interest rate r_H):

$$r = H \cdot \log\left(1+\frac{r_H}{H}\right)$$

or

$$r_H = H \cdot \left(e^{\frac{r}{H}} - 1\right)$$

Example 1.2 Figure 1.5 above illustrated the development of a capital amount of EUR 100 over a 20-year period at a continuously versus an annually compounded interest rate of 10%.

For a given annually compounded interest rate R_H of 10%, the continuously compounded rate r according to the first of the two formulas (we set $H = 1$ for the annual compounding) would have to be

$$r = \log\left(1+\frac{10}{100}\right) = \log(1.1) = 0.0953$$

to obtain the same capital development. This means that an annually compounded interest rate of 10% is equal to a continuously compounded interest rate of 9.53%. Conversely:

At a given continuously compounded interest rate R of 10%, the annually compounded interest rate r_H according to the second of the two formulas (we set $H = 1$ for the annual compounding) would have to be

$$r_H = \left(e^{\frac{10}{100}} - 1\right) = 0.1052$$

to obtain the same capital development. We see that a continuously compounded interest rate of 10% equals an annually compounded interest rate of 10.52%.

We already noted above that in financial mathematics, the continuous compounding method is the version that is practically always used, as it is the most expedient for mathematical analyses. We will now give you an example bearing out this statement by calculating the development of a given capital using continuous compounding at a time-varying interest rate.

1.6 Continuous Compounding at a Time-Varying Interest Rate

Let $R(t)$ be an interest rate that varies over the period of time t.

An example of such an interest rate would be one that increases at regular intervals. For example, an interest rate $R_1(t)$ starts at a value of 1% at the time 0 and increases every 6 months by a value of 0.2% until year 10 (see Fig. 1.7, in red)

However, other examples can also be given using an explicit closed formula. For example, an interest rate $R_2(t)$ starts at a value of 1% at the time 0 and then increases until year 10 according to the formula $R(t) = 1 + \frac{t^2}{20}$ (see Fig. 1.7, in blue)

The development of a capital K on which interest is compounded continuously at the time-varying rate $R(t)$ over a time period from 0 to T can be calculated using the following formula (for the proof see Theorem 1.4):

When interest is compounded continuously, an initial capital K on which interest accrues at a time-varying interest rate $R(t)$ for a period of T years will, over the

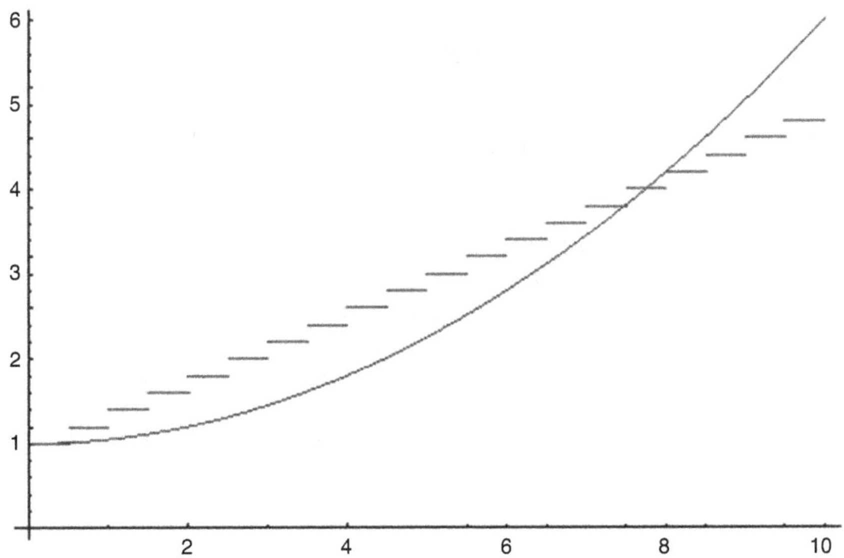

Fig. 1.7 Time-varying interest rates

course of T years, become

$$K \cdot e^{\int_0^T r(t)dt}$$

(here again we used the notation $r(t) = \frac{R(t)}{100}$).

Example 1.3 We calculate the development of a capital of EUR 100 over a 10-year period using the two interest rates R_1 and R_2 from the above examples.

To calculate this, we need the value of $\int_0^T r_1(t)dt$ and of $\int_0^T r_2(t)dt$, respectively.

The first value is easily calculated by dividing the area under the graph of the staircase function representing the interest rate development (i.e. $1 \times 0.5 + 1.2 \times 0.5 + \ldots + 4.8 \times 0.5$) by 100, which gives us 0.29.

The second value is easily calculated through simple integration:

$$\int_0^T r_2(t)dt = \int_0^{10} \frac{1+\frac{t^2}{20}}{100}dt = \int_0^{10} \frac{1}{100} + \frac{t^2}{2000}dt = \frac{t}{100} + \frac{t^3}{6000}\Big|_0^{10} =$$

$$= \frac{1}{10} + \frac{1000}{6000} = 0.267.$$

Thus, when applying continuous compounding, the development with respect to the first interest rate R_1 results in a final value of $100 \cdot e^{0.29} = 133.64$, while the development with respect to the second interest rate R_2 results in a final value of $100 \cdot e^{0.267} = 130.56$.

We will now prove this development formula for continuous compounding at a time-varying interest rate.

So what we have to demonstrate is the following.

Theorem 1.4 *When interest is compounded continuously, an initial capital K on which interest accrues at a time-varying interest rate $R(t)$ for a period of T years will, over the course of T years, become*

$$K \cdot e^{\int_0^T r(t)dt}$$

(here again we used the notation $r(t) = \frac{R(t)}{100}$).

Proof When applying continuous compounding to a short period from time t to time $t + dt$ (i.e. dt is assumed to be a very small time increment), capital $K(t)$ (at time t) develops to capital $K(t + dt)$ (at time $t + dt$) approximately

(continued)

like

$$K(t+dt) \approx K(t) \cdot e^{r(t)dt}.$$

Since dt becomes arbitrarily small, $e^{r(t)dt}$ can be approximated through $1 + r(t)dt$. We thus get

$$K(t+dt) \approx K(t) \cdot e^{r(t)dt} \approx K(t) \cdot (1 + r(t)dt).$$

Using rearrangement we get

$$\frac{K(t+dt) - K(t)}{dt} \approx K(t)r(t),$$

and taking the limit for dt to 0, we get

$$K'(t) = K(t)r(t)$$

$$\Leftrightarrow \frac{K'(t)}{K(t)} = r(t)$$

$$\Leftrightarrow (\log(K(t)))' = r(t)$$

$$\Leftrightarrow \log(K(T)) = \int_0^T r(t)dt + C$$

$$\Leftrightarrow K(T) = e^{\int_0^T r(t)dt} \cdot e^C$$

and since we have $K = K(0) = e^C$, the result follows. □

Example 1.5 The development of capital, as seen, e.g. in Example 1.3 for both variable interest rates, can also be visualized dynamically over time. To do so, we need to calculate $G(t) = K \cdot e^{\int_0^t r(u)du}$ for both interest rates r_1 and r_2 for an arbitrary variable t between 0 and T. $G(t)$ then gives us the accumulated total capital at the time t and can be visualized in a time/capital diagram. If we do this for the variable interest rates given in Example 1.3, we can easily state the integral explicitly for the second interest rate by analogy to above:

$$\int_0^t r_2(u)du = \int_0^t \frac{1 + \frac{u^2}{20}}{100} du = \int_0^t \frac{1}{100} + \frac{u^2}{2000} du = \frac{u}{100} + \frac{u^3}{6000} \Big|_0^t =$$

$$= \frac{t}{100} + \frac{t^3}{6000}$$

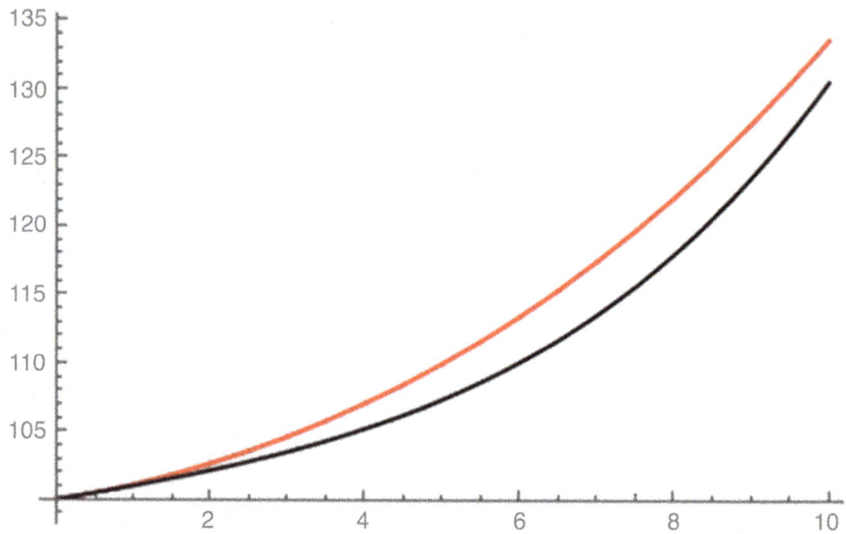

Fig. 1.8 Capital development at time-varying interest rates

Calculating the integral over the first interest rate is somewhat more intricate if you aim for an explicit representation of the integral. We leave this task to the motivated reader and only present the result here:

$$\int_0^t r_1(u)du = \frac{1}{100}\left(t\left(\frac{[2t]}{5}+1\right) - \frac{[2t]+[2t]^2}{20}\right)$$

Here, the expression $[x]$ denotes the largest integer lower than or equal to x.

The dynamics of the two interest rate developments are then plotted using these integral values in the graph of Fig. 1.8. The red line represents the development with respect to the first interest rate and the blue line the development with respect to the second interest rate.

1.7 Euribor, Libor, Swap Rates, and Key Interest Rates

Many financial products are defined by certain "benchmark interest rates" such as one of the Euribor rates, one of the Libor rates, or an ICE swap rate (formerly: ISDAFIX swap rate). We will need such "ideal interest rate" at a later point for comparative purposes and to determine fair values of financial products.

Put simply, and purely informally, a current Euribor interest rate, a Libor interest rate, or a swap rate for a certain currency W and a certain maturity T is an "ideal interest rate" that is currently demanded or offered between "ideal financial partners" (e.g. between top major banks) for borrowing or investing funds in the currency W for a time period T as from now.

The Euribor and Libor rates are generally more short-term interest rates of up to 1 year. Swap rates are longer-term interest rates of 1 year or more. In essence, Euribor and Libor interest rates differ only in the way they are calculated and determined on the financial market. Note that Euribor rates refer only to interest rates in euro, while Libor rates are also calculated for other currencies.

For nonprofessional investors, these benchmark interest rates are also relevant in that they are often used as a reference in loan offers or investment offers from banks, or as a basis for negotiations. Euro loans, for instance, are sometimes offered with variable interest rates, for example, in the form *based on Euribor 6 months + 2%, reset semi-annually*.

Euribor and Libor interest rates as well as the ICE swap rates are described in more detail below (full details can be found on the websites of the institutions calculating these rates. See emmi-benchmarks.eu or de.euribor-rates.eu and reuters.com or theice.com).

Euribor is short for "European Interbank Offered Rate".

Euribor rates are provided for maturities of 1 week, 2 weeks, and for 1, 2, 3, 6, 9, and 12 months. There is also a Euribor overnight rate for shortest-term investments. This overnight Euribor is called "Eonia" (= Euro OverNight Index Average). The Euribor rates are considered to be the most important benchmark rates for euro-denominated loans. Euribor rates have been calculated and published since 1 January 1999. Each business day, at 11:00 am Brussels time, the so-called panel banks selected by the Advisory Committee of the European Banking Association report the interest rates they currently offer for short-term euro loans in the interbank market to an information provider. Based on these reported rates, the information provider calculates an average value for each Euribor rate, according to certain rules.

Libor rates are calculated in the same way as Euribor rates, yet on the basis of daily (11:00 am London time) reports from the major international banks operating in London. Currently, Libor rates are calculated and published for five currencies: the euro (EUR), the British pound (GBP), the US dollar (USD), the Swiss franc (CHF), and the Japanese yen (JPY) and for maturities of 1 week and 1, 2, 3, 6, and 12 months. A Libor overnight rate is also calculated and reported. As a benchmark interest rate for euro loans, Euribor is more important than Libor. Libor on the other hand also provides benchmark rates for other major currencies.

Whenever we assume "an ideal benchmark interest rate" in the following without any further specification, we informally refer to one of these benchmark rates, and we denote such benchmark rates for a term from now (time 0) to time T as $f_{0,T}$. Even if the official benchmark rates are given only for certain periods of time, we assume the existence of benchmark rates $f_{0,T}$ for any time periods T. Often linear interpolations from adjacent official benchmark rates are used for this purpose.

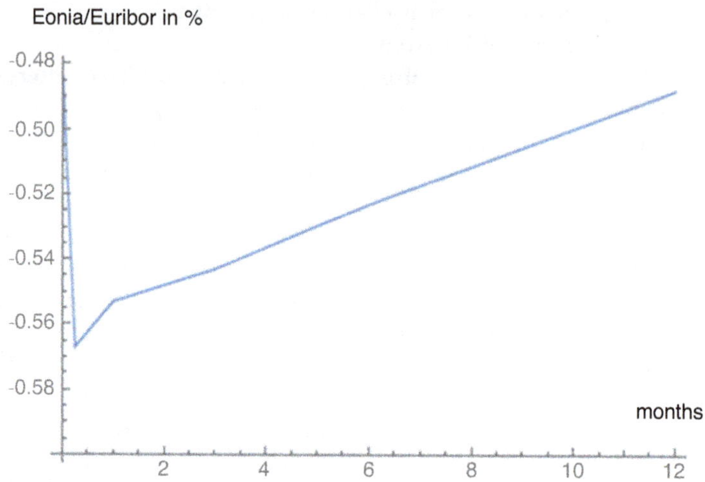

Fig. 1.9 Linearly interpolated Euribor interest rates on 24 September 2021

Example 1.6 On 24 September 2021, the Euribor interest rates had the following values (see, e.g. the Bloomberg abbreviations: EONIA, EUR001W Index, EUR001M Index, EUR003M Index, EUR006M Index, EUR0012M Index):

$$\begin{array}{ll} Eonia & \ldots -0.484\% \\ Euribor\ 1\ week & \ldots -0.567\% \\ Euribor\ 1\ month & \ldots -0.553\% \\ Euribor\ 3\ months & \ldots -0.543\% \\ Euribor\ 6\ months & \ldots -0.523\% \\ Euribor\ 12\ months & \ldots -0.488\% \end{array}$$

In Fig. 1.9, these values were entered in a chart (the x axis being labeled in months) and linearly connected. The interpolated values of $f_{0,T}$ for a T (in this example, which is an exception purely for better illustration, we specify the value T in months) between, e.g. 6 and 12 months can be calculated as follows:

$$\mathbf{f_{0,T}} = \text{Euribor 6 months} + \left(\frac{(T-6)}{(12-6)}\right) \times (\text{Euribor 12 months}-$$

$$- \text{Euribor 6 months}) = \mathbf{-0.523} + \left(\frac{\boldsymbol{(T-6)}}{\mathbf{6}}\right) \times \mathbf{0.035}$$

For example:
$f_{0,10\text{ months}} = -0.523 + \left(\frac{10-6}{6}\right) \times 0.035 = -0.499\%$

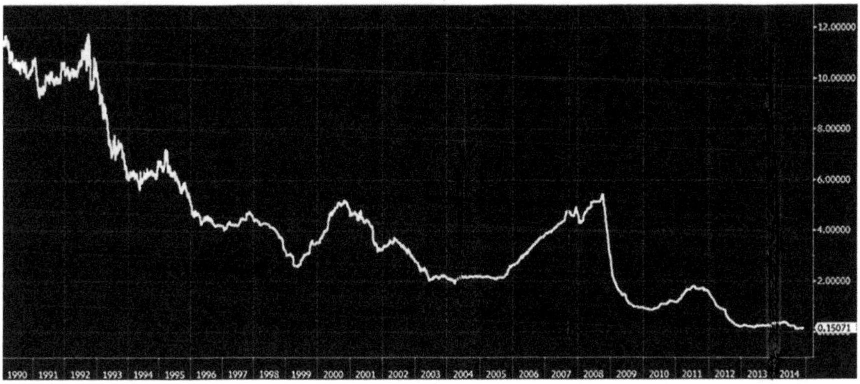

Fig. 1.10 Historical development of the 6-month euro Libor rate until November 2014 (source: Bloomberg)

Programs for determining interpolated benchmark rates can be found on this book's website (see https://app.lsqf.org/book/interest-rate-interpolation).

The general formula for calculating interpolated benchmark rates $f_{0,T}$ from the known actual benchmark rates $f_{0,A}$ and $f_{0,B}$, where A denotes the next shorter horizon and B denotes the next longer horizon for which actual benchmark rates are available, is as follows:

$$f_{0,T} = f_{0,A} + \left(\frac{(T - A)}{(B - A)} \right) \times \left(f_{0,B} - f_{0,A} \right) \tag{1.1}$$

An overview of the historical development of benchmark rates for the main currencies is presented in Figs. 1.10, 1.11, 1.12, and 1.13, plotting the 6-month Libor values for the euro, US dollar, and Japanese yen since January 1990. As the 6-month Libor for the euro currency was only calculated until 28 November 2014, we used the 6-month Euribor from 1 December 2014 onwards. Looking at these charts, we immediately note some striking peculiarities:

- Euro: The essentially continuous downward movement of the interest rates from their high-level period in the early 1990s (with interest rates around 10%) into substantially negative territory
- US dollar: The much more irregular decline in interest rates starting in the first half of the 1990s, from around 8% to around 0.1%–0.2%
- Japanese yen: The rapid fall in interest rates in the mid-1990s from around 8% to a range just slightly above 0%, where interest rates have remained largely unchanged since around 1996.

Figure 1.14 depicts the euro Libor interest rate movements for different maturities (1-week Euribor (green), 6-month Euribor (white), 12-month Euribor (red)).

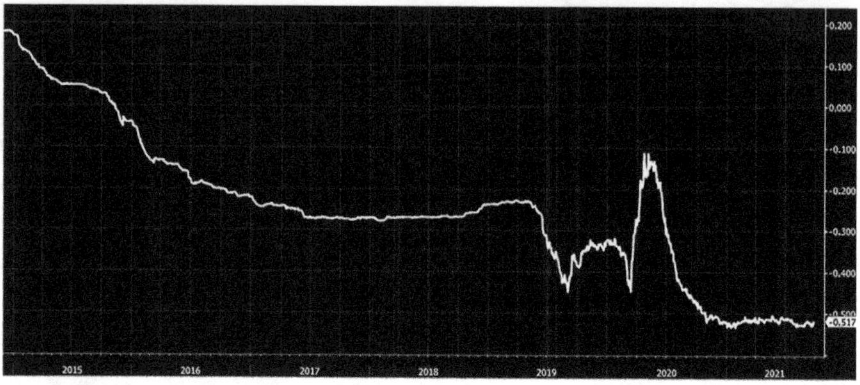

Fig. 1.11 Historical development of the 6-month euro Libor rate as from December 2014 (source: Bloomberg)

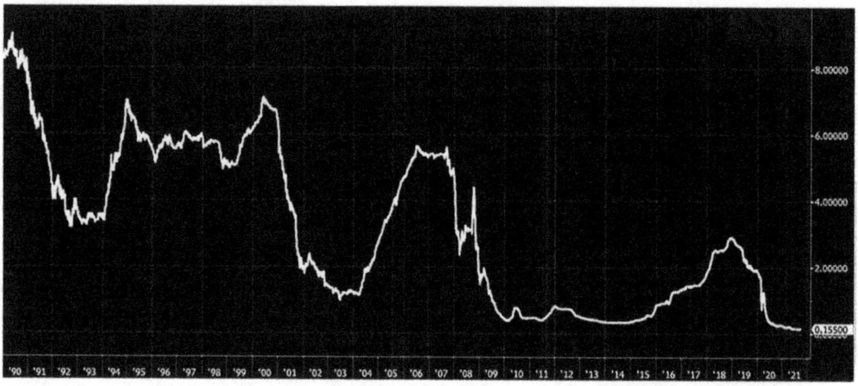

Fig. 1.12 Historical development of the 6-month US dollar Libor rate (source: Bloomberg)

Libor and Euribor Manipulation Scandal

The Libor and Euribor benchmark rates are determined on the basis of daily reports from major international banks. Historically, these reports were usually prepared by only a few responsible persons at each of these banks. This fact was taken advantage of by various parties at different banks in the period from about 2005 to 2013. Agreements were reached between these parties with the intention of circulating slightly falsified data and thus manipulating the values of the benchmark rates. At the same time, or shortly before that, those bankers had used intermediaries to place financial bets on the short-term development of the benchmark rates, enabling them to take enormous profits. As we will see later, due to the leverage effect of financial

(continued)

derivatives, moving an interest rate by even just one tenth of one percent (i.e. one per mil) can already lead to large gains—or losses for the counterparty—in the derivative business. The uncovering of these manipulations resulted in several convictions and high fines for the banks involved in the collusions. Subsequently, the rules for determining and calculating the benchmark rates were substantially tightened and partly transferred to new entities.

The **ICE Swap Rates** (formerly known as “ISDAFIX swap rates”) are also benchmark rates like the Euribor or Libor rates, yet for longer maturities. These swap rates are also calculated and published on the basis of bank reports. ICE swap rates for the EUR, CHF, GBP, and USD currencies are published once to twice a

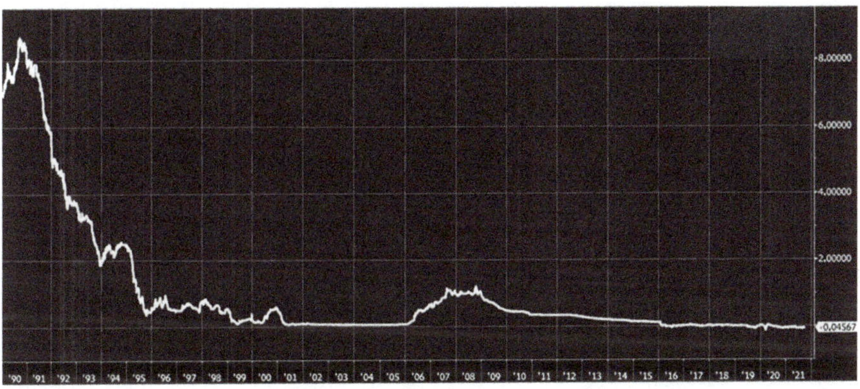

Fig. 1.13 Historical development of the 6-month Japanese yen Libor rate (source: Bloomberg)

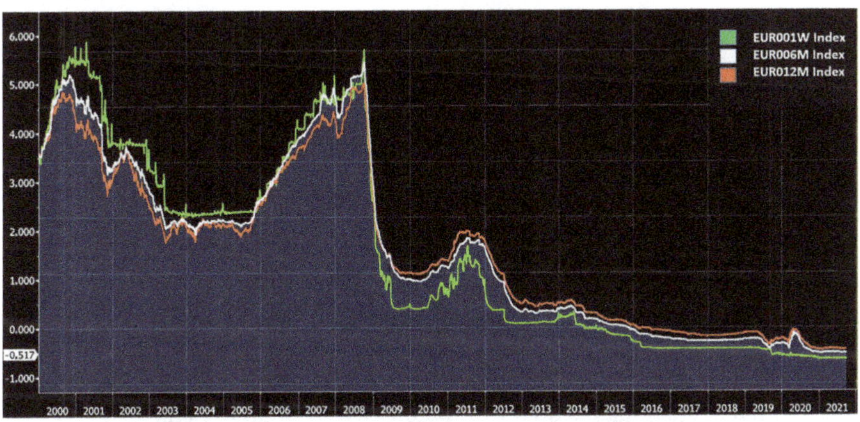

Fig. 1.14 Comparison of EUR benchmark interest rates for different maturities, 6-month Euribor (white), 1-week Euribor (green), 12-month Euribor (red) (source: Bloomberg)

Fig. 1.15 Comparison of euro swap rates for different maturities, 1-year swap rate (white), 10-year swap rate (red) (source: Bloomberg)

day (all currencies at 11:00 am GMT, and EUR additionally at 12:00 noon and USD at 3:00 pm GMT).

The maturities for which the swap rates are published are 1–10 years for all currencies, as well as 12, 15, 20, 25, and 30 years for the EUR, GBP, and CHF, and 15, 20, and 30 years for the USD. Historical data (with a 1-day delay) can be found, for example, on https://www.theice.com/marketdata/reports/180.

An overview of the typical development of ICE swap rates in the euro area over the past few years is presented in Fig. 1.15 below, where we illustrate the Euro ICE swap rates for 1 year (white) and 10 years (red) for the period from 2000 to 2021.

Key Interest Rates

Key interest rates are set by the central banks of the respective currency areas (ECB, European Central Bank for the euro area; Federal Reserve, the US central bank system). The key interest rate is the rate at which large banks can borrow money in the respective currency from the central banks—subject to deposition of collateral. Figure 1.16 contains the following information:
Key interest rate euro area (blue) Key interest rate USA (yellow)
Key interest rate Japan (red) Key interest rate UK (green)
Source: http://www.finanzen.net/leitzins/

1.8 Two Comments Regarding Loans: Calculation of Continuously Redeemable Loan Instalments as Well as Foreign Currency Loans and Interest Rate Parity Theory

Calculation of Instalments on a Continuously Redeemable Loan

Purchasing a bond means granting a loan to a borrower. That borrower then repays the loan amount by paying a fixed coupon (the interest rate on the loan)

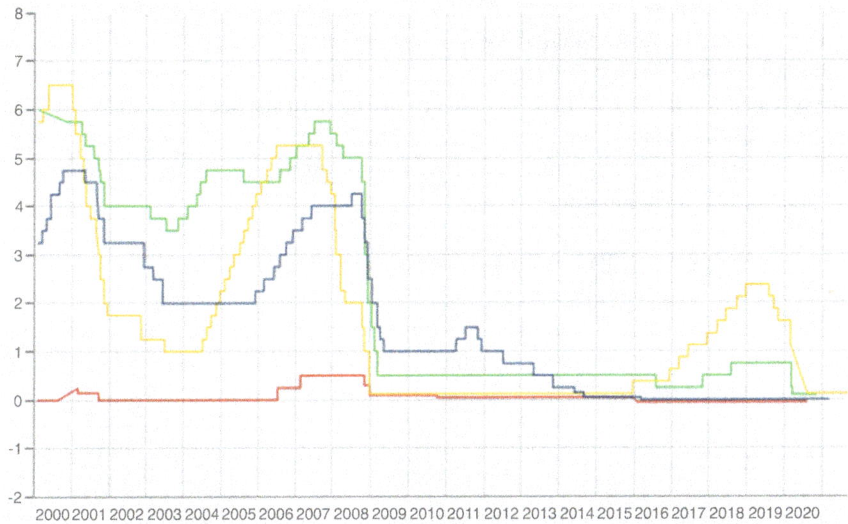

Fig. 1.16 Comparison of historical key rate developments

at regular intervals and the entire loan amount (more precisely: the principal or nominal amount) at the end of the term. Traditional loans often offer the possibility (or the requirement) that the regular (constant) payments already include partial repayments of the principal, so that when the principal becomes due, only a previously agreed residual balance (if any) needs to be redeemed.

We want to digress briefly here to calculate the amount of such fixed instalments. For this, we assume the following given parameters:

- Principal K .... e.g. $K = \text{EUR } 1{,}000{,}000$
- Term of the loan T .... e.g. $T = 30$ years
- Lending rate r .... e.g. Lending rate 5% p.a., so $r = 0.05$
- Periodicity of instalments h .... e.g. $h = \frac{1}{12}$, i.e. monthly instalments (we assume that T is an integral multiple of h)
- Residual balance Z .... e.g. $Z = 200{,}000$

The question we want to answer is:

- What is the constant regular repayment amount X that is required in this setting?

In the following, we assume lending rates with annual compounding (for other compoundings, the procedure is analogous).

The initial loan amount is K.

When the first instalment becomes due for payment at time h, the outstanding loan amount will have risen to $K \cdot (1 + r)^h$ due to the compounded interest. In the following, we denote $(1+r)^h$ as y. At time h, the outstanding loan amount is reduced to $K \cdot y - X$ following payment of the first instalment X. Until payment of the second instalment at time $2h$, the outstanding loan amount rises to $(K \cdot y - X) \cdot y$ but is in turn reduced to $(K \cdot y - X) \cdot y - X$ due to payment of the second instalment. This continues until the last instalment is paid at time T, i.e. for a total of $\frac{T}{h}$ times (which is the number of times that instalment payments are made). At time T, therefore, the outstanding loan amount is

$$(((\ldots((K \cdot y - X) \cdot y - X) \cdot y - X) \cdot y - X) \cdot y - \ldots - X) \cdot y - X$$

(Here the summand X is subtracted $\frac{T}{h}$ times in total.) This amount, which is the principal that is still outstanding at time T, is to be equal to the residual balance Z. So we get the equation

$$(((\ldots((K \cdot y - X) \cdot y - X) \cdot y - X) \cdot y - X) \cdot y - \ldots - X) \cdot y - X = Z$$

which we can use, after performing some rearrangements and applying the geometric sum formula $1 + y + y^2 + y^3 + \ldots + y^{n-1} = \frac{y^n - 1}{y-1}$, to calculate the instalment X as follows:

$$\begin{aligned}
(((\ldots((K \cdot y - X) \cdot y - X) \cdot y - X) \cdot y - X) \cdot y - \ldots - X) \cdot y - X = Z &\Leftrightarrow \\
K \cdot y^{\frac{T}{h}} - X \cdot y^{\frac{T}{h}-1} - X \cdot y^{\frac{T}{h}-3} - \ldots - X \cdot y - X = Z &\Leftrightarrow \\
X \cdot \left(y^{\frac{T}{h}-1} + y^{\frac{T}{h}-2} + y^{\frac{T}{h}-3} + \ldots + y + 1\right) = K \cdot y^{\frac{T}{h}} - Z &\Leftrightarrow \\
X \cdot \frac{y^{\frac{T}{h}} - 1}{y - 1} = K \cdot y^{\frac{T}{h}} - Z &\Leftrightarrow \\
X = \frac{y - 1}{y^{\frac{T}{h}} - 1} \cdot \left(K \cdot y^{\frac{T}{h}} - Z\right)
\end{aligned}$$

By using the definition for y, we have thus obtained:

The **constant repayment instalment X** is

$$X = \frac{(1+r)^h - 1}{(1+r)^T - 1} \cdot \left(K \cdot (1+r)^T - Z\right)$$

So, for our example above, we get a monthly instalment of EUR 5055. The total of 360 payments of this monthly instalment plus the balance of EUR 200,000 result in a total amount to be repaid of EUR 2,019,900. The development of the outstanding loan amount over the 30-year period is shown in Fig. 1.17.

Given the same parameters, but a residual balance of 0, i.e. in the case of full repayment through instalments, the monthly repayment amount is EUR 5301. This means that the total repayment amount is approximately EUR 1,908,200.

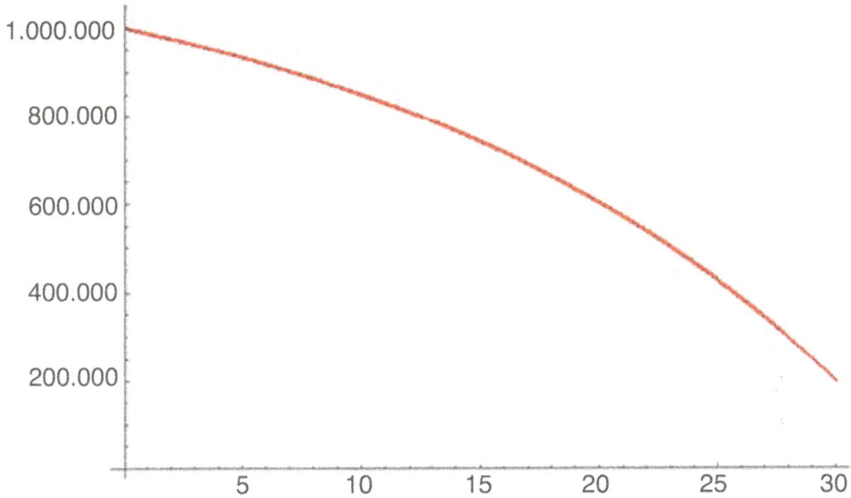

Fig. 1.17 Development of the outstanding loan amount

Assuming the same parameters as in the initial example but with annual payments, the borrower has to repay an annual amount of EUR 62,041. This results in a total repayment amount of approximately EUR 2,061,230.

Foreign Currency Loans and Interest-Rate Parity Theory

Taking out a loan in a foreign currency is worth considering if the lending rates in that other currency are significantly lower than the lending rates in the home currency (EUR).

This was the case, for example, in the early to mid-2000s, when the Japanese yen (JPY) and the Swiss franc (CHF) rates compared favourably to the euro. At the time, this led to a boom in demand for loans in JPY and CHF.

The 10-year swap rates for the euro stood at 4.917% on 31 July 2008 and the 10-year swap rate of the Swiss franc at 3.555%. The exchange rate was 1.6337 Swiss francs for 1 euro.

The procedure for taking out a CHF loan, for example, was basically the following: A German borrower who, e.g. needed a loan of EUR 300,000 for building a house in 2008, would—with assistance from the borrower's bank—take out a CHF loan at the time (we stick to the date used in our previous example of 31 July 2008) in an amount equivalent to approximately EUR 300,000.

At the then exchange rate of 1.6337, this meant taking out a loan in the amount of CHF 490,110. Let us assume (by way of example) a term of 10 years and that the above swap rates were fixed as the lending rates for the entire term, with annual interest payments and repayment of the principal at the end of the term. This means that the borrower would make annual payments of CHF $490{,}110 \times 0.03555 = 17{,}423$ in loan interest and repay 490,110 CHF at the end of the term on 31

July 2018. For this purpose, at each interest pay date as well as at the principal repayment date, the borrower has to come up with the corresponding amount in EUR and convert that amount into CHF before effecting the payment. However, due to fluctuations in the exchange rate, this EUR amount varies from date to date and is not known in advance.

If the **exchange rate of the CHF to the EUR remained unchanged throughout the term**, each interest payment would have a value of EUR 17,423/1.6337 = 10,655, and the principal repayment at maturity would have the value of EUR 300,000. The cost of the loan, which was simply added up here, would therefore be 10 × 10,655 = 106,550 euros. A loan purely in EUR, on the other hand, would have required annual interest payments of EUR 300,000 × 1.04917 = 14,751. The cost of a EUR loan (when simply added up) would thus have been EUR 147,510. **With the CHF loan, the borrower in this case would have saved EUR 40,860.**

Exchange rates do not generally remain unchanged, however. In fact, the exchange rate of the euro to the Swiss franc changed massively over the course of the 10-year period from 31 July 2008 to 31 July 2018. Such changes in exchange rate parity can have both positive and negative effects for the borrower, of course. The effects are positive for the borrower if the foreign currency weakens against the euro over the term of the loan, and they are negative if the foreign currency strengthens against the euro.

The actual CHF-EUR development in the period from 31 July 2008 through 31 July 2018 is shown in Fig. 1.18.

Fig. 1.18 Exchange rate development CHF EUR from 31 July 2008 to 31 July 2018

Table 1.1 Exchange rates and interest amounts from 2008 to 2018

Date	CHF EUR	Interest instalment
31 July 2008	1.6341	10,662
31 July 2009	1.5233	11,438
30 July 2010	1.3582	12,828
29 July 2011	1.1311	15,404
31 July 2012	1.2012	14,505
31 July 2013	1.2317	14,145
31 July 2014	1.2168	14,319
31 July 2015	1.0608	16,424
31 July 2016	1.0833	16,083
28 July 2017	1.1382	15,308
31 July 2018	1.1569	15,060

The exchange rates and thus the payable interest amounts in EUR at each of the interest pay dates and the principal repayment date are given in Table 1.1.

Thus, in reality, the total interest burden would have been EUR 156,175, i.e. slightly higher than with a euro-only loan. The strongest negative effect, however, would have resulted from the exchange rate of 1.1569 applicable to the repayment of the principal on 31 July 2018. At that exchange rate, the amount to be repaid would have been EUR $490{,}110/1.1569 = 423{,}641$. The loss compared to the initial loan amount of EUR 300,000 would therefore have been EUR 123,641.

As noted above, this exchange rate risk of a foreign currency loan could also have worked to the borrower's advantage, namely, if the CHF had weakened against the euro in the period from 2008 to 2018.

The **interest rate parity theory** postulated by various economic theories speaks against the likelihood of such a positive effect.

Without going into further details of this approach, I just want to briefly outline the idea: investors who recognize that investments in a foreign currency (FC) offer the prospect of higher profits due to higher market interest rates r_f than investments in their home currency (HC) with market interest rates r will generally tend to buy foreign currency and invest at such high interest rates. However—according to theory—the dynamics of the financial markets will always seek to achieve equilibrium, in the sense that expected returns are assumed to be essentially equal around the globe, meaning there are no investment opportunities with a very high likelihood of generating above-average returns.

For the above example, however, where we compare an investment K in a home currency HC at a lower interest rate r (for a certain fixed term T) with an investment in the same amount (after conversion) in a foreign currency FC at a higher interest rate r_f, this would mean: the principal amount in HC would have the same expected (!) value after investing in HC as it would have after investing in FC, i.e.

$$K \cdot (1+r)^T \approx K \cdot w_0 \cdot \frac{\left(1+r_f\right)^T}{w_T}.$$

(where w_0 denotes the price of 1 HC in FC at time 0 and w_T denotes the expected (!) price of 1 HC in FC at time T).

Thus,

$$\boldsymbol{w_T \approx w_0 \cdot \frac{(1+r_f)^T}{(1+r)^T}}.$$

According to the interest rate parity theory, we thus obtain, at time 0, an estimate of the expected value of the exchange rate w_T at time T.

For our example of the 10-year CHF loan from 2008 to 2018, the interest rates at the end of July 2008 would have resulted in the following estimate for the development of the CHF-EUR exchange rate until 2018 (where w_T denotes the price of 1 euro in Swiss francs on 31 July 2018):

$$w_T \approx \frac{1.6341 \cdot 1.03555^{10}}{1.04917^{10}} = 1.4339$$

According to the interest rate parity theory, a profitable transaction through a foreign currency loan would not have been foreseeable on 31 July 2008, since the exchange rate expected for 31 July 2018 was 1.4339. In fact, the actual development of the exchange rate would have resulted in an even substantially higher loss to the borrower.

1.9 Bond Yields

The **yield of a bond** is a measure representing the interest rate at which the buyer of the bond earns interest on that investment.

For illustration purposes, we will again use the example of the VW bond presented in Sect. 1.2. To recap, the relevant parameters of this bond were as follows:

Issuer Volkswagen International Finance N.V.
Interest rate 1.625%
Interest accruing as from 16 January 2015
Next interest payment 16 January 2022
Maturity 16 January 2030
Spot rate 108.76
Quotation time 22 September 2021 1:14:30 pm

On 22 September 2021, the date at which the above data were published, this bond represents an investment until its maturity on 16 January 2030, i.e. for a period of approximately 8 years and 4 months, or 8.33 years.

This bond generates a series of cash flows (expressed as a percentage) over the remaining term (provided the issuer remains solvent throughout the term).

The first cash flow consists in payment of the bond's total purchase price in the amount of its price plus its pro rata coupon (this is a negative cash flow from the investor's point of view).

This is followed by positive cash flows in the amount of the coupons at the coupon pay dates on 16 January 2022, 16 January 2023, ..., 16 January 2030.

And finally comes a positive cash flow consisting in payment of the bond's face value (= 100%).

For the example of the VW bond, this means:

- negative cash flow now (at quotation time) of $108.76 + \frac{2}{3} \cdot 1.625 = 109.84$
- positive cash flows of 1.625 in 0.33 years, 1.33 years, 2.33 years, ..., 8.33 years
- a final cash flow of 100 in 8.33 years

We now want to compare this bond with a fixed-rate savings account with the same maturity, and ask ourselves the following:

What interest rate x (compounded annually) would this savings account need to offer in order for it to generate the same cash flows as the bond does?

If we succeed in calculating this interest rate x, we can use x to denote the yield of the bond. The investment in the bond would then correspond to an investment in a savings account bearing interest at rate x.

To summarize:

The yield of a bond is the interest rate at which a savings account would have to bear interest (compounded annually) in order for it to generate the same cash flows as the bond does.

In the following, we determine this interest rate x, i.e. the yield of a bond, for a general case of a bond, meaning: We will derive a general formula for calculating yields and then apply this formula to our example of the VW bond.

For this purpose, we will use the notations

$A \ldots$	to denote the current (at time 0) total price of the bond (= price + pro rata coupon), as a percentage
$t_1, t_2, t_3, \ldots, t_{n-1}, t_n$	to denote the times at which coupon payments take place (in years from now), where$t_n = T$ denotes the remaining maturity of the bond
Ngiven in %	to denote the nominal amount to be repaid at time T (usually $N = 100$)

The balance in a savings account with interest x (expressed in absolute numbers, not as a percentage), which is opened at this moment—at time 0—and has the same cash flows as the bond in the time period $[0, T]$, would thus evolve as follows:

We deposit amount A in the savings account at this moment, i.e. at time 0.

This amount bears interest until time t_1, so that by the time t_1, the balance will be $A \cdot (1 + x)^{t_1}$. Additionally, an amount equal to the first coupon C_1 is withdrawn at time t_1. The balance in the savings account will then be $A \cdot (1 + x)^{t_1} - C_1$.

This amount will now bear interest until the time t_2 (i.e. for the time period $t_2 - t_1$), which means that by the time t_2, the balance in the account will be $\big(A \cdot (1+x)^{t_1} - C_1\big) \cdot (1+x)^{t_2-t_1}$.

Additionally, an amount equal to the second coupon C_2 is withdrawn at time t_2. The balance in the account will then be

$$\big(A \cdot (1+x)^{t_1} - C_1\big) \cdot (1+x)^{t_2-t_1} - C_2.$$

The following cycles continue in the same way: this amount will in turn bear interest until the time t_3 (i.e. for the time period $t_3 - t_2$), which means that by the time t_3 the balance in the account will be

$$\big(\big(A \cdot (1+x)^{t_1} - C_1\big) \cdot (1+x)^{t_2-t_1} - C_2\big) \cdot (1+x)^{t_3-t_2}.$$

Additionally, an amount equal to the third coupon C_3 is withdrawn at time t_3. The balance in the account will then be

$$\big(\big(A \cdot (1+x)^{t_1} - C_1\big) \cdot (1+x)^{t_2-t_1} - C_2\big) \cdot (1+x)^{t_3-t_2} - C_3.$$

$$\ldots$$

Finally, we get: the balance accrued in this way up to the time t_{n-1} of

$$\begin{aligned}(((\ldots \big(\big(A \cdot (1+x)^{t_1} - C_1\big) \cdot (1+x)^{t_2-t_1} - C_2\big) \cdot (1+x)^{t_3-t_2} - C_3) \cdot \ldots \\ \ldots - C_{n-2}) \cdot (1+x)^{t_{n-1}-t_{n-2}} - C_{n-1}\end{aligned}$$

will now bear interest until the time t_n (i.e. for the time period $t_n - t_{n-1}$), which means that by the time t_n the balance in the account will be

$$\begin{aligned}((((\ldots \big(\big(A \cdot (1+x)^{t_1} - C_1\big) \cdot (1+x)^{t_2-t_1} - C_2\big) \cdot (1+x)^{t_3-t_2} - C_3) \cdot \ldots \\ \ldots - C_{n-2}) \cdot (1+x)^{t_{n-1}-t_{n-2}} - C_{n-1}\,) \cdot (1+x)^{t_n-t_{n-1}}.\end{aligned}$$

Furthermore, an amount equal to the last coupon C_n and the nominal amount N are withdrawn at time t_n. The balance in the savings account will then (at the time $t_n = T$) be

$$\begin{aligned}((((\ldots \big(\big(A \cdot (1+x)^{t_1} - C_1\big) \cdot (1+x)^{t_2-t_1} - C_2\big) \cdot (1+x)^{t_3-t_2} - C_3) \cdot \ldots \\ \ldots - C_{n-2}) \cdot (1+x)^{t_{n-1}-t_{n-2}} - C_{n-1}\,) \cdot (1+x)^{t_n-t_{n-1}} - C_n - N.\end{aligned}$$

At time T, however, all of the bond's cash flows have been settled, so the balance remaining in the savings account should also be equal to 0. Consequently, we get the equation

$$((((\ \dots((A\cdot(1+x)^{t_1}-C_1)\cdot(1+x)^{t_2-t_1}-C_2)\cdot(1+x)^{t_3-t_2}-C_3)\cdot\dots$$
$$\dots-C_{n-2})\cdot(1+x)^{t_{n-1}-t_{n-2}}-C_{n-1}\ )\cdot(1+x)^{t_n-t_{n-1}}-C_n-N=0$$

with the unknown x. The yield x can now be calculated using this equation.

Before we proceed to calculating x based on this equation, however, we first simplify the equation somewhat further:

First we replace the expression $(1+x)$ by y and multiply the terms of the equation to obtain the much nicer-looking equation

$$A\cdot y^T-C_1\cdot y^{T-t_1}-C_2\cdot y^{T-t_2}-C_3\cdot y^{T-t_3}-\dots-C_{n-1}\cdot y^{T-t_{n-1}}-C_n-N=0$$

with the unknown y. This equation is then to be solved for y. The interest rate x finally results from $x=y-1$.

To summarize:

The **yield** x **of a bond** with the following parameters:

A	current total price of the bond as a percentage
$t_1, t_2, t_3, \dots, t_{n-1}, t_n$	coupon pay times
$C_1, C_2, C_3, \dots, C_{n-1}, C_n$	coupons paid at the above times
N	the nominal value to be repaid at time $T=t_n$

is calculated by solving the equation

$$A\cdot y^T-C_1\cdot y^{T-t_1}-C_2\cdot y^{T-t_2}-C_3\cdot y^{T-t_3}-\dots-C_{n-1}\cdot y^{T-t_{n-1}}-C_n-N=0$$

for the variable y and through $x=y-1$.

Normally, such an equation cannot be solved explicitly but has to be solved by approximation using a computer (e.g. Mathematica, Maple, etc.).

Brief Account on Solving Equations and Niels Henrik Abel

At first glance, the above equation for calculating the yield might give the impression that we are dealing with a polynomial equation in the form $3x^2-5x-4=0$, or $7x-2=0$, or $x^6-3x^5-x^3-2x^2-7x-15=0$. This is generally not the case, as the exponents T or $T-t_i$ in the equation need not necessarily be natural numbers. Thus, the equation for calculating the yield can also be of the form $2x^{3.176}-5x^{1.749}-x^{0.351}-7=0$. Such equations are much more intricate than polynomial equations with positive integral exponents and can practically never be solved explicitly.

However, this is not a problem at all, since arbitrarily accurate approximation solutions of such equations can be instantly obtained using numerical methods (such as Newton's approximation method) and the computer.

(continued)

But even with polynomial equations, we cannot always hope for explicit solutions y. Linear equations, i.e. first-degree polynomial equations of the form $ax + b = 0$, are of course explicitly solvable through $x = -\frac{b}{a}$.

The formula $x = \frac{-b\pm\sqrt{b^2-4ac}}{2a}$ for solving quadratic equations $ax^2 + bx + c = 0$ will be familiar to any high school student.

There are also formulas for solving third- and fourth-degree polynomial equations $\left(ax^3 + bx^2 + cx + d = 0 \text{ und } ax^4 + bx^3 + cx^2 + dx + e = 0\right)$ but they are very complex. The formula for solving general polynomial equations of the third degree was first published by Gerolamo Cardano in 1545. Cardano's formula was based on important preparatory work by the Venetian mathematician Niccolo Tartaglia. In the same year, Cardano also published a general formula for solving fourth-degree equations, which, as Cardano himself pointed out, had been developed by his student Lodovico Ferrari. The formulas for solving polynomial equations up to and including the fourth degree give us explicit solutions of these equations in the form of radicals. This means that the solutions x of the equations can be written as combinations of integers, coefficients $a, b, c, \ldots$, a finite number of additions, multiplications, divisions, and nth roots (with integral n).

Thus, the solution can be a radical like this $x = \frac{-b+\sqrt{b^2-4ac}}{2a}$ but also like this $\sqrt[5]{1 + \sqrt[3]{a - \sqrt[6]{\frac{b^5-1}{\sqrt{c+1}}}}}$.

For equations of the fifth or higher degree, various mathematicians, including the great Leonhard Euler, tried in vain to find formulas for solving them by radicals. In 1824, the then 22-year-old Norwegian mathematician Niels Henrik Abel succeeded in proving that for polynomial equations of the fifth degree or higher, there can never be any such general solution formula in radicals. There are, of course, polynomial equations of the fifth degree or higher that can be solved by radicals, but generally that is not the case. For instance, the solutions of the equation $x^5 - x + 1 = 0$ are not radicals.

Note the qualitative difference between the two mathematical results delivered by Cardano versus Abel: Cardano demonstrated superb mathematical craftsmanship by developing a formula for solving equations up to the fourth degree. Abel showed that no matter how much ingenuity and technical skill you apply, it will never be possible to find a formula for solving higher-degree equations. Propositions and proofs of this kind are what define the sublime art of pure mathematics, which goes far beyond the appreciation of mathematics as a technical craft. Nils Henrik Abel died of tuberculosis, in complete poverty and without a job, at age 26.

As noted above, the fact that we have to approximate the solutions of the equations for determining the yield of a bond is, of course, no problem at all and can easily be accomplished with the use of a mathematical software.

As an example, we calculate the yield of the VW bond that we presented in Sect. 1.2:

Example 1.7 The VW bond data were taken from 22 September 2021.

Future interest payments of 1.625% were scheduled for 16 January of each year, starting on 16 January 2022 and ending on 16 January 2030. If we set the period from 22 September of a year to 16 January of the following year at roughly 4 months, i.e. at $\frac{1}{3}$ of a year, then the time intervals until the future interest payments $t_1, t_2, t_3, \ldots, t_8, t_9$ equal $0.33, 1.33, 2.33, \ldots, 8.33$.

The coupon payments $C_1, C_2, C_3, \ldots, C_8, C_9$ each have the value 1.625.

The (total) purchase price for the bond (dirty price) was 109.84.

To calculate the bond's yield x, we must therefore solve the following equation in y and then set $x = y - 1$.

$$
\begin{aligned}
109.84 \cdot y^{8.33} - 1.625 \cdot y^{7.33} - 1.625 \cdot y^{6.33} - 1.625 \cdot y^{5.33} - \\
- \ldots - 1.625 \cdot y^{0.33} - 101.625 = 0
\end{aligned}
$$

Mathematica produces the approximate solution $y = 1.0054$ (using the "FindRoot" command).

For the yield x, we thus get $x = 0.0054$, i.e. 0.54%.

If we compare this result with the 0.4% yield stated in the bond's fact sheet in Sect. 1.2, we see that it differs slightly from our result. This is because the intervals until the future interest payments that we used in our calculation were not entirely exact but rounded to $\frac{1}{3}$ each.

1.10 Forward Interest Rates

In Sect. 1.7, we defined benchmark interest rates $f_{0,T}$ for any time periods T starting now (time 0). These benchmark rates are also referred to as "spot rates".

In this chapter we will look at forward interest rates of the form $f_{k,k+T}$.

Imagine the following hypothetical situation: you know that, exactly 1 year from today, you are going to receive a cash amount of EUR 1 million. You also know that you will be required to keep that amount for 2 years and that, after those 2 years—i.e. in 3 years from today—you will have to return it.

For purposes of this example, we will again assume ideal benchmark rates $f_{0,T}$ for investments and loans. We also assume that you are not going to stash those EUR 1 million in a drawer at your home but want to invest it risk-free at a bank. Anticipating a further decline in interest rates (which, in the current situation, means that you fear interest rates will move even further into negative territory), you decide to ask the bank already now to agree on the interest rate for the investment of the one million euros for the period $[1, 3]$, i.e. from year 1 from now until year 3 from now. So you set up an appointment with your banker and start negotiating. A key reference for those negotiations will obviously be the currently prevailing benchmark rates for the next 3 years, i.e. $f_{0.1}$, $f_{0.2}$, $f_{0.3}$.

For purposes of this exercise, we use the current (28 September 2021) ICE swap rates (see https://www.theice.com/marketdata/reports/180):

$$f_{0,1} = -0.526\%$$
$$f_{0,2} = -0.429\%$$
$$f_{0,3} = -0.350\%$$

Your banker also expects interest rates to decline further. In particular, he believes that in 1 year from now, the interest rate for then 2-year investments will be even lower than at present, i.e. below -0.429%. Nevertheless, as a gesture of goodwill to you as a regular customer, he decides to offer you an even slightly better interest rate of -0.4% on the EUR 1 million investment for the period $[1, 3]$.

It would appear that you negotiated well and should therefore happily accept this offer . . .

But should you really?

Accepting the offered interest rate would mean:

Over the period $[1, 3]$, your cash amount would go from EUR 1 million to $1 \cdot (1 - 0.004)^2 = 992{,}016$ EUR. You would **make a loss of EUR 7984.**

Let us compare this with the following approach (or "strategy") (always assuming ideal benchmark rates for investments and loans):

We take out a loan now (at time 0) in the amount of EUR $\frac{1{,}000{,}000}{(1-0.00526)} = 1{,}005{,}288$ for a period of 1 year at interest rate $f_{0.1} = -0.526\%$, and we immediately invest this amount at the benchmark interest rate $f_{0.3} = -0.350\%$ for a period of 3 years.

One year from now, we can repay our loan with the 1 million euro amount that we receive (the amount to be repaid is now exactly EUR $1{,}005{,}288 \cdot (1 - 0.00526) = 1{,}000{,}000$).

Over the course of the 3-year period, the money invested at interest rate $f_{0,3} = -0.350\%$ develops to EUR $1{,}005{,}288 \cdot (1 - 0.0035)^3 = 994{,}769$. If we then, after 3 years, return the one million euros as agreed, we will even have made a smaller **loss of EUR 5230.**

Let us also calculate what interest rate x we devised through this alternative strategy for the time period $[1, 3]$. The equation

$$1{,}000{,}000 \cdot (1 + x)^2 = 994{,}769 \tag{1.2}$$

gives us the value $x = -0.0026$, i.e. $x = -0.26\%$

The strategy therefore enabled us to secure an interest rate of $x = -0.26\%$ instead of the interest rate of -0.4% that was offered to us by the bank.

Now let's see how we arrived at the above amount of 994.769:

$$994{,}769 = 1{,}005{,}288 \cdot (1 - 0.0035)^3 = \\ = \left(\frac{1{,}000{,}000}{(1 - 0.00526)}\right) \cdot (1 - 0.0035)^3 = \\ = \left(\frac{1{,}000{,}000}{(1 + f_{0,1})}\right) \cdot (1 + f_{0,3})^3 .$$

If we now set this expression $\left(\frac{1{,}000{,}000}{(1+f_{0,1})}\right) \cdot (1 + f_{0,3})^3$ instead of 994,769 in Eq. (1.2) and solve the equation for x, we get the following expression for x:

$$x = \left(\frac{(1 + f_{0,3})^3}{(1 + f_{0,1})}\right)^{\frac{1}{2}} - 1 \tag{1.3}$$

Consequently, by using the strategy set out above, we are able to secure the interest rate $x = \left(\frac{(1+f_{0,3})^3}{(1+f_{0,1})}\right)^{\frac{1}{2}} - 1$ for the time period $[1, 3]$.

Obviously, we can apply the same approach to any future period of time $[k, k + T]$ and thus secure an interest rate of the form $x = \left(\frac{(1+f_{0,k+T})^{k+T}}{(1+f_{0,k})^k}\right)^{\frac{1}{T}} - 1$.

We refer to this interest rate as $\boldsymbol{f_{k,k+T}}$ and call it the **forward interest rate for the time period $[k, k + T]$**:

$$\boldsymbol{f_{k,k+T} = \left(\frac{(1 + f_{0,k+T})^{k+T}}{(1 + f_{0,k})^k}\right)^{\frac{1}{T}} - 1} \tag{1.4}$$

It is important to note that $f_{k,k+T}$ is **not** the interest rate that will apply to investments made in the year k for a period of T years (as we don't know it yet!); rather,

$f_{k,k+T}$ is the interest rate that we can secure **now** for the period $[k, k + T]$, using the above approach.

That is the best possible strategy!

Hence, the strategy set out above has proved to be advantageous compared to the interest result achieved in the negotiations with the bank. This immediately begs the question, of course, whether there might be another strategy with the potential of even greater success?

The short answer is: No, there isn't! We are going to show below that it is not possible, not by any strategy whatsoever, to create a safe interest rate for the time period $[k, k + T]$ that is higher than $f_{k,k+T}$.

Let us for a moment assume the opposite, that is, that we will indeed be able to find a strategy allowing us to secure an interest rate z for the time period $[k, k+T]$ that is higher than $f_{k,k+T}$. Thus, $z > f_{k,k+T}$.

In this case, I would immediately do the following (to illustrate my approach, let us first look at the initial example in the time period $[1, 3]$ again and then argue the general case):

I would take out a loan in the highest possible amount (for the sake of simplicity: EUR 1 million) for 3 years at the interest rate $f_{0,3} = -0.350\%$ and initially invest it for 1 year at the interest rate $f_{0.1} = -0.526\%$. In this first year, the invested amount becomes $1{,}000{,}000 \cdot \left(1 + f_{0,1}\right) = 1{,}000{,}000 \cdot (1 - 0.00526) = 994{,}740$ euros. I would now invest this amount at the interest rate $z > f_{1.3}$ for the second and third years. This brings the initial amount to a total of:

final amount $= 1{,}000{,}000 \cdot \left(1 + f_{0,1}\right) \cdot (1+z)^2$ euros.

At the end of this third year, I have to redeem (repay) the loan. The total amount to be redeemed is:

redemption amount $= 1{,}000{,}000 \cdot \left(1 + f_{0,3}\right)^3$

The formula (1.3) above, or the general formula (1.4) for forward rates, gives us the equation $\left(1 + f_{0,3}\right)^3 = \left(1 + f_{0,1}\right) \cdot \left(1 + f_{1,3}\right)^2$.

If we replace $\left(1 + f_{0,3}\right)^3$ by $\left(1 + f_{0,1}\right) \cdot \left(1 + f_{1,3}\right)^2$ in the formula for the redemption amount, we get:

redemption amount $= 1{,}000{,}000 \cdot \left(1 + f_{0,1}\right) \cdot \left(1 + f_{1,3}\right)^2$.

Since we have assumed that the interest rate z is greater than $f_{1,3}$, it follows that the final amount of our investment process is greater than the redemption amount that we have to pay.

Therefore, after repayment of the loan, we have made a safe risk-free profit equal to the difference between the final amount and the redemption amount, without having to invest any funds of our own at any time (remember: we started the process by taking out a loan)!

At this point in our argumentation, and for the first time in this book, we apply THE axiom of financial mathematics, the so-called **no-arbitrage principle**,—albeit in a rather informal, not strictly mathematical form.

We will provide a strict definition of this axiom in the chapter on "Frictionless Markets" and also discuss it there.

Here, as noted above, we are going to provide an informal introduction only and accept it as a given for the time being:

The **no-arbitrage principle** states: ***It is not possible in the financial markets to create a safe and completely risk-free positive profit in a certain period of time through any strategy whatsoever without the use of capital.***

If it were indeed possible, a large number of market participants would massively employ such a strategy. Due to the enormous demand for or supply of certain financial products on which the strategy is based, the prices of these products would change so rapidly and to such an extent that any potential arbitrage strategy (= arbitrage opportunity) would quickly become impossible to execute.

The no-arbitrage principle is sometimes referred to, rather informally, as *there is no such thing as a free lunch.*

Above we noted as follows: if we were able to secure, for the future period $[1, 3]$, an interest rate z that is higher than the forward rate $f_{1,3}$, we would have an arbitrage opportunity, which however—as per the no-arbitrage principle—is not possible. It follows then—as argued above—that the forward interest rate $f_{1,3}$ is indeed the best possible interest rate that we can secure right now for the time period $[1, 3]$.

The same observations can of course be applied analogously to any future time periods, so that we get the following:

Theorem 1.8 *The forward rate* $f_{k,k+T} = \left(\frac{(1+f_{0,k+T})^{k+T}}{(1+f_{0,k})^{k}}\right)^{\frac{1}{T}} - 1$ *is the maximum interest rate for investments, or the minimum interest rate for loans that can be secured at time 0 for the future period* $[k, k+T]$ *based on the ideal benchmark interest rates* $f_{0,t}$ *.*

A simple program for calculating the forward rate can be found on the book's website.

See https://app.lsqf.org/book/forward-rates

1.11 The Fair Value of a Future Payment, Discounting

In this short subsection, we will once again apply the no-arbitrage principle in its informal version in order to strictly justify something so seemingly self-evident that it is often not questioned, namely, the process of discounting future payments.

What is the present value (at time 0) of a payment of, say, 1 million euros that is going to be made to us in, for example, 3 years?

Any reader with basic knowledge of handling cash flows will answer:

That one million euros that we are going to get in 3 years – that is, in the future – has a different value right now, namely: The value that we get when we "discount" the payment to now.

The **discounted value of a payment of K euros in T years** is the value

$$\text{discounted value} = \frac{\boldsymbol{K}}{\left(\mathbf{1} + \boldsymbol{f}_{\mathbf{0},\boldsymbol{T}}\right)^{\boldsymbol{T}}}.$$

In our example, the discounted value would therefore be

$$\text{discounted value} = \frac{1{,}000{,}000}{\left(1 + f_{0,3}\right)^{3}} = \frac{1{,}000{,}000}{(1 - 0.0035)^{3}} = 1{,}010{,}574 \text{ EUR}.$$

Again, in times of negative short-term interest rates, we need to get used to the fact that discounted values can be greater than the nondiscounted value!

As noted above, the need for discounting values is often accepted without question. But how can we strictly justify the proposition that the actual value of a future payment is the discounted value? Again, the reason for what appears so unquestionably obvious is the no-arbitrage principle (in the following occasionally referred to as the NA principle).

The justification for the argument that the actual present value of a payment of K euros at a future time T is precisely the discounted value $\frac{K}{(1+f_{0,T})^T}$ is:

If the present value of a payment in the amount of K euros at a future time T were greater than $\frac{K}{(1+f_{0,T})^T}$, meaning, if someone were willing to pay a price M greater than $\frac{K}{(1+f_{0,T})^T}$ in exchange for a promised payment of K euros at time T, then I would sell this promise at price M and invest the purchase price I receive for a period up to time T at interest rate $f_{0,T}$. Over the period until time T, the money I invested will become $M \cdot \left(1 + f_{0,T}\right)^T$ euros. This value is greater than $K \left(\text{since } M > \frac{K}{(1+f_{0,T})^T} \Rightarrow M \cdot \left(1 + f_{0,T}\right)^T > K\right)$. I can therefore pay out the promised amount K and keep the difference $M \cdot \left(1 + f_{0,T}\right)^T - K$ as my safe positive profit. This means that we would have an opportunity for arbitrage, which runs counter to the no-arbitrage principle.

If the present value of a payment in the amount of K euros at a future time T were less than $\frac{K}{(1+f_{0,T})^T}$, meaning, if someone were willing to sell the promise of making a payment of K euros at time T at a price M that is below $\frac{K}{(1+f_{0,T})^T}$, then I would take out a loan of M euros for T years at interest rate $f_{0,T}$ and use that money to buy this promise. At time T, I would receive K euros and can redeem my loan with that.

I can do so of course because the redemption amount is $M \cdot \left(1 + f_{0,T}\right)^T$ and that is less than $K \left(\text{since } M < \frac{K}{(1+f_{0,T})^T} \Rightarrow M \cdot \left(1 + f_{0,T}\right)^T < K\right)$. I would thus make a safe positive profit in the amount of the difference $K - M \cdot \left(1 + f_{0,T}\right)^T$. Here again we would have an opportunity for arbitrage, which contradicts the no-arbitrage principle.

And with that we have positively concluded the argument!

To summarize:

A future payment in the amount of K EUR at time T currently—at time 0—has the value $\frac{K}{(1+f_{0,T})^T}$ EUR.

For any other value, there would be an opportunity for arbitrage.

We refer to the amount $\frac{K}{(1+f_{0,T})^T}$ EUR, i.e. the **discounted value** of the payment, as the **fair value** of that payment.

However, all of the above considerations are only valid if I have 100% certainty that the future payments will actually take place, i.e. that the promises will definitely be kept!

1.12 The Fair Value of a Bond

The most important fixed technical parameters of a bond are the maturity date T, the coupon C, and the future payment dates $t_1, t_2, t_3, \ldots, t_{n-1}, t_n$.

The price of a bond is a variable parameter that changes over the course of the bond's life as a result of supply and demand dynamics.

We want to know, and thus ask the question: What is the "fair price", i.e. the fair value of a bond? For example, is the current (22 September 2021) price of 108.61 a fair price for our VW bond?

The question as to how to calculate the fair price of a bond is easy to answer **in principle**, based on the arguments that we outlined in the previous chapter.

You know that, as a holder of a bond, you will receive certain future payments on that bond. **The fair value of a bond is therefore simply the sum of the fair values of these payments, i.e. the sum of the discounted payments.** The formula for calculating this sum of discounted payments is shown below.

But first, I want to clarify why I wrote that we can calculate the value "in principle".

I say that because this approach—as pointed out above—is only applicable if I can be absolutely certain that the payments that I expect to receive will actually be made in full. However, only bonds with the highest credit quality can (largely) guarantee that.

For bonds of the highest quality with the following fixed parameters:

$t_1, t_2, t_3, \ldots, t_{n-1}, t_n = T$ coupon pay dates,
$C_1, C_2, C_3, \ldots, C_{n-1}, C_n$ coupons paid at the stated dates

the fair value (FV) of the bond is given by

$$FV = \frac{C_1}{\left(1+f_{0,t_1}\right)^{t_1}} + \frac{C_2}{\left(1+f_{0,t_2}\right)^{t_2}} + \frac{C_3}{\left(1+f_{0,t_3}\right)^{t_3}} + \ldots + \frac{C_{n-1}}{\left(1+f_{0,t_{n-1}}\right)^{t_{n-1}}} + \frac{C_n + 100}{\left(1+f_{0,t_n}\right)^{t_n}}. \tag{1.5}$$

This value is indeed a benchmark for the actual price to be paid for the bond, i.e. for the bond's dirty price (price + pro rata coupon)!

In other words, the "fair price" of the bond is given by the "fair value FV" of the bond minus the pro rata coupon.

Now, if we apply this new insight to our VW bond, we will immediately fail, for the following reason: as we know from its fact sheet, the VW bond has a Standard& Poor's rating of BBB+, meaning it is far from having the highest credit quality rating.

We will discuss "valuation", i.e. the determination of a fair price for bonds with lower credit ratings, in more detail in Volume III Chapter 1. Such bonds require a slightly different approach.

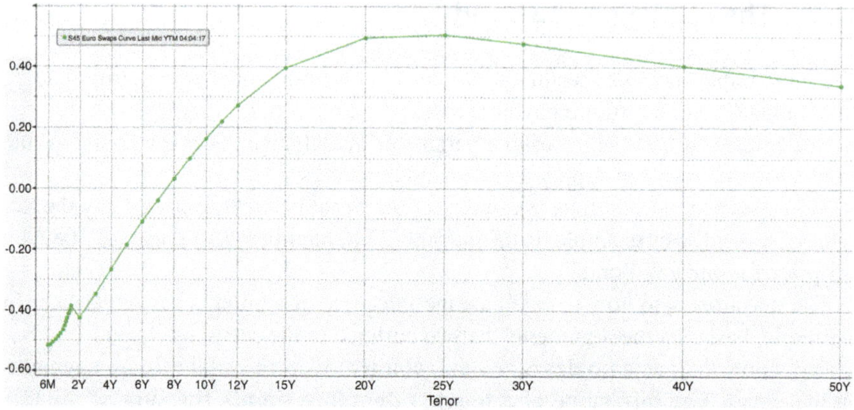

Fig. 1.19 Euro swaps curve on 22 September 2021 (source: Bloomberg)

Example 1.9 For purposes of a practical exercise, we want to calculate the fair value and the fair price for the VW bond—as if it had the highest credit rating—based on the above formula:

The VW bond data were from 22 September 2021.

Future interest payments of 1.625% were scheduled for 16 January of each year, starting on 16 January 2022 and ending on 16 January 2030.

If we set the period from 22 September of a year to 16 January of the following year at roughly 4 months, i.e. at $\frac{1}{3}$ of a year, then the time intervals until the future interest payments $t_1, t_2, t_3, \ldots, t_8, t_9$ are given by 0.33, 1.33, 2.33, ..., 8.33. In order to determine the fair value, we also need the interest rates f_{0,t_i} where $i = 1, 2, \ldots, 9$ on 22 September 2021. We will extrapolate these values from the EONIA ($f_{0,0}$) the 6-month Euribor $\left(f_{0,0.5}\right)$, the 9-month Euribor $\left(f_{0,0.75}\right)$, and the ICE swap rates for 1–9 years. We go to Bloomberg to get the historical data for the Euribor rates and the ICE swap rates and find the following data for 22 September 2021 (see also Fig. 1.19)

$$
\begin{aligned}
f_{0,0} &= -0.488\% \\
f_{0,0.5} &= -0.517\% \\
f_{0,0.75} &= -0.496\% \\
f_{0,1} &= -0.464\% \\
f_{0,2} &= -0.426\% \\
f_{0,3} &= -0.3479\% \\
f_{0,4} &= -0.267\% \\
f_{0,5} &= -0.1871\% \\
f_{0,6} &= -0.112\% \\
f_{0,7} &= -0.04\% \\
f_{0,8} &= 0.0304\% \\
f_{0,9} &= 0.098\%
\end{aligned}
$$

By interpolation using the formula (1.1) in Sect. 1.7 (see also Fig. 1.19 for illustration of the interest rate structure on 22 September 2021 and the interpolation), we obtain

$$
\begin{aligned}
f_{0,0.33} &= -0.507\% \\
f_{0,1.33} &= -0.451\% \\
f_{0,2.33} &= -0.400\% \\
f_{0,3.33} &= -0.321\% \\
f_{0,4.33} &= -0.240\% \\
f_{0,5.33} &= -0.162\% \\
f_{0,6.33} &= -0.088\% \\
f_{0,7.33} &= -0.017\% \\
f_{0,8.33} &= 0.053\%
\end{aligned}
$$

We then get the fair value using Formula (1.5) (assuming highest credit rating):

$$
\begin{aligned}
FV = {} & \frac{1.625}{\left(1+f_{0,0.33}\right)^{0.33}} + \frac{1.625}{\left(1+f_{0,1.33}\right)^{1.33}} + \ldots + \\
& + \frac{1.625}{\left(1+f_{0,7.66}\right)^{7.33}} + \frac{1.625+100}{\left(1+f_{0,8.33}\right)^{8.33}} = 114.27
\end{aligned}
$$

Thus, the "fair price" is
FV—pro rata coupon = 114.27 – 1.08 = 113.19

The bond's actual price of 108.61 is much lower than the calculated fair price of 113.19. Remember, however, that this would be the bond's fair price if the bond had the highest credit rating. Yet because it only has a BBB+ rating, the actual price is considerably lower, so the bond is significantly less valuable than the above fair value would suggest.

Example 1.10 Let us therefore calculate another example, this time for an actual AAA-rated bond.

WKN A180TW
Issuer: Cie de Financement Foncier
Statistics on 29 September 2021
Maturity date: 4 September 2024
Coupon: 0.5%
Interest payments on 4 September of each year
The pro rata coupon on 28 September is thus almost equal to 0.
The actual price on 28 September was 102.5 EUR.

The next coupon payments will take place almost exactly in 1, 2, and 3 years. We will therefore discount the following payments at the ICE swap rates:

$$f_{0,1} = -0.526\%$$
$$f_{0,2} = -0.429\%$$
$$f_{0,3} = -0.350\%$$

These values (which were current at the time of writing this example) were again retrieved from the website https://www.theice.com/marketdata/reports/180.

Using that data to calculate the fair value FV, we get:

$$\begin{aligned} FV &= \frac{0.5}{(1+f_{0,1})} + \frac{0.5}{(1+f_{0,2})^2} + \frac{0.5+100}{(1+f_{0,3})^3} = \\ &= \frac{0.5}{(1-0.00526)} + \frac{0.5}{(1-0.00429)^2} + \frac{0.5+100}{(1-0.0035)^3} = \\ &= 102.57 \end{aligned}$$

The calculated fair value therefore almost perfectly coincides with the actual current price of 102.5 EUR!

If interested, you can use the software provided on this book's website to calculate and chart various bond metrics (see https://app.lsqf.org/anleihen).

1.13 Some Examples of Junk Bonds

In general, bonds—especially those with top ratings—are purchased primarily because they represent a relatively safe haven for investments and offer coupon earnings. Speculations on price fluctuations are only of minor importance (as opposed to the stock market).

However, bond price speculations can become quite interesting when it comes to so-called junk bonds. These are bonds that have suffered severe, far above-average price losses as a result of massive downgradings. Bargain hunters in the bond market may buy such bonds in the hope that the issuer will recover and the bonds that they bought at exceedingly low prices during the crisis can then be sold at prices close to 100. Here are just two examples of such bonds from recent years.

The first example is an Argentine government bond issued in 1996 with a coupon of 11.75% and maturity on 13 November 2026.

The price chart of the bond for the past 23 years is shown in Fig. 1.20.

Its lowest price was 2.00 at the beginning of 2009. If we had bought the bond then, trusting that Argentina would not default, we could have sold it between 2016 and 2019 at a price of around 125. Currently, it is again traded at a very low price of around 30.

TRADE PLATFORM	BÖRSE STUTTGART (XSTU)	
LAST PRICE / YIELD	30 G / 64.15	0 nom.
PRICE DETERMINATION TIME	09/29/2021 / 12:10:47 PM	
DAILY VOLUME (NOMINAL)	0	
YIELD SPREAD	64.9008	
SPOT PRICE	84	
DAILY HIGH / LOW	30	30
PREV. DAY'S PRICE	30	(28.09.)
CHANGE DAY BEFORE	→ +0	→ 0.00 %
52 WEEK HIGH / LOW	34 (27/07)	5 (06/01)

Fig. 1.20 Price curve of an Argentine government bond (source: Stuttgart Stock Exchange)

The second example is a Greek government bond.
Its key data are:
Coupon: 3%
Maturity: 24 February 2029

The bond's price performance since 2012 is illustrated in Fig. 1.21.

Here, too, we would have made a significant profit if we had bought this bond at the beginning of 2012 at prices between 11 and 20 and sold it at the currently prevailing bid price of around 123.

Griechenland EO-Bonds 2012(28) Ser.6

WKN A1G1UF ISIN GR0133006198

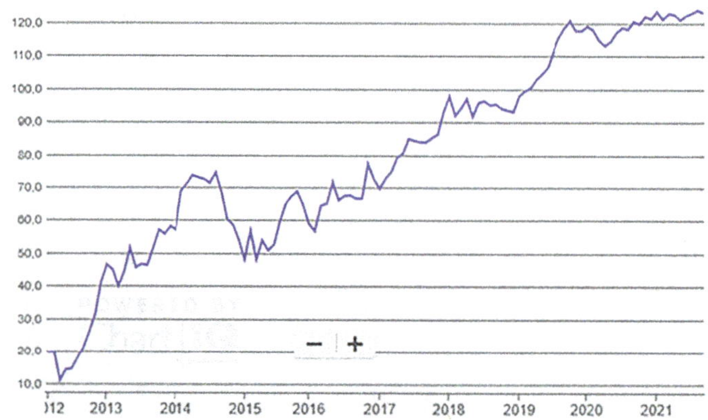

Price data

TRADE PLATFORM	BORSE STUTTGART (XSTU)	
LAST PRICE / YIELD	123.476 G	0 nom.
PRICE DETERMINATION TIME	09/29/2021 / 12:13:30 PM	
DAILY VOLUME (NOMINAL)	16	
YIELD SPREAD	1.3318	
SPOT PRICE	64.5	
DAILY HIGH / LOW	123.448	122
PREV. DAY'S PRICE	123.196	(28.09.)
CHANGE DAY BEFORE	↗ +0.28	↗ +0.23%
52 WEEK HIGH / LOW	125.6 (05/02)	117 (30/09)

Fig. 1.21 Price curve of a Greek government bond (source: Stuttgart Stock Exchange)

1.14 Stock (Basics)

This section will only provide a brief overview of the basic characteristics and determinants of stocks (or equities, as they are also called). Other related issues, such as stock price modelling, will be addressed in other sections of this book.

When you purchase stock, you buy part ownership of a company.

(Remember: When you buy bonds from a company, you merely grant that company a loan.)

The key *defining parameters* of a stock are:

the *issuer*: a publicly traded stock corporation (such as Coca Cola, Deutsche Bank, etc.)

the *currency* in which the stock is denominated (e.g. EUR, USD, etc.)

issued *volume* (shares or units) of stock (e.g. 150 million shares)

variable parameters are

the *share price* (e.g. 87.59, etc.)

market capitalization in EUR (e.g. EUR 2257 billion, etc.)

amount of the periodic (annual) *dividend distribution* (e.g. EUR 5.10 per share, etc.)

For illustration purposes, we once again use an actual example and choose the Allianz share traded on the Frankfurt Stock Exchange. All relevant data pertaining to this share can be found at:
http://www.boerse-frankfurt.de/aktie/Allianz-se

Issuer: Allianz

WKN/German securities identification number: 840400

Currency: euro

Volume: 412,290,000 shares

Closing price on 29 September 2021: EUR 195.28

Market capitalization (= number of shares × price): EUR 80,512,000,000

Last dividend payouts:

2017: EUR 7.60
2018: EUR 8.00
2019: EUR 9.00
2020: EUR 9.60
2021: EUR 9.60

Figure 1.22 displays the Allianz stock price performance for the last 31 years, whereby Fig. 1.23 shows the intraday performance on 29 September 2021.

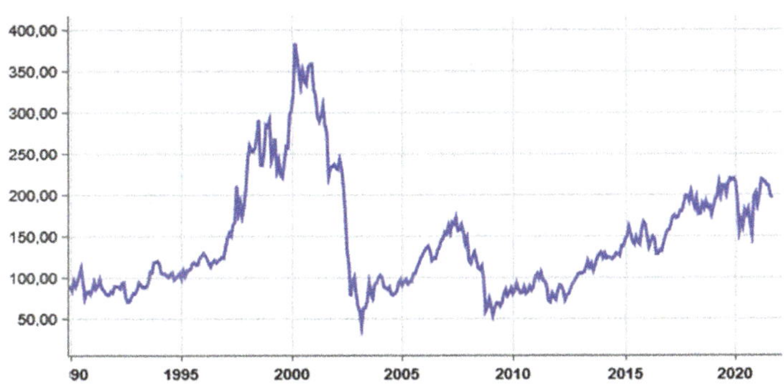

Fig. 1.22 Allianz stock price performance, last 31 years (source: Frankfurt Stock Exchange website)

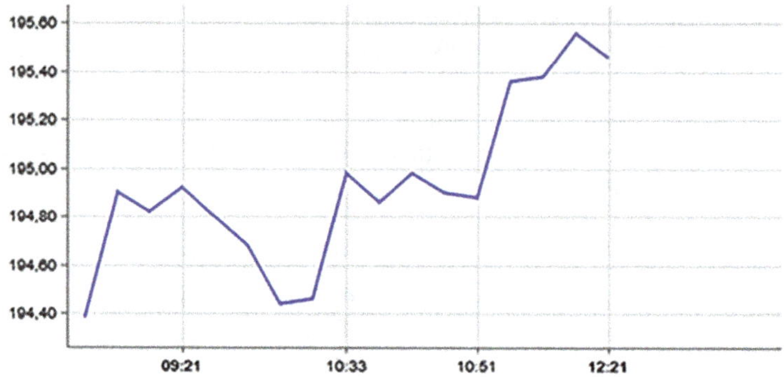

Fig. 1.23 Allianz stock price performance, intraday, 29 September 2021 (source: Frankfurt Stock Exchange website)

As explained above, when you buy stock from a company, you buy a certain ownership share in that company. Your part ownership is determined in relation to the total number of issued shares.

For example, by purchasing one Allianz share, you get to own $\frac{1}{412,290,000}$ of the company. With 206,145,000 Allianz shares, you would own half of the company.

When the shares of stock are first issued, they are purchased at an **initial offering price**. This initial offering price is determined by the bank in charge of coordinating the IPO (initial price offering) on the basis of demand from interested investors. Immediately after the initial offering, trading of the stock begins on the stock exchange(s) on which the stock is listed (admitted to trading). Major price fluctuations may already occur immediately after trading has begun.

Shares are traded at the prevailing stock price. This price is the actual purchase price (as opposed to bond trading, where the buyer also has to pay a pro rata coupon).

The price of a stock is generally subject to much stronger short-term and long-term fluctuations. It is in no way linked to any fixed benchmark metric. (Bond prices, on the other hand, are usually close to the value of 100 and practically never significantly higher than 100.)

At a company's general meeting, which is usually held annually and to which all shareholders are invited, and at which all shareholders have voting rights (in proportion to the number of shares they hold), the company's management reports on the company's business performance over the past financial year.

One of the main topics that management reports on at the annual general meeting (AGM) is the company's earnings performance. The company owners—i.e. the shareholders—then decide by vote what proportion of those earnings should be paid to the shareholders (e.g. and what proportion should be retained by the company for new investments or as reserves).

The annual profit distribution is referred to as the annual dividend. The amount of the **dividend** is stated in EUR per share. The above-mentioned dividend distribution

Historical prices and volumes Allianz SE

1 2 3 4 5 6

Date	Open	Close	High	Low
10/05/21	215.25	215.10	215.50	214.00
07/05/21	213.95	215.05	215.80	213.35
06/05/21	213.20	213.45	215.30	212.00
05/05/21	218.80	221.15	221.90	218.00
04/05/21	220.00	218.90	221.00	216.10
03/05/21	218.45	220.25	220.25	217.30
30/04/21	216.60	217.30	218.05	216.50

Fig. 1.24 Historical stock price data for Allianz (source: Frankfurt Stock Exchange website)

of EUR 9.60 for the year 2021 therefore means that a total profit share of EUR $412{,}290{,}000 \times 9.60 = 3{,}957{,}984{,}000$ was distributed to the shareholders for that year.

The dividend distribution for the year 2020 took place on 5 May 2021.

The share price before distribution of the dividend (the closing price on 5 May 2021) was 221.15 (see Fig. 1.24 with the historical Allianz share prices around 5 May 2016). At 213.20, the opening price on 6 May 2021—i.e. after payment of the dividend—was significantly lower than on the day before. This reflects a typical and very logical phenomenon: upon distribution of the dividend, the share price usually falls instantly by approximately the amount of the dividend. This is logical in that the stock's value decreases by the amount of the dividend payout.

This is a phenomenon that **does not** occur with bonds when a coupon is paid! That's because, when you buy bonds, you pay the pro rata coupon in addition to the bond's price. So, when a coupon is paid, the bond's **purchase price** is reduced by the amount of that coupon (before the payout, the pro rata coupon = coupon amount, after the payout the pro rata coupon = 0), **but not the price** of the bond per se.

The concept of a "dirty price" to prevent price fluctuations around a dividend distribution date cannot be applied to stock, of course, since the amount of the dividend to be paid is not known in advance (as opposed to the fixed coupon of a bond).

Let us see how the amount of the dividend—for 2020, for instance—correlates with the stock's purchase price 1 year before the dividend is paid, hence, in May 2020:

On 5 May 2020, for instance, the stock's closing price was 165.40.

The dividend on 5 May 2021 was 9.60. So, if you bought a share at EUR 165.40 and held it for 1 year, you would have earned a payout of EUR 9.60. This payment can be interpreted as "a variable return" on the invested capital in the amount of $100 \times \frac{9.60}{165.40} = 5.80\%$.

1.15 Stock Market Dynamics

In addition to income generated through dividend payouts, many investors are also interested in actively trading stock to profit from price movements, as prices can fluctuate quite strongly. We will therefore briefly discuss the dynamics of stock trading below.

As an example, we will again use the Allianz stock as it is traded on the Frankfurt Stock Exchange. Figures 1.25 and 1.26 are screenshots of the Allianz stock traded on the Frankfurt Stock Exchange at a particular date. The two screenshots were taken approximately one and a half hours apart.

The current price of a stock is the price at which it was last traded. In the screenshots above, we see the price of the Alliance stock at each of the two points in time in the last but one row (i.e. 196.16 at 3:33:13 pm and 196.30 at 4:49:44 pm). If the price of a stock is the price at which it was last traded, it would seem to allow for all kinds of price manipulations: all you would need to do is make an offer to buy an Allianz share at, say, 300 euros. Any number of shareholders would be delighted to offer you shares at this price (which is well above the current price of 196.30). The

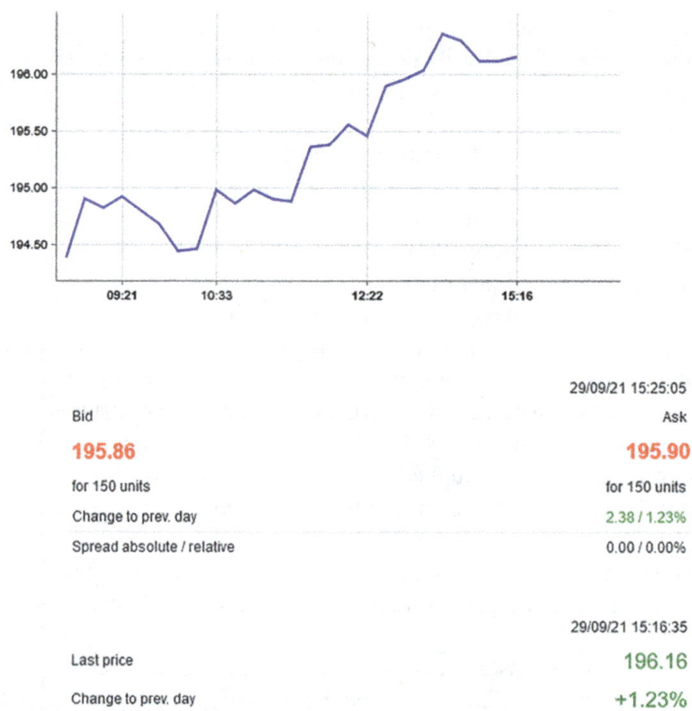

Fig. 1.25 Screenshot of the Allianz stock as traded on the Frankfurt Stock Exchange on 29 September 2021 at 3:33:13 pm (source: Frankfurt Stock Exchange website)

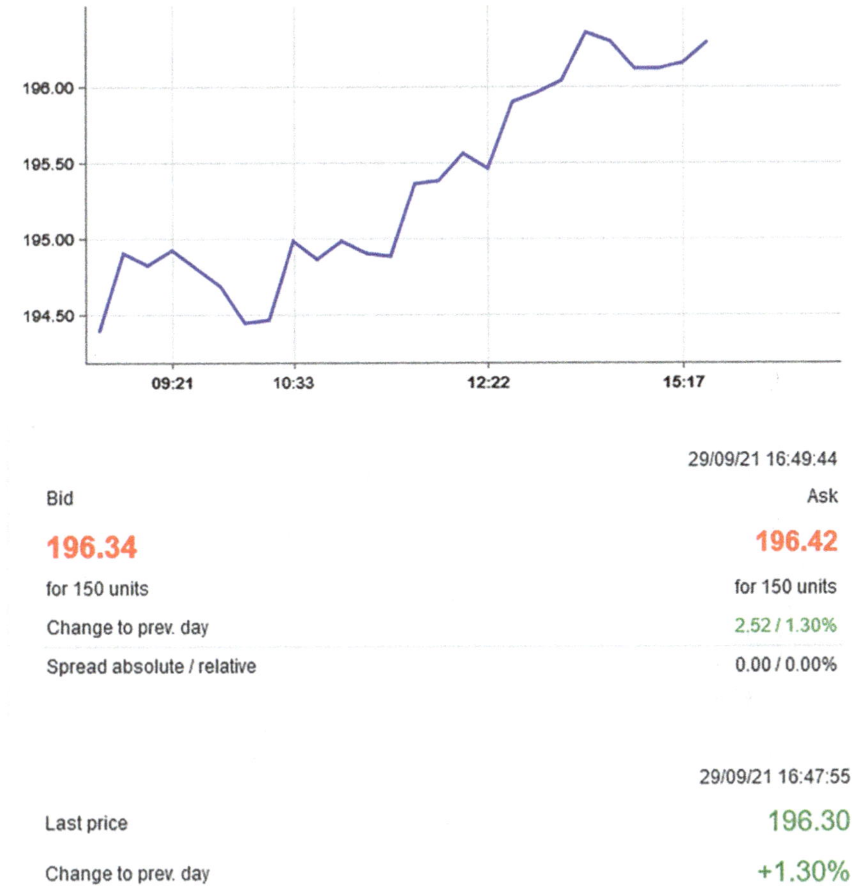

Fig. 1.26 Screenshot of the Allianz stock as traded on the Frankfurt Stock Exchange on 29 September 2021 at 4:49:44 pm (source: Frankfurt Stock Exchange website)

trade would be immediately executed at said price of 300 euros, and the new price would then be 300 euros. Subsequently, a shareholder would offer a share for sale at the price of 1 euro, again purely for the purpose of manipulating the price. Interested parties would immediately report interest in the share, it would then be traded at 1 euro, and the price (= last price at which a trade was made) would immediately drop to 1 euro.

The dynamics of stock exchange tradings, however, prevent this from happening: in the screenshots of the Frankfurt Stock Exchange for the Allianz stock, you see the stock's current top of the **order book** with the **highest bid and lowest ask prices** (196.34/196.42), each for a certain volume of shares (150 shares/150 shares).

Table 1.2 Fictional example of an order book up to depth 4

Buy orders // bid price				Last	Sell orders // ask price			
28.60	29.00	29.10	29.50	**30.00 €**	30.20	30.40	30.50	30.70
2000 shares	50 shares	1300 shares	150 shares		20 shares	550 shares	1000 shares	120 shares

These values tell us that:

There is currently one (or more) market participant(s) willing to buy up to (a total of) 150 Allianz shares at the price of EUR 196.34 per share (bid price).

And:

There is one (or more) market participants willing to sell up to (a total of) 150 Allianz shares at the price of EUR 196.42 per share (ask price).

These are the best prices that are currently offered on the stock exchange for the purchase and sale of Allianz shares. However, the order book normally has a much greater **depth** on both sides. That is to say, on the left side it lists further purchase offers (all lower than 196.34, however) for different volumes, sorted by size, while the right side lists further selling offers (all higher than 196.42) for different volumes, sorted by size.

To illustrate the order book process and the trading process, consider the theoretical example of an order book up to depth 4 as given in Table 1.2, whereby the data is somewhat clearer and easier to handle than the Allianz stock.

The price at which this stock A was traded ("Last"), was EUR 30.00.

Currently, there are parties interested in purchasing
150 shares of stock A at EUR 29.50
1300 shares of stock A at EUR 29.10
50 shares of stock A at EUR 29.00
2000 shares of stock A at EUR 28.60
(etc., as the order book may have a much greater depth).

In addition,

There are currently parties interested in selling
20 shares of stock A at EUR 30.20
550 shares of stock A at EUR 30.40
1000 shares of stock A at EUR 30.50
120 shares of stock A at EUR 30.70
(etc., as the order book may have a much greater depth).

As long as this situation persists, no shares will be traded.

Table 1.3 Limit order of a potential seller in the given order book from Table 1.2

Buy orders // bid price				Last	Sell orders // ask price			
28.60	29.00	29.10	29.50	30.00€	30.20	30.40	30.50	30.70
2000 shares	50 shares	1300 shares	150 shares		20 shares	550 shares	1000 shares	120 shares
28.60	29.00	29.10	29.50	**30.00 €**	29.90	30.20	30.40	30.50
2,000 shares	50 shares	1,300 shares	150 shares		100 shares	20 shares	550 shares	1,000 shares

Table 1.4 Order book from Table 1.3 with an actual trade

Buy orders // bid price				Last	Sell orders // ask price			
28.60	29.00	29.10	29.50	30.00€	30.20	30.40	30.50	30.70
2000 shares	50 shares	1300 shares	150 shares		20 shares	550 shares	1000 shares	120 shares
28.60	29.00	29.10	29.50	30.00€	29.90	30.20	30.40	30.50
2000 shares	50 shares	1300 shares	150 shares		100 shares	20 shares	550 shares	1000 shares
28.60	29.00	29.10	29.50	**29.90€**	29.90	30.20	30.40	30.50
2,000 shares	50 shares	1,300 shares	80 shares		30 shares	20 shares	550 shares	1,000 shares

Let us now assume that a potential seller *B* places a new order in the form of a limit order for the sale of 100 shares of stock *A* at a minimum price of EUR 29.90. The order book given in the previous Table 1.2 then changes as follows:

However, there is still no trade taking place.

The change in the ask price might, however, induce an investor *C*, who was previously logged into the order book given in the Table 1.3 with, say, 70 shares at 29.50 on the bid price side, to increase that limit to 29.90. This will immediately result in a trade: *C* buys 70 shares from *B* at a price of 29.90. The new price is 29.90. *B* remains listed on the order book's ask-price side with 30 shares on offer.

Now comes yet another investor *D*, who wants to buy 500 shares of stock A and is willing to pay a maximum of EUR 30.50 per share. The decisive factor now is the following: the ask price side is processed according to the offers on the ask price side. So, *D* does not get 500 shares at the price of 30.50 but will automatically buy 30 shares at 29.90, 20 shares at 30.20, and 450 shares at 30.40. The new order book situation is presented in Table 1.5 (the two new entries on the far right were previously depths 5 and 6 in the old order book given in the Table 1.4).

The new price is now 30.40. This would have been the case even if investor *D* had been willing to pay EUR 100 per share, or had been willing to pay any price (in which case he would have placed a so-called at-best order). A massive manipulation

Table 1.5 Final order book

Buy orders // bid price				Last	Sell orders // ask price			
28.60	29.00	29.10	29.50	30.00€	30.20	30.40	30.50	30.70
2000 shares	50 shares	1300 shares	150 shares		20 shares	550 shares	1000 shares	120 shares
28.60	29.00	29.10	29.50	30.00€	29.90	30.20	30.40	30.50
2000 shares	50 shares	1300 shares	150 shares		100 shares	20 shares	550 shares	1000 shares
28.60	29.00	29.10	29.50	29.90€	29.90	30.20	30.40	30.50
2000 shares	50 shares	1300 shares	80 shares		30 shares	20 shares	550 shares	1000 shares
28.60	29.00	29.10	29.50	**30.40 €**	30.40	30.50	30.70	30.80
2000 shares	50 shares	1300 shares	150 shares		100 shares	1000 shares	120 shares	3500 shares

of a stock's price would therefore only be possible if an entire side of the order book were processed (Table 1.5).

1.16 Stock Indices

A stock index (equity index) reflects the average performance of a certain number of stocks that have been selected according to geographical markets, industry sector, or on the basis of other common characteristics. The criteria used for selecting stocks may vary from index to index.

Their performance is usually averaged on the basis of weighting, where "larger" stocks are given more weight than "smaller" stocks. The way in which the "size" of a stock is defined (capitalization, turnover, etc.) depends on the index in question.

Some indices only measure how the value of the stocks in the index is changing (**price index**), while others add the dividends paid on the stocks in the index (**performance index**). The question as to whether a particular index is a price index or a performance index will also be relevant when we later discuss valuation of derivatives on an index.

If it is relevant for investment decisions to know exactly how a given index is composed and calculated, the investor will have to consult the specifications of that index (on the relevant website).

Stock indexes are usually defined, calculated, and regularly published by a stock exchange (provided all stocks of the index are listed on that exchange) or by an information provider (e.g.: Reuters, Standard & Poor's, etc.).

In principle, indexes can be traded indirectly by trading the stocks of the index according to their relative weighting. However, this requires a great deal of effort and usually a great deal of money. In most cases, investors can trade indices more directly by investing in smaller units, either through **index futures** (see Sect. 2.17) or through **index certificates**, which replicate the relevant index one-to-one. These are issued by banks and investment firms and offered for purchase and sale on their trading platforms. We will have a look at two examples of index certificates in the following subsections.

To give you an idea, we will present a few select examples of some of the main international stock indices in the following and describe them in a few keywords. The only stock index that we will discuss in more detail is the American S&P500 Index, as it will form the basis of most of our observations on derivative trading strategies.

An excellent overview of international indices and their creation, history, and components can be found at http://www.finanzen.net/indizes/.

Europe

Euro Stoxx 50 tracks the average price movements of the 50 largest listed companies in the euro area—price index (there is also a performance index version of Euro Stoxx 50, which is explicitly identified as such). Its development for the past 21 years is given in Fig. 1.27.

Stoxx Europe 50 tracks the average price movements of the 50 largest listed companies in Europe—price index (a separately identified performance index version also exists).

DAX tracks the average price movements of the 30 largest listed companies in Germany; performance index (a separately identified price index version also exists). The development is given in Fig. 1.28.

FTSE 100 is short for Financial Times Stock Exchange and reflects the average price movements of the 100 largest UK stocks listed on the London Stock Exchange—price index. For a historical development, compare Fig. 1.29.

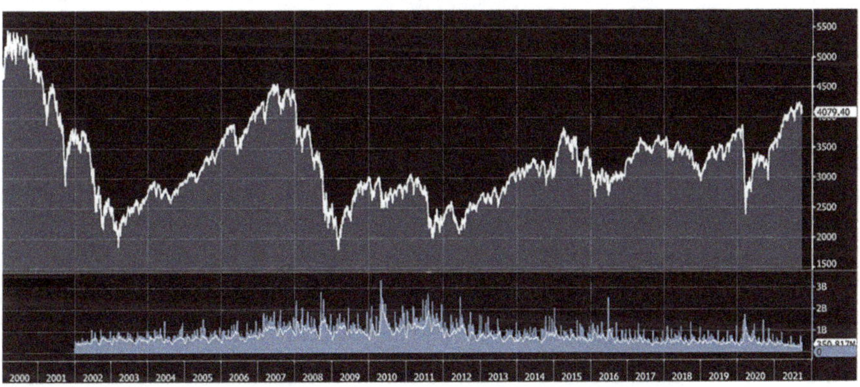

Fig. 1.27 Euro Stoxx development from January 2000 to September 2021

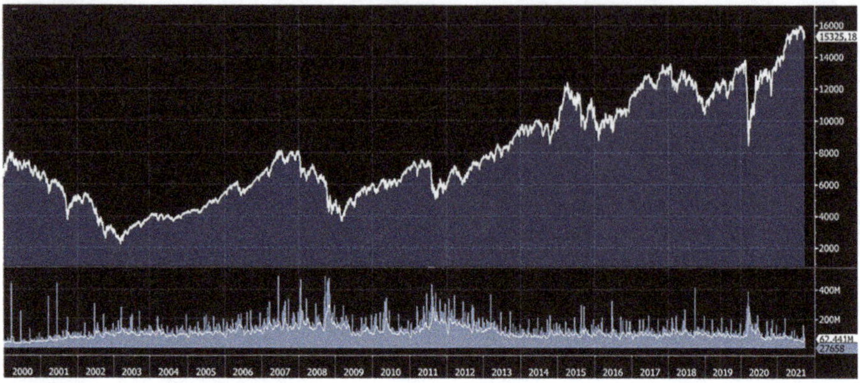

Fig. 1.28 DAX development from January 2000 to September 2021

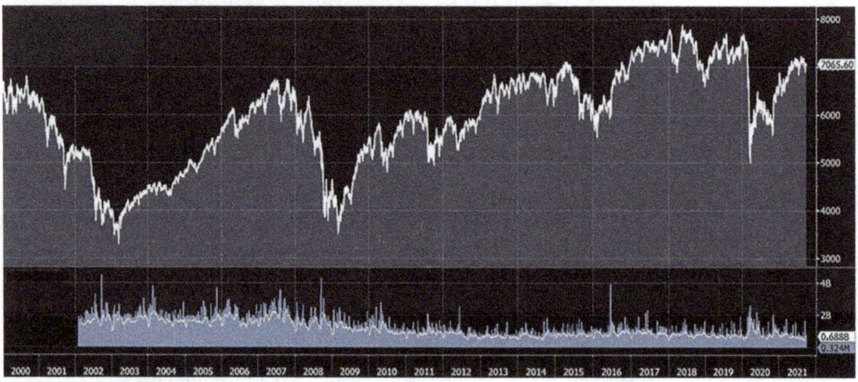

Fig. 1.29 FTSE development from January 2000 to September 2021

CAC 40 tracks the average price movements of the 40 largest and most actively traded French companies listed on the Paris Stock Exchange—price index

SMI tracks the average price movements of the 20 largest and most liquid companies listed on the Swiss Stock Exchange (SIX)—price index

IBEX 35 tracks the average price movements of the 35 largest and most actively traded Spanish companies listed on the Madrid Stock Exchange—price index

ATX short for Austrian Traded Index which reflects the average price movements of the 20 largest listed Austrian companies—price index

AEX is the stock index of the Amsterdam Stock Exchange and is composed of up to 25 Dutch stock corporations—price index

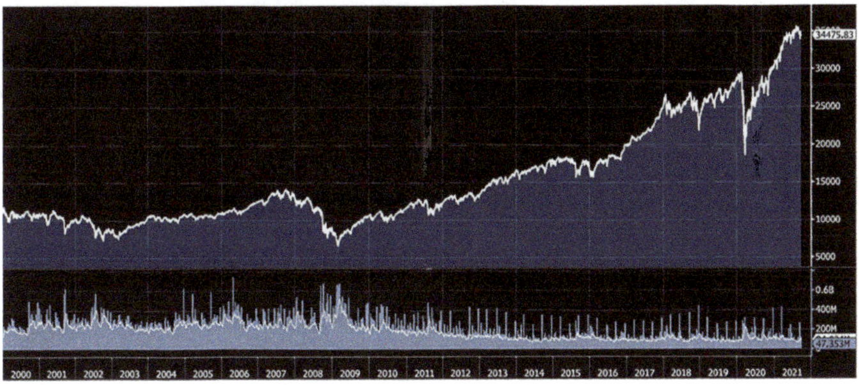

Fig. 1.30 Dow Jones development from January 2000 to September 2021

RTS Index is composed of up to 50 of Russia's largest companies listed on the Moscow Stock Exchange and is considered the indicator of Russian securities trading and price index

USA

Dow Jones The Dow Jones Industrial Average consists of 30 large US companies listed on the New York Stock Exchange. The index was first compiled and published in the 1880s by the American publishing firm Dow Jones—performance index. The development from 2000 to 2021 is given in Fig. 1.30.

NASDAQ Comp. This index is the largest stock index on the NASDAQ (the largest electronic exchange in the USA). It is comprised of up to 3000 stocks of companies mainly from the technology sector—price index.

S&P500 This index—which will be the most relevant one in our further discussions—includes the 500 most actively traded listed US corporations and is calculated and published by Standard & Poor's. We will look at this index in more detail in a later subsection—price index.

Asia

Nikkei 225 tracks the average price movements of the 225 largest Japanese companies traded on the Tokyo Stock Exchange—price index (see Fig. 1.31 for a historical development).

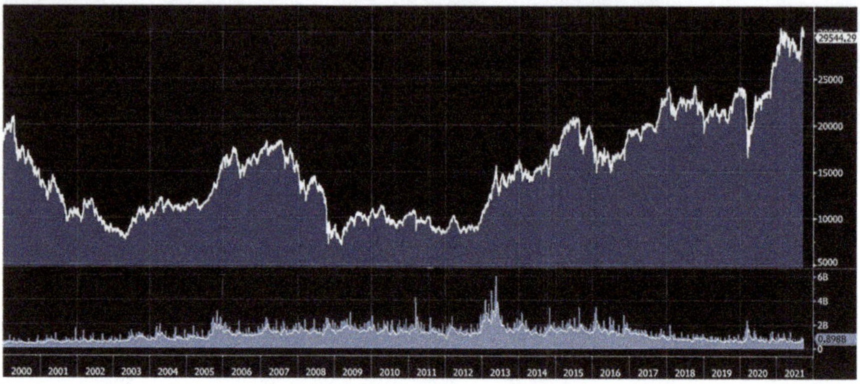

Fig. 1.31 Nikkei development from January 2000 to September 2021

Hang Seng The Hang Seng Index consists of the 45 largest companies listed on the Hong Kong Stock Exchange (these are predominantly but not exclusively Chinese companies).

Kospi The KOSPI Index reflects the average performance of all South Korean equities listed on the Korea Stock Exchange in Seoul—price index.

SSE Composite The Shanghai Stock Exchange Composite Index is considered the most important equity index in mainland China and includes all companies listed on the Shanghai Stock Exchange—price index.

Further Examples

Bovespa the Bovespa Index consists of all Brazil-based companies listed on the São Paulo Stock Exchange—performance Index

Australia All Ordinaries This index—Australia's most important stock index—contains almost all of the stocks listed on the Sydney Stock Exchange—price index.

MSCI World the MSCI World Index (see Fig. 1.32) is a global stock index compiled by Morgan Stanley Capital International based on the price movements of more than 1500 stocks from more than 20 industrialized countries, making it one of the most important international stock indices—price index (performance-based versions also exist)

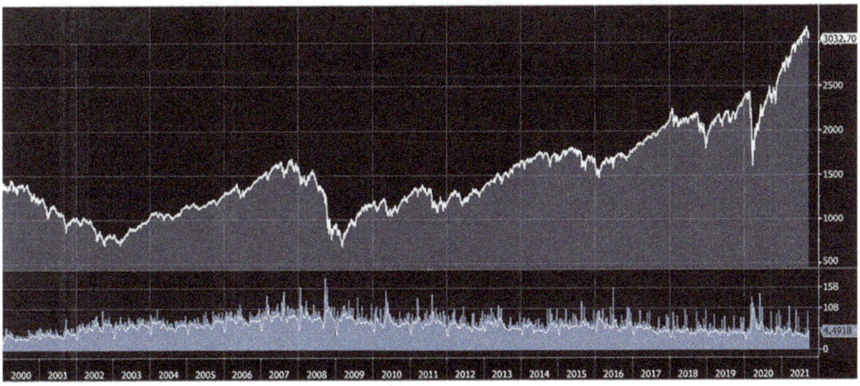

Fig. 1.32 MSCI World development from January 2000 to September 2021

1.17 Trading in Indices Through Index Certificates

As has been noted above, a wide range of financial services providers offer the possibility to trade in indices in the form of index certificates. Here are two examples of such certificates.

See, for example, www.sg-zertifikate.de where the French Société Générale offers a number of index certificates for trading.

In Fig. 1.33 we see the data of an index certificate on the Dutch AEX Index that can be purchased or sold through Société Générale. These data were current on 30 September 2021 at 10:19:43 am and represent the certificate's defining parameters.

The "multiplier" ratio of 10:1, for instance, tells you that by buying an index certificate, you acquire one tenth of the index (meaning that you have to buy 10 index certificates in order to hold 1 unit of the index).

The information "Index 781.96" tells us that the underlying instrument, i.e. the AEX, had a value of 781.96 points when these data were extracted.

The first two rows show the current bid/ask prices of the certificate of EUR 78.190 and EUR 78.220, respectively. This means as follows.

One unit of the certificate (i.e. one tenth of the index) can currently be purchased from Société Générale at the price of EUR 78.22, while Société Générale currently offers EUR 78.19 to buy 1 unit of the certificate.

These prices should be compared with the current value of one tenth of the index, which is 78.196.

1.18 The ShortDAX and Index Certificates on the ShortDAX

In many cases you also have the possibility of "shorting" indices, again via suitable index certificates. When you "short" indices, you trade indices in such a way that, if the index increases by $y\%$, you make a loss of $y\%$ and vice versa. Shorting is

Fig. 1.33 AEX index certificate

Unlimited Index Certificate on AEX	
Bid	78.190
Ask	78.220
% daily change	0.84
High	78.51
Low	77.91
Last update	10:19:43
Index	781.96
Static data	
ISIN	DE000CJ8V3R6
WKN	CJ8V3R
Product type	Unlimited Index Certificate
Underlying	AEX
Multiplier	10 : 1
Issue Date	10 / 01 / 2019
Maturity date	unlimited

a strategy that you use when you anticipate that prices will fall. The possibility of creating "short positions" is an essential component of various trading strategies. A short position in a product A is a financial product D whose value will fall by $y\%$ when the value of A rises by $y\%$.

When we apply shorting strategies in the following, we will primarily use derivatives of various types. In most cases, this is the most efficient method, provided the required derivatives are traded on certain exchanges (more details will follow in subsequent chapters).

At this point we are going to look at an example of how to create a short position in an index using suitable index certificates.

More specifically, let us assume that we are interested in creating a short position on the DAX.

In order to short the DAX via an index certificate, we make use of the so-called ShortDAX index.

The ShortDAX index is an index calculated and published by the Deutsche Börse which essentially reflects changes in the DAX in an inversely proportional way. Hence, if the DAX goes up $y\%$ during a trading day, the ShortDAX goes down $y\%$ during that day and vice versa. Changes always refer to the closing price of the previous day.

Figure 1.34 is a chart of the ShortDAX in the period from January 2002 to September 2021 (blue) compared to the development of the DAX (green) in the same period, expressed as percentages.

Due to how it is calculated, the ShortDAX reflects the DAX in a way that is essentially but not entirely inversely proportional. Noticeable discrepancies occur especially over longer periods of time, the main reason for that being the daily value adjustment based on the closing price of the previous day. (This is simply due to

Fig. 1.34 Relative development ShortDAX versus DAX, from January 2002 to September 2021

Historische Kurse zu DAX Export als CSV

Jahr 2017 Monat November Datum von bis Börsenplatz Xetra

Datum	Eröffnung	Hoch	Tief	Schluss	Performance	Volumen	Umsatz	Hoch-Tief
21.11.17	13.042,94	13.209,01	13.026,77	13.167,54	+0,83 %	125.832.391	5.493.936.627 EUR	182,24 PKT
20.11.17	12.932,81	13.085,17	12.926,13	13.058,66	+0,50 %	110.335.518	4.549.470.373 EUR	159,04 PKT
17.11.17	13.051,71	13.089,72	12.984,67	12.993,73	-0,41 %	126.980.479	5.192.241.015 EUR	105,05 PKT
16.11.17	13.024,39	13.071,94	13.008,02	13.047,22	+0,55 %	124.108.813	4.531.664.443 EUR	63,92 PKT
15.11.17	12.963,09	12.996,12	12.847,88	12.976,37	-0,44 %	156.313.065	5.241.853.710 EUR	148,24 PKT
14.11.17	13.101,09	13.139,27	13.000,15	13.033,48	-0,31 %	111.694.332	4.426.241.199 EUR	139,12 PKT
13.11.17	13.150,78	13.163,94	12.960,65	13.074,42	-0,40 %	124.080.429	4.740.905.276 EUR	203,29 PKT
10.11.17	13.206,35	13.216,97	13.111,65	13.127,47	-0,42 %	145.310.727	5.712.012.249 EUR	105,32 PKT
09.11.17	13.378,96	13.402,05	13.175,22	13.182,56	-1,49 %	171.398.376	6.783.466.104 EUR	226,83 PKT
08.11.17	13.404,58	13.419,77	13.345,11	13.382,42	+0,02 %	129.475.309	5.166.432.586 EUR	74,66 PKT
07.11.17	13.517,98	13.525,56	13.369,85	13.379,27	-0,66 %	107.987.751	4.696.218.250 EUR	155,71 PKT
06.11.17	13.459,42	13.481,23	13.441,66	13.468,79	-0,07 %	111.743.438	4.114.910.653 EUR	39,57 PKT
03.11.17	13.476,53	13.505,01	13.430,22	13.478,86	+0,28 %	104.620.442	4.302.925.511 EUR	74,79 PKT
02.11.17	13.448,52	13.460,86	13.405,91	13.440,93	-0,18 %	116.326.477	4.762.673.682 EUR	54,95 PKT
01.11.17	13.342,44	13.488,59	13.341,30	13.465,51	+1,78 %	157.592.893	6.929.994.154 EUR	147,29 PKT

Fig. 1.35 Historical prices of DAX

the fact that an increase of $x\%$ on one day, followed by a decrease of $y\%$ on the following day, generally does not produce the same result as a decrease of $y\%$ on one day, followed by an increase of $x\%$ on the next day.)

In the tables with historical daily data on the development of the DAX (Fig. 1.35) and the ShortDAX (Fig. 1.36), for example, from 1 to 21 November 2017, the numbers in the performance column are essentially inversely proportional when compared on a day-to-day basis. However, if we calculate and compare the changes

Historische Kurse zu ShortDax-Index (Performance-Index) Export als CSV

Jahr 2017 Monat November Datum von [illegible] bis [illegible] Börsenplatz Xetra

Datum	Eröffnung	Hoch	Tief	Schluss	Performance	Volumen	Umsatz	Hoch-Tief
21.11.17	2.276,21	2.279,02	2.247,29	2.254,51	-0,84 %	0	0 EUR	31,73 PKT
20.11.17	2.295,65	2.296,82	2.268,85	2.273,52	-0,51 %	0	0 EUR	27,97 PKT
17.11.17	2.274,98	2.286,67	2.268,35	2.285,09	+0,41 %	0	0 EUR	18,32 PKT
16.11.17	2.279,84	2.282,73	2.271,46	2.275,81	-0,55 %	0	0 EUR	11,27 PKT
15.11.17	2.290,68	2.310,82	2.284,91	2.288,36	+0,44 %	0	0 EUR	25,91 PKT
14.11.17	2.266,69	2.284,22	2.260,05	2.278,43	+0,31 %	0	0 EUR	24,17 PKT
13.11.17	2.258,21	2.290,98	2.255,94	2.271,37	+0,40 %	0	0 EUR	35,04 PKT
10.11.17	2.248,89	2.265,08	2.247,08	2.262,38	+0,42 %	0	0 EUR	18,00 PKT
09.11.17	2.220,43	2.254,23	2.216,60	2.253,01	+1,49 %	0	0 EUR	37,63 PKT
08.11.17	2.216,23	2.226,10	2.213,71	2.219,91	-0,03 %	0	0 EUR	12,39 PKT
07.11.17	2.197,76	2.222,03	2.196,52	2.220,48	+0,66 %	0	0 EUR	25,51 PKT
06.11.17	2.207,40	2.210,31	2.203,84	2.205,87	+0,07 %	0	0 EUR	6,47 PKT
03.11.17	2.204,76	2.212,37	2.200,07	2.204,37	-0,28 %	0	0 EUR	12,30 PKT
02.11.17	2.209,42	2.216,40	2.207,39	2.210,66	+0,18 %	0	0 EUR	9,01 PKT
01.11.17	2.227,58	2.227,78	2.202,76	2.206,68	-1,79 %	0	0 EUR	25,02 PKT

Fig. 1.36 Historical prices of ShortDAX

in the DAX and the ShortDAX for the period from 1 to 21 November 2017 as a whole, we get:

Changes in ShortDAX:

$$\frac{(2254.51-2206.68)}{2206.68} = 0.0216751 \text{ corresponding to } 2.17\%$$

Changes in DAX:

$$\frac{(13167.54-13465.51)}{13465.51} = -0.221284 \text{ corresponding to - } 2.21\%$$

Below is an index certificate on the ShortDAX offered by BNP Paribas (see https://www.derivate.bnpparibas.com/zertifikat/details/shortdax-index-open-end-zertifikat/de000aa1sdx3). The key parameters of that index certificate are presented in Fig. 1.37.

Upon extraction of this data sheet, the ShortDAX index stood at 1567.75 points.

The multiplier for this certificate is 0.00967438. Therefore, when you buy this certificate, you buy 0.967438% of 1 unit of the ShortDAX index.

The price of the certificate should therefore be $1,567.75 \times 0.00967438 = 15.167$.

The bid/ask prices were actually EUR 15.15 and EUR 15.17, respectively.

Note that the bank also charges a management fee of 0.25% p.a. for the certificate!

1.19 The S&P500 Index

The S&P500 Index is one of the most important stock indices in the world and is widely regarded as a key economic indicator of the US economy. The S&P500 is of particular interest to us in this book, as most of our discussions of derivatives and derivative trading strategies will be based on it. When we present examples of option valuations, we will mostly use options on the S&P500.

SHORTDAX Index Open End Certificate

WKN: AA1SDX
ISIN: DE000AA1SDX3

Bid: 15.15
Ask: 15.17

Last update: 30 September 2021 11:38:40 a.m.

Underlying: 1,567.75

Type: Open End Certificate
Multiplier: 0.00967438
Issue Date: 04/10/2013
Management Fee: 0.25% p.a.

Fig. 1.37 BNP Paribas, ShortDAX index certificate

The main reason for this is that the options market on the S&P500 (these options are traded on the CBOE [Chicago Board Options Exchange]) is arguably the largest and most liquid market of exchange-traded options in the world. It also allows investors to trade large volumes of options in very short time spans at mostly narrow bid-ask spreads. But more about that later.

The S&P500 is also especially relevant as a basis for calculating the volatility index (VIX) of the S&P500, which plays an essential role in a wide range of trading and hedging strategies. Again, there will be much more about that later.

As with the DAX, there is also a short version of the S&P500, the S&P500 Short Index.

The S&P500 index is calculated and published by the rating agency Standard and Poor's. It reflects the average performance of the 500 most actively traded stocks of large-cap US corporations. The index has been published since 4 March 1957. However, there are also data for the S&P500 that go back as far as 1789.

The S&P500 index is a price index, meaning that dividend payments are not taken into account when calculating index value changes.

Traditional statistical data, such as the highest and lowest daily or monthly yields, all-time highs, etc., as well as the weighting of the most important stocks in the index can be found on a multitude of websites (including the associated Wikipedia page).

Here we merely want to provide a brief overview of the development of the S&P500 index, followed by some additional, but still basic, statistical observations.

The first of the following graphs shows the historical development of the S&P500 since its inception in 1957. Then, in the next graph, we see the development in the period from 1990 to today. This is a period that we will analyse more specifically in the following, as the time around 1990 marks the beginning of a truly liquid and modern options market. Figure 1.40 illustrates the dramatic price developments during the financial crisis from the summer of 2008 to spring 2009 and Fig. 1.42 shows the sharp decrease of the index during the first COVID wave in spring 2020.

In the next four paragraphs, we will take a closer look at four very special trading days in the history of the S&P500: 17 October 1987 (Black Monday), 11 September 2001 (attacks on the World Trade Center), 10 October 2008 (stock markets crashing at the height of the financial crisis), and 6 May 2010, when a flash crash caused by a trading error briefly shook the S&P500. The first published value of the S&P500 on 4 March 1957 was 44.06 points. Its all-time high (based on closing prices) until 30 September 2021 was 4524.66 points on 2 September 2021 (see Fig. 1.38). Converted to percentages, the S&P500 thus recorded an increase from 100% to 10,269.31% between March 1957 and September 2021 (i.e. in the course of 64 and a half years). Capital invested in the S&P500 in 1957 would therefore have increased more than 100-fold by 2021. In percent per annum, this corresponds to approximately 7.45% per annum. An investment over this period in a savings account with a fixed interest rate of 7.45% would have yielded the same amount.

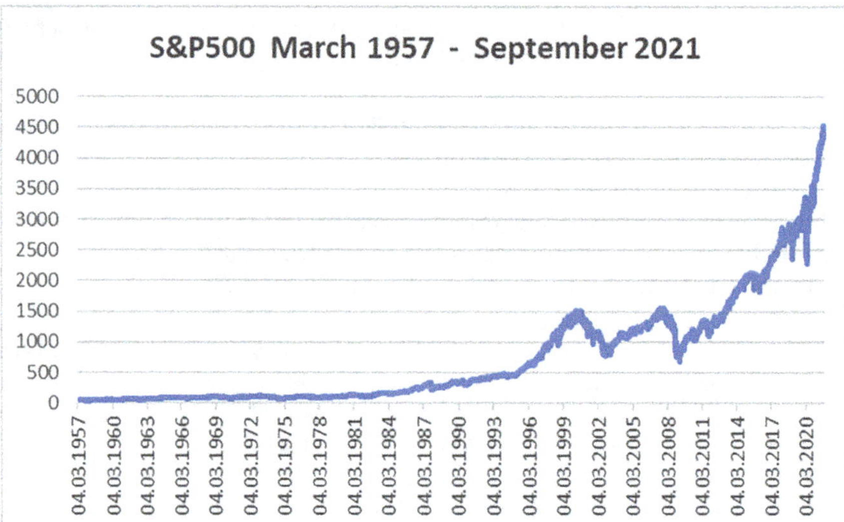

Fig. 1.38 S&P500 Index, 1957–2021

Excursus: Determining the Per Annum Interest Rate of a Capital Development

Although this explanation is probably superfluous for most readers, we include it here for the sake of completeness.

If we have a capital development from an initial capital A to a final capital E over a period of T years, and we want to calculate its annual return x, i.e. the annual interest rate x on which this development is based, then that interest rate x results from the equation

$$E = A \cdot (1 + x)^T \quad \text{als} \quad x = \left(\frac{E}{A}\right)^{\frac{1}{T}} - 1$$

In our example, we chose $E = 4524.66$, $A = 44.06$, and $T = 64.5$.

Between 1 January 1990 and September 2021, the index moved from 359.69 points to 4524.66 points (compare Fig. 1.39). This corresponds to an increase from 100% to 1257.93%, which in turn results in an annual interest rate of 8.30% p.a. from the beginning of 1990 to September 2021.

Three major slumps are strikingly noticeable in the otherwise fairly steadily positive development of the index: in the period 2000/2002 (when the Internet bubble burst), 2007/2009 (financial crisis), and in the period February/March 2021 during the first corona virus wave.

The maximum drop (drawdown) of the S&P500 in the 2000/2002 period occurred when it fell from the then all-time high of 1520.77 points on 1 September

Fig. 1.39 S&P500 Index, 1990–2021

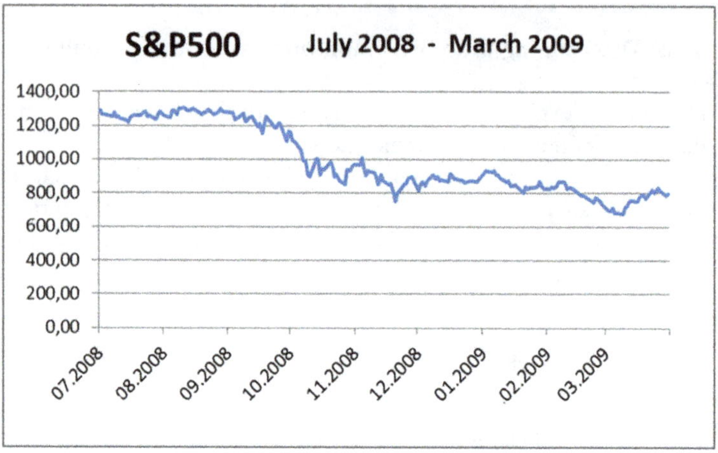

Fig. 1.40 S&P500 Index at the time of the 2008/2009 financial crisis

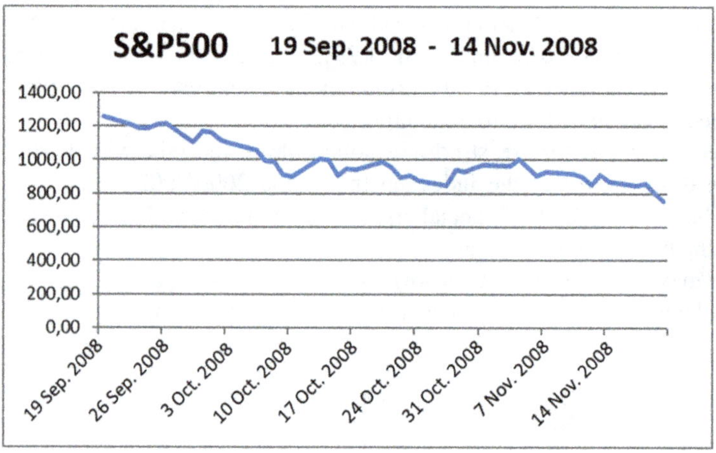

Fig. 1.41 S&P500 Index at the height of the financial crisis in fall 2008

2000 to its lowest level of 776.76 points on 9 October 2002. (This corresponds to a plunge of 48.92%, meaning an annual price loss of 27.38% p.a. over the slightly more than 2 years of decline.)

The development during the financial crisis in 2008/2009 was even more dramatic (see also Figs. 1.40 and 1.41). The maximum drawdown of the S&P500 in the 2007/2009 period occurred when it fell from the then all-time high of 1563.15 points on 9 October 2007 to its lowest level before rebounding at 676.53 points on 9 March 2009.

(This corresponds to a plunge of 56.72%, meaning an annual price loss of 44.63% p.a. over the period of decline of slightly less than one and a half years.)

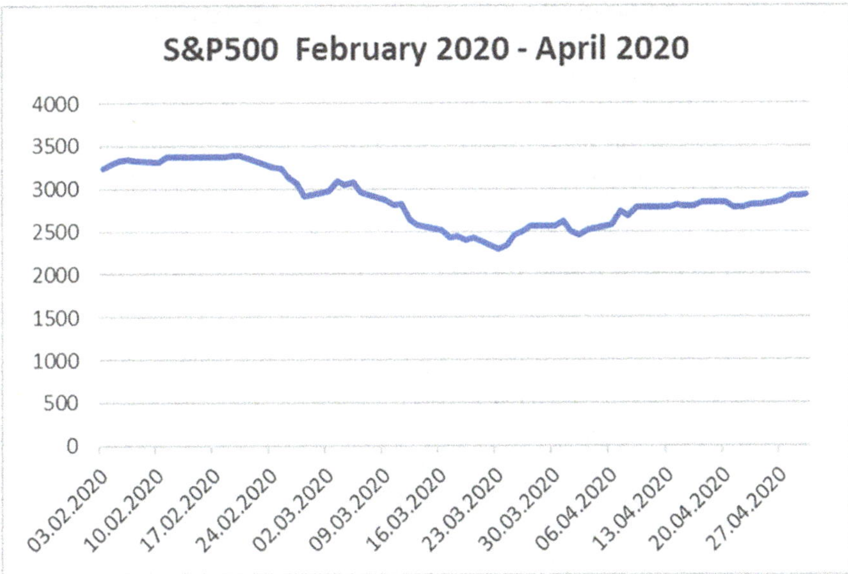

Fig. 1.42 S&P500 Index during the first coronavirus wave

Equally striking is the rapid recovery of the price from 676.53 points on 9 March 2009 to the new all-time high of 4524.66 points on 2 September 2021, which far exceeded all previous highs. This is an increase from 100% to 668.80%, corresponding to an average annual increase of 16.42% p.a.

From 19 February 2020 until 23 March 2020, the S&500 Index fell from 3393.52 (high on 19 February) to 2191.86 (low on 23 March). This is a dramatic decrease of 35.4% in just slightly more than 1 month. This abrupt movement was caused by the irritations of the first COVID wave (Fig. 1.42).

Let us now have a look at the best and worst 10-year periods of the S&P500.

To get the most out of a 10-year investment in the S&P500, you would have had to buy at a price of 304.59 points on 24 September 1990. By 22 September 2000, the S&P500 had climbed to 1448.72 points. That is an increase from 100% to 475.63%, corresponding to an annual return of 16.88%. If your 10-year investment started on 10 March 1999 at 1286.84 points, however, you were definitely out of luck, as it would have ended on 9 March 2009 at 676.53 points. That is a performance from 100% to 52.57%, corresponding to an annual loss of 6.22%.

We could of course list many more statistics of a rather simple nature, such as the worst and the best results in the course of a trading day or a month. However, there are plenty of Internet sites where you will find such information. If any such data become relevant in the course of the following discussions, we will of course provide them there and then.

1.20 The S&P500 on the Black Monday of 19 October 1987

Four days in the history of the S&P500 since 1957 stand out more than any others, and we want to look at these 4 days in some more detail, if only briefly. Anyone intending to actively engage in derivatives trading in particular should know which short-term price distortions and jitters can occur even with such large and broad indices as the S&P500 and what enormous impact they can have on real trading and accurately timed executions.

The first of the 4 days that we want to examine more closely is the so-called Black Monday of 19 October 1987, given in Fig. 1.43.

The 20.47% drop in the S&P500 index on 19 October 1987 happened almost out of the blue. As we can see in the index chart from the beginning of September 1987 to the end of November 1987 and in Table 1.6 of the daily movements of the index in the weeks before and after 19 October 1987, the development before and after that Black Monday was relatively stable and calm. Only on the immediately preceding trading day, i.e. on Friday, 16 October 1987, had there been a rather substantial decrease of 5.16%. The steep drop on 19 October of course triggered irritations and caused major fluctuations to persist for several days afterwards. But as early as 27 October 1987, price swings were back to normal.

As can be seen from the tick data of 19 October 1987, the price decline occurred gradually over the trading day without any really significant temporary recoveries in between (see Fig. 1.44).

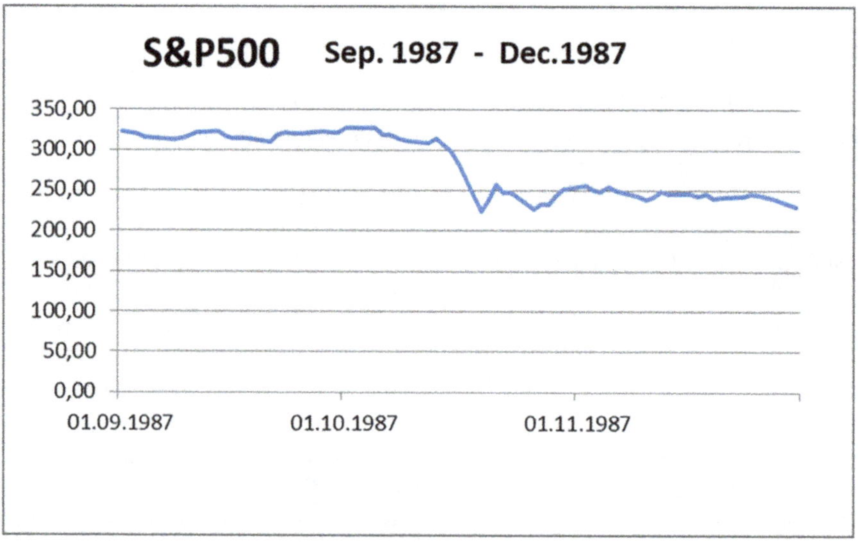

Fig. 1.43 S&P500 Index around the Black Monday in October 1987

Table 1.6 Daily movements of the S&P500 from 1 October 1987 to 11 November 1987

Trading day	S&P500 price	Change on trading day in %
01.10.1987	327.33	1.71
02.10.1987	328.07	0.23
05.10.1987	328.08	0.00
06.10.1987	319.22	−2.70
07.10.1987	318.54	−0.21
08.10.1987	314.16	−1.38
09.10.1987	311.07	−0.98
12.10.1987	309.39	−0.54
13.10.1987	314.52	1.66
14.10.1987	305.23	−2.95
15.10.1987	298.08	−2.34
16.10.1987	282.70	−5.16
19.10.1987	**224.84**	**−20.47**
20.10.1987	236.83	5.33
21.10.1987	258.38	9.10
22.10.1987	248.25	−3.92
23.10.1987	248.22	−0.01
26.10.1987	227.67	−8.28
27.10.1987	233.19	2.42
28.10.1987	233.28	0.04
29.10.1987	244.77	4.93
30.10.1987	251.79	2.87
02.11.1987	255.75	1.57
03.11.1987	250.82	−1.93
04.11.1987	248.96	−0.74
05.11.1987	254.48	2.22
06.11.1987	250.41	−1.60
09.11.1987	243.17	−2.89
10.11.1987	239.00	−1.71
11.11.1987	241.90	1.21

In fact, there was no readily apparent fundamental economic reason for the crash on 19 October 1987. A plausible explanation is that a coincidence of several circumstances (including speculations on higher US interest rates) led to an initial wave of sell-offs, which triggered computer-controlled hedging systems, much en vogue back then, of large fund companies and consequently triggered sell signals, leading to a cascade of sell-offs and, finally, panic sales.

A very significant consequence of the events on 19 October 1987 was the introduction of so-called circuit breakers especially on the most important US stock exchanges. Circuit breakers require stock exchanges to suspend trading in the event of extreme price swings in order to calm the market and prevent panic selling.

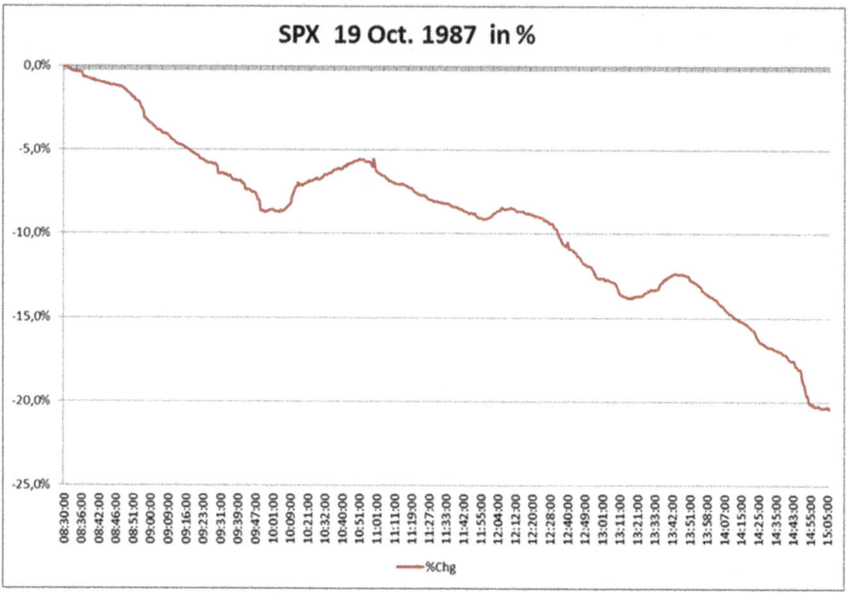

Fig. 1.44 S&P500 index on Black Monday of 1987, losses in percent, tick data

For example, trading on US stock exchanges is halted for 15 min if the S&P500 index falls by 7% or more before 3:25 pm (the NYSE stock exchange is open from 9:30 am to 4:00 pm local time, which corresponds to 3:30 pm to 10:00 pm CET). If the price continues to fall massively after that suspension, resulting in a loss of 13% or more before 3:25 pm, trading will be suspended for another 15 min. Upon reaching a daily loss of 20% or more in the S&P500, the exchange will close for the remainder of the trading day.

Trading can also be suspended on the stock exchanges in Frankfurt, Zurich, and Vienna, but only for large individual stocks and not for the entire market. In the event of large price fluctuations of stocks of a certain size, trading for that particular stock is suspended for a certain period of time.

1.21 11 September 2001

In principle, the board of most stock exchanges (particularly in the USA) has the right to suspend all trading if maintaining "fair and orderly" markets is no longer possible. The closing of the New York Stock Exchange (NYSE) for several days following the 9/11 attacks on 11 September 2001 was such a case.

The NYSE remained closed on Tuesday 11 September 2001 and was not reopened until the following Monday, 17 September 2001.

On 10 September 2001, the index had closed at 1092.54 points.

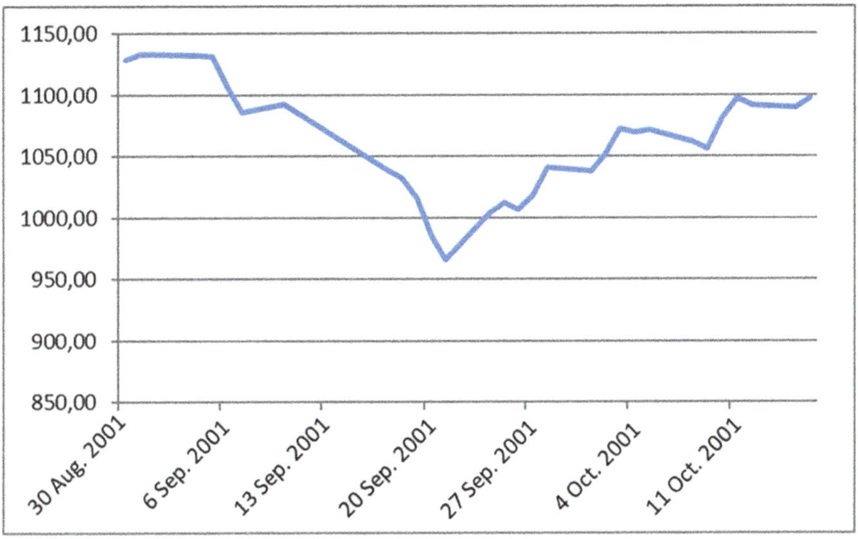

Fig. 1.45 The S&P500 around 11 September 2001

On 17 September 2001, the index fell rapidly to around 1040 points and closed at 1038.77 points (close to its lowest value on that day). This corresponds to a daily loss of 5.18%.

Over the next 3 trading days, the index fell to 965.8 points on 21 September 2001 (daily low at 944.75 points). Between 10 September 2001 and 21 September 2001, the index had therefore lost 11.6% (13.53% if we take into account the daily lows).

Prices recovered rapidly in the following days, and as early as 11 October 2001, prices were again back above the 1092 points recorded on 10 September 2001 as shown in Fig. 1.45.

The Euro Stoxx, DAX, SMI, and ATX indices—where trading was not suspended—showed a basically similar picture: Sharp price declines on 11 September 2001, further declines until 21 September 2001, and recovering prices from then on. Incidentally, the disruptions in the ATX were far less pronounced, as can be seen from Table 1.7.

Table 1.7 Price decline of the Euro Stoxx, DAX, ATX, and SMI on 11 September 2001 and until the 21 September 2001

Index	Price decline on 11 Sept 2001	Decline until 21 Sept 2001
Euro Stoxx	−6.40%	−16.36%
DAX	−8.49%	−18.90%
ATX	−1.12%	−9.18%
SMI	−7.07%	−16.61%

1.22 The S&P500 Index Around 10 October 2008, Overnight Gaps

During the financial crisis from 2007 to 2009, the S&P500 fell 56.72% between 9 October 2007 and 9 March 2009. For many traders in that period, especially derivative traders, the extreme volatility of the index in a single trading day, within very short time intervals and often overnight, had an even greater impact than the total fall in prices.

As an example of an extreme trading day, we are going to take a brief look at Friday, 10 October 2008. Anyone trading in the derivatives market should be aware that circumstances like the ones on that day can potentially occur on the financial market. Figure 1.46 shows a chart with tick data of the S&P500 from the time shortly before the close of trading on Thursday, 9 October 2008, to shortly after trading opened on Monday, 13 October 2008.

On 9 October 2008, the S&P500 had closed at 909.92 points. Within a matter of just a few minutes after the stock exchange opened on 10 October 2008, the index tumbled to 839.80 points. This meant a drop of 7.7% within an extremely short period of time.

Later in the day, the index peaked at 936.36 points, resulting in a difference of 11.50% between its lowest and highest level on that day. It finally closed the day

Fig. 1.46 The S&P500 Index around 10 October 2008, tick data (source: Bloomberg)

Table 1.8 Largest positive and negative overnight gaps of the S&P500 history before 2017

Largest negative overnight gaps	
Date	Gap size
Before 1990	
From 17 April 1961 to 18 April 1961	−3.61%
From 1990 to 2017	
From 23 October 2008 to 24 October 2008	−1.42%
From 13 February 2009 to 14 February 2009	−1.00%
From 18 January 2008 to 19 January 2008	−0.92%
Largest positive overnight gaps	
Date	*Gap size*
Before 1990	
From 22 October 1957 to 23 October 1957	+4.49%
From 1990 to 2017	
From 10 October 2008 to 11 October 2008	+1.50%
From 7 May 2010 to 8 May 2010	+1.03%
From 29 October 2008 to 30 October 2008	+1.00%

at 899.22 points. The following day, the index opened with an "overnight gap" of +1.50% at 912.75 points.

Such massive fluctuations within 1 trading day, either overnight or immediately after opening of trading, can make it immensely difficult to implement adequate trading strategies.

Until the end of 2017, all three of the largest negative overnight gaps and two of the three largest positive overnight gaps after 1990 occurred during the 2008/2009 financial crisis, as displayed in the Table 1.8. The second largest positive overnight gap occurred from 7 May 2010 to 8 May 2010, 1 day after another extreme event in the history of the S&P500.

As from the beginning of 2018 (and especially during the first COVID wave in 2020), large positive and negative overnight gaps became much more common. In Table 1.9, we see the largest positive and negative overnight gaps in the period from January 2018 until September 2021.

1.23 The "Flash Crash" on 6 May 2010

On 6 May 2010, after a relatively unremarkable activity in the first 5 h, the S&P500 suddenly tumbled from about 1140 points to 1065.79, i.e. by about 6.5% within just a few minutes. However, it only took about 15 min for the index to recover back into the 1130 points range. The closing price on 6 May 2010 was 1128.15 points (see Fig. 1.47).

There were conflicting assumptions about what had triggered this sudden "flash crash". It was probably a combination of "mistrades" and deliberate price manipulations.

Table 1.9 Largest positive and negative overnight gaps of the S&P500 history after 2018

Largest negative overnight gaps	
Date	Gap size
16.03.2020	−8.07
12.03.2020	−4.20
18.03.2020	−3.80
09.03.2020	−3.79
01.04.2020	−3.46
27.03.2020	−2.90
24.02.2020	−2.46
06.03.2020	−2.36
11.06.2020	−2.13
28.02.2020	−2.12
11.03.2020	−2.00
Largest positive overnight gaps	
Date	Gap size
24.03.2020	4.57
13.03.2020	3.48
06.04.2020	3.48
07.04.2020	2.74
10.03.2020	2.38
12.06.2020	2.24
21.07.2021	2.22
20.07.2021	2.08
16.06.2020	2.06
09.11.2020	2.05

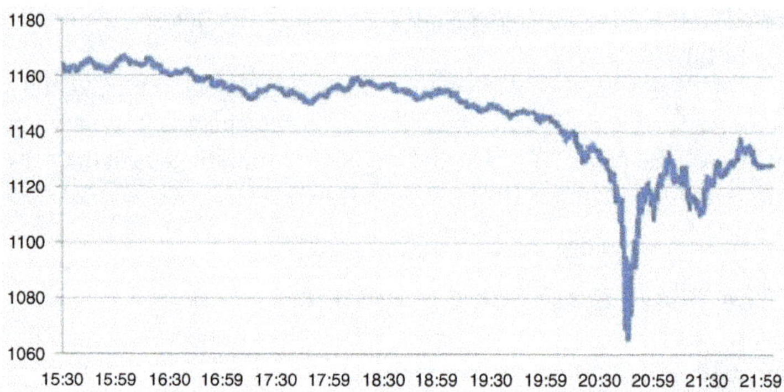

Fig. 1.47 The S&P500 index around 6 May 2010, tick data

1.24 Price Forecasting and Run Analyses of Stock Prices and Index Prices

In my presentations on trading strategies and in my seminars, and also in personal conversations or interviews, I am often asked how I assess the future development of certain indices or commodity prices, exchange rates, or interest rates. I invariably answer that I don't know and have absolutely nothing to say about that. I am not a market analyst nor a financial economist; I am a mathematician. Strictly, mathematically oriented traders and analysts such as myself are interested in the present state and present correlations between prices of products, not in a presumed future state of the markets. We will explain in detail how we do this in the course of this book.

Many other traders try their luck using a wide variety of price forecasting methods. It is in no way our intention to underestimate or dismiss these approaches; they are just not the subject of our discourse. It may occasionally still become necessary to touch on these topics, however.

Most of these price forecasting methods are based on various indicators (intersections with moving averages, chart pattern analyses, etc.), on certain trend-following or mean-reversion strategies, or on identifying and exploiting various actual or perceived market anomalies ("sell-in-May effect", "year-end rally", "weekend effect", "winner-loser effects", etc.).

Of course, I am often confronted with the view that dealings on stock exchanges are ultimately nothing more than gambling. (Or in the words of a mathematician colleague, who once said, in all seriousness, that what I was doing was really the "devil's work".) If I have a bit more time on my hands and am speaking to someone with a genuine interest in the topic, I briefly explain how to make insightful use of derivatives through so-called statistical arbitrage. The latter enables me to turn something that is indeed a mere game of chance (if one approaches the markets in complete ignorance of its workings) into a strategy that, while still being a game of chance, **brings a significant portion of skill into play**. And as soon as skill can impact the outcome of a game, it is no longer just a game of chance!

The question as to which (actual) games are entirely of chance and which games allow skill (to what extent?) to determine the outcome in addition to chance is a very interesting, challenging, and a quite relevant mathematical problem. Maybe we will have time to draw on my extensive experience with such challenges and explore them in a later chapter.

For now, let us get back to the frequently expressed view that dealings on stock exchanges are basically pure gambles. If there is little time, I usually give the (peripherally) interested food for thought, making them reconsider their apodictic views, with the following information.

Toss a—demonstrably—fair coin and observe if it lands heads or tails up. A fair coin means that in an average of 50% of all cases it lands heads up and in an average of 50% of cases it lands tails up. Suppose you got heads 5 times in a row and are now about to flip the coin for the 6th time. Can we expect that the sixth toss—given that it is a fair coin—is more likely to show tails rather than heads once again? Of course not. The coin is completely unaffected by the past; the probability of showing heads or tails is still 50%, no matter how often the coin has shown heads in a row.

(This is a fact that occasionally comes up for discussion, of course. And—let's be honest here—aren't there situations where even we in the know may succumb to such thoughts, against our better judgement? If red comes up seven times in a row at the roulette table, wouldn't we be inclined to think that it's finally time for a black number, making us bet on black?) Most people I speak with, however, do understand that the heads-tails sequence resulting from multiple tossings of a fair coin is a purely random sequence.

I would then continue by saying something like, let us apply that to the S&P500 stock index, for example. What do you think? Is this index a purely random sequence?

Or more specifically:

Let us look at the daily closing prices of the S&P500 over a certain period of time and mark the days on which the index rose as "1" and the days on which it fell as "0". This gives us a 0-1 sequence analogous to the heads-tails sequence of the coin toss, which is given in Table 1.10.

Then, among all the S&P500 data that we want to analyse, we search for all the blocks of 4 consecutive trading days on which the index rose (i.e. that have a 1 in the third column as displayed in Table 1.11) and look at the following fifth trading day.

Similarly, among all the S&P500 data that we want to analyse, we then search for all the blocks of 4 consecutive trading days on which the index fell (i.e. that have a 0 in the third column) and look at the following fifth trading day. A corresponding example is given in Table 1.12.

Table 1.10 Development of the S&P500 from 24.03.2017 to 13.04.2017 whereby "1" denotes an increase and "0" a decrease

Date	Daily closing price	Change from previous day
24.03.2017	2343.98	
27.03.2017	2341.59	0
28.03.2017	2358.57	1
29.03.2017	2361.13	1
30.03.2017	2368.06	1
31.03.2017	2362.72	0
03.04.2017	2358.84	0
04.04.2017	2360.16	1
05.04.2017	2352.95	0
06.04.2017	2357.49	1
07.04.2017	2355.54	0
10.04.2017	2357.16	1
11.04.2017	2353.78	0
12.04.2017	2344.93	0
13.04.2017	2328.95	0

Table 1.11 Four consecutive days with an increase of the S&P500

Date	Daily closing price	Change from previous day
01.11.2017	2579.36	1
02.11.2017	2579.85	1
03.11.2017	2587.84	1
06.11.2017	2591.13	1
07.11.2017	2590.64	**0**

Table 1.12 Four consecutive days with an decrease of the S&P500

Date	Daily closing price	Change from previous day
16.03.2017	2381.38	0
17.03.2017	2378.25	0
20.03.2017	2373.47	0
21.03.2017	2344.02	0
22.03.2017	2348.45	**1**

And then we ask ourselves:

"Is it more likely that, with 1 occurring 4 days in a row, the fifth day shows a 0 or a 1, or is there no correlation?" And analogous to that, of course:

"Is it more likely that, with 0 occurring 4 days in a row, the fifth day will show a 0 or a 1, or is there no correlation here?"

We are going to "answer" that question by simply counting the historical frequencies of the conditional occurrence of "zeros and ones on the fifth days" and derive empirical probabilities from that.

Note of course that the relative frequencies of the conditional occurrence (i.e. after a block of four occurrences of the same digits) of 0 or 1 in any one time period must not be compared with the 50% probability in the case of the fair coin. This is because the index is not necessarily fair in those time periods. Rather, the comparison has to be made with the total number of ones or zeros occurring in a time period.

Instead of looking for blocks of 4 consecutive days with the same tendency (0 or 1) and then asking for the following day, we could just as well of course look for blocks of 2, 3, or 5 consecutive days and ask for the following trading day.

Let's do this with the S&P500 index for different time periods and different block lengths and note the results in Tables 1.13, 1.14, 1.15, and 1.16, which we will then discuss.

If our hypothesis is that after the occurrence of several 1s in a row, the occurrence of a 0 becomes more likely and

that after the occurrence of several 0s in a row, the occurrence of a 1 becomes more likely, then we will see this hypothesis confirmed for the periods 1990–1999, 2000–2009, and 2010–2017 (bold blue results in the last column).

And for most cases, we can state that the longer the preceding block of 1s or 0s is, the more clearly the hypothesis is confirmed.

Let's take an example that illustrates this more clearly:

Table 1.13 Various block lengths of a consecutive increase/decrease of the S&P500 from 2010–2017 and their occurrences

Period	Block length s	Number of blocks of 1 with length s	Frequency of 1 in total	Frequency of 1 after s times 1
2010–2017	2	572	0.546	**0.517**
	3	296	0.546	**0.480**
	4	142	0.546	**0.465**
	5	66	0.546	**0.424**
Period	Block length s	Number of blocks of 0 with length s	Frequency of 1 in total	Frequency of 1 after s times 0
2010–2017	2	389	0.546	**0.576**
	3	165	0.546	**0.562**
	4	69	0.546	**0.551**
	5	31	0.546	**0.613**

Table 1.14 Various block lengths of a consecutive increase/decrease of the S&P500 from 2000 to 2009 and their occurrences

Period	Block length s	Number of blocks of 1 with length s	Frequency of 1 in total	Frequency of 1 after s times 1
2000–2009	2	633	0.523	**0.472**
	3	298	0.523	**0.470**
	4	139	0.523	**0.453**
	5	63	0.523	**0.460**
Period	Block length s	Number of blocks of 0 with length s	Frequency of 1 in total	Frequency of 1 after s times 0
2000–2009	2	519	0.523	**0.584**
	3	216	0.523	**0.639**
	4	78	0.523	**0.666**
	5	26	0.523	**0.769**

In the trading period from 2000 to 2009 (1 January 2000 to 31 December 2009), the proportion of days with a price increase was 52.3%. The proportion of days with a *price increase* after *4 days with a price increase* was only 45.3%. (Such a block of four price increases in a row occurred 139 times.)

On the other hand, the percentage of days with a *price increase* after *4 days with a drop in prices* was 66.6%. (Such a block of four price drops in a row occurred 78 times.)

This phenomenon is much more pronounced in the periods 2000–2009 and 2010–2017 than in the period 1990–1999.

Table 1.15 Various block lengths of a consecutive increase/decrease of the S&P500 from 1990 to 1999 and their occurrences

Period	Block length s	Number of blocks of 1 with length s	Frequency of 1 in total	Frequency of 1 after s times 1
1990–1999	2	741	0.536	**0.533**
	3	394	0.536	**0.485**
	4	191	0.536	**0.497**
	5	95	0.536	**0.474**
Period	Block length s	Number of blocks of 0 with length s	Frequency of 1 in total	Frequency of 1 after s times 0
1990–1999	2	562	0.536	**0.543**
	3	257	0.536	**0.603**
	4	102	0.536	**0.608**
	5	40	0.536	**0.650**

Table 1.16 Various block lengths of a consecutive increase/decrease of the S&P500 from 1957 to 1989 and their occurrences

Period	Block length s	Number of blocks of 1 with length s	Frequency of 1 in total	Frequency of 1 after s times 1
1957–1989	2	2517	0.524	**0.564**
	3	1419	0.524	**0.567**
	4	803	0.524	**0.567**
	5	455	0.524	**0.587**
Period	Block length s	Number of blocks of 0 with length s	Frequency of 1 in total	Frequency of 1 after s times 0
1957–1989	2	2124	0.524	**0.477**
	3	1112	0.524	**0.477**
	4	582	0.524	**0.486**
	5	299	0.524	**0.482**

The exact opposite is true for the trading period 1957–1989.

However, in all cases, we observe the opposite tendency, contrary to our hypothesis. The results that contradict the hypothesis are shown in red.

It seems that the S&P500 before 1990 (i.e. before the development of highly efficient and liquid derivatives markets) shows stronger trend-following tendencies, followed by stronger mean-reversion tendencies after 1990.

Similar results are also notable for DAX and Euro Stoxx (although we only have data from 1988 and 1992 respectively), as can be seen in Tables 1.17 and 1.18 (for the value $s = 4$ as an example).

Table 1.17 Analysis of the trend following tendencies of the DAX

Period	Block length s	Frequency of 1 in total	Frequency of 1 after s times 1	Frequency of 1 after s times 0
1988–1999	4	0.537	**0.512**	**0.563**
2000–2009	4	0.524	**0.509**	**0.619**
2010–2017	4	0.541	**0.529**	**0.566**

Table 1.18 Analysis of the trend following tendencies of the Euro Stoxx

Period	Block length s	Frequency of 1 in total	Frequency of 1 after s times 1	Frequency of 1 after s times 0
1988–1999	4	0.559	**0.545**	**0.525**
2000–2009	4	0.501	**0.444**	**0.508**
2010–2017	4	0.510	**0.469**	**0.575**

Table 1.19 Updated analysis of the trend following tendencies of the DAX

Period	Block length s	Number of blocks of 1 with length s	Frequency of 1 in total	Frequency of 1 after s times 1
2018–Sep 2021	2	294	0.566	**0.548**
	3	161	0.566	**0.453**
	4	73	0.566	**0.466**
	5	34	0.566	**0.412**

DAX

Euro Stoxx

The results for the hypothesis are less evident than for the S&P500, however.

Yet again the results for the hypothesis are more convincing for the post-2000 period than for the pre-2000 period.

On our website you can use our software to conduct your own run tests for any price data and any periods (see https://app.lsqf.org/book/run-analysis).

Note that you cannot base trading strategies for long-term success on such observations, of course! But they can be viewed as an explanation of the fact that, at least over certain periods of time, certain trends can be seen in the development of stock indices or in the prices of individual stocks which distinguish them from purely random developments.

The preceding facts are based on an extensive investigation conducted in early 2018. For the new edition (English publication) of this monograph, we have added the corresponding results for the S&P500 for the time period 1 January 2018 until 30 September 2021. The results are shown in Tables 1.19 and 1.20.

Table 1.20 Updated analysis of the trend following tendencies of the DAX

Period	Block length s	Number of blocks of 0 with length s	Frequency of 1 in total	Frequency of 1 after s times 0
2018–Sep 2021	2	170	0.566	0.553
	3	76	0.566	0.553
	4	34	0.566	0.559
	5	15	0.566	0.533

1.25 Notes on a Simple Trading Strategy by Signals Using Exponential Moving Averages

This chapter and the related software on the website were largely developed and written by my team member Alexander Brunhuemer.

A fundamentally obvious strategy to succeed in the stock market would be to buy a stock whenever its price is low and to then sell it (possibly opening a short position in addition) whenever the price is high again. However, if we want to answer the question when the price of a stock should be considered low or high, things start to become a little trickier. A case as illustrated in Fig. 1.48 would be ideal.

There is no way of knowing, of course, how an underlying is going to perform in the future, which is why the above scenario remains pure utopia. However, strategies have been developed that aim to get close to this scenario. One often-used strategy consists in trading based on exponential moving averages (EMA). To calculate such

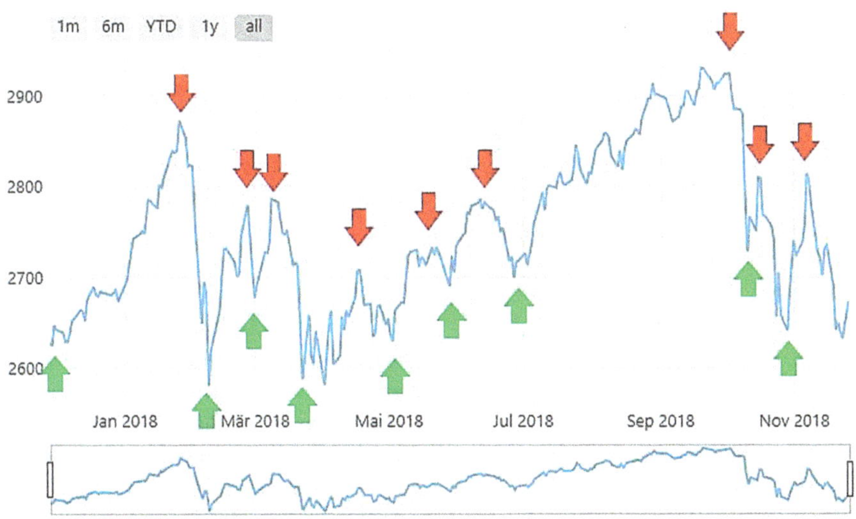

Fig. 1.48 Ideal case of a market-timing strategy applied to daily S&P500 prices. We want to buy at the points with green arrows and sell at the red arrows peaks

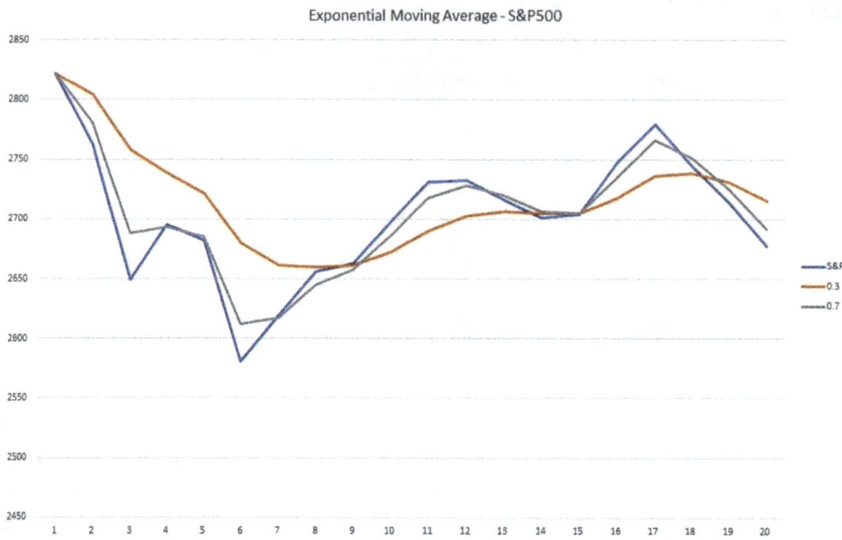

Fig. 1.49 Illustration of the exponential moving average where $\alpha = 0.3$ (orange) versus $\alpha = 0.7$ (gray)

exponential moving averages, the following formula is applied to any time series S_t (e.g. daily stock prices).

We define an arbitrary starting point 0 in the past and set

$$E_0 = S_0$$

$$E_{t+1}(\alpha) = \alpha S_{t+1} + (1 - \alpha)E_t, \quad t \geq 0$$

This leads to a smoothed time series E_t, where the strength of the smoothing can be controlled using the α parameter (see Fig. 1.49).

Let us now take a closer look at the intersections of the underlying time series and the associated EMA. In Fig. 1.50, we divided the intersections into two categories. The red dots are those where the exponential moving average crosses the time series coming from below and the green dots where the exponential moving average crosses the time series coming from above. Looking at that chart, one could easily get the impression that each of the red dots is above the following green dot and that each of the green dots is below the following red dot. If that were really the case, it would mean that, in terms of stock prices, you would have a safe way to make a profit by selling at the red crossovers and buying at the green crossovers. However, this approach ignores essential limitations of the strategy. First of all, the assumption that the red dots are always above the green dots is not necessarily correct. There may well be scenarios where this is not the case (e.g. given in Fig. 1.51). In addition, it will not necessarily be possible to trade directly at the points at which the price/average intersection occurs; in fact, it may not be possible

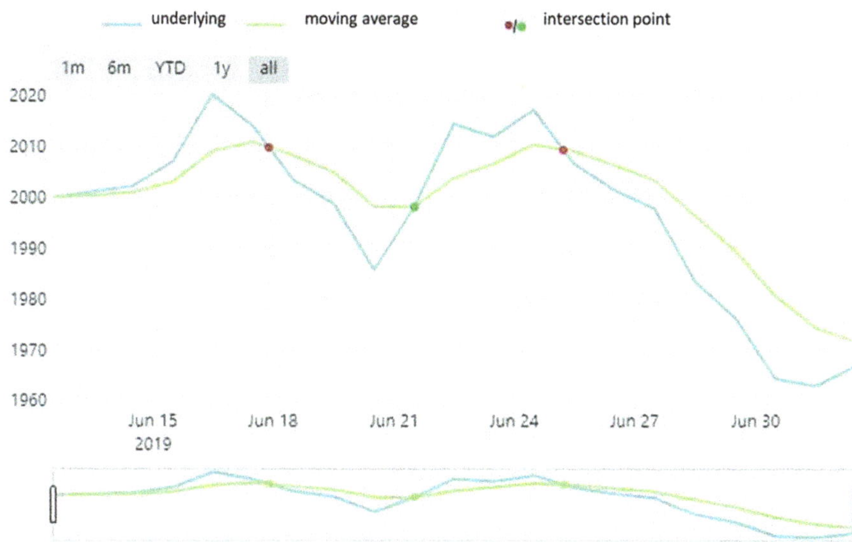

Fig. 1.50 Intersections of the underlying with the exponential moving average

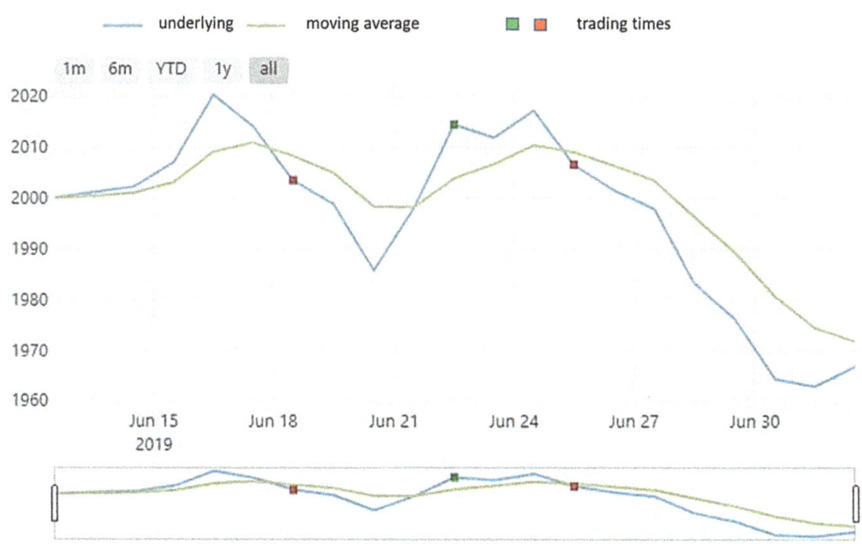

Fig. 1.51 Divergences between intersections and actual trading times lead to potential losses

to do so before the time when prices are looked into (at the daily closing time, see Fig. 1.52) plus a certain response time until the trade is actually executed. It is therefore necessary to analyse whether these problems are of a purely formal nature and the strategy can still be successfully implemented in the long term or whether these divergences actually prevent the success of the strategy.

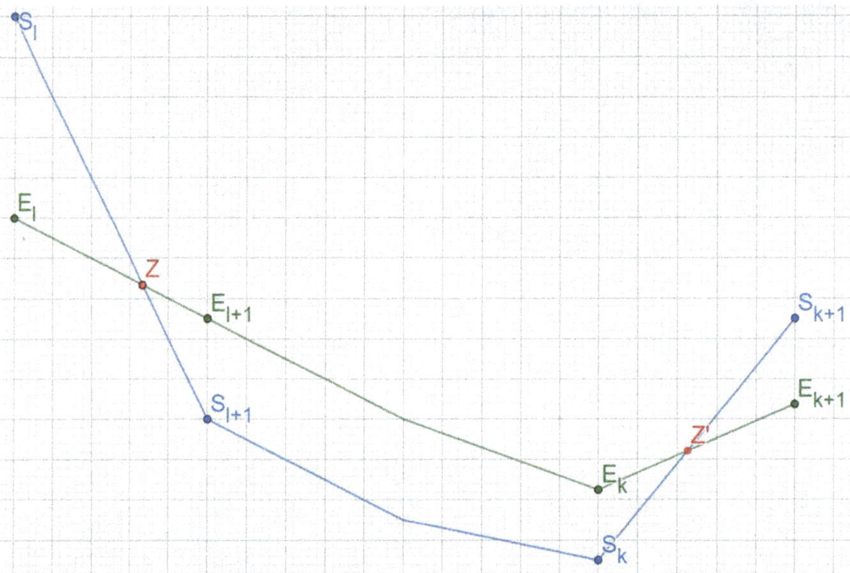

Fig. 1.52 Illustration of the basic mechanism of the strategy

The best way to find the answer would be if we could show the following: Let Z and Z' be two consecutive intersections of the exponential moving average (E_i) with the time series of the underlying (S_i). Let us further assume that the exponential moving average crosses the underlying from below in Z (the other case can be viewed analogously). Using this strategy we could make a guaranteed profit if the value of the underlying at the first time point after Z is greater than the value of the underlying at the first time point after Z' (this corresponds to $S_{l+1} > S_{k+1}$ in the chart). However, this cannot be shown, as we cannot rule out the possibility that the value of the underlying may be higher at the point in time directly after Z' than directly after Z. What we can do, however, is show this relationship for the time points directly before Z and directly before Z', i.e. $S_l > S_k$ in the above illustration.

To do so, we can proceed as follows:

S_{l+1} is by definition smaller than E_{l+1} and therefore

$$E_{l+1} = \alpha S_{l+1} + (1-\alpha)E_l < \alpha E_{l+1} + (1-\alpha)E_l.$$

Hence,

$$E_{l+1} < \alpha E_{l+1} + (1-\alpha)E_l,$$

and after rearrangement we get

$$E_{l+1} < E_l.$$

If S_{l+2} is also smaller than E_{l+2}, then we will show analogously that $E_{l+2} < E_{l+1} < E_l$ holds true. We can continue this for as long as the price is smaller than the associated average value, i.e. up to S_k and E_k. This then gives us

$$E_k < \ldots < E_{l+2} < E_{l+1} < E_l$$

and consequently

$$S_k < E_k < E_l < S_l,$$

which is what we needed to show.

Conclusion: At the last point before the price crosses the exponential moving average from above, the price is always higher than at the subsequent last point before the price crosses the exponential moving average from below.

But as stated above, this statement is no longer valid if we replace the expression ... *at the last point before* ... above with ... *at the first point after* ....

This fact leads to a risk of loss, however, since we do not yet know at times S_l and S_k that the two time series will soon intersect. It is therefore only possible to act afterwards, which can lead to said losses. The question now is whether the potential profits outweigh these losses or if there is no prospect of success in the long term.

To shed more light on this, the team around the author has developed a program that makes it possible to automatically test this strategy for any given price data and smoothing parameters α . The program is available at https://app.lsqf.org/book/exponential-moving-average and can be used free of charge. With this program, we use price data from 30 different stocks in the Dow Jones Index to review the performance of the trading strategy over two different periods, 1999–2008 and 2009–2018. On average, we got the following disastrous result when choosing different parameters (from 0.1 to 0.5 in one tenth increments) for α, even without considering transaction costs (see Fig. 1.53).

Theoretical observations also show an average negative return of this strategy. When applying the strategy to simulated stock prices according to the Wiener model, a noticeably negative return can be seen on average after carrying out Monte Carlo simulation (again without taking account of transaction costs). If you would like to perform this analysis for yourself with continuously updated datasets, you can do so using a program on this book's website: https://app.lsqf.org/book/exponential-moving-average-simulation

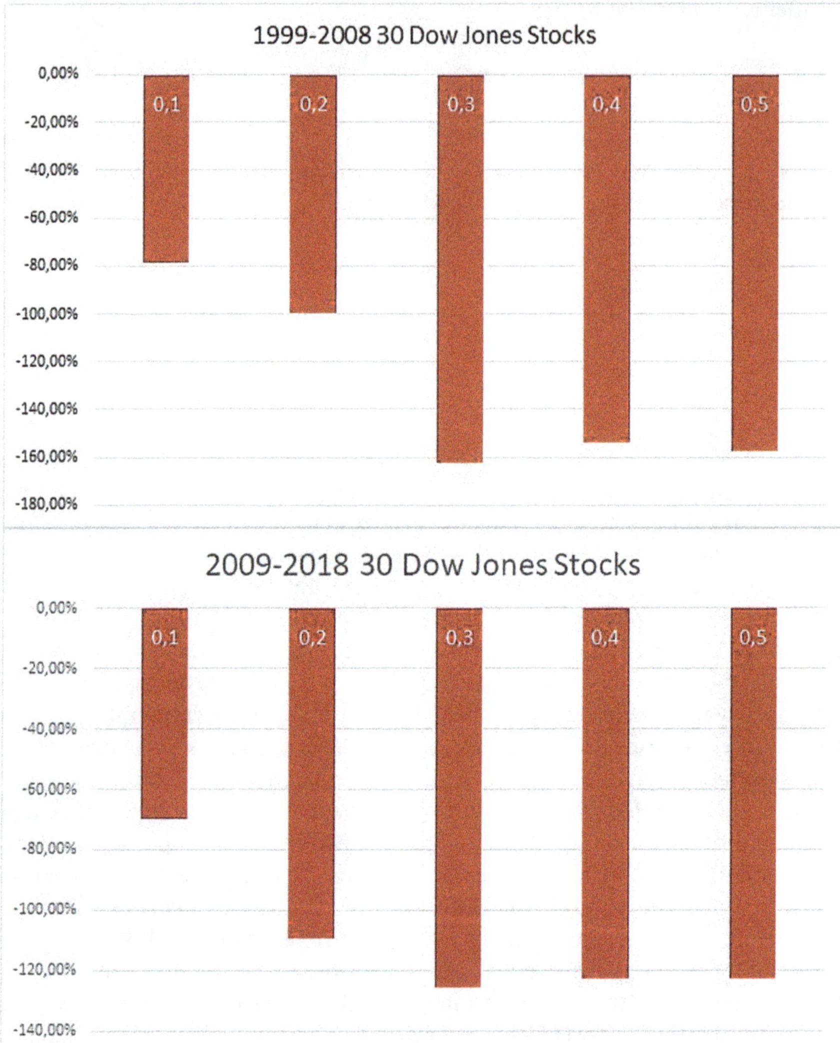

Fig. 1.53 Performance of the trading strategy using exponential moving averages as calculated for 30 stocks of the Dow Jones Index using different α values between 0.1 and 0.5. Calculation of the relative profit or loss was based on the price of the stock that was paid on the first trade

2 Derivatives and Trading in Derivatives, Basic Concepts and Strategies

Abstract

We define the basic types of derivatives, namely, European and American plain-vanilla call- and put-stock-options and futures and we discuss fundamental properties. We introduce the most basic option-strategies like straddles, protective puts, and covered calls. We discuss the question which type of investor is especially interested in which type of application of derivatives. We also give details on how to trade options and futures on an electronic trading platform and on margin requirements for short positions in derivatives.

Keywords

Financial derivatives · Plain-vanilla options · American options · Basic option strategies · How to trade options · Option combinations · SPX options · Futures · Euro-Bund Futures

At the centre of our interest in this book are financial derivatives and trading in such derivatives.

Options, futures, swaps, caps, floors, etc. are all derivatives. Derivatives are the key elements of a wide range of complex structured financial products.

In addition, derivatives can often occur in a less obvious, not immediately recognized form, as part of other structures, e.g. in call rights.

And for mathematicians, i.e. investors who base their trading strategies on mathematics, the derivative market is their real sphere of activity. It is only in the derivative market that mathematics can be essentially applied. Only there can mathematical methods be used to provably identify opportunities in certain financial market situations.

And the mathematical methods used in this field often are highly sophisticated.

G. Larcher, *The Art of Quantitative Finance Vol.1*, Springer Texts in Business and Economics, https://doi.org/10.1007/978-3-031-23873-4_2

In this—very elementary—chapter, we will introduce the basic types of derivatives and their basic properties. We are going to ask ourselves who uses these derivatives, in what ways, and for what reasons and thus gain insights into some elementary derivative strategies.

And finally, after defining "non-frictional markets" and discussing the "no-arbitrage principle", we will explore some basic relationships between different derivatives and attempt to exploit very simple arbitrage opportunities.

2.1 What Is a Derivative?

A derivative, a derivative financial product, is a financial product that derives its value from another financial product. In the following, therefore, being precise in our choice of words, we will always refer to an "option on the Allianz share", an "option on the euro/dollar exchange rate", a "futures contract on a German government bond", an "option on the VIX volatility index", a "swap between the 6-month Euribor and the 10-year EUR swap rate", . . . etc. And we will even look at derivatives of derivatives, such as "options on futures on the S&P500 index".

The financial product A from which a particular derivative D is derived is referred to as the "underlying asset" of D.

Derivative D is always given by defining which cash flows will be generated by that derivative in the future as a result of how the underlying asset performs.

A derivative can also be based on several underlying assets, i.e. it can generate cash flows that depend on the price movements of several other financial instruments.

Many derivatives are traded on derivatives exchanges (e.g. on two of the major international derivatives exchanges, the CBOE and the CME, both based in Chicago). However, many derivatives are also traded off the board between investors and banks or other financial firms. This is referred to as OTC ("over-the-counter") trading.

Derivatives can be highly complex. We will start with the simplest type of derivatives, so-called plain-vanilla options and simple futures contracts.

2.2 European Plain-Vanilla Options, Definition and Basic Characteristics

"Plain-vanilla" options are standard options with no special features. The term "plain vanilla" is said to have its origin in the Southern USA, where the most common ice cream flavour used to be vanilla. Therefore, if you wanted a scoop of just plain, ordinary ice cream, you would order the "plain-vanilla" variety.

Plain-vanilla options are therefore the most basic and simplest type of options.

In the following, we will focus specifically on:

Call Options and **Put Options**
These options can be either **American** or **European**. (Note that this classification has nothing to do with geography!)
You can take a **long position** or a **short position**, i.e. go long or short, in each of these two types of options.

All in all, we are therefore dealing with a total of eight different possible variants: from **American Long Call** to **European Short Put**.

Instead of saying: *I take a **long position** in a European call option*, we often say *I **buy** a European call option.*

And instead of saying: *I take a **short position** in a European call option*, we often say *I **sell** a European call option.*

We do so because:
When taking a **long position** or “going long”, we always have to **pay a premium** (i.e. the price of the option), at the moment of taking that long position; hence, we “buy” the option.
When taking a **short position** or “going short”, we always **receive payment of a premium** (i.e. the price of the option) at the moment of taking that short position; hence, we “sell” the option.
This kind of usage is practical and widespread but can be confusing when you first start out in the field of options trading, especially when it comes to short positions.
The reason why this can be confusing is that you can go short on an option at any time, regardless of whether you are already holding long or short positions, whereas normally, you need to actually own a product before you can “sell” it.
Still, even though it may appear confusing at first, we will often adopt this usage in the following. The important thing to keep in mind, however, is that: you can always take a short position (go short) on an option (i.e. “sell” an option), even if you do not hold that option!
All of the options listed above, of all types and positions, have the following defining parameters.

Option parameters:

*the **underlying** (underlying asset)* **A** *(a stock, an index, an exchange rate, an interest rate, etc.)*
*the **expiration date*** **T** *(a specifically defined date in the future)*

the ***strike price*** **K**

All of the options listed above, of all types and positions, have the following variable parameter:

the ***option price*** *(hereinafter denoted by* **C** *for call options and* **P** *for put options)*

Every option refers to the underlying asset A, and every option "lives" until time T, which is when it expires.

The option's **time to expiration** is defined as the period of time **from now until the expiry date T**. Often (if there is no risk of confusion), the time to expiration is also denoted by the letter T, in which case we denote the **time interval from now (= time 0) to expiration by** $[0, T]$.

We assume that the options we are considering can be traded throughout the entire time to expiration, either on the exchange on which they are traded (for exchange-traded options) or directly with the financial firm with which the option contract was entered into (for OTC options).

For any point in time t from $[0, T]$, we denote the **price of a call option at time t by** $C(t)$ and the **price of a put option at time t** by $P(t)$. (For $C(0)$ or $P(0)$, we often write C or P for short.)

If we take a long position in an option at a certain time t_1 from $[0, T]$, and take a short position in the same option at a certain other time t_2 from $[0, T]$, then the two positions cancel each other out. We also say: **the option is closed or closed out.** We then no longer have any position in that option.

At time t_1 we paid the price of option ($C(t_1)$ or $P(t_1)$), and at time t_2 we received the price of option ($C(t_2)$ or $P(t_2)$).

Let us look at **European** options first.

Call options:

(a) If you, at time t, took a **long position** in a **European call option** on the underlying asset **A** with **expiration T** and **strike price K**:

Then you paid the option's price $C(t)$ *at time* **t**.
And you thus acquired the ***right*** *to buy a unit of* **A** *at time* **T** *at the strike price* **K** *(regardless of the price that the underlying asset A has at time T).*

It is important to note that you acquire the **right, not the obligation**, to make that purchase! When will you exercise that right? If and only if the actual price of the underlying A at time T is greater than the strike price K. Because then you can immediately resell the underlying A at the higher actual price and have thus created a positive payoff at time T.
In the following, we denote **the actual price of the underlying asset A at time t by S(t)**.

- So, if at time T, $S(T) > K$, you would exercise the right to buy A at the price K and immediately resell A at the actual price $S(T)$. The payoff that you created at time T is thus the difference $S(T) - K$.
- However, if at time T, $S(T) \leq K$, then of course you would not exercise this right. In that case you would have a cash flow of 0 at time T. We say: "The option expires worthless".

The **payoff resulting from a long position in a call option at time T** therefore is

$$\mathbf{max}\,(\boldsymbol{S}\,(\boldsymbol{T}) - \boldsymbol{K}, \mathbf{0})$$

(i.e. the greater of the two values $S\,(T) - K$ and 0).

However, in reality, this **process of buying and selling the underlying asset does not actually take place** (except in a few exceptional cases that we will not go into here)!

Instead, at time T, this payoff $\mathbf{max}\,(\boldsymbol{S}\,(\boldsymbol{T}) - \boldsymbol{K}, \mathbf{0})$ is automatically paid to the holder of the long position.

To summarize:
The holder of a **long position in a European call option** on the underlying asset A with expiration date T and strike price K receives payment of a **payoff** at time T of $\mathbf{max}\,(\boldsymbol{S}\,(\boldsymbol{T}) - \boldsymbol{K}, \mathbf{0})$. Since that holder initially purchased the option at the price $C(t)$, the **profit (loss) from the option position is** $\mathbf{max}\,(\boldsymbol{S}\,(\boldsymbol{T}) - \boldsymbol{K}, \mathbf{0}) - \boldsymbol{C}\,(\boldsymbol{t})$.

Some notes on this:

- The obvious question as to who pays the payoff will be answered right below.
- And remember: The long position can be closed at any time by taking a short position before expiration T (at the option's then current price)!

It is helpful and important to visualize and memorize the payoff function and the profit function of a long position in a call option, as shown in Fig. 2.1.
Payoff/Profit Function of a Long Call Option

The possible values of the underlying asset's price $S(T)$ at time T are plotted on the horizontal x-axis. The **red graph** shows the **payoff of a call option** with strike price K (where $K = 100$). The payoff is 0 if $S(T)$ is less than K and increases linearly (with a gradient of 1) if $S(T)$ is greater than K.

The **green graph** shows the **profit/loss** generated through a call option with strike price K (where $K = 100$). The green curve of course corresponds exactly to the payoff curve translated downward by $C(t)$. You can see that the loss range is bounded below by $C(t)$ and that the profit range principally does not have any

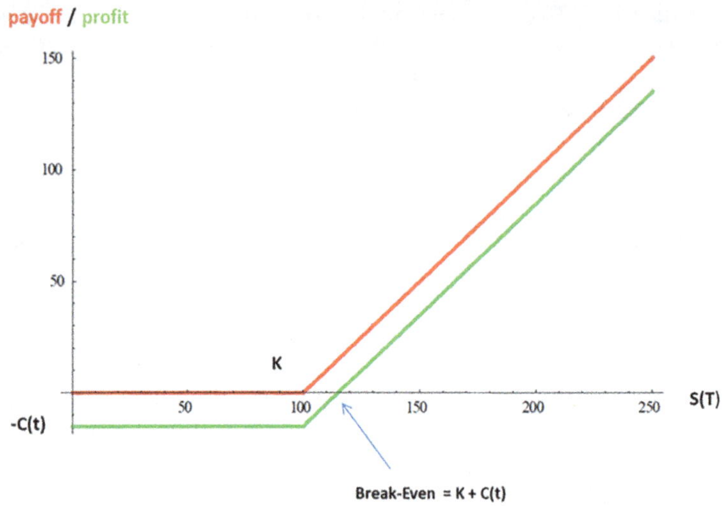

Fig. 2.1 Payoff and profit function of a long call option

upside limit. As long as $S(T)$ is smaller than the **break-even value** $\boldsymbol{K + C(t)}$, the long position in the call option causes a loss. It is only when $S(T)$ is greater than $K + C(t)$ that the call option generates a profit—which increases linearly as $S(T)$ increases.

At the moment when an investor W takes a long position in a particular call option, i.e. pays the option price $C(t)$ and thus buys a right, at exactly that moment, another investor M has to be willing to sell precisely this right to W, i.e. enter into an obligation towards W. As consideration for selling this right, investor M receives the premium $C(t)$ in advance from W. In other words, investor M goes short on the call option in question. Otherwise, no transaction will take place.

(b) If you, at time t, took a **short position** in a **European call option** on the underlying asset **A** with **expiration T** and **strike price K**:

Then you were paid the option's price **C(t)** *at time* **t**. *In return, you entered into the* ***obligation*** *to* ***sell at time*** **T** ***a unit of*** **A** ***at the strike price*** **K** *if called upon to do so (by the holder of the long position).*

Note that you **are obligated** to make this sale if called upon to do so!

When will you have to meet this obligation? If and only if the actual price of the underlying A at time T is greater than the strike price K. Because that is when the holder of the long position will want to exercise his or her option. In theory, you would then have to buy the underlying asset on the market at $S(T)$ and sell it to the long position holder at K. This would cost you a total of $S(T) - K$. Otherwise, the option would expire worthless. You would not have to make any payment. However, as explained in (a) above, this is not how the process is handled; instead you have to make a payment of $\mathbf{max}\,(S\,(T) - K, 0)$ to the holder of the long position.

To summarize:
The holder of a **short position in a European call option** on the underlying asset A with expiration date T and strike price K has to pay out a **payoff** of $\mathbf{max}\,(S\,(T) - K, 0)$. Since that holder initially was paid the option price $C(t)$, the **profit (loss) from the option position is** $-\,\mathbf{max}\,(S(T) - K, 0) + C\,(t)$.

One comment:

- Note that, in addition, the short position can be closed out at any time by taking a long position before expiration T (at the option's then current price)! In that case the obligation is passed on to another investor.

The **options exchange** now has the following tasks:

- Bring together parties interested in taking long and short positions in the options and ensure adequate pricing.
- Handle "purchase" and "sale" of the option.
- Ensure upfront that the holder of the short position will actually be able to meet his or her obligation (to make any requisite payment) at time T. **Every investor shorting an option must deposit collateral (margin)!** We will discuss the amount of this margin further below.
- Process the payout of the payoff between the holders of the long and short positions at time T.

The shape of a short position's payoff curve and profit/loss curve is obviously exactly the opposite of the long position's curves.
Payoff/Profit Function of a Short Call Option (Compare Fig. 2.2)
You can see that the profit range is bounded above by $C(t)$ and that the loss range principally does not have any downside limit. As long as $S(T)$ is smaller than the **break-even value** $K + C(t)$, the short position in the call option will generate a profit. If $S(T)$ is greater than $K + C(t)$, the short position in the call option will cause a loss that increases linearly as $S(T)$ increases.

Put options are defined by exactly the same concept. The only difference is that the right to buy (for the long position) is replaced by the right to sell.

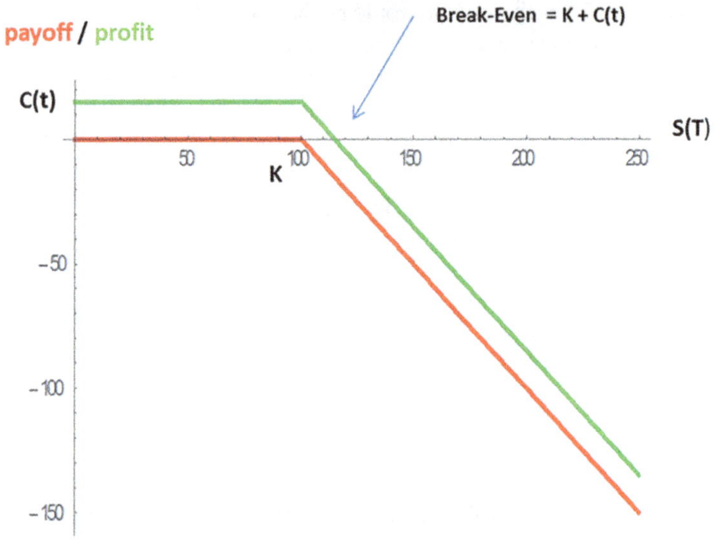

Fig. 2.2 Payoff and profit graph of a short call option

PUT Options

(c) If you, at time t, took a **long position** in a **European put option** on the underlying asset **A** with **expiration date T** and **strike price K**:

Then you paid the option's price **P(t)** *at time* **t**.
You thus acquired the ***right*** *to* ***sell*** *one unit of* **A** *at time* **T** *at strike price* **K** *(regardless of the price that the underlying asset A has at time T).*

The **payoff resulting from a long position in a put option at time T** is therefore $\mathbf{max}\,(K - S\,(T)\,,0)$ (i.e. the greater of the two values $K - S\,(T)$ and 0). However, in reality, this **process of selling and buying the underlying asset does not actually take place**.
Instead, at time T, this payoff $\mathbf{max}\,(K - S\,(T)\,,0)$ is automatically paid to the holder of the long position.

To summarize:
The holder of a **long position in a European put option** on the underlying asset A with expiration date T and strike price K **receives** a **payoff** at time T of $\mathbf{max}\,(K - S\,(T)\,,0)$. Since that long position holder initially purchased the option at the price $P\,(t)$, his or her **profit (loss) from the option position is**
$\mathbf{max}\,(K - S(T), 0) - P(t)$.

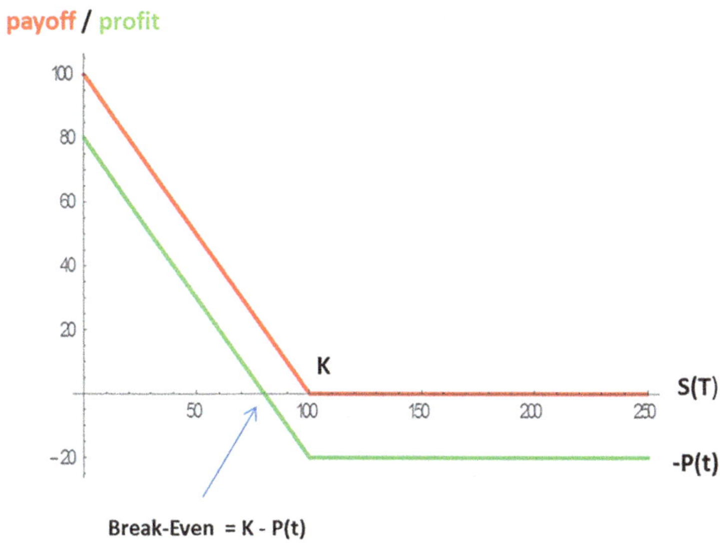

Fig. 2.3 Payoff and profit graph of a long put option

Payoff/Profit Function of a Long Put Option (Compare Fig. 2.3)
The **red curve** plots the **payoff of a put option** with strike price K (where $K = 100$). The payoff is 0 if $S(T)$ is greater than K and increases linearly with decreasing $S(T)$ (with a slope of 1) if $S(T)$ is smaller than K.

The **green curve** illustrates the **profit/loss** generated by a put option with strike price K (where $K = 100$). The green curve of course corresponds exactly to the payoff curve translated downwards by $P(t)$. You can see that the loss range is bounded below by $P(t)$ and that the profit range, while limited, has great upside potential. As long as $S(T)$ is greater than the **break-even value** $\boldsymbol{K - P(t)}$, the long position in the put option will result in a loss. It is only when $S(T)$ is smaller than $K - P(t)$ that the put option generates a profit—which increases linearly as $S(T)$ decreases.

(d) If you, at time t, took a **short position** in a **European put option** on the underlying asset **A** with **expiration date T** and **strike price K**:

Then you were paid the option's price $\boldsymbol{P(t)}$ *at time* **t**. *In return, you made the* ***commitment*** *to* ***buy at time*** **T** ***a unit of*** **A** ***at the strike price*** **K**, *if called upon to do so (by the holder of the long position).*

Yet again, this is not how the process is handled; instead, you have to make a payment of $\mathbf{max}\,(\boldsymbol{K - S(T)}, \mathbf{0})$ to the holder of the long position.

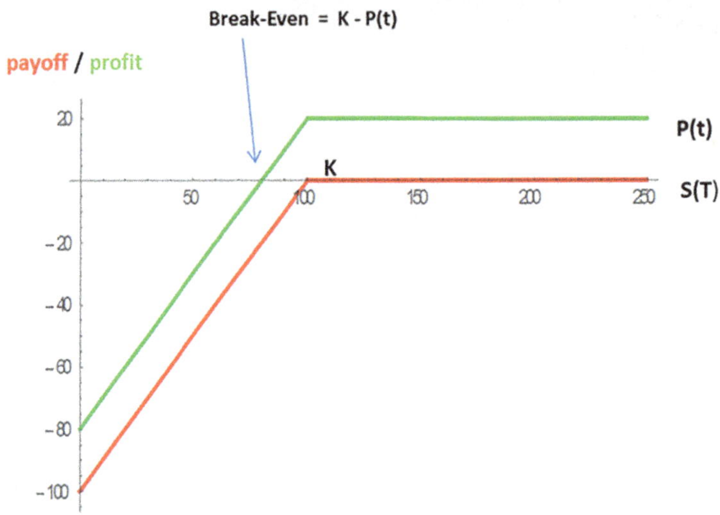

Fig. 2.4 Payoff and profit function of a short put option

To summarize:
The holder of a **short position in a European put option** on the underlying asset A with expiration T and strike price K has to pay a **payoff** at time T of $\mathbf{max}\,(K - S\,(T)\,,0)$. Since that short position holder was initially paid the option price $P\,(t)$, the **profit (loss) from the option position is** $-\mathbf{max}\,(K - S\,(T)\,,0) + P\,(t)$.

Payoff/Profit Function of a Short Put Option (Given in Fig. 2.4)
You can see that the profit range is bounded above by $P\,(t)$ and that the loss range, while limited, has high downside potential.

As long as $S\,(T)$ is greater than the **break-even value** $K - P\,(t)$, the short position in the put option will generate a profit. If $S\,(T)$ is smaller than $K - P\,(t)$, the short position in the put option will cause a loss that increases linearly as $S\,(T)$ falls.

Note on terminology used: For an option that would result in a "significant" payoff when exercised immediately, we say, "The option is in the money" (ITM).

For a call option, this is the case when the current price $S\,(t)$ of the underlying asset is "significantly" above the strike price K (the current payoff would then be $S\,(t) - K$, hence significantly positive).

For a put option, this is the case when the current price $S\,(t)$ of the underlying asset is "significantly" below the strike price K (the current payoff would then be $K - S\,(t)$, hence significantly positive).

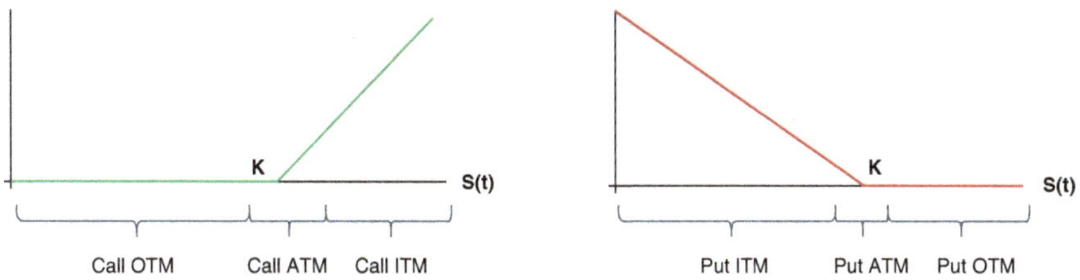

Fig. 2.5 OTM, ATM, and ITM options

For an option that would result in no "significant" payoff when exercised immediately, we say, "The option is out of the money" (OTM).

For a call option, this is the case when the current price $S(t)$ of the underlying asset is "significantly" below the strike price K (the current payoff would then be clearly 0).

For a put option, this is the case when the current price $S(t)$ of the underlying asset is "significantly" above the strike price K (the current payoff would then be clearly 0).

For an option where the current price $S(t)$ of the underlying asset is close to the strike price K, we say, "The option is at the money" (ATM) (see, e.g. Fig. 2.5).

2.3 American Options

Let us recall:

> The holder of a **long position in a European call option** on the underlying asset A with expiration date T and strike price K receives a **payoff** at time T of $\max(S(T) - K, 0)$.

In the two figures below, we have illustrated the situation of a long position in a call option for potential price movements of the underlying asset (blue) over the life of the option contract (x-axis). In Fig. 2.6, the price curve of the underlying asset ends significantly above the strike price K. The difference $S(T) - K$ is therefore paid out to the holder of the long position.

In Fig. 2.7, the price curve of the underlying asset ends below the strike price K. There is no payout, and the option expires worthless.

Holders of an **American** call option, however, have an additional choice.

If you hold an **American** call option, you can choose **to exercise the option at any point until expiration T**, i.e. you have the right to demand a payout in the amount of $S(t) - K$ at any time t you choose (yet only once!) until expiration T.

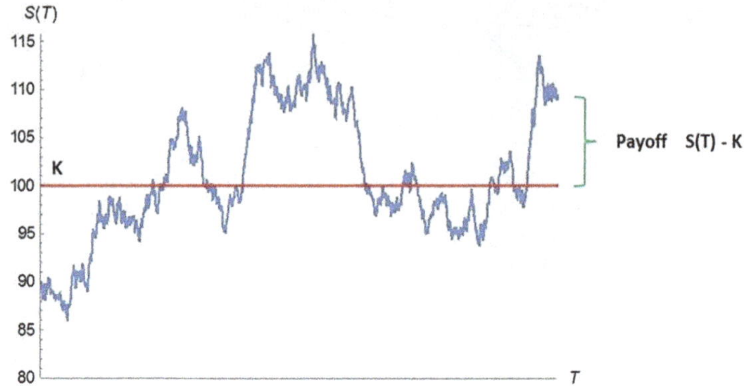

Fig. 2.6 Example of a long call payoff

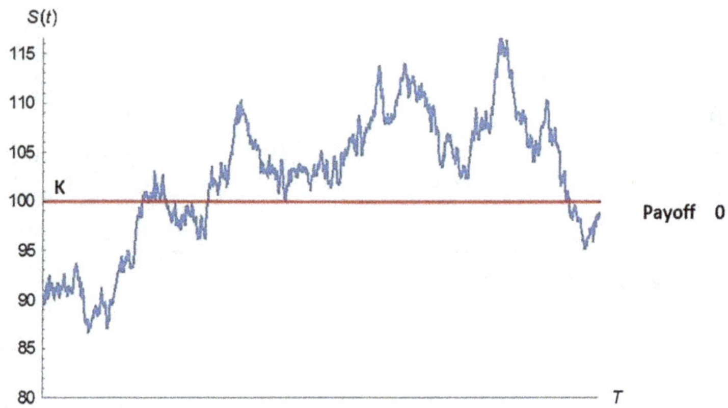

Fig. 2.7 Example of a long call payoff

This means:

If you hold a **long position in an American call option** on the underlying asset A with expiration T and strike price K, you once will **receive** a **payoff** in the amount of $\mathbf{max}\,(S\,(t) - K, 0)$ at any time **t** that you choose during the life of the option.

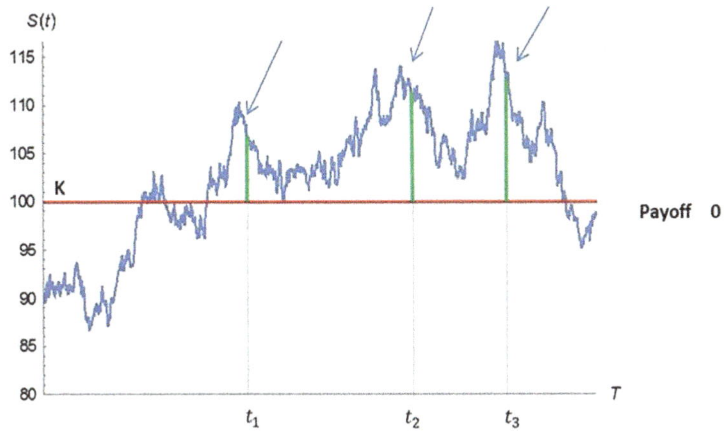

Fig. 2.8 Example of a long call payoff, American style

And conversely:

> If you hold a **short position in an American call option** on the underlying asset A with expiration T and strike price K, you have to pay out a **payoff** of $\mathbf{max}\,(S\,(t) - K, 0)$ as soon as the holder of the corresponding long position requires you to do so at a certain date t during the life of the option.

So, if you were the holder of an American call option in the scenario shown in Fig. 2.7, you could well have considered exercising the option at an earlier point (e.g. at dates t_1, t_2 or t_3 (see Fig. 2.8)). You would then have received one of the payoffs shown in bold green.
The situation is analogous for the American put option, of course.

> If you hold a **long position in an American put option** on the underlying asset A with expiration T and strike price K, you once will **receive** a **payoff** in the amount of $\mathbf{max}\,(K - S\,(t)\,, 0)$, at any date **t** that you choose during the life of the option.

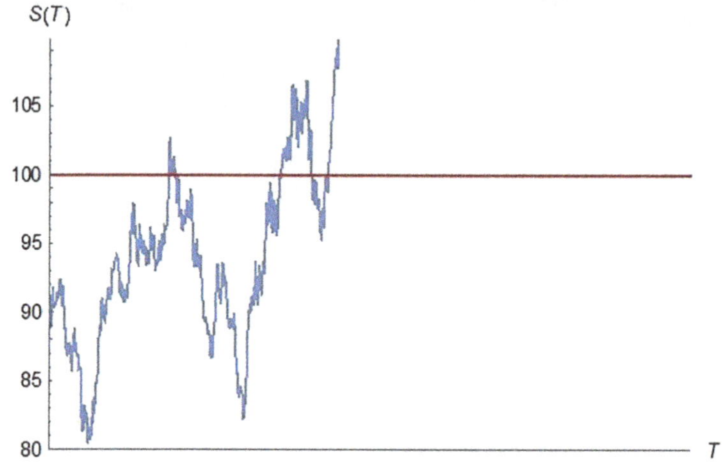

Fig. 2.9 American call, how to decide?

And conversely:

> If you hold a **short position in an American put option** on the underlying asset A with expiration T and strike price K, you have to pay out a **payoff** in the amount of $\mathbf{max}\,(K - S\,(t)\,,0)$ as soon as the holder of the corresponding long position requires you to do so at a certain date t during the life of the option.

So, the holder of an American option can additionally choose to exercise the option at an earlier date. However, with that also comes the pressure of having to take the best possible decision! How should we decide in the situation in Fig. 2.9? Should we exercise or wait? And if we wait, how are we going to decide a bit later (Fig. 2.10)? How to make the right decision? How can you know if you are making the right decision or not? In Figs. 2.9, 2.10, and 2.11, waiting was the wrong decision. But in Fig. 2.10, the price of the underlying asset could just as well have risen sharply, so that an early exercise would have prevented a potentially much higher profit.

Can there even be a strategy that would result in an optimal outcome?

We will deal with this question in depth later. For now, let us simply say: there is, of course, no magic recipe with an inbuilt guarantee for identifying the optimum exercise time (i.e. in our example, the moment during the life of the option at which the price of the underlying asset is at its highest). But there are some "best strategies", and you will get to know them!

To give you just a small taste of what's to come, I am including a section on strategies below. Feel free to skip it if not interested; it has no bearing on the discussions that follow.

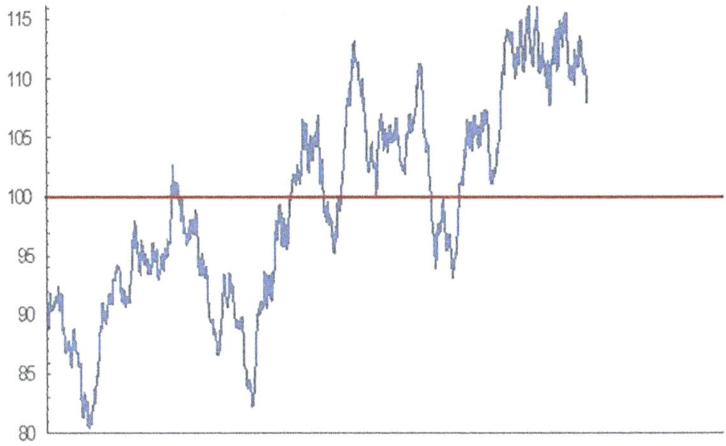

Fig. 2.10 American call, what now?

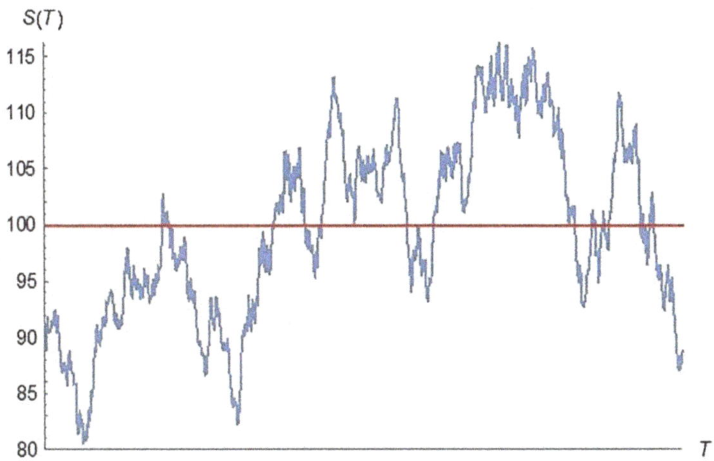

Fig. 2.11 Too late!

2.4 Any Strategy Is Better than No Strategy and "The Secretary Problem"

We will briefly discuss a **very** simple game and then apply that to a real-life everyday situation—which in a sense is an extension of the game.

Let us start with the rules of the **game (beginner's version)**:

The game is played by two people, player A against player B.

Player A secretly writes one arbitrary number each on two pieces of paper. Any numbers are allowed, be they small, huge, immensely negative, irrational, rational, integral, $-5, \pi, 2, 812, 745, -10,000^{1,000,000^{10,000,000^{10,000,000,000}}}$*, or whatever A chooses. The only rule is that the two numbers be different.*

The two pieces of paper are shuffled, and one of them (let's call it Note 1) is revealed—visible to all.

Now it's B's turn: B needs to decide whether to choose Note 1 or the second piece of paper (Note 2), which has not yet been uncovered.
If B chooses Note 1, then A gets Note 2 and vice versa.

The winner is the player with the bigger number on their note.
An Illustration of the game is given in Fig. 2.12.

We now ask ourselves: **Is this purely a game of chance or is there a strategy that B can apply to increase the odds of winning from 50:50 to a more favorable ratio?**
B can of course always achieve a chance of winning of at least 50%.

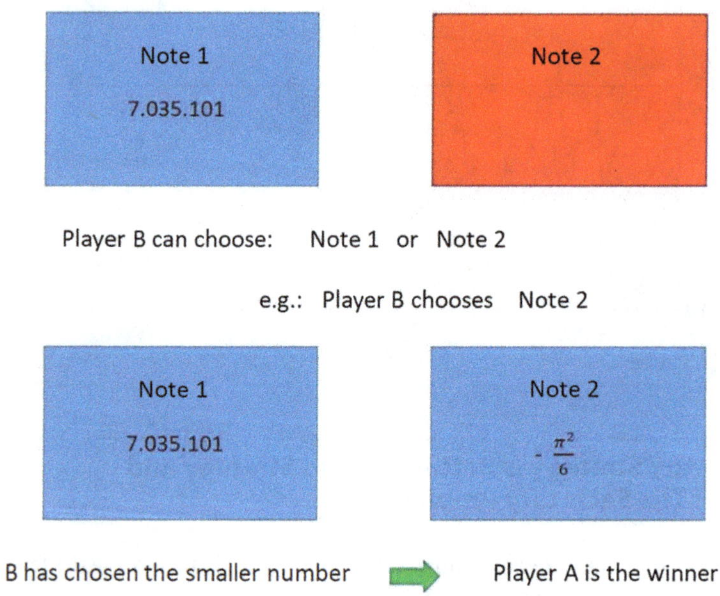

Fig. 2.12 Illustration of the number notes game

For example:

The strategy for player B to "always choose the first note" will of course result in an exactly 50% chance of winning for both players just like, for example, the strategy "choose alternately the first and then the second note" would.

Another, albeit somewhat, vaguely defined "strategy" for player B would be: "If the number on Note 1 appears rather large, then pick Note 1, otherwise choose Note 2". This begs two questions:

First: *What does: "rather large" mean, exactly?*

Does 1,000,000 qualify as large, or is 100 already a large number, or is only something like $10{,}000^{1{,}000{,}000^{10{,}000{,}000^{10{,}000{,}000{,}000}}}$ actually large?

But as you may well know, with any real number there is just as much space "to the right" of

$10{,}000^{1{,}000{,}000^{10{,}000{,}000^{10{,}000{,}000{,}000}}}$ as there is "to the left" of it!

So, in the infinite universe of numbers, "large" is a highly relative qualifier!

In any case, player B has to decide in advance when to consider a number as large and when not. Suppose B chooses a certain bound X. If the number is greater than X, player B will consider it to be large, if it is smaller than X, player B will consider it small.

Second: *Does this strategy improve B's odds of winning, or rather, Which bounds X would improve the odds?*

What do you think?

The—maybe—surprising answer is: **Whatever number B chooses as bound X, B's odds of winning will always improve!** In other words: *any kind of strategy is better than no strategy.*

And why is that?

Let us denote the number on Note 1 by $Z1$ and the number on Note 2 by $Z2$.

That gives us six different cases:

Case 1: $Z1 < Z2 \leq X$
Case 2: $Z2 < Z1 \leq X$
Case 3: $X < Z1 < Z2$
Case 4: $X < Z2 < Z1$
Case 5: $Z1 \leq X < Z2$
Case 6: $Z2 \leq X < Z1$

We'll say as much right away:

Case 1 is as likely to occur as Case 2 (if both numbers are smaller than X, then it is equally likely that either one of these two numbers is the larger one.) Similarly, Case 3 is as likely to occur as Case 4.

If we apply our strategy, then:

- we win in Case 1 (we do not choose $Z1$, since it is smaller than X, so we take $Z2$, which is larger than $Z1$, which means we have won)
- we lose in Case 2

- we lose in Case 3
- we win in Case 4

Since Case 1 (we win) is as likely to occur as Case 2, their odds of winning cancel each other out. The same goes for Case 3 and Case 4. They cancel each other out. Hence, we have neither won nor lost anything so far. Up to now, the odds of winning are still 50:50.

We still have Cases 5 and 6. If we apply our strategy, then:

- we will always win when Case 5 occurs
- we will always win when Case 6 occurs

Meaning: When Cases 5 and 6 occur, our odds of winning are 100%. And so we have increased our overall odds of winning to more than 50%!

Yet we cannot say anything about **how high** our actual odds are of winning. That depends on how likely it is for Cases 5 and 6 to occur, and we have no information about that.

It is interesting to note, however: no matter what kind of bound X player B chooses, it will always increase B's odds of winning to more than 50%!

Let us now explore the **advanced version** of the game (in addition to that, there is also a "professional version"):

In the advanced version, player A writes a different number on each of ***three*** *pieces of paper (the Notes). These numbers* $Z1$, $Z2$, $Z3$ *are arbitrary, but each is different from the others.*

The notes are shuffled.

Note 1 is uncovered. Player B decides whether or not to take Note 1.

If not, then Note 2 is uncovered. B decides whether or not to take Note 2.

If not, player B gets Note 3.

Player A will get the two notes that B decided not to take.

Player B can only choose ***one*** *note!*

The winner of the game is the player who has the note with the largest number on it.

At first glance, it doesn't look good for player B, given that he or she ends up with just one note, while A gets two notes. A's odds of winning thus seem to be twice as great as B's odds. And that's actually right, if B's "strategy" were, for example, to always take the first note, or always take the third note, then the likelihood for player B to get the largest number is $\frac{1}{3}$. B's odds of winning would be 33.33%.

Now, when I mentioned player B choosing the **first** or the **third** note above, did I do so intentionally or was that purely coincidental? What about choosing the second note?

The situation here is a little different:

Let us assume I am player B. I did not choose the first note with the number $Z1$. Now I look at the second note with the number $Z2$. If $Z2$ is smaller than $Z1$, I will certainly not choose $Z2$, since $Z2$ cannot be the largest of the three numbers. I will discard $Z2$ therefore and choose $Z3$.

So this would be a possible **strategy for B**:
Discard $Z1$.
If $Z2 > Z1$ *then choose* $Z2$.
If $Z2 < Z1$ *then choose* $Z3$.

What is the likelihood of winning with this strategy?
If $Z1$ is the largest number, I will definitely lose (the probability of that occurring is $\frac{1}{3}$).
If $Z2$ is the largest number, then I will definitely win (the probability of that happening is again $\frac{1}{3}$).
If $Z3$ is the largest number (probability $\frac{1}{3}$), then $Z1 > Z2$ in half of the cases (and I win) and $Z2 > Z1$in half of the cases and I lose).

All in all, this means:
Using this strategy, the odds for B to win is exactly $\frac{1}{2}$!

So, what seemed like a rather hopeless situation for player B at first has at least become a 50:50 game for B. And perhaps there is an even better strategy!?
No, there isn't: lacking any further information, this is the best possible strategy for B.
By now, you will probably have guessed what the **professional version** of the game looks like:
In the professional version, player A writes a different number on each of ***n pieces*** *of paper.*
These numbers $Z1, Z2, Z3, \ldots, Zn$ *are arbitrary, but each is different from the others.*
The notes are shuffled.
Note 1 is uncovered. Player B decides whether or not to take Note 1.
If not, then Note 2 is uncovered. B decides whether or not to take Note 2.
If not, then Note 3 is uncovered. B decides whether or not to take Note 3.
etc.
If B also decides against taking $n - 2$, *then Note* $n - 1$ *is uncovered. B decides whether or not to take Note* $n - 1$.
If not, player B gets Note n.
Player B can only choose ***one*** *note!*
Player A gets all the notes that B decided not to take.
The winner of the game is the player who has the note with the largest number on it.

If n is large (e.g. $n = 1000$), the outlook for player B would seem catastrophic! How can player B, with just one (!) note against $n - 1$ (e.g. 999) notes, expect to end up with the largest number?

This game situation is often illustrated using the following real-life example:

An office manager is urgently looking for **one** new secretary. There are ***n* candidates** outside, waiting to be interviewed for this job. The office manager is in such a hurry that she does not actually intend to interview all n applicants. Hence, she starts briefly interviewing one candidate after another (in random order) and decides immediately after each interview for or against that candidate. She does want to get the best candidate, of course.... Is there a suitable strategy that the office manager can apply to ensure the highest likelihood of getting the best candidate despite the time constraints? In the literature, this—admittedly hypothetical—example is called the "secretary problem" and represents the same predicament that we saw in the professional version of our game. The office manager has the role of player B.

So, let's get back to the professional version, find a solution to our question, and apply that to the secretary problem.

After some analysis, we see that in the general case, we should proceed just the same way that we did when we had three notes:

We start with a certain **observation phase** of, say, length m.

This means that we will look at the first m numbers $Z1, Z2, \ldots,$ Zm but will reject them.

From then on, however, after that observation phase, i.e. from $Z\,(m+1)$, we pick the first number that is greater than all previously uncovered numbers (if no such number comes up anymore, then there is only Zn left and we have lost anyway).

The questions that arise are:

- How can we ensure that the observation phase m we choose is of optimum length?
- Using this strategy (and assuming optimal m), how likely is player B to win the game?

For readers with a deeper interest in mathematics, we develop the answers to both questions below. If less interested, you can just skip this part (grey box), as we will summarize the answers anyway.

If we proceed as described, then we will win a game when **both** of the following conditions are met (this is easy to think through, we encourage you to do it!):

(a) The largest number, let us denote it by Z, is not among the first m numbers (but occurs, e.g. in spot k which is greater than m (i.e. $k = m + 1$ or $m + 2$ or $m + 3$ or ... or n)).

(continued)

(b) The largest number that occurred up to $k-1$ was among the first m numbers (and not among the numbers $Z(m+1), Z(m+2), \ldots, Z(k-1)$).

The probability that Z will occur exactly at k is, of course, $\frac{1}{n}$.
The probability that condition (b) is met for a given k is obviously $\frac{m}{(k-1)}$ (m "good" spots for the largest number to occur among $k-1$ possible spots so far).
The probability that Z occurs at k and condition (b) is met is (since the two events are independent of each other) equal to the product of the probabilities, i.e. equal to $\frac{1}{n} \cdot \frac{m}{(k-1)}$.
Since k can attain any of the values $m+1, m+2, \ldots, n$, the probability of conditions (a) and (b) occurring therefore equals

$$\sum_{k=m+1}^{n} \frac{1}{n} \cdot \frac{m}{k-1} = \frac{m}{n} \cdot \sum_{k=m+1}^{n} \frac{1}{k-1}$$

Now we only have to select m such that we get the highest possible probability. Based on this, we can get an exact result for any given n of course.

For example, with $n = 4$ we get:
From among these probabilities, we get the highest one for $\mathbf{m = 1}$ as $\frac{11}{24} =$ **0.458333**. . ., as displayed in Table 2.1.

Hence, the best strategy with 4 numbers is: pass on the first number and then select the first one that is larger than all previous ones. The probability of choosing the largest number with this method is almost 46%.

The case of five numbers, i.e. $n = 5$, is illustrated in Table 2.2:
From among these probabilities, we get the highest one for $m = 2$ as $\frac{13}{30} =$ 0.43333 . . .

Table 2.1 Odds of winning the game with four numbers for $m = 1, \ldots 4$

m	Odds of winning $= \frac{m}{4} \cdot \sum_{k=m+1}^{4} \frac{1}{k-1}$
1	$\frac{11}{24}$
2	$\frac{5}{12}$
3	$\frac{1}{4}$

Table 2.2 Odds of winning the game with five numbers for $m = 1, \ldots 4$

m	Odds of winning $= \frac{m}{5} \cdot \sum_{k=m+1}^{5} \frac{1}{k-1}$
1	$\frac{5}{12}$
2	$\frac{13}{30}$
3	$\frac{7}{20}$
4	$\frac{1}{5}$

Hence, the best strategy with five numbers is: pass on the first number and the second number and then select the first one that is larger than all previous ones. The probability of choosing the largest number with this method is more than 43%.

It would seem that, as n grows, player B's odds of winning get smaller and smaller and go to zero. Is that actually so?
To find the answer to that question, we analyze the sum $\sum_{k=m+1}^{n} \frac{1}{k-1}$ more closely for a large n (and thus also for an increasing m). It is almost exactly $\log(n/m)$, where log denotes the natural logarithm. So if we want to maximize the probability $\frac{m}{n} \cdot \sum_{k=m+1}^{n} \frac{1}{k-1}$ (for a large n), we need to maximize $\frac{m}{n} \cdot \log\left(\frac{n}{m}\right)$.
Let us set $\frac{n}{m} = x$.
This leaves us with the task of maximizing the function $f(x) := \frac{\log(x)}{x}$.
We do this by differentiating the function and setting the derivative equal to zero:
$f'(x) := \frac{1-\log x}{x^2} = 0$; hence $1 - \log x = 0$ and thus $x = e = 2.71828$.
This gives us

$$\frac{n}{m} = x = e$$

and therefore $m = \frac{n}{e} = 0.3679 \cdot n$.
The probability of winning the game is then close to $f(e) = \frac{\log(e)}{e} = \frac{1}{e} = 0.3679\ldots$.

To summarize:
In a game with n numbers (or n applicants in the secretary problem), where n is a large number, choose the observation period as integer m that is the closest to $0.3679 \cdot n$. Discard the first m numbers and then choose the first number that is larger than all previous ones. Using this method, your likelihood of winning the game (choosing the best candidate) is about 36.7%.

This holds for arbitrarily large n! Hence, the odds of winning with this strategy are surprisingly high!

Further questions in this context can of course be developed ad infinitum, for example, how likely is it to get one of the two biggest numbers, when using an optimal strategy?

What has this got to do with our problem? Can this type of strategy be applied to our problem of finding the most favorable moment in time to exercise American options?

The answer is "no", not directly. That is because, in the case of American options, the stock prices do not occur in a sequence of completely independent numbers. A high share price is most likely followed by a share price that is close to the previous price rather than one that is suddenly much lower.

Nevertheless, this section about the relevance of having a strategy rather than no strategy at all was important to get a sense of, and gain some confidence in, what a strategy is and what it can do for you—even in seemingly hopeless situations.

2.5 How Do You Trade Options? Trading Through a Bank

Options are traded on options exchanges or directly with financial institutions that issue options (in which case we usually refer to them as "warrants") (OTC trading).

In all of the following—wherever we execute trades for ourselves—we will deal practically exclusively with exchange-traded options.

In our strategies, we always strive to discover options prices that we can potentially profit from and then make the most of those opportunities for our purposes.

Prices of options issued by financial institutions are set by the institutions themselves and are therefore designed from the outset to work in their favor and less so in ours—we being their counterparties in a transaction.

Prices of options on exchanges (in which we can always take both long and short positions), on the other hand, are created purely through supply and demand on the exchange. This can often result in options prices that can be advantageous for us.

Furthermore, we will deal almost exclusively with options on the S&P500 index for the following reasons: The options market on the S&P500 index is probably the world's most liquid market of exchange-traded options. The author himself has been trading intensively in this market for decades and can provide the most competent information about trading specifically in this market. There are also other important highly liquid derivative markets that are closely related to the S&P500 options market and can therefore be included in the creation of trading strategies (S&P500 futures, S&P500 futures options, VIX futures, VIX options, etc.).

What we are aiming for is not for you to get to know a wide range of different markets but rather for you to learn to act, analyse, and trade in a fully confident and circumspect manner in a select market.

In practice, there are two different ways to trade options on an exchange. In both cases, you need a trading account with a bank on which trading in derivatives is permitted.

Trades can then be executed either by means of orders transmitted by telephone (or by fax or e-mail) to a broker or can be placed directly on the respective options exchange itself via an electronic trading platform.

The author of this book trades for his own account, as well as for clients of his investment management company, including via the German Baader Bank in Munich and the Zurich-based UBS through brokers, as well as via the electronic trading platform Interactive Brokers.

When you trade through brokers with Baader Bank, UBS, or any other bank:

You open a securities account (also referred to as a trading account or a brokerage account) with one of these banks and transfer an investment amount into that account. Upon opening the securities account, you state that you (or the asset manager who will be authorized to use the account for trading purposes) have experience in trading derivatives. Once the bank has given the green light to allow trading, orders for option trades can be placed with the brokers at that bank (during their brokerage service hours).

Especially when trading options on the US options market, which is open from 15:30 to 22:00 Central European Time, you want to make sure you work with a bank whose trading desk is actually open for business until 22:00. Not every bank offers that (but Baader Bank and UBS do, for example).

One problem that many investors have, especially private investors trading for their own account, is to obtain relevant **real-time market data**. While real-time (bid/ask) data for basic investments such as stocks, indices, bonds, etc. are provided free of charge on many Internet sites, reliable real-time price data (including bid/ask prices) for derivatives can rarely be accessed at a low cost.

Access to professional financial data information systems such as Bloomberg or Reuters is highly advantageous but very cost-intensive. The author has access to Bloomberg (at a cost of currently about 1700 euros per month), so screenshots of Bloomberg pages will be shown quite often below. Your broker will of course also provide data on real-time prices of derivatives that you specifically request information on. For example, a free page with—not always all data on—option prices for the S&P500 can be found at http://bigcharts.marketwatch.com/quickchart/options.asp?sid=3377&symb=SPX.

Let us take a look at Bloomberg screenshots, as shown in Figs. 2.13 and 2.14, of options for example on 10 September 2019.

Let us pick one of these options that we would like to trade, for example, the put option on the S&P500 with expiration date 18 October 2019, a strike price of 2950 and bid/ask prices = USD 39.10 // 40.20.

The bid price of 39.10 means that we could currently go short (i.e. take a short position) on this option at a guaranteed price of 39.10.

The ask price of 40.20 means that we could currently go long (i.e. take a long position) on this option at a guaranteed price of 40.20.

With decades of experience in trading S&P500 options, the author can confirm that it is in fact mostly possible to make a trade close to the midpoint between bid and ask price. In our case, this means that the trade is highly likely to occur even if we place an order to sell (take a short position) at a price of approximately 39.50 or if we place an order to buy (take a long position) at a price of approximately 39.80.

S&P500 options cannot be purchased or sold piecewise, but only in the form of contracts. A single contract of S&P500 options covers 100 units of the option.

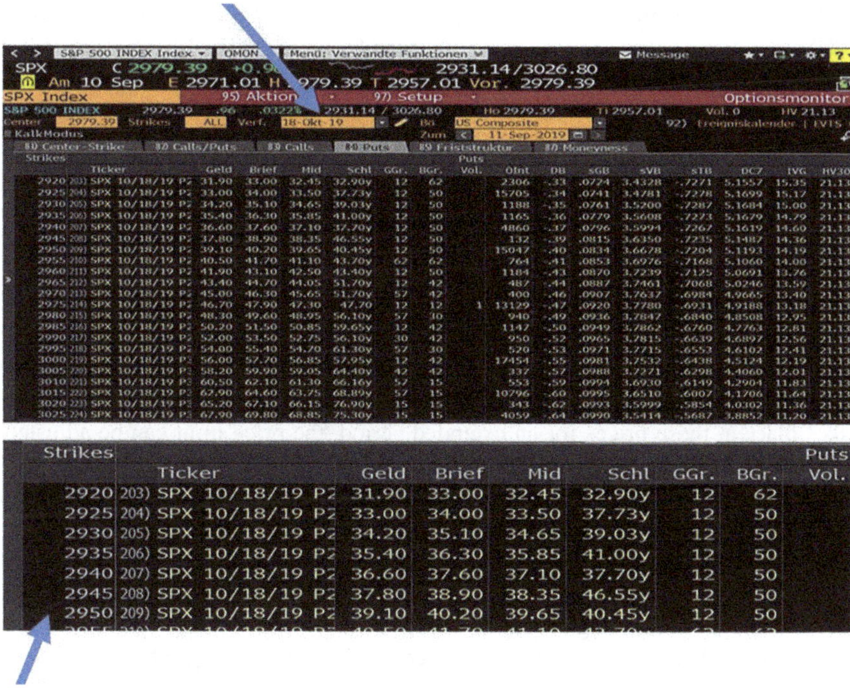

Fig. 2.13 Bloomberg screenshot: put options on 10 September 2019 (source: Bloomberg)

Fig. 2.14 Bloomberg screenshot: call options on 10 September 2019 (source: Bloomberg)

The limit order that you would then place for trading a contract of the above S&P500 option in these two cases would then look something like this ("SPX" is the ticker symbol for the S&P500 index):

Sell to open a put contract on SPX, expiration October 2019, strike 2950, limit 39.50 dollars.
and
Buy to open a put contract on SPX, expiration October 2019, strike 2950, limit 39.80 dollars.

The broker places these orders, which are then either executed immediately at the specified prices or—if the trade does not take place immediately—are shown as new bid/ask prices in the real-time information systems (e.g. Bloomberg). If no trade takes place at these prices for an extended period, the price limit can be adjusted, for example, to 39.40 for the sell order or to 39.90 for the buy order.

As soon as the trade has been executed, the sale proceeds of, for example, 3950 dollars will be credited to your trading account, or the purchase amount of, for example, 3980 dollars will be debited to your account. In addition, your broker will calculate the necessary margin in the event of a sale (a new short position) and block that amount in your trading account.

We will look at the margin that you need when taking short positions in options at a later point.

If you already have a long position in an option and then take a short position in the same option, the two positions cancel each other out and both positions expire. We say: "The long position was **closed (out)**". Conversely, an already existing short position in an option is closed out by taking a matching long position.

The order would then be phrased like this:

Sell to close a ... contract on SPX, ... strike ... limit ... dollars. and
Buy to close a ... contract on SPX, ... strike ... limit ... dollars.

Trading through an electronic trading platform has many advantages, as we will see. However, one sometimes major advantage of trading via a bank's brokerage service is the following: as collateral, you have to deposit margin in cash with the options exchange. If an investor A trades via a bank B, then bank B does not normally transfer money from A into the options exchange's margin account on a case-by-case basis. Instead, bank B maintains a permanent and always sufficiently funded margin account with the option exchange of which it is a member. Any loss resulting from option trades by investor A is thus debited to the margin account of bank B. Bank B is covered against any such losses through the investor's margin account with the bank. In this setup, the question of the margin deposit required from A is therefore solely a matter between investor A and bank B. Many banks therefore also offer investors the possibility to deposit securities or securities portfolios (or bank guarantees) as collateral. This means that investor A does not necessarily have to put up cash to engage in options trading. However, in the event of a loss due to options trading, the investor may have to liquidate part of the securities portfolio to settle that loss with the bank. This is an advantage that investors appreciate especially when their options strategies consist mainly of short positions, since the latter create positive premiums (the prices of the options sold) that the investor receives upfront. This allows the investor to use an existing securities portfolio in a twofold capacity: on the one hand, as a portfolio that generates income and, on the other hand, as collateral for options trading.

However, the bank does not accept the entire current value of the portfolio as margin deposit. The reason for this is, of course, that the value of a securities portfolio is not fixed but can fluctuate. Some banks assign a marginable value of 90% to bonds with the highest credit rating, a value in the range of 50%–70% to bonds with a lower credit rating and of 50% to stable stocks listed in major indices.

2.6 How Do You Trade Options? Trading Through an Electronic Trading Platform

The major advantages of trading through an electronic trading platform are:

you are much more flexible in your trades, you usually get the real-time data you need relatively inexpensively, you are not dependent on the availability of a broker, and the transaction fees are usually much lower than when trading through a broker at a bank.

To show you how an electronic trading platform works, we are going to use the example of the trading platform at Interactive Brokers.

This provider offers an excellent platform for derivatives trading. There are several other trading platforms through which derivatives can most probably be traded just as comfortably and professionally. The author's extensive experience in online trading however is above all with Interactive Brokers, which is why this platform was chosen for the following representations.

Opening a trading account at Interactive Brokers (hereinafter IB) is a matter of simply a few formalities on their website at

www.interactivebrokers.com .

After you have wired an investment amount (only cash transfers, of course, not securities portfolio transfers as described above) to Citibank (in Germany, Switzerland, or in the USA) and your account has been activated, you can start trading.

When logging into IB, you first see the Portfolio page of the Trader Workstation. Here you can see all the open positions that you currently hold in your portfolio. In the portfolio shown in Fig. 2.15 (with two zooms or excerpt for better readability), we begin our explanations with the line marked with the blue arrow.

EUR cash

You currently have a cash amount of EUR 680,339 in your portfolio.

USD cash

You currently have a cash amount of USD 205,651 in your portfolio.

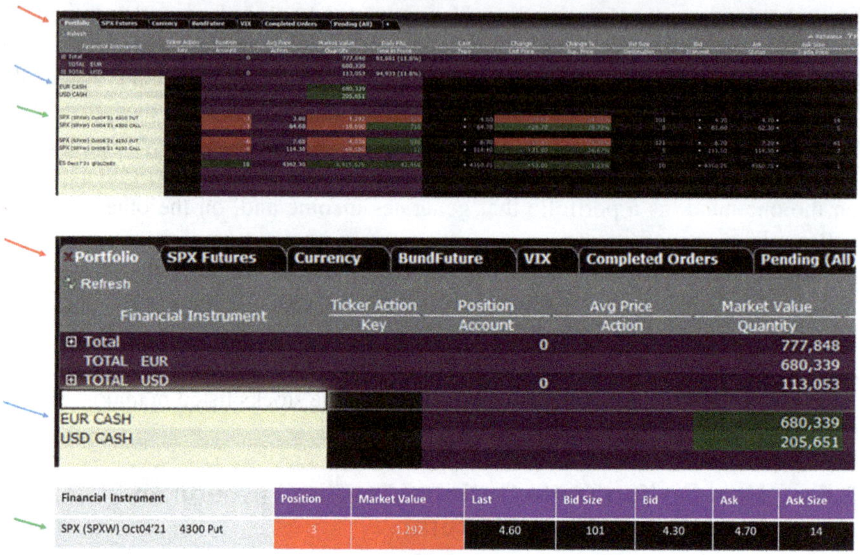

Fig. 2.15 IB Trader Workstation Portfolio window

This is followed by five rows of products that you are currently holding in your portfolio:

SPX (SPXW) Oct 04' 21 4300 Put

This is a put option on the S&P500 Index that expires on 4 October 2021 and has a strike price of 4300.

Further to the right in the same row, we have:

−3 (highlighted in red) means that 3 **short**-position contracts have been taken on this option.

The last four columns in that row are:

Bid size 101 and **bid price 4.30** as well as **ask price 4.70** and **ask size 14**.

So, right now, the best buy offer is USD 4.30 **per unit** (i.e., USD 430 per contract) for a total of 101 contracts of this option and the best sell offer is USD 4.70 per unit (i.e., USD 470 per contract) for a total of 14 contracts of this option.

These are the most important data that you need for trading this option.

The column labelled **Last** tells us that this option was last traded at a price of USD 4.60.

The market value column is essentially the number of positions (times 100 contract size) multiplied by the current mean value of bid and ask price. The value **−1292** thus roughly reflects the **actual current value** of these three short option contracts in US dollars.

The other columns provide additional statistical (daily) information.

The following lines list further short positions in call and put options expiring on 4 October 2021 and 6 October 2021 that are held in the portfolio.

The last line

ES Dec 17' 21 @ GLOBEX

shows **18 long positions** of (mini)**futures on the S&P500 Index** expiring on 17 December 2021. We will look at the details of this line later in the context of futures. For now, let us simply say: the value of 3,915,675 in "the market value column" has nothing to do with the actual value of these 18 positions. The value of futures (as we will explain later) is always 0.

Let us now move on to the first two lines of the overview:

Total USD

The stated value of USD 113,053 is composed of the USD 204,651 in cash and the total (at the moment negative) cash values of the various options contracts denominated in USD.

Total EUR

The stated value of EUR 680,339 would be composed of the EUR 680,339 in cash and the total cash values of the EUR-denominated financial products in the portfolio (however, these do not exist in this portfolio).

In addition to the Portfolio page that we logged in to, the Trader Workstation allows you to set up any number of additional pages. The red arrow pointing to the tabs in Fig. 2.15 indicates that, in our case, we have six more pages besides the currently opened Portfolio page: SPX Futures, Currency, Bund Future, VIX, Completed Orders, and Pending (All). On these pages, we can save the products we want to watch, trade, and track.

In our example, the "SPX Futures" page lists futures on the S&P500 with three different expiration dates (see Fig. 2.16).

To display an overview of the (mostly numerous) options offered and their real-time data on a given underlying asset, you can use a separate tool called the "OptionTrader", which can be accessed by clicking on the "Option Trader" button in the toolbar at the top of the IB Trader Workstation.

Figure 2.17 shows a small selection of options on the S&P500 traded on the CBOE. These are options expiring on 26 November 2021 (see red arrow in Fig. 2.17). Any amount of other expiry dates can be selected by clicking on the relevant tabs in the top bar. The blue column (labelled Strike) in the centre of the panel displays the strike prices of the options shown here. In this case they range from 4220 to 4370. Scrolling down, you will be able to see all existing strike prices (from 1000 to 6400 in this case). The call options are listed to the left of the Strike

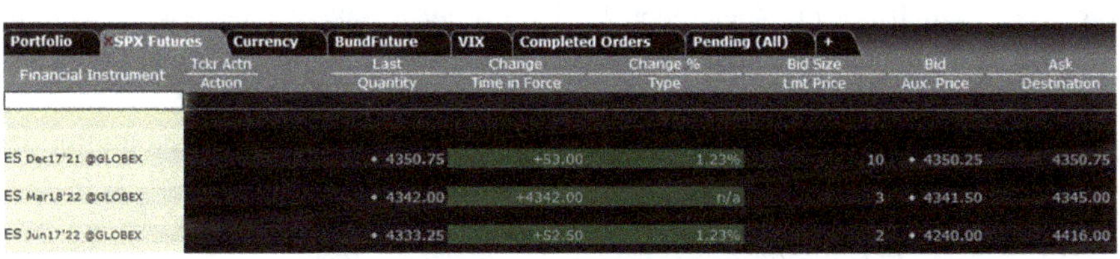

Fig. 2.16 es window in IB Trader Workstation

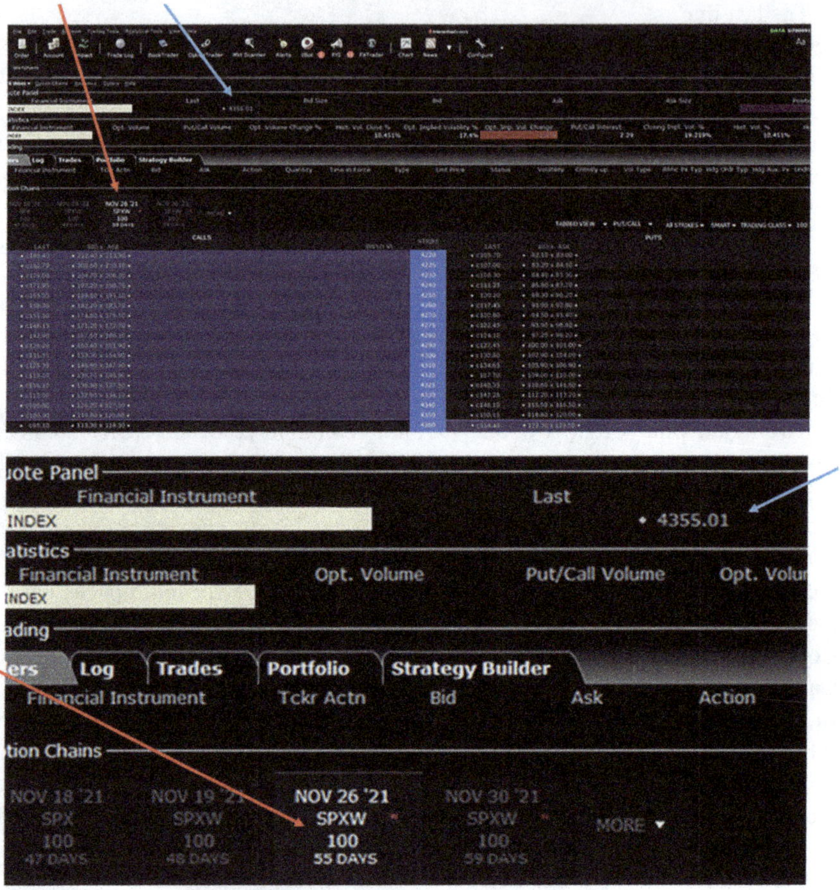

Fig. 2.17 IB OptionTrader, selection of S&P500 Options

column, and the put options are listed to the right. The products listed on these pages can be traded directly from any of these pages.

If you want to trade products that you are already holding in your portfolio, you can do so directly from the IB Trader Workstation's Portfolio page.

For example, if we want to trade the first option in our portfolio shown in Fig. 2.18, which is from 4 October 2021: in order to take a long position, i.e. to buy, we move the cursor to the ask price of the option (8.40, red arrow in Fig. 2.18) and left-click on it. (If we want to take a short position, i.e. sell, we left-click on the option's bid price (8.00).) An order line opens directly below the option line, automatically suggesting the following:

BUY 1 contract, time in force DAY, with limit USD 8.40.

When you click the *Transmit* button, the order is placed (and would also be immediately executed at the price of 8.40).

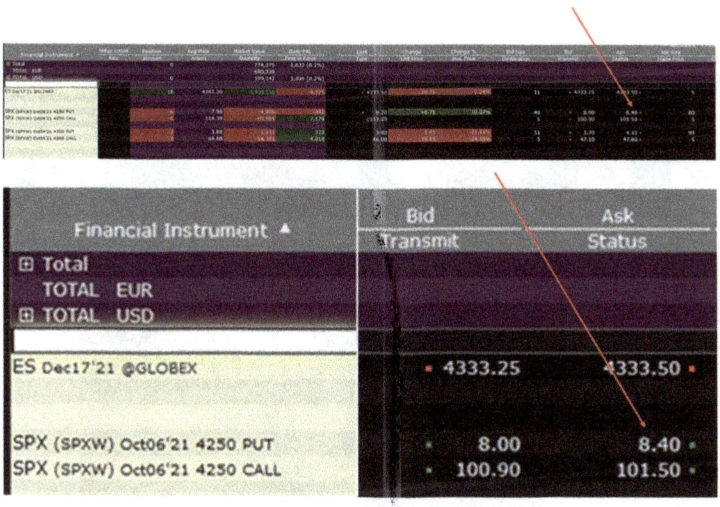

Fig. 2.18 IB Trader Workstation, Trading Activity I

DAY means that the order will expire if it is not executed by the end of the current trading day. By clicking the *DAY* button, this setting can be changed to *GTC* (good till cancelled). GTC means that the order will remain active until it is executed or deleted.

In the following, however, we do not want to buy just **one** contract but **two** contracts, and we are not willing to pay 8.40 but just a maximum of 8.40.

To do this, we click on the current quantity of "1", which opens a window where we can select the number of contracts we want to trade.

We choose the number "2" (see red arrow in Fig. 2.19).

Furthermore, to change the limit, we click on the suggested price of "8.40". Again, a window opens in which we can select the desired limit.

The order is then placed by clicking on the "Transmit" button.

The order we just placed is now shown in the Order line in Fig. 2.20.

Additionally, in the option line to the right, we see the new Bid size (now 2) and Bid price (now 8.10), which have both changed as a result of the order we placed.

Shortly afterwards, the order has already been executed—despite the fact that we set the limit "left" of the midpoint between bid and ask price (see Fig. 2.21). The number of positions (red arrow in Fig. 2.21) in this option has changed from six short positions to four short positions (as we closed out two short position contracts by taking long positions) and the last price has changed from 8.20 to 8.10.

Click "Trade Log" in the header of the IB Trader Workstation to view all the details of that trade (and of previous transactions, if you wish). For the trade that we just transacted, the Trade Log would look like this (Fig. 2.22):
The entry on the far right of this line provides information about the transaction fees. In this case, the fees were 2.54$, i.e. 1.27$ per contract.

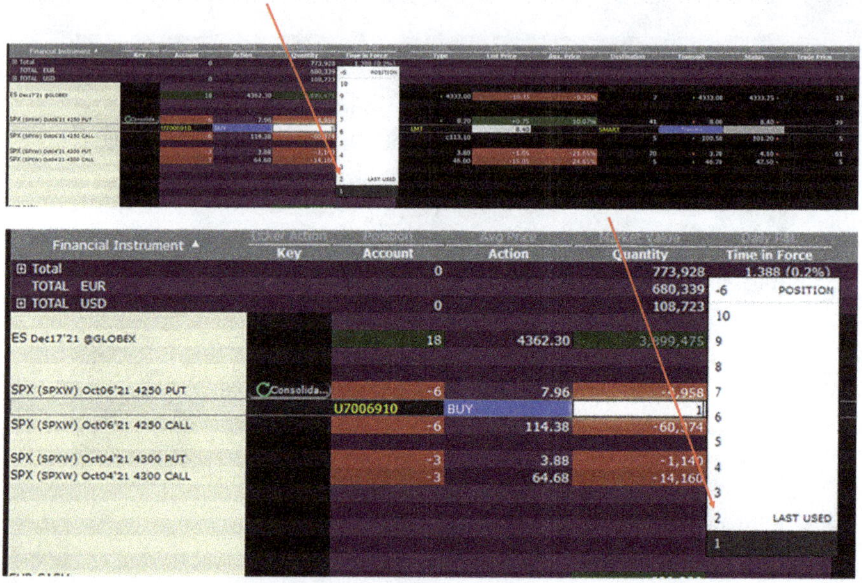

Fig. 2.19 IB Trader Workstation, Trading Activity II

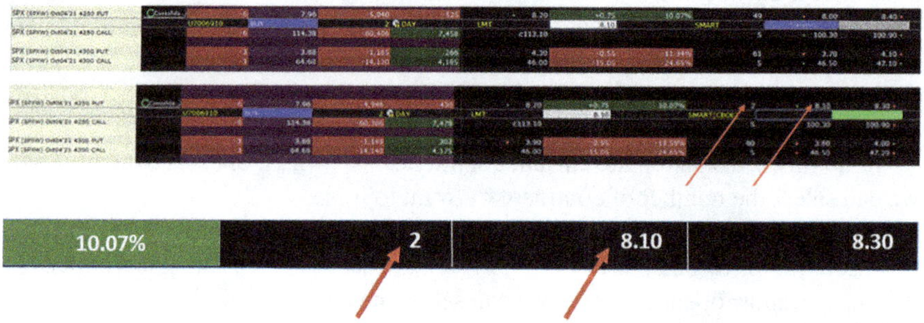

Fig. 2.20 IB Trader Workstation, Trading Activity III

Considering that by trading one options contract we essentially trade 100 units of the S&P500 at a current value of 4340 points, meaning we can trade a value of 434,000$ at 1.27$ in transaction fees, we see once again the efficiency of trading options also in terms of transaction costs.

In general, trading through a bank's brokerage service is substantially more costly of course. Here are two examples of transaction fees per order charged by banks that the author is also using:

Bank A: Cost 6.42$ per options contract
Bank B: 3$ per options contract +
+ Max (25, number of contracts $\times 0.875 \times$ price per option)

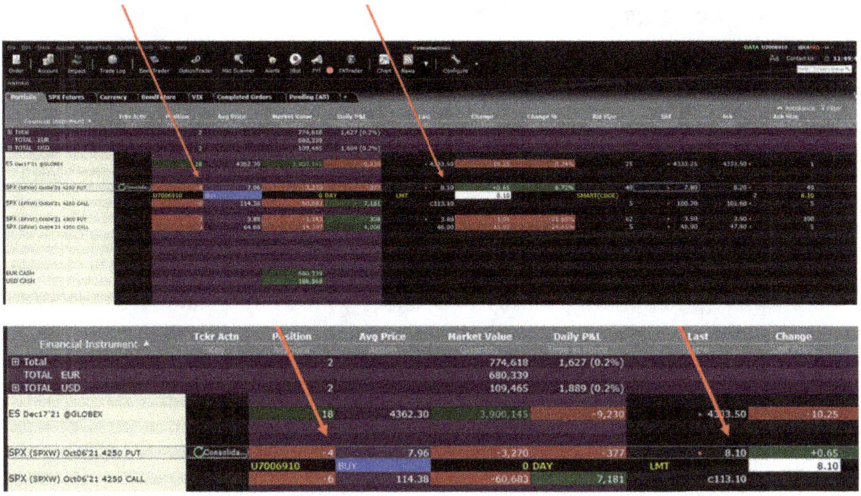

Fig. 2.21 IB Trader Workstation, Trading Activity IV

+/-	Action	Quantity	Comb.	Fin Instrument	Price	Currency	Exch.	Time	Order Ref.	Commission
	BOT	2	■	SPX (SPXW) Oct06'21 4250 PUT	8.10	USD	CBOE	11:49:47		2.54

Fig. 2.22 Trade Log I

+/-	Action	Quantity	Comb.	Fin Instrument	Price	Currency	Exch.	Time	Order Ref.	Commission
	BOT	2	■	SPX (SPXW) Oct06'21 4250 PUT	8.10	USD	CBOE	11:49:47		2.54
⊞	SLD	2	■	SPX (SPXW) Oct06'21 4250 PUT	8.3	USD	CBOE	11:54:15		2.54

Fig. 2.23 Trade Log II

Especially the second fee structure can result in high transaction costs when trading more expensive options.

A little later, we reopened the two options contracts that we closed for demonstration purposes. As we can see in the corresponding trade log (second line in Fig. 2.23), they could be sold at an even slightly better price of 8.30. This is yet another indication of the high liquidity of the S&P500 option market on the CBOE.

As already mentioned further above, decades of experience in trading S&P500 options have shown that it is in fact mostly possible to make a trade close to the midpoint between the bid/ask prices. This was again confirmed by the transaction that we just described. Here are a couple of additional comments on this: if the current bid/ask prices are due to market-maker quotes, it is often possible to achieve a buy price in the range of "Mid +0.05" or "Mid +0.10" and a sell price in the range of "Mid −0.05" or "Mid −0.10". Occasionally, the price on one of the two sides, bid or ask, is set by an investor's limit order. In such cases, it may sometimes even be possible to achieve a price "on the side of the midpoint that's better for me". For example:

The market-maker quotes for a given option are 1.00//2.00. So realistically, it should be possible to sell at a price of 1.45 and buy at a price of 1.55. If an investor places a sell order with a sell limit of 1.80, the order will normally not (immediately) result in a trade. The new quotes will then be 1.00//1.80. If you place a sell order now with a limit of 1.45, it is quite possible that the order will indeed be executed, even though the limit is "to the right" of the current midpoint at 1.40.

Conversely, if an investor places a buy order with a buy limit at 1.20, the order will normally not (immediately) result in a trade. The new quotes will then be 1.20//2.00. If you place a buy order now with a limit at 1.55, it is quite possible that the order will indeed be executed, even though the limit is "to the left" of the current midpoint at 1.60.

The same applies analogously if we place a new option order from within the OptionTrader page or if we trade another product listed on one of the other pages of the IB Trader Workstation.

For example:

We place a sell order for the put option with expiration 26 November 2021, strike 4325, by clicking the corresponding bid price of 110.60$ (see Fig. 2.24, blue arrow). Clicking on it opens the order line (see Fig. 2.24, red arrow, or Fig. 2.25).
In this order line, we can again modify the number of contracts we want to trade (blue arrow in Fig. 2.25), the time in force (DAY or GTC, red arrow), and the limit (green arrow). The order is then placed by clicking the "Transmit" button (violet arrow).

2.7 Who Trades in Options? Long Positions in Call Options, Leverage

The following sections are intended to provide only some preliminary insights into the possible reasons for taking basic positions in options (rather than investing directly in the underlying assets).

Before we start, it is important to note that options are highly fine-tuned tools in the financial market, which we can use, in suitable combinations and often in conjunction with other products (futures, the underlying asset itself, etc.), to pursue a wide range of investment objectives. Here, however, we are only going to describe a few basic positions and how to proceed when you first start out in options trading.

Let us assume that today is 1 October 2021. If we look at the chart of the S&P500 of the last 12 months in Fig. 2.26, we might be inclined to assume a very clear upward trend in the index. And since there are no fundamental indications that anything could stop this trend, we might start to think that this trend is likely to continue for some time and that we should therefore invest in this index. (The validity of this assumption is not the issue at this point; for purposes of the following, let us simply accept it as a working hypothesis.)

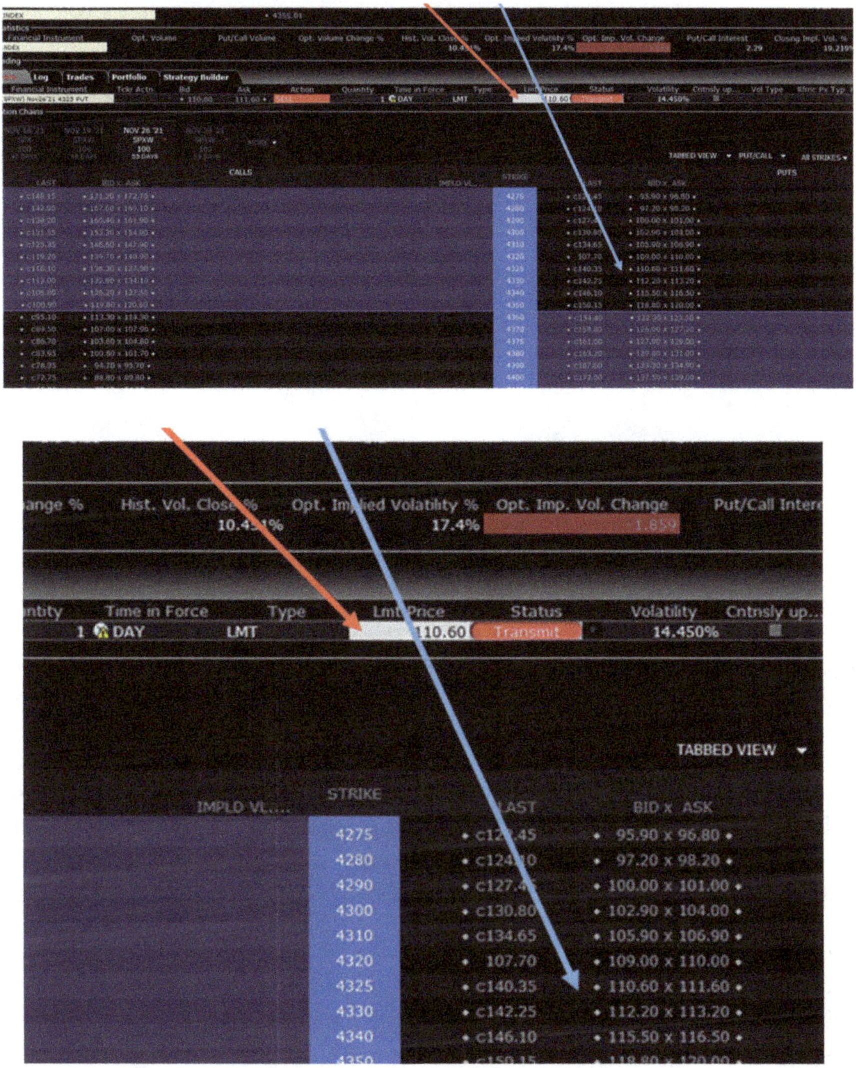

Fig. 2.24 Trading from within the OptionTrader page

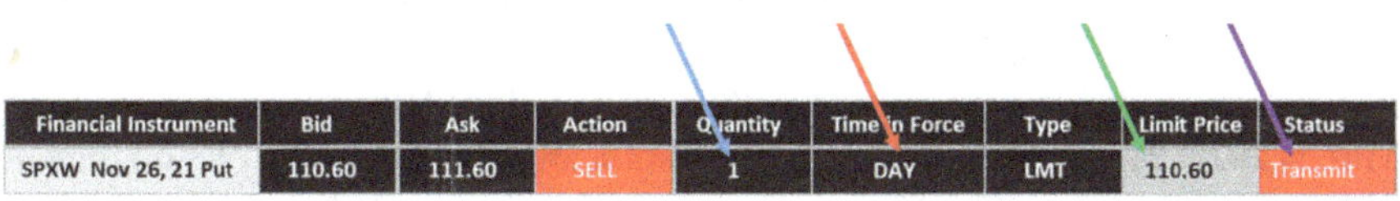

Fig. 2.25 Order line on the OptionTrader page

Suppose we have 100,000 US dollars at our disposal that we can use for investments. (We assume cash in dollars here, as the prices of the S&P500 and the prices of options on the S&P500 are quoted in dollars.)

Fig. 2.26 S&P500 index chart: October 2020 to October 2021 (with "trend line")

If, as shown in Fig. 2.26, we also draw an approximate "trend line", we believe that the S&P500 could be expected to rise from 4355 points now to around 5200 points by mid-March 2022.

An **obvious approach** to capitalizing on this forecast would, of course, be to buy the S&P500 in the form of an index certificate. With a total investment of 100,000 dollars, we could buy 22 units of the index at a price of $22 \times 4355 = 95,810$ dollars. At this point, we are not taking into account transaction costs or bid/ask spreads (we are only interested in the basic principle for now). An amount of 4190$ of the 100,000$ are still at our disposal. If the S&P500 index indeed rises to around 5200 points by 17 March 2022, we can sell the 22 index certificates at a price of 5200 dollars each on the very same day. We would receive $22 \times 5200 = 114,400$ dollars. Together with the leftover (unused) amount of 4190 dollars, we now have a total of 118,590 dollars, which means that we made a profit of 18.59%.

An **alternative approach** would be the following:

We open the Option Trader window with options on the SPX expiring on 17 March 2022 (see Fig. 2.27).

As an example, let us pick the **call option with strike 4310 and expiration on 17 March 2022**. The bid/ask quotes for this option are 244.00//246.60. We go long on four contracts (100 units) of this option. A trade (purchase) should be possible at the price of 246.00 dollars per option, i.e. at 24,600 dollars per contract. That leaves us with a balance of 1600 dollars. If the SPX rises to 5200 points by 17 March as assumed, we will receive a payout of $5200 - 4310 = 890$ dollars per option. The total payout for our contracts is therefore of $400 \times 890 = 356,000$ dollars. Including

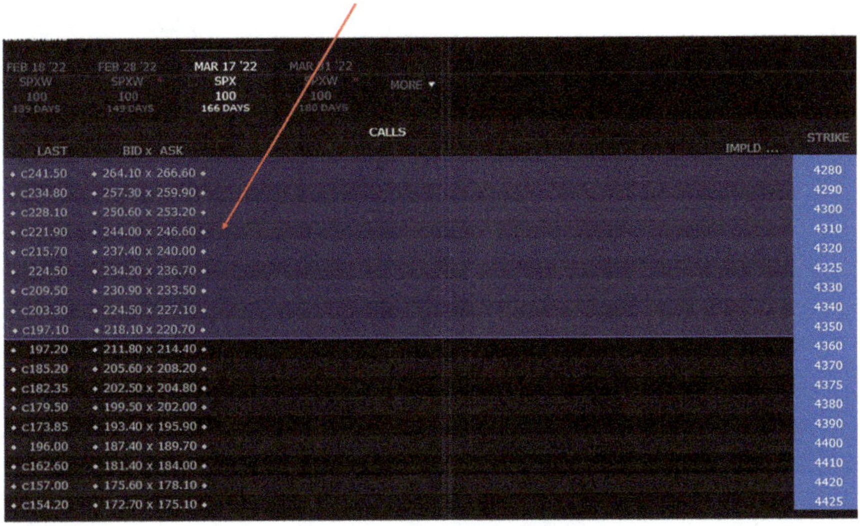

FEB 18 '22 SPXW 100 139 DAYS	FEB 28 '22 SPXW 100 149 DAYS	MAR 17 '22 SPX 100 166 DAYS	MAR 31 '22 SPXW 100 180 DAYS	MORE ▾

CALLS

LAST	BID x ASK	IMPLD ...	STRIKE
• c241.50	• 264.10 x 266.60 •		4280
• c234.80	• 257.30 x 259.90 •		4290
• c228.10	• 250.60 x 253.20 •		4300
• c221.90	• 244.00 x 246.60 •		4310
• c215.70	• 237.40 x 240.00 •		4320
• 224.50	• 234.20 x 236.70 •		4325
• c209.50	• 230.90 x 233.50 •		4330
• c203.30	• 224.50 x 227.10 •		4340
• c197.10	• 218.10 x 220.70 •		4350
• 197.20	• 211.80 x 214.40 •		4360
• c185.20	• 205.60 x 208.20 •		4370
• c182.35	• 202.50 x 204.80 •		4375
• c179.50	• 199.50 x 202.00 •		4380
• c173.85	• 193.40 x 195.90 •		4390
• 196.00	• 187.40 x 189.70 •		4400
• c162.60	• 181.40 x 184.00 •		4410
• c157.00	• 175.60 x 178.10 •		4420
• c154.20	• 172.70 x 175.10 •		4425

Fig. 2.27 IB OptionTrader on 1 Oct 2021, options on SPX expiring 17 March 2022

our unused balance of 1600 dollars, we then have 357,600 dollars, i.e. we have made a profit of 257.6% (compared to 18.59% with the conventional strategy).

This is called the "leverage effect" of an option: using the same amount of money, the leverage effect allows you to make a much higher profit than if you traded directly in the underlying asset.

Comments

(a) The second approach has a major drawback, of course: the leverage effect can also work to your disadvantage. Hence, if the SPX dropped to, say, 4300 points, i.e. by 1.26%, you would not get any payout from the options. Once the options expire, all you have is the 1600 dollars that you didn't invest, meaning you have made a loss of practically 100%.

(b) Do I depend on the behaviour of the SPX on 17 March 2022 for the second approach, i.e. when using options? Do I have to stand idly by while a situation unfolds as shown in Fig. 2.28 (SPX increases to 5300 points by mid-February, then stagnates, and, within just a few days until the beginning of March, drops to 4600 points, with a further downward trend)?
If we were dealing with American options, we could, of course, exercise the option and take the profit at any time before the option expires, if we consider that preferable. However, as mentioned above, the SPX options on the CBOE are European style. So we can't exercise them, but we can sell (close) them on the exchange at any time.
The only question is: At what price?
This question—how the price of an option changes over time—is an aspect that we will look at in detail over the course of this book. To give you a rough idea

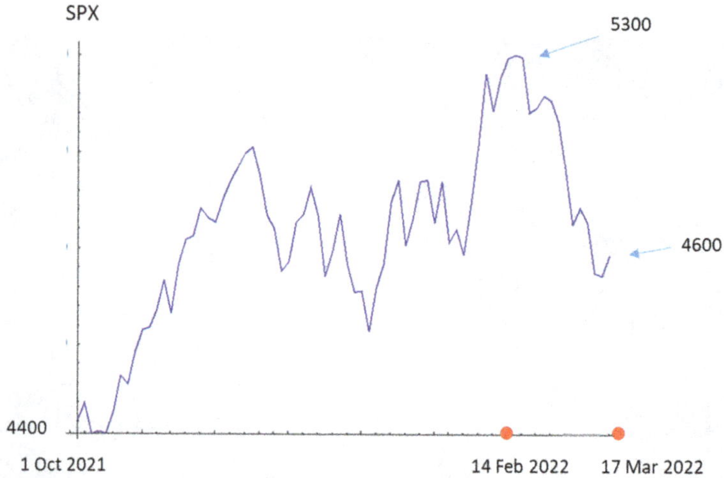

Fig. 2.28 Possible forecast for the S&P500 until mid-March 2022 from the point of view of 1 October 2021

upfront: throughout its life until expiration, the sell price of a call option is above—or at most only very slightly below—the payoff that would be achieved if the option were exercised immediately.
So, in our example, if in mid-February the SPX is at 5300 points and we were to sell the option at that time (i.e. closing it), we would earn a premium of at least $5300 - 4310 = 990$ dollars per contract (or just slightly less).
Given the possibility of trading the option at any time, we are therefore not dependent on the behaviour of the index exclusively on 17 March 2022!

(c) Very similar results could of course be achieved using call options with the same term to expiration and other strike prices. We could, for example, consider a call option with a strike price of 4390. The bid/ask quotes are 193.40//195.90 and we therefore trade five contracts for our 100,000 dollar budget. If the index rises to 5200 points by 17 March 2022, we will receive a payout of $5 \times 100 \times (5300 - 4390) = 455{,}000\$$, making a profit of 355%. What then is the disadvantage of choosing strike 4390 instead of strike 4310? Right: If we choose a strike price of 4390, we will sustain a total loss if the index stays below 4390 points (as there will be no payout). With a strike price of 4310 and an index level of 4390 on 17 March 2022, we would at least receive a payout of $500 \times (4390 - 4310) = 40{,}000$ dollars, i.e. incur a loss of only 60%.

(d) It is very important to develop very good intuition when trading in options. With that in mind, we are going to think through two further extreme cases, purely as an exercise and for you to become familiar with options trades.
So, we are convinced that the index will rise to 5200 points by 17 March 2022, and we will again assume that this is actually going to be the case. Which call option, which strike price would give us the highest profit, if we invest the entire amount of 100,000 dollars? We will of course not consider options with a strike

Table 2.3 Possible profit obtained from options with different strike prices

Strike	Quotes	Approx. buy price	Qty	Unused bal.	Payoff	Total
3970	495.50//498.00	496.50	2	700	246, 000	246, 700
4000	471.50//474.10	473.00	2	5400	240, 000	245, 400
4050	432.30//434.80	433.70	2	13, 260	230, 000	243, 260
4100	393.80//396.40	395.40	2	20, 920	220, 000	240, 920
4150	356.30//358.90	357.80	2	28, 440	210, 000	238, 440
4200	319.90//322.40	321.30	3	3610	300, 000	303, 610
4250	284.60//287.10	286.40	3	14, 080	285, 000	299, 080
4275	267.40//270.00	268.90	3	19, 330	277, 500	296, 830
4300	250.60//253.20	252.00	3	24, 400	270, 000	294, 400
4310	244.00//246.60	245.50	4	1800	356, 000	357, 800
4320	237.40//240.00	238.80	4	4480	352, 000	356, 480
4330	230.90//233.50	232.30	4	7080	348, 000	355, 080
4340	224.50//227.10	226.00	4	9600	344, 000	353, 600
4350	218.10//220.70	219.50	4	12, 200	340, 000	352, 200
4360	211.80//214.40	213.30	4	14, 680	336, 000	350, 680
4370	205.60//208.20	207.00	4	17, 200	332, 000	349, 200
4380	199.50//202.00	200.90	4	19, 640	328, 000	347, 640
4390	193.40//195.90	194.80	5	2600	405, 000	407, 600
4400	187.40//189.70	188.70	5	5650	400, 000	405, 650
4420	175.60//178.10	177.00	5	11, 500	390, 000	401, 500
4440	164.10//166.60	165.50	5	700	456, 000	456, 700
4460	153.00//155.50	154.50	5	7300	444, 000	451, 300
4480	142.30//144.70	143.60	5	13, 840	432, 000	445, 840
4500	132.50//134.20	133.50	7	6550	490, 000	496, 550
4550	108.40//110.10	109.40	9	1540	585, 000	586, 540
4600	86.70//88.70	87.90	11	3310	660, 000	663, 310
4650	68.40//70.30	69.60	14	2560	770, 000	772, 560
4700	53.10//54.80	54.10	18	2620	900, 000	902, 620
4750	40.70//42.10	41.50	24	400	1, 080, 000	1, 080, 400
4800	30.90//32.10	31.60	31	2040	1, 240, 000	1, 242, 040
4850	23.30//24.30	23.90	41	2010	1, 435, 000	1, 437, 010
4900	17.50//18.50	18.10	55	450	1, 650, 000	1, 650, 450
4950	13.10//14.10	13.70	72	1360	1, 800, 000	1, 801, 360
5000	9.90//10.60	10.30	97	90	1, 940, 000	1, 940, 090
5100	5.60//6.30	6.00	166	400	1, 660, 000	1, 660, 400
5200	3.10//3.70	3.50	285	250	0	250

price greater than or equal to 5200, as such options would not deliver any payoff if the index rose to 5200 points. Let us draw up a Table 2.3 with results for various strike prices between 3970 and 5200.

The total final value tends to grow with growing strike (until 5200). But of course the danger of losing the whole investment if the SPX does not increase enough grows drastically.

It is obvious therefore that long positions in call options are used mainly to exploit the leverage effect when speculating on rising prices of the underlying asset.

2.8 Who Trades in Options? Long Positions in Put Options, Protective Put

Long positions in put options can of course be used analogously to long positions in call options, i.e. to speculate on falling prices with a leverage effect.

Here is just a quick example without further analysis: if we had purchased a put option contract with expiration 17 March 2022 and strike 4350 (with quotes according to Fig. 2.29 of 228.40 // 230.90 respectively) at the price of 23,000 dollars on 1 October 2021, and the price of the S&P500 had fallen to 3900

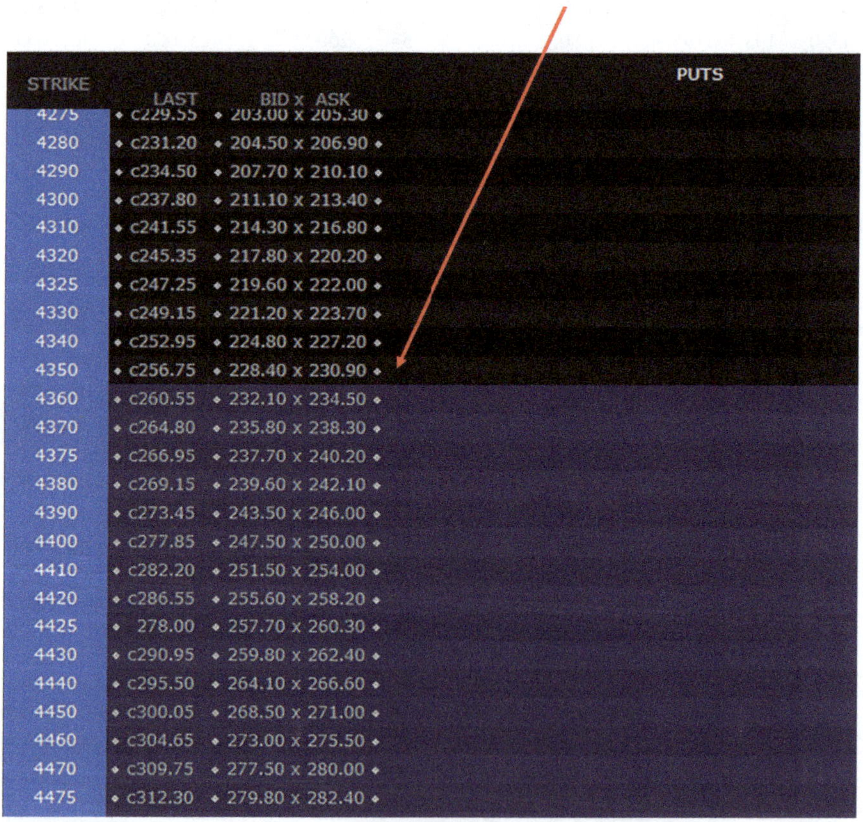

PUTS

STRIKE	LAST	BID x ASK
4275	• c229.55	• 203.00 x 205.30 •
4280	• c231.20	• 204.50 x 206.90 •
4290	• c234.50	• 207.70 x 210.10 •
4300	• c237.80	• 211.10 x 213.40 •
4310	• c241.55	• 214.30 x 216.80 •
4320	• c245.35	• 217.80 x 220.20 •
4325	• c247.25	• 219.60 x 222.00 •
4330	• c249.15	• 221.20 x 223.70 •
4340	• c252.95	• 224.80 x 227.20 •
4350	• c256.75	• 228.40 x 230.90 •
4360	• c260.55	• 232.10 x 234.50 •
4370	• c264.80	• 235.80 x 238.30 •
4375	• c266.95	• 237.70 x 240.20 •
4380	• c269.15	• 239.60 x 242.10 •
4390	• c273.45	• 243.50 x 246.00 •
4400	• c277.85	• 247.50 x 250.00 •
4410	• c282.20	• 251.50 x 254.00 •
4420	• c286.55	• 255.60 x 258.20 •
4425	• 278.00	• 257.70 x 260.30 •
4430	• c290.95	• 259.80 x 262.40 •
4440	• c295.50	• 264.10 x 266.60 •
4450	• c300.05	• 268.50 x 271.00 •
4460	• c304.65	• 273.00 x 275.50 •
4470	• c309.75	• 277.50 x 280.00 •
4475	• c312.30	• 279.80 x 282.40 •

Fig. 2.29 Quotes for put options on SPX with expiration date 17 March 2022

points (i.e. by 10.33%) by 17 March 2022, we would have received a payoff of $100 \cdot (4350 - 3900) = 45{,}000$ dollars on 17 March 2022. Assuming an investment of 23,000 dollars, we would have made a profit of 95.6% (instead of 10.33% for a direct short investment, e.g. via a short index certificate).

Anticipating falling prices of certain indices, stocks, or commodities and speculating on such price movements is, in and of itself, not at all reprehensible but rather part of a functioning and highly liquid financial market. Speculation based on insider knowledge, however, rightly brings the use of such put option strategies into disrepute. And this is exactly what happened, for example, just before and around the time of the 9/11 attacks on the World Trade Center in New York on 11 September 2001.

On 6 and 7 September 2001, CBOE dealers recorded the purchase of 4744 put options contracts on shares of United Airlines (UAL), one of the two airlines whose planes were hijacked for the attack. This volume was many times higher than put option trades in this stock on other trading days. There was no such increase for other airlines. Likewise, there was a sharp increase in put options on the stock of the investment bank Morgan Stanley, which operated offices on 22 floors of the World Trade Center. That trading volume increased to 20 times the usual volume. Other banks, insurance companies, and airlines also saw significantly stepped-up trading in put options.

On 10 September 2001, a total of 1535 put options contracts (at 100 options each) on UAL with expiration October 2001 and strike 30 were bought at a price of 2.15 dollars per option.

If we look at the performance of the UAL stock in September 2001 (see Fig. 2.30 and Table 2.4), we find that the stock was trading in the range of just over 30 dollars before 11 September 2001. When the exchange reopened after having closed around 11 September, the price was 18.25 dollars. The stock price had dropped by around 40%.

On 17 September 2001, following the reopening of the exchange, these put options were closed at a price of 18 dollars per option. The profit per option was thus 15.85 dollars, which (based on the investment of 2.15 each) translates into a return on investment of 637%! In absolute numbers, this means a profit of $1535 \times 100 \times 15.85 = 2{,}432{,}975$\$ solely from the put options on the UAL stock traded on 10 September.

Suspicion of insider trading by individuals who had been privy to information about the attack and used their knowledge on the exchanges had been raised by stock market observers shortly after 11 September 2001. This suspicion was refuted in a few sentences by the US Congress' 500-page official investigation report in 2004, which did not however disclose the stock market movements or even the names of the parties involved in those transactions. A comprehensive scientific analysis was published by Zurich financial professor Marc Chesney and co-authors in the "Journal of Empirical Finance" [1].

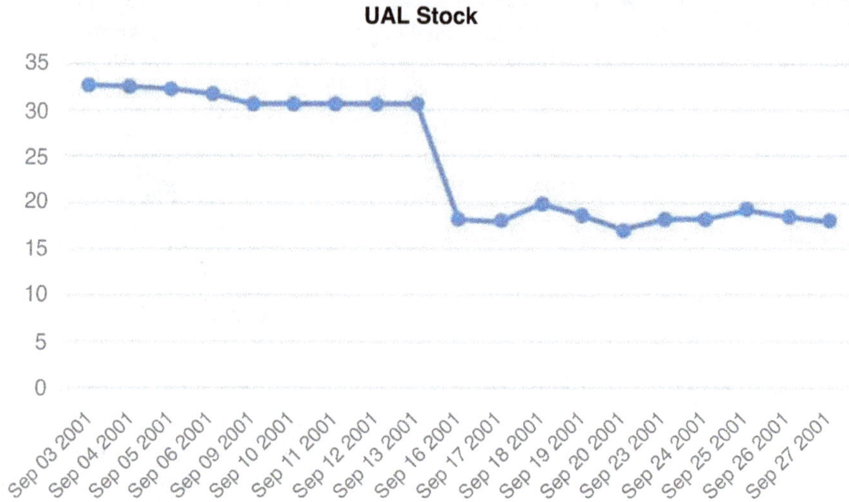

Fig. 2.30 Development of UAL stock in September 2001

Table 2.4 Development of the UAL stock in Sep 2001

Date	UAL
Sep 03 2001	32.69
Sep 04 2001	32.60
Sep 05 2001	32.25
Sep 06 2001	31.70
Sep 09 2001	30.64
Sep 10 2001	30.64
Sep 11 2001	30.64
Sep 12 2001	30.64
Sep 13 2001	30.64
Sep 16 2001	18.25
Sep 17 2001	18.00
Sep 18 2001	19.89
Sep 19 2001	18.72
Sep 20 2001	17.10
Sep 23 2001	18.25
Sep 24 2001	18.30
Sep 25 2001	19.25
Sep 26 2001	18.48
Sep 27 2001	17.98

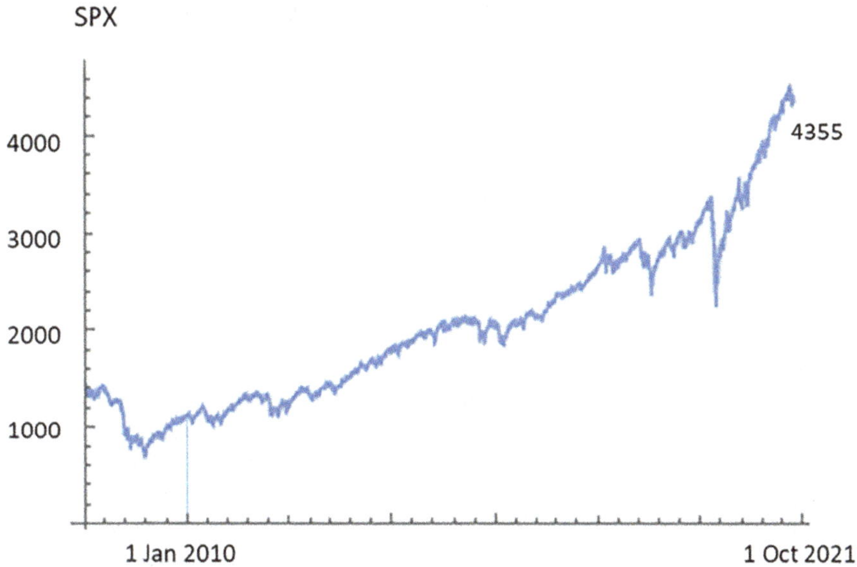

Fig. 2.31 S&P500 index chart from 2010 to October 2021

The much more important and much more frequent use of put options is, however, to hedge equity portfolios.

To explain how put options are used for hedging purposes, let us again assume a special scenario, analogous to the previous section on call options. We are still assuming that today is 1 October 2021.

For the following, we will again consider the S&P500, but this time we will go further back into the past (see Fig. 2.31).

Investor *A* had the savviness to buy a total of 100 S&P500 index certificates at the end of 2009 at a price of 1000 dollars each. The total investment was therefore 100,000 dollars. Today, on 1 October 2021, with the S&P500 at 4355 points, the certificate portfolio has a value of 435,500 dollars. *A* intends to close the portfolio in March 2022 for the purchase of an apartment. The purchase price of the apartment—which has already been agreed—is 400,000 dollars. If *A* were to close the portfolio now, the apartment could be financed with the payoff.

On the other hand, analyses—like the ones we saw in the previous section—suggest that the current strong rise in the index is likely to continue into the region of around 5200 until mid-March 2022.

If this forecast turns out to be correct, it would have been a pity to have closed the index portfolio already in October 2021, given that the money was not actually needed before March 2022.

On the other hand, one is never immune to sudden falls in the stock markets. For instance, we cannot rule out that the index might decline by around 20% over the next 5 or 6 months, i.e. in the range of approximately 3500 points. The index portfolio would then only have a value of 350,000 dollars, and the purchase of that apartment could be jeopardized.

One possible course of action in such a situation or a comparable situation would be as follows: We will return to our screenshot as shown in Fig. 2.29 and retrieve the quotes for the put option on the S&P500 with expiration 17 March 2022 and strike 4350.

The quotes are 228.40//230.90. We then go long on this option and pay 23,000 dollars for the contract.

Table 2.5 below illustrates a few scenarios that could potentially arise between now and 17 March 2022. We will then proceed to analyse the general case.

Let us present the results of Table 2.5 in Fig. 2.32:

The chart clearly shows:

As long as the index stays above the strike price of the put, i.e. above 4350, we fully benefit from the increase of the S&P500. We only need to subtract the initially paid option premium of 23,000 dollars. However, if the S&P500 falls below the strike price, the value of the portfolio remains constant at 412,000 dollars, no matter how far the S&P500 falls. Even in the event of a catastrophic crash, the value of the entire portfolio remains unchanged! The losses in the index certificates are offset by the payoffs from the put option. The total portfolio is therefore perfectly *hedged* at a value of 412,000!

The buyer of the put option has acquired "insurance" for the index certificate portfolio by purchasing the put option. This is called a "**protective put**".

The counterparty in the options trade, the one who took the short position in the put option, has therefore sold insurance, thus playing the role of an insurer in the financial market. The initial selling price collected for the put option is the insurance premium.

Table 2.5 Potential scenarios for the time span from now until 17 March 2022

SPX on 17 Mar 2022	Cost put option	Index value	Option payoff	Total value
6000	23,000	600,000	0	577,000
5200	23,000	520,000	0	497,000
4800	23,000	480,000	0	457,000
4400	23,000	440,000	0	417,000
4375	23,000	437,500	0	414,500
4350	23,000	435,000	0	412,000
4325	23,000	432,500	2500	412,000
4300	23,000	430,000	5000	412,000
4200	23,000	420,000	15, 000	412,000
4000	23,000	400,000	35, 000	412,000
3000	23,000	300,000	135, 000	412,000
1000	23,000	100,000	335, 000	412,000

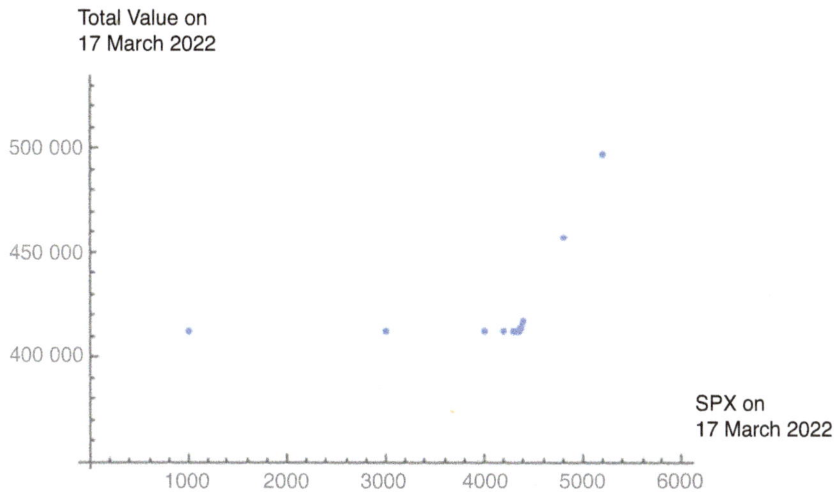

Fig. 2.32 Potential scenarios until 17 March 2022

This use of put options generates a great demand for such options. Large pension funds, for example, tend to hedge their portfolios in the short term by buying put options. Such pension funds are even required to provide proof that they are hedging their investments in order for them to qualify as pension funds.

The insights that we gained above using some tabulated examples and their illustration can of course also be generalized. This is best visualized in a graph by superimposing the profit functions of each of the two parts of the portfolio, i.e. by superimposing the profit function of the put option (which we already know and which is shown in red in Fig. 2.33 and whose formula is given by $\max(0, K - x) - P(0)$, and the profit function of the index.

The shape of the profit function of the index, or more generally of a derivative's underlying asset, is very simple and shown in blue in Fig. 2.33. The line accurately reflects the fact that the index certificate generates no profit if the index stays at 4355. If the index goes up by x points, the profit from the certificate also increases by x dollars and vice versa. So, the profit function of the underlying asset—which we will often need in the following—always consists of a straight line that intersects the x-axis in the current price of the underlying asset and has a slope of 1. The formula of the profit function is thus simply $f(x) = x - 4355$, and generally $f(x) = x - S(0)$, where $S(0)$ denotes the current price of the underlying instrument.
The profit function—shown in green in Fig. 2.34—of the combination of index certificate and put option is obtained through addition of the two profit functions. The addition gives us the result shown in Fig. 2.34.

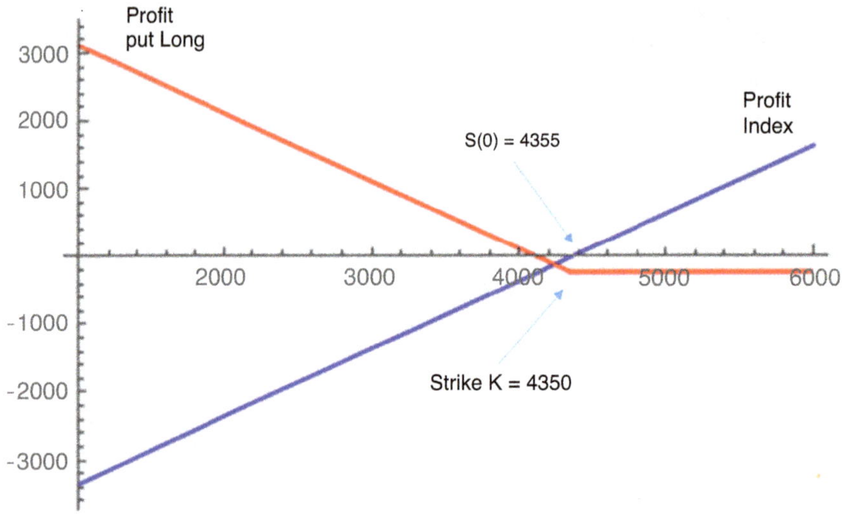

Fig. 2.33 Profit function of put option (red) and underlying asset (blue)

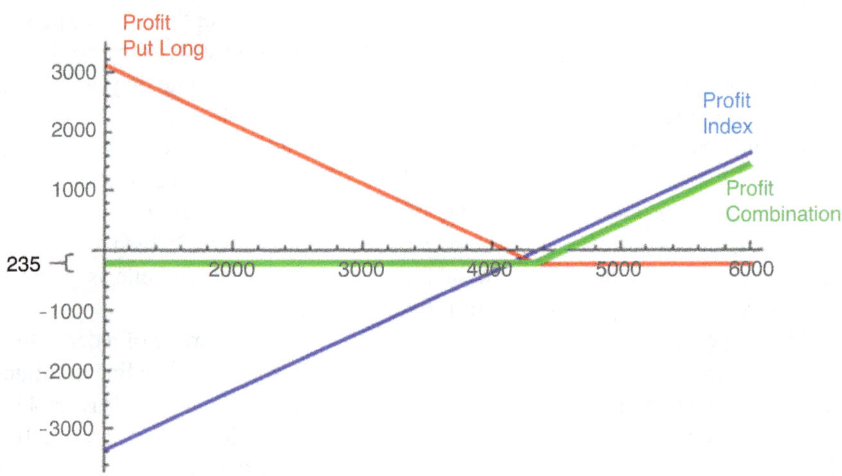

Fig. 2.34 Profit function: put option (red), underlying asset (blue), and combination (green)

The profit function shows exactly what we had assumed: a maximum potential loss of 235 dollars per option (23,500 dollars per contract) and 1:1 participation in price gains of the S&P500 (minus the option premium).

In addition to the graphic illustration, we can also easily corroborate this through calculation.
To get the profit function G of the combination, we simply add the two individual profit functions:

$$\boldsymbol{G(x)} = x - S\,(0) + \max\,(0, K - x) - P\,(0) =$$

$$\mathbf{max}\,(\boldsymbol{x} - \boldsymbol{K}, \mathbf{0}) - (\boldsymbol{S}(\mathbf{0}) - \boldsymbol{K}) - \boldsymbol{P}\,(\mathbf{0})$$

In our example $-(S(0) - K) - P\,(0) = -(4,355 - 4,350) - 230 = -235$. The maximum possible loss is thus the sum of the option price and the difference between the initial value of the index and the strike price of the option.

Of course it is also possible to hedge with trades using put options with the same term to expiration but different strike prices.

As can easily be seen from the quotes of the put options in Fig. 2.29, hedging becomes cheaper when using a lower strike price, the flipside being that in this case the trade is only protected at a lower index value and vice versa. Figure 2.35 shows the profit functions of the overall portfolio hedged with put options with a strike price of 4350 (green), 4300 (blue), and 4400 (red).

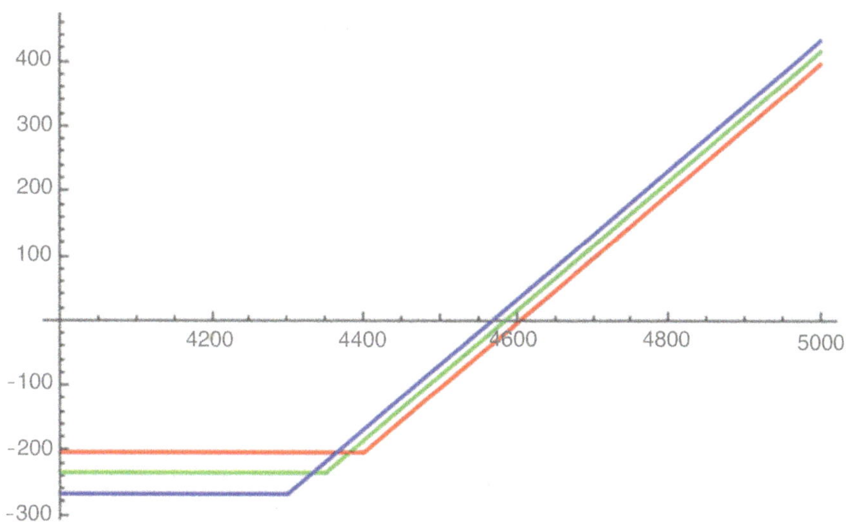

Fig. 2.35 Hedging with put options with strike price 4300 (blue), 4350 (green), and 4400 (red)

2.9 Who Trades in Options? Short Positions in Put Options, Selling Insurance, Put Spreads

So, we have seen that long positions in put options can be used to hedge portfolios. Yet every long position requires a short position as its counterpart. But who could possibly be interested in taking short positions in put options, and why? After all, taking a short position in a put option comes at a considerable risk: if the SPX were to suddenly nosedive, a short put could cause massive losses. And the maximum profit we can expect is only the initial option premium we received. We will look at this using the profit chart of our example above, only this time from the perspective of the holder of the short put (Fig. 2.36).

This does not look very promising at all. Selling a put option contract will earn you 23,000 dollars. But if the price were to fall to 2000 points, for example, you would incur a loss of $100 \cdot (4350 - 2000) = 235{,}000$ dollars. We could of course argue: "We can close the option at any time if the index falls further or if any other greater risk seems imminent". However, caution should be exercised for at least two reasons:

First, at this point, we still know far too little (in fact nothing) about the future movements of put option prices as they get closer to expiration or when market conditions change.

Second, in principle, it is of course true that the option can be traded at any time before expiration. However, it should not be taken as a given that you will always be able to make your trade precisely when you want to, especially in situations

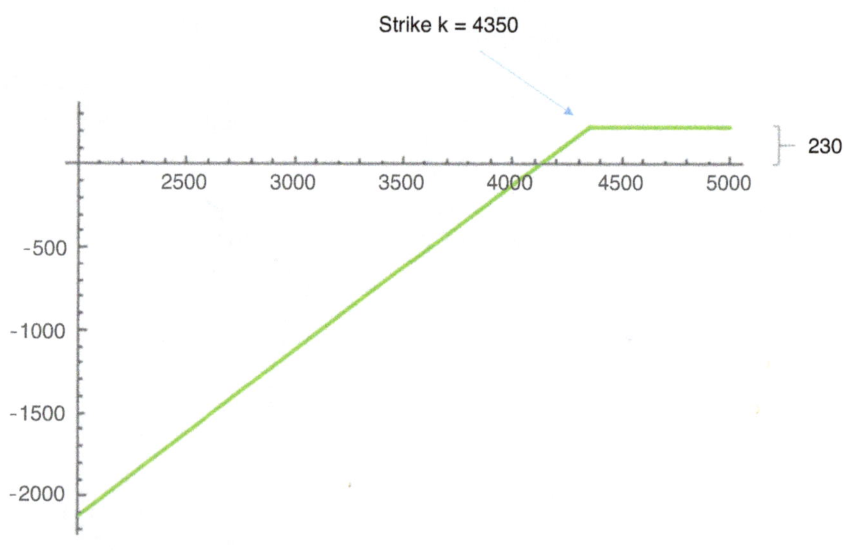

Fig. 2.36 Profit function of a short put

of turmoil in the financial markets. Let us just remind you of two situations that we discussed further above:

First, on the days around 11 September 2001, when the NYSE remained closed and no trading was possible after the terrorist attack. Second, on 10 October 2008 (see Fig. 1.46), where no options could be traded with any precision immediately after the exchange opened

In fact, only very few sellers of put options would get themselves into the situation of a so-called naked short position. The following approach would be much more advisable.

Sell a put option on the S&P500 with expiration date 17 March 2022 and strike price 4350 at the price of 230 dollars and simultaneously buy a put option on the S&P500 with expiration date 17 March 2022 and strike price of—for example—4300 at the price of 213 dollars (see quotes in Fig. 2.29). The resulting premium earned is 17.00 dollars.

The effect of combining a long put and a short put with different strike prices is again illustrated by plotting the profit functions of each of the two positions and the addition of the two profit functions, which gives us the profit function (green) of the combination (see Fig. 2.37).

We see a profit range with an upper bound at 17.00 dollars (in case the S&P500 stays above 4350), a loss range with a lower bound at 33.00 dollars (in case the S&P500 falls below 4300), and a linearly increasing loss/profit level if the S&P500 moves in the range between 4300 and 4350 points in mid-March. This also gives us the value of the maximum loss amount, which is the premium income of 17.00 minus

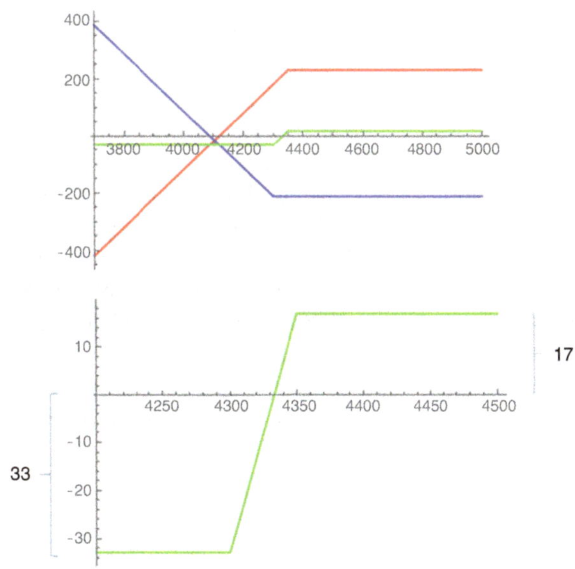

Fig. 2.37 Profit function of a put spread (green)

50, the difference between the two strike prices. This simple option combination is generally referred to as a **"put spread"**.

The investor in the short position has therefore "sold insurance" and got reinsurance at the same time, by buying another put option. (This is in fact a common and necessary approach used by the traditional insurance industry to limit excessive losses.) As both the naked position and the spread contain short positions, the investor needs to provide a margin. Another major advantage of the combined spread strategy compared to a naked short position is that the necessary margin for the spread is generally significantly lower than for the naked position and that the required margin amount for the combination is fixed from the outset, while the amount that is needed for the naked position can increase over time and may have to be adjusted. But more on this will follow in the section on margin requirements.

Some investors, especially institutional ones, systematically run these types of strategies and aim to take long-term profits through occasionally (in certain situations and areas) slightly overpriced put options. We will discuss this in much more detail later (e.g. in Volume III Section 3.4).

2.10 Who Trades in Options? Short Positions in Call Options, Covered Call Strategies

And finally, who would be interested in being the counterpart to call option buyers, i.e. to go short on call options?

Short call positions are often used in "covered call" strategies. Let us look at a case in point: We'll assume it is still 1 October 2021, the SPX is at 4355 points, and we still hold 100 index certificates or ETFs on the S&P500 Index, just like we did in Sect. 2.9. For further forecasts, we will again use a previously used graphic, shown below as Fig. 2.38 (now without "trend line").

After studying the chart, we assume (again, this is just a hypothesis, the validity of which we are not going to question any further at this point) that the S&P500 will increase further by March 2022, though likely not beyond 4600 points. As a way to further increase our payoff (and slightly cushion any unexpected stagnation or price declines), we might be tempted to go short on a call options contract on the S&P500 with expiration date 17 March 2022 and a strike price of—for example—4600.

On 1 October 2021, the quotes for this option were as follows: 86.70//88.70. We can therefore assume that we will receive a premium of 87.00 (i.e. 8700 dollars for the contract) when shorting this call contract.

The resulting profit function for the combination of S&P500 and the shorted call options contract as of 17 March 2022 is plotted graphically (Fig. 2.39) and also calculated in the following.

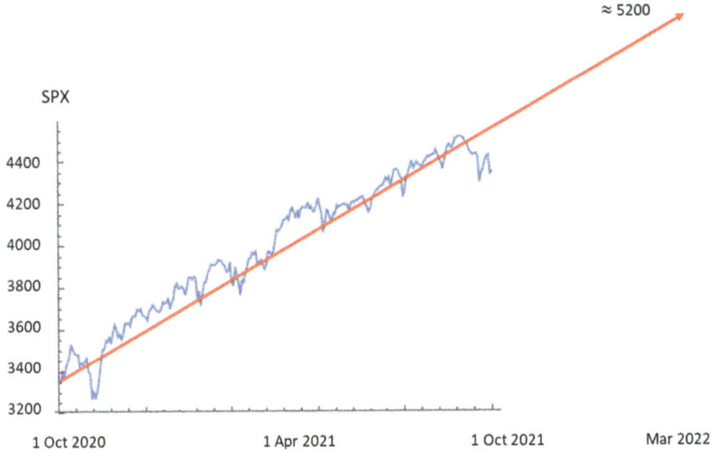

Fig. 2.38 S&P500 index chart: October 2020 through October 2021

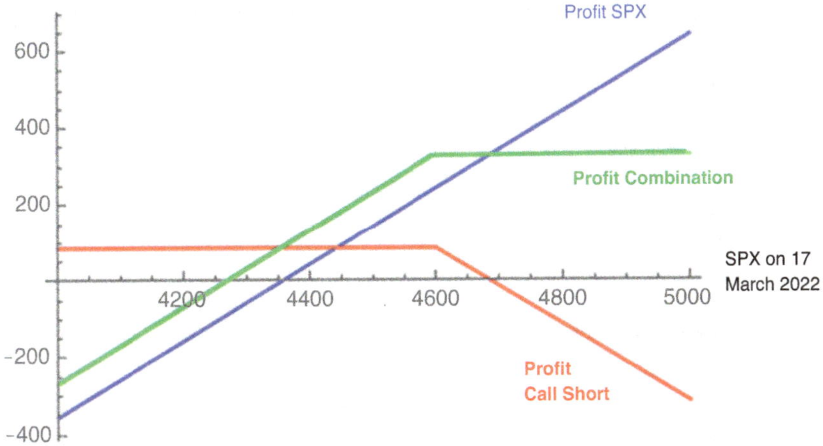

Fig. 2.39 Profit function of a covered call strategy

In addition to the graphic illustration, this can again be easily calculated.
To get the profit function G of the combination, we simply add the two individual profit functions:

$$\begin{aligned} \boldsymbol{G(x)} &= x - S(0) - \max(0, x - K) + C(0) = \\ &= -(\max(0, x - K) - x) - S(0) + C(0) = \end{aligned}$$

(continued)

$$= -(\max(-x, -K)) - S(0) + C(0) =$$
$$= \mathbf{\min(x, K) - S(0) + C(0)}$$

Compared to the profit function $x - S(0)$ of just the index certificate, x is thus replaced by $\min(x, K)$, and $C(0)$ is added to compensate for this "impairment". In our case this gives us the profit function $G(x) = \min(x, 4600) - 4268$.
The maximum possible profit in this strategy is the sum of the option's price and the difference between the option's strike price and the initial value of the index, which is 332 per option in our case (i.e. 33,200 per contract).

The covered call strategy therefore generates a profit consisting of the collected option premium of 8700 dollars in addition to the result of the index in every case where the index falls, stagnates, or does not rise too sharply. On the other hand, the possible profit in the event of a very strong increase in the S&P500 by March 2022 is limited by the option's strike price (plus the option premium). The covered call strategy therefore always generates additional income, except in the event that the S&P500 rises sharply (in which case we will have reaped strong profits due to the index increase anyway).

Of course, we are again flexible in choosing the strike price of the call option. A higher strike will result in lower premium income, yet in return, the potential profit is limited only in the higher range.

2.11 Discount Certificates

The so-called certificates of various types are widely offered to investors by various banks and investment firms. You can find such certificates on the BNP Paribas website, for example
https://www.derivate.bnpparibas.com/startseite
or Societe Generale
https://www.sg-zertifikate.de
and many other providers.

In this chapter, we want to pick out a specific certificate of this type, a discount certificate from Societe Generale, as an example, and show that such a certificate can easily be replicated using basic products and derivatives. The product we are going to look at is described in Fig. 2.40 from the "basic information" provided by Societe Generale.
The quotes for the product on 4 October 2021 at 15:44:37 were 36.940//36.950. The indicative value of the underlying asset at that time was 4340.44 points.

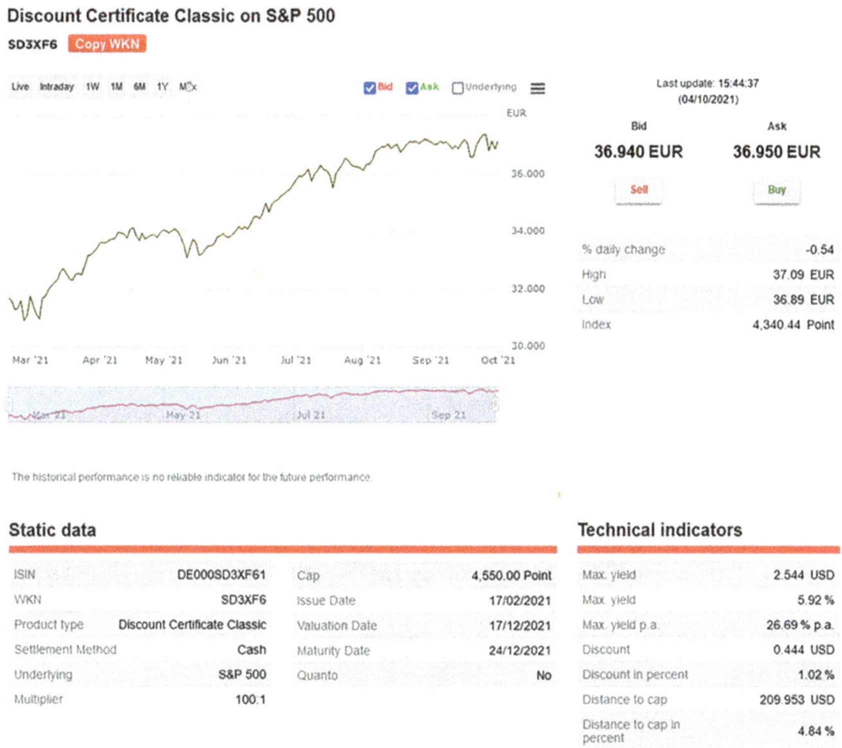

Fig. 2.40 Excerpt from SG's basic information on a discount certificate

In short, the certificate has the following properties:

It was possible to buy the discount certificate on 4 October 2021 at 15:44:37 pm at a price of 36.95 euros.

To better understand the product, it is helpful to state this price in dollars (as the S&P500 is quoted in dollars). The dollar exchange rate at that time was 1.163 (i.e. the price of 1 euro was 1.163 dollars). Hence, it was possible to buy the discount certificate on 4 October 2021 at 15:44:37 pm at the converted price of 42.97 dollars.

If the S&P500 reaches or moves beyond the "cap" of 4550 points on the valuation day, 17 December 2021, the certificate holder will receive a payout of 45.50 dollars. If the S&P500 is below 4550 points on the valuation day, 17 December 2021, the certificate holder will receive one hundredth (multiplier 100:1) of the value of the SPX in dollars.

This essentially means that I can buy now the SPX at a price of 4297 dollars, although its present value is 4340.44 dollars. However, the maximal profit I can make with this SPX is bounded by 4550 dollars. Figure 2.41 illustrates the profit function (in dollars) of 100 units of the discount certificate compared to a direct investment in the S&P500.

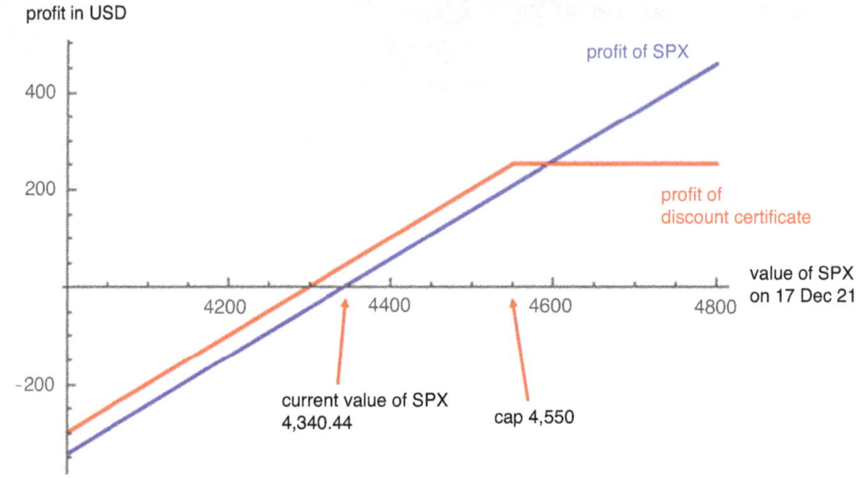

Fig. 2.41 Profit function for 100 units of the discount certificate vs. profit function for S&P500

Table 2.6 Performance of the discount certificate and the S&P500 in percent for some hypothetical values of the index on 17 December 2021

S&P500 on 17 Dec 2021	Profit S&P500	Profit discount certificate
4200	−3.23	−2.26
4250	−2.08	−1.09
4275	−1.51	−0.51
4297	−1.00	0
4325	−0.35	0.65
4350	0.22	1.23
4400	1.37	2.40
4500	3.68	4.72
4593	5.82	5.89
4650	7.13	5.89
4800	10.59	5.89

We see that the profit from the discount certificate is always somewhat higher than the profit from a direct investment in the S&P500, except when the S&P500 rises sharply, to more than $4550 + (4340.44 - 4,297) = 4593.44$ points.

Table 2.6 shows the performance of the discount certificate, expressed in percent, compared to the performance of the S&P500 for some potential values of the S&P500 on 17 December 2021, based on the respective initial investment.
We recognize a certain similarity to a covered call strategy.

And in fact, we can indeed replicate the discount certificate by investing directly in the S&P500 (by investing in an SPX index certificate) and by taking a short position in a call option on the S&P500 with expiration date 17 December 2021 and strike price 4550.

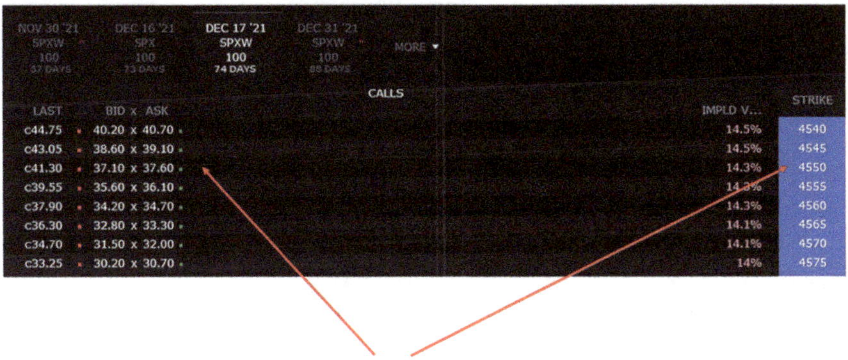

Fig. 2.42 Quotes for S&P500 call options on 4 October 2021 with expiration date 17 December 2021

Let us see how we can accomplish this, by looking at the quotes of the call options on the S&P500 on 4 October 2021 at 15:44:37 (see Fig. 2.42).
The quotes for the call option with strike price 4550 are 37.10//37.60. It can therefore be shorted at a price of approximately 37.50 dollars. The combined investment in the S&P500 and this call option, which now has exactly the same payoff function as the discount certificate, thus comes at a total price of $4340.44 - 37.50 = 4302.94$ dollars. This is slightly more than the current price of 4297 dollars for 100 units of the discount certificate. We can therefore conclude that in this case, we would be better off with a direct investment in the discount certificate than tracking the index with an exchange-traded call option.

2.12 Who Trades in Options? Long Straddle, Short Straddle

Combinations of a wide variety of options with a wide variety of strike prices and expiration dates can be used to create and generate any number of possible profit functions. Many basic combinations have become so popular that names have been coined for them. You can look them up in many websites and other sources. However, we do not consider an extensive discussion of these combinations to be really fruitful. Instead, we would like to present an example of another basic use of options combinations. To this end we are again going to explore a hypothetical case: Let us look at the price movements of a US stock, which we will simply call XXX.

The price movements of the XXX stock in the 2 weeks prior to 6 December 2021 are shown in Fig. 2.43.
Although the stock had proved fairly stable throughout the year, moving steadily in a range above 80 points, it dropped sharply to 62.19 by 6 December 2021. This fall in the stock's performance had been triggered by a recent news flow regarding severe turmoil within the XXX corporation, which had led to serious concerns about the survival of XXX. The fall in the company's stock price had slowed down somewhat

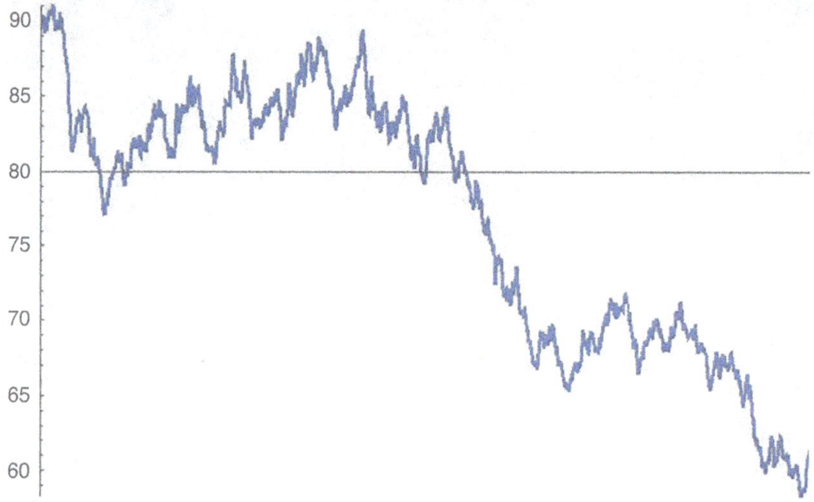

Fig. 2.43 Price movements of XXX stock from 22 November 2021 to 6 December 2021

in the last 3 days before 6 December due to the fact that another company YYY had expressed interest in a possible takeover of company XXX and had begun negotiations in this regard. Furthermore, it was announced by the management of the two companies that a decision regarding the takeover (and thus more or less the survival of XXX) would be taken before Christmas.

An observer and analyst of the XXX stock price would interpret this as follows: *The stagnation of the last 3 days shows that the market is waiting to see whether these negotiations will be successful or not. As soon as a clearly positive outcome of those negotiations is in sight, investors will again speculate on sharply rising prices, buy the stock, and thus drive up the prices. However, if and as soon as it becomes evident that the negotiations will definitely fail, investors will attempt to get rid of the last remaining shares in XXX and engage in heavy selling, causing prices to plummet further.* Based on these considerations, it is therefore highly likely that the price of the XXX stock is in for either a strong rise or a strong fall in the next few days. The direction it will take seems, however, completely unpredictable.

One possible strategy to respond adequately to this situation would be as follows: we buy both a put contract and a call contract on the XXX stock with expiration date 22 December 2021 and strike price 62.

The relevant quotes can be found on the corresponding Bloomberg page:

Put 1.32//1.55
Call 1.50//1.74

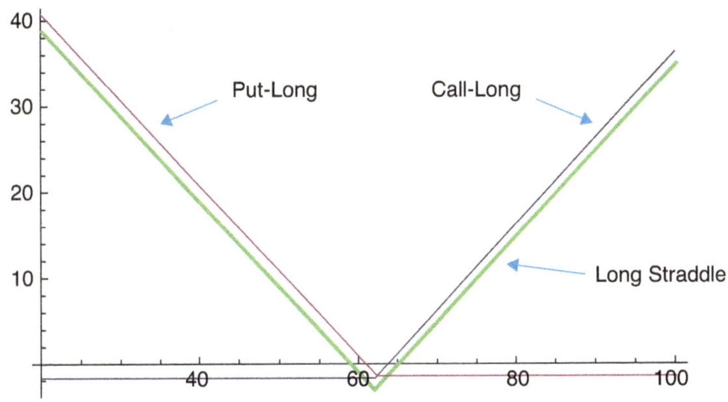

Fig. 2.44 Profit function: "long straddle"

We assume purchase prices of 1.45 for the put and 1.65 for the call. We pay a total of 3.10 for the two options combined (i.e. 310 dollars per combination contract). The resulting profit function (**"long straddle"**) is plotted in Fig. 2.44.
Both a sharp drop and a sharp rise in prices over the next 2 weeks would therefore result in considerable gains through this strategy.

For example:

If the XXX stock price were to rebound to the previous level of 80 points (i.e. go up by 29%), the call option would earn us a payoff of $80 - 62 = 18$ dollars per option. After subtraction of 3.10 dollars for the premium we paid, we would end up with 14.90 dollars per option, thus 1490 dollars per contract. Based on our investment of 310 dollars, that means a profit of 380.6%. We would achieve the same level of profit if the index fell by 18 points to 44 points. The profit function is symmetrical around the price of 62.

We would suffer a loss if the price moved sideways between the two breakeven points $62 - 3.10 = 58.90$ and $62 + 3.10 = 65.10$ until 22 December.

Again, it would be possible to exploit strong price swings even before 22 December of course, if we close the two options early. To be able to analyse such opportunities more closely, we need insights into price movements of options over their life. This is something we will gain in coming chapters.

(We must confess that, for this particular example, we manipulated history a little: the stock whose option prices we used here did not really experience such a sharp decline as shown in Fig. 2.43. If that had been the case, the actual option prices would have been significantly higher.)

The diametrically opposed strategy to the above would be to take a short position in the above call and put option ("short straddle"). Regarding the prices for taking these two short positions, we need to assume slightly modified sell prices, of course (e.g. sell price of 1.41 instead of buy price of 1.45 for the call and 1.61 instead of 1.65 for the put). The corresponding profit/loss curve (Fig. 2.45) strongly suggests that we hedge the two short positions with long positions.

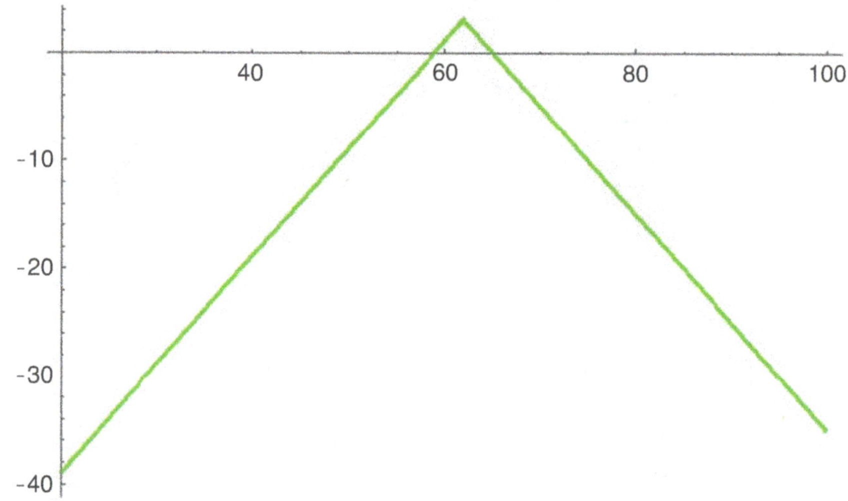

Fig. 2.45 Profit function: short straddle

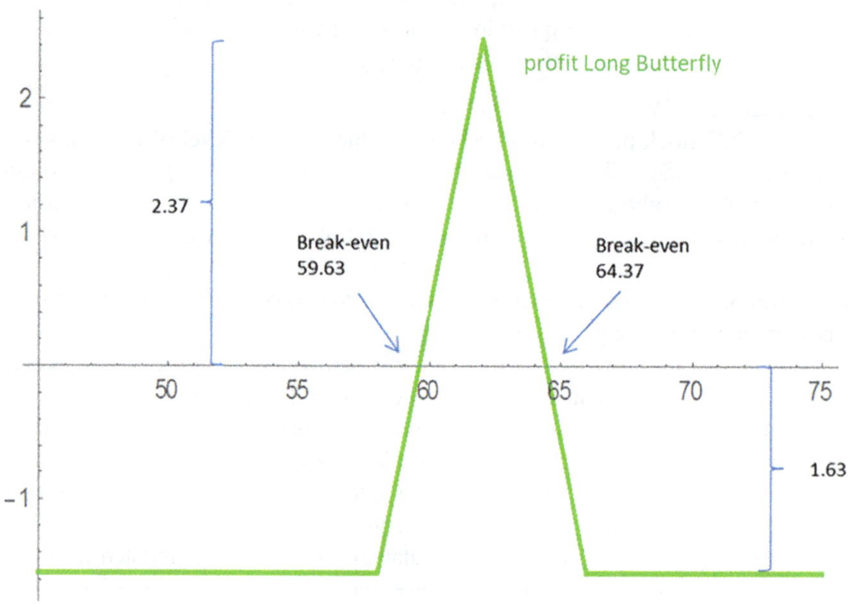

Fig. 2.46 Profit function: long butterfly spread

To that end, we take a long put option with strike 58 (quotes 0.23//0.38, price approx. 0.32) and a long call with strike 66 (quotes 0.26//0.39, price approx. 0.33), which gives us the profit function shown in Fig. 2.46 (**"long butterfly spread"**).

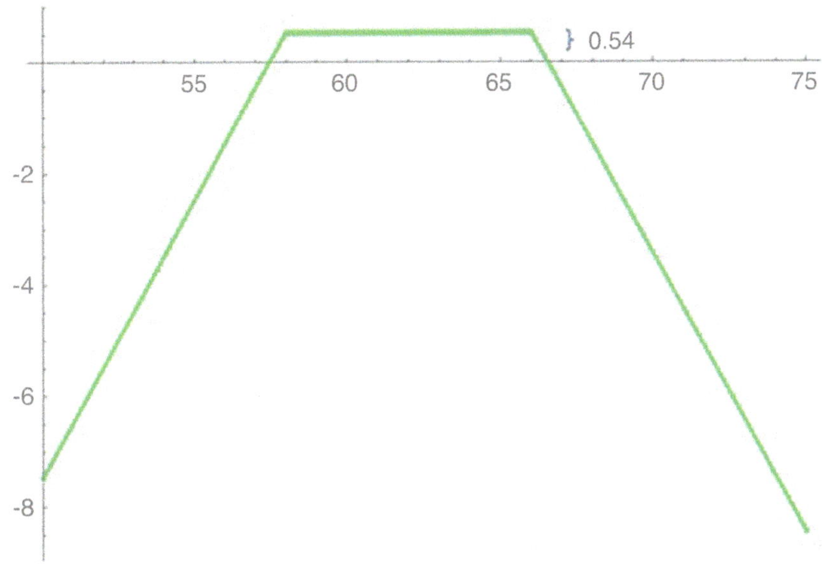

Fig. 2.47 Profit function: short strangle

Again, however, it is important to reiterate that we should not ponder this strategy only in terms of the payoff chart or the profit chart at expiration! For us to close the options and take the profit, it would suffice if the stock price moved only briefly into a range that's clearly around 62 points a few days before expiration. As noted above, these price movements during the life of an option will be dealt with in detail later and can also be analysed more closely using our software.

For all of these strategies (long straddle, short straddle, long butterfly, . . .), we can easily generate variants in which the highest point—the vertex—of the respective profit function is replaced by a "broader base".

As an example, let us have a closer look at the short straddle: by selecting a call with a higher strike and a put with a lower strike (e.g. call, strike 66, premium 0.32; put, strike 56, premium 0.22), we get the profit function shown in Fig. 2.47, a **short strangle**.

The analogous opposite strategy is a "long strangle" (Fig. 2.48) or a "long condor spread" (Fig. 2.49).

Many standard combinations can be traded directly using the OptionTrader tool on the Interactive Brokers trading platform. We describe how you do this below, again using the S&P500 options.

Open the Options Trader screen, click on "Option Chains" from the top menu bar (blue arrow in Fig. 2.50), and select "Option Spreads" from the dropdown menu (red arrow in Fig. 2.50).

Another window opens (Fig. 2.51), where we can select the parameters for our straddle (expiration and strike). In our example, we choose a straddle with expiration

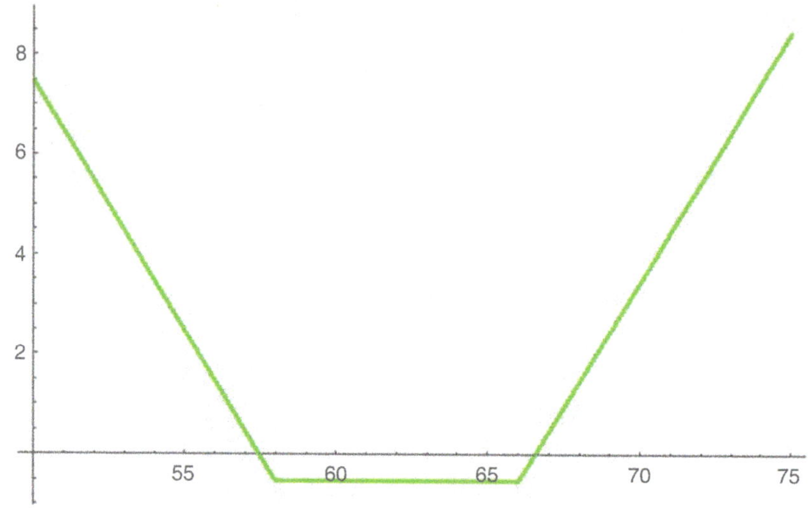

Fig. 2.48 Profit function: long strangle

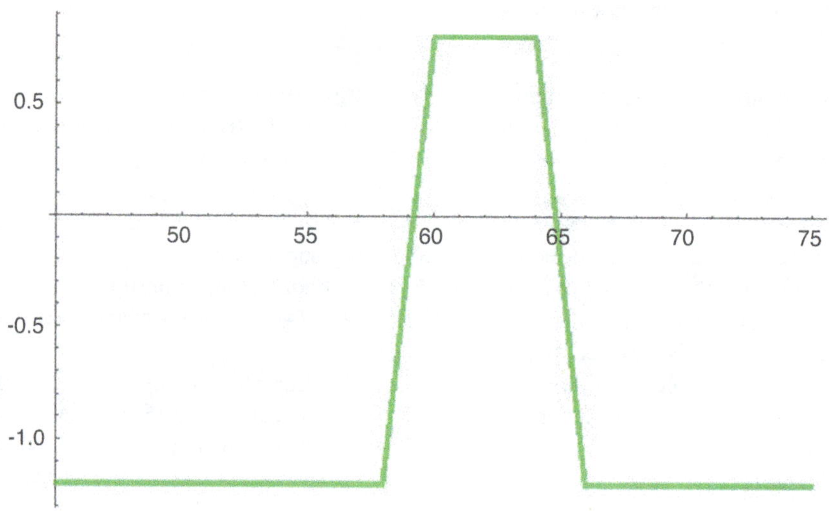

Fig. 2.49 Profit function: long condor spread

date 30 November 2021 and strike price (vertex of the straddle) at 4400 points (red arrow in Fig. 2.51). The OptionTrader screen will now display some additional rows that contain all the relevant information for the selected combination and where we can trade the selected combination directly—by placing just one single order! (See Fig. 2.52.)

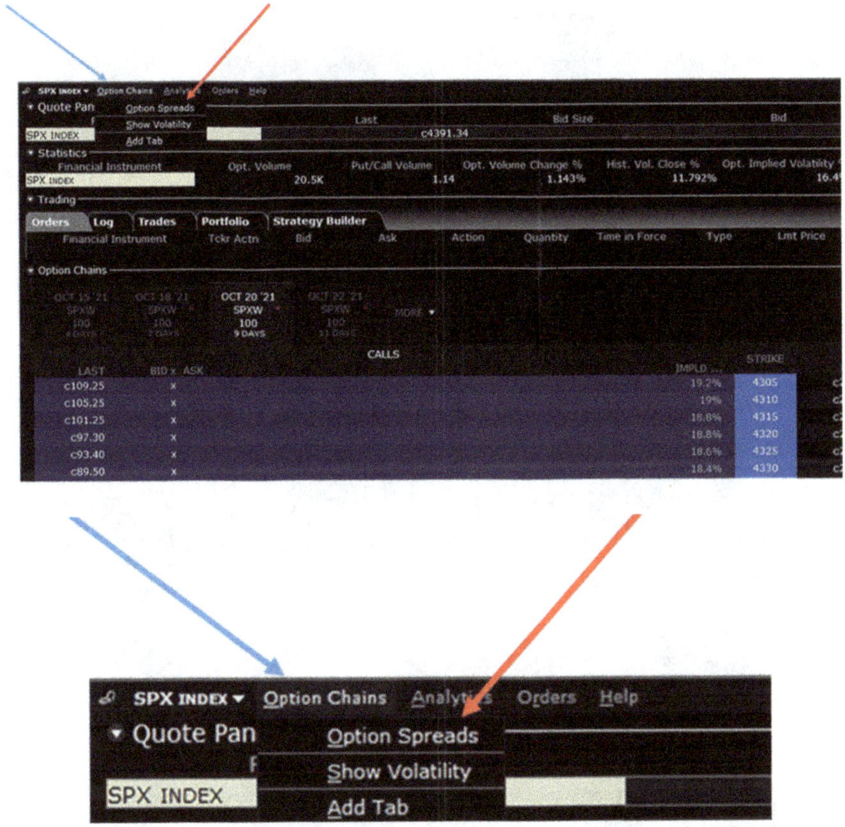

Fig. 2.50 Trading option combinations via IB

Before actually placing the order, we first click on the name of the straddle to activate it (blue arrow in Fig. 2.52) and select "Contract Statistics" → "Description" to view the data on the selected combination, as shown in Fig. 2.53.

As further information an extra screen shows the selected combination (see Fig. 2.53)!

That description tells us that the straddle is composed of a call and a put with the same strike and that "to buy 1 straddle" means we buy both positions (payoff function: V-shaped, open at the top) and that "to sell 1 Straddle" means we sell both positions (payoff function: reverse V-shaped, open at the bottom).

Further, some information on the necessary margin is given there. Figure 2.52 lists the two positions of the straddle with their respective quotes as well as the quotes of the straddle. The bid prices of the straddle are simply made up of the bid prices of the two positions.

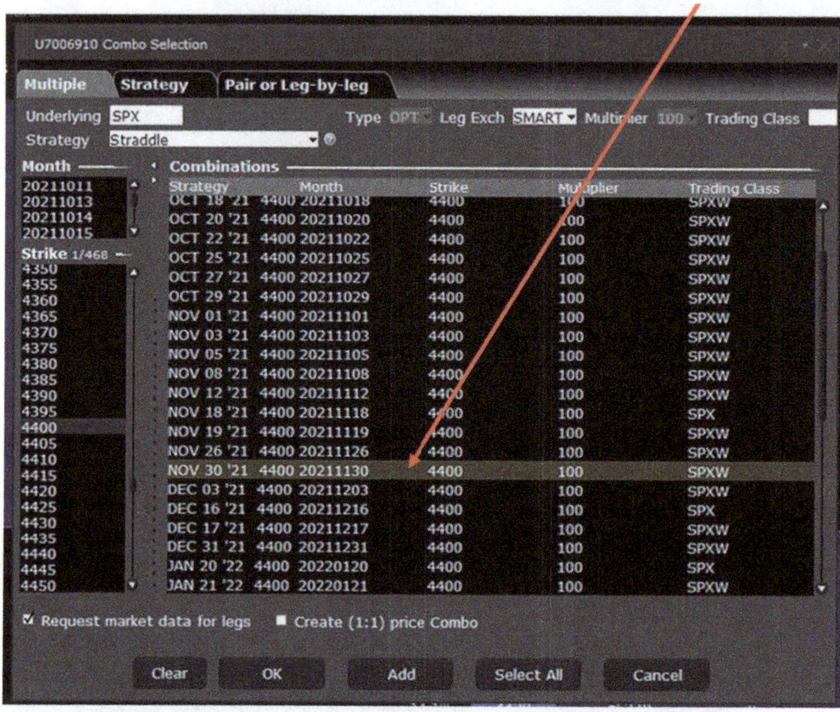

Fig. 2.51 Trading option combinations via IB, list of parameters for a straddle

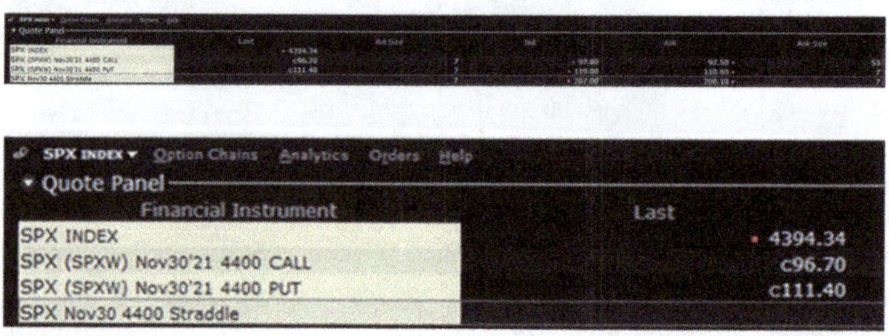

Fig. 2.52 Trading a straddle via IB

2.13 Relationship Between the Payoffs of Puts, Calls, and Underlying Asset

At this point we want to address, if only very briefly, the fairly obvious relationship between the payoff functions of put and call options on the same underlying asset with the same expiration date and the same strike price and the payoff of the associated underlying asset. This relationship will be dealt with in more depth

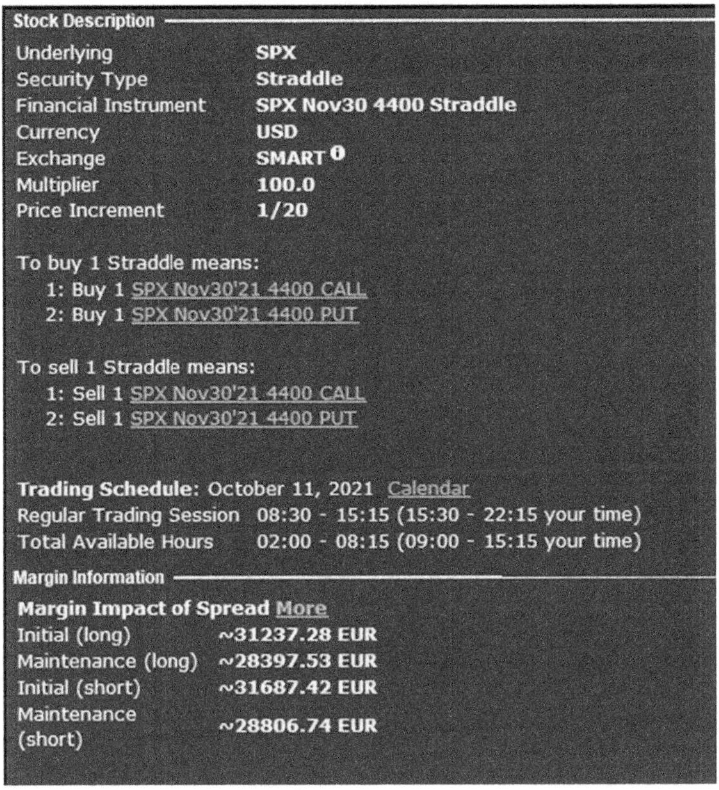

Stock Description

Underlying	**SPX**
Security Type	**Straddle**
Financial Instrument	**SPX Nov30 4400 Straddle**
Currency	**USD**
Exchange	**SMART**
Multiplier	**100.0**
Price Increment	**1/20**

To buy 1 Straddle means:
1: Buy 1 SPX Nov30'21 4400 CALL
2: Buy 1 SPX Nov30'21 4400 PUT

To sell 1 Straddle means:
1: Sell 1 SPX Nov30'21 4400 CALL
2: Sell 1 SPX Nov30'21 4400 PUT

Trading Schedule: October 11, 2021 Calendar
Regular Trading Session 08:30 - 15:15 (15:30 - 22:15 your time)
Total Available Hours 02:00 - 08:15 (09:00 - 15:15 your time)

Margin Information

Margin Impact of Spread More

Initial (long)	**~31237.28 EUR**
Maintenance (long)	**~28397.53 EUR**
Initial (short)	**~31687.42 EUR**
Maintenance (short)	**~28806.74 EUR**

Fig. 2.53 Description of the selected straddle

later—especially in relation to the profit function—and will lead us, among other things, to the so-called put-call parity equation.

Let us first recall the payoff and profit functions of an underlying asset S at time T, which evidently look like the ones shown in Fig. 2.54. In this case, $S(0)$ denotes the current price of the underlying asset, while $S(T)$ denotes the price of the underlying asset at time T.

Let us now combine (green line in Fig. 2.55) a long call (blue) and a short put (red) with the same strike K.

We see that the payoff function of the combination (green line in Fig. 2.55) exactly matches the payoff function of the underlying asset (red line in Fig. 2.54) but translated downward by the value K.

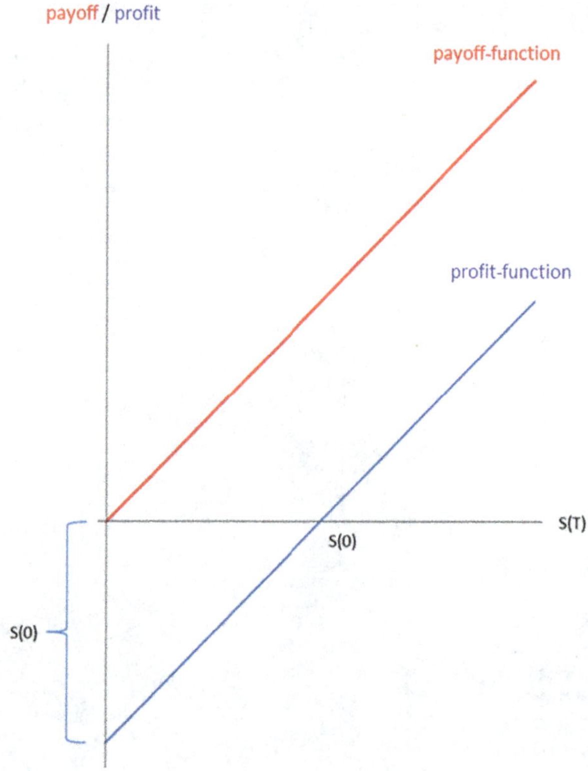

Fig. 2.54 Payoff and profit functions of an underlying asset

This can also be easily verified by adding the payoffs:
payoff long call + payoff short put =
$= \max(S - K, 0) - \max(K - S, 0)$
$= \max(S - K, 0) + \min(S - K, 0) = S - K + 0 = \boldsymbol{S - K}$

This means:
The payoff of a long call plus a short put plus the amount K exactly matches the payoff of the underlying asset sold at time T.

Schematically, we write this relationship as follows:

$$S = C - P + K$$

And this equation can now be rearranged in any way, for example, into

$$C = S + P - K$$

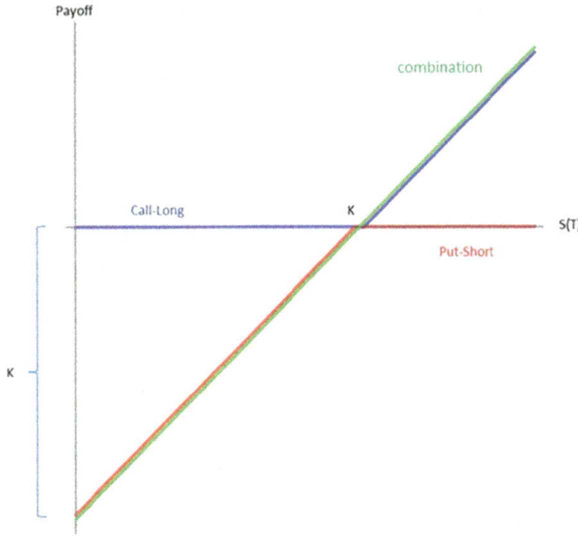

Fig. 2.55 Combination of a long call and a short put

(the payoff of a long call can be generated through the payoff of the underlying asset, a long put, and a fixed payout amount $-K$)

or

$$P = C - S + K$$

(the payoff of a long put can be generated through the payoff of a short position in the underlying asset, a long call, and a fixed payout amount $-K$)

or

$$-S = P - C - K$$

(the payoff of a short position in an underlying can be generated through the payoff of a long put, a short call, and a fixed payout amount $-K$)

Similarly, short positions in call and put options can of course be generated by the respective other option type and the underlying asset plus a fixed payout amount.

2.14 More Option Combinations

As we mentioned at the beginning of Sect. 2.12, there are endless ways to combine options (and subsequently options, underlying assets, futures, etc.). So far, we have only dealt with combinations of options with the same expiration date, whose payoff and profit functions can be visualized graphically.

Yet combinations of options with different expiration dates are also of great practical interest. Our goal here is not to provide a detailed discussion of the advantages and risks of all of these combinations, however. Such analyses can be found in [2] or in [3], for example.

In the course of this book, we will continue to look at a number of specific combinations to attain certain investment objectives, and we will also discuss them in detail.

The website associated with this book also provides a program which you can use to visualize any combinations of payoff and profit functions of any number of options (futures and underlying assets) (see https://app.lsqf.org/book/option-payoffs).

Below is just one example of a combination created with this program, which shows that options can be combined for even the most absurd payoff functions.

Example 2.1 Using the program on our website in Fig. 2.54, we combine the following options in combination with the associated underlying asset and generate the associated payoff function:

50 short calls, strike 50
10 long calls, strike 30
20 short calls, strike 60
70 long calls, strike 55
12 short calls, strike 100
30 short puts, strike 45
80 long puts, strike 47
50 short puts, strike 40
1 underlying short
−200 cash

The resulting combination is shown in Fig. 2.56.

This of course takes us to the question: **What kind of** given payoff functions can be represented by a finite combination of options? It is clear that any such combination will basically have the following shape: it must be a continuous line composed of a finite number of straight-line segments (as shown in the example in Fig. 2.56), with the rightmost segment continuing to infinity.

But can any given payoff function of this basic shape also be represented as an option combination? **The answer is "Yes"**—provided we are allowed to use not just **integral quantities** of options (or options contracts) but any real-valued quantities of options and provided further that there are options with strike K for each given value K. We want to show this in the following and even provide an algorithm for identifying the desired options combination.

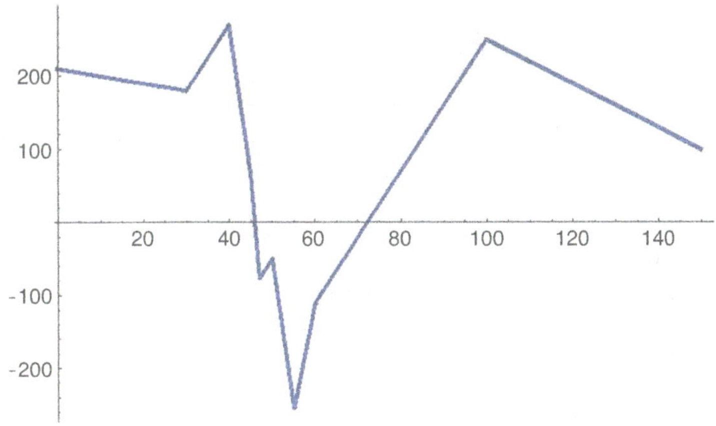

Fig. 2.56 Example of a combination, payoff function

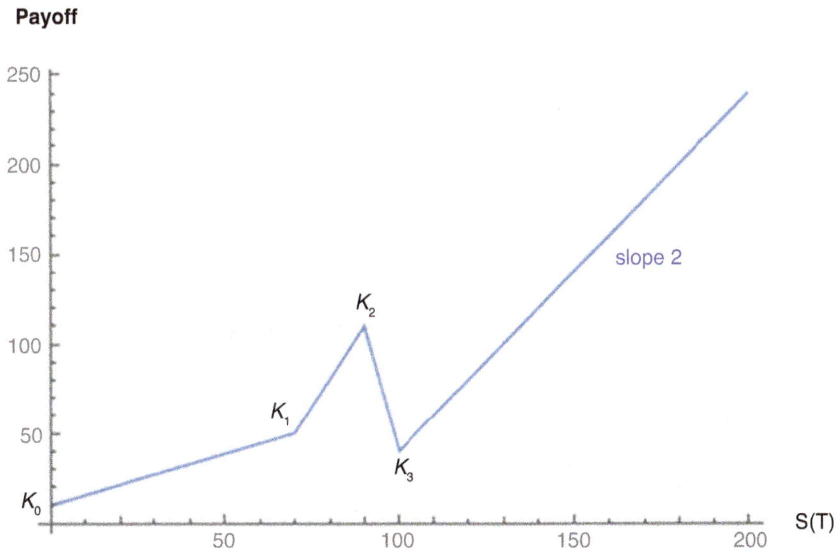

Fig. 2.57 Desired payoff function

A payoff function composed of n segments is given by n "breaking points" $K_0, K_1, \ldots, K_n$ (including the "starting point on the y-axis") and the slope of the last line segment, which continues to infinity. We are going to illustrate this with an example. The payoff function in Fig. 2.57 is given by the points

(continued)

$K_0 = \{0, 10\}$, $K_1 = \{70, 50\}$, $K_2 = \{90, 110\}$ and $K_3 = \{100, 40\}$ as well as Slope 2 for the segment farthest to the right.
We denote the slope of the line segment between K_i and K_{i+1} by a_i (the slope is obtained simply by dividing the difference of the y-axis coordinates of K_i and K_{i+1} by the difference of the x-axis coordinates of K_i and K_{i+1}).
In our example, $a_0 = \frac{4}{7}$, $a_1 = 3$, $a_2 = -7$, and $a_3 = 2$ (as specified).
To construct the payoff function through options, we proceed step by step as follows. We start with the first segment at the left and construct the payoff function segment by segment from left to right. Once we have constructed the function all the way to point K_i, we have a preliminary payoff function, which we are going to denote by Pf_i.
Between the points K_0 and K_i, this function Pf_i has basically the same shape as the given payoff function PF but may be higher or lower than the given payoff function PF. To the left of K_i, the preliminary function PF_i has breaking points at the x-values of $K_{i-1}, \ldots, K_0$. The y-values of the breaking points of PF_i will differ from the y-values of $K_{i-1}, \ldots, K_0$.
Once we have constructed the payoff function up to point K_i (i.e. PF_i), we construct the next segment, from K_i to K_{i+1}, in the following way: we add a certain quantity A_i of long calls, or a certain quantity A_i of short calls with strike at K_i to Pf_i such that the resulting new preliminary payoff function PF_{i+1} has a slope of a_i between the x-values of K_i and K_{i+1} and to the right beyond that. Since our call option to the left of K_i is constant, it does not change the basic shape of Pf_i to the left of K_i. Pf_{i+1} therefore has the same basic shape as PF up to the x-value of the point K_i.
How do we choose the quantity A_i?
As a result of its construction, the slope of PF_i between the x-values of K_i and K_{i+1} is equal to a_{i-1}. We therefore need to choose $a_i - a_{i-1}$ long calls if $a_i - a_{i-1}$ is positive or $a_{i-1} - a_i$ short calls if $a_i - a_{i-1}$ is negative.
We continue to proceed in this way until we get to the last segment (which goes to infinity). This gives us a preliminary payoff function PF_{n+1}, which basically matches the given payoff function PF but may be situated slightly higher or lower.
Finally, we want to correct this constant difference d in the level between the functions, which warrants a brief parenthesis remark (see Fig. 2.58).
The combination of a long call and a short put with the same strike K and of a short call and a long put with the same strike L evidently results in a constant payoff $L - K$.
In order to correct the level difference, therefore, we select two strikes K and L with distance d between them and add or subtract the described combination of four options to PF_{n+1} to obtain the given payoff function PF.
Let us now apply this step by step to our case in point.

(continued)

We start with $a_0 = \frac{4}{7}$ long calls, strike 0; this is our PF_0 (see Fig. 2.59, red graph).
Then we continue with $a_1 - a_0 = 3 - \frac{4}{7} = \frac{17}{7}$ long calls, strike 70, which we add to PF_0. This results in PF_1 (see Fig. 2.59, green graph).
The slope on the right edge of PF_1 is $a_1 = 3$. We therefore need to add $a_1 - a_2 = 10$ short calls with strike 90 to obtain PF_2 (see Fig. 2.60, red graph).
Finally, we add $a_3 - a_2 = 9$ long calls with strike 100 to get PF_3 (see Fig. 2.60, green graph).
The value of the function PF_4 for $x = 0$ is given by $PF_4(0) = 0$ (as PF_4 is the sum of long/short call options with a non-negative strike). The difference between the given payoff function and PF_4 is therefore $d = 10$, and we minimize this difference as described above.
Using this method, it is of course possible to approximate payoff functions arbitrarily well to **any** given continuous functions $f : \mathbb{R} \to \mathbb{R}$ on given finite intervals (!) through a suitable combination of options.
To do this, we simply approximate the given function (e.g. blue function $f(x) = \sin \frac{x}{30}$ in Fig. 2.61) arbitrarily well on the given interval (e.g. [0, 200] in Fig. 2.61) by means of a line of segments (e.g. red function in Fig. 2.61) and then represent this approximation by options, as described above.

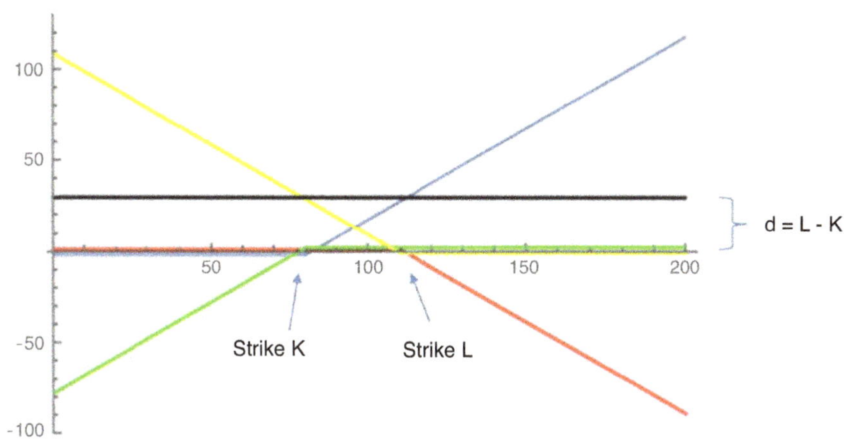

Fig. 2.58 Combination of long call/short put with strikes K and short call/long put with strikes L

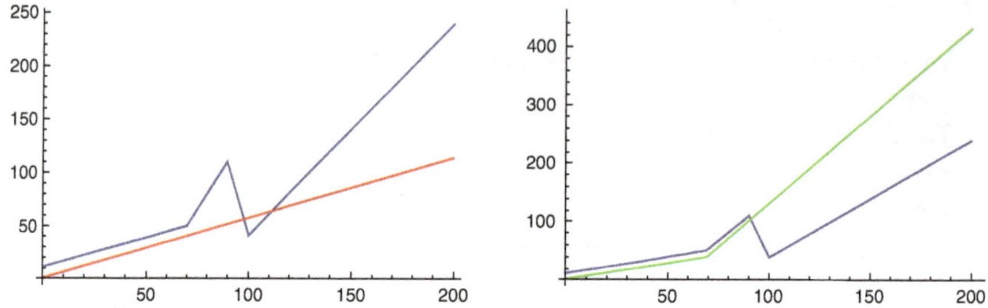

Fig. 2.59 Example PF_0 (red) and PF_1 (green)

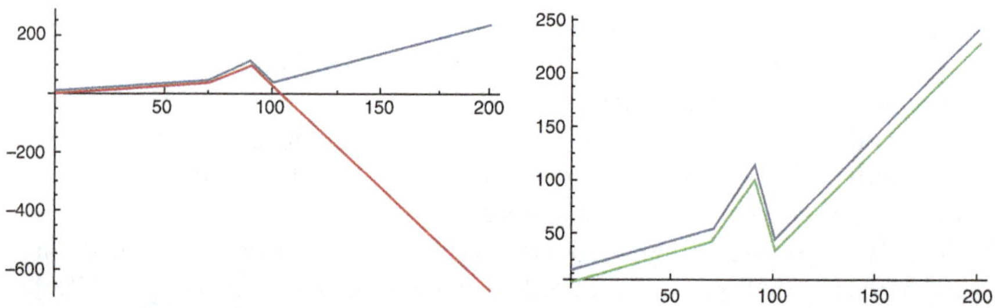

Fig. 2.60 Example PF_2 (red) and PF_3 (green)

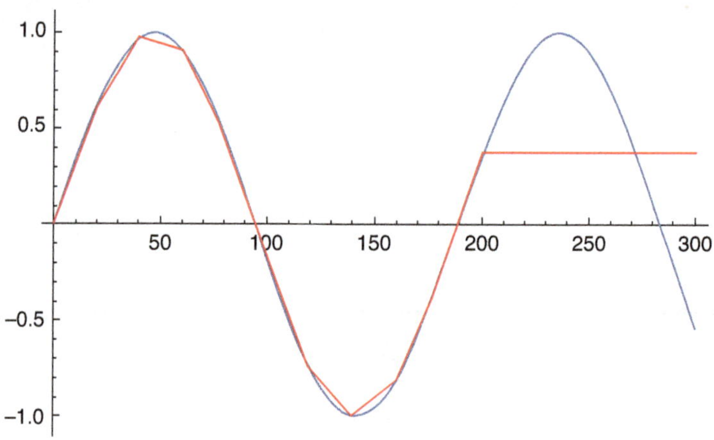

Fig. 2.61 Arbitrary payoff function (blue) and approximation (red)

2.15 Margin Rules for Short Positions in (CBOE S&P500) Options

For short positions in options, collateral, called **margin**, must be deposited at the broker (and by the broker at the respective derivative exchange).

With this margin, the exchange ensures that investors will be able to meet the payment obligations they entered into upon taking the short position. Note that the required margin amount may change over the life of the option. If the required margin exceeds the deposited margin—due to price movements of the underlying instrument—the broker makes a "margin call" to the investor, demanding that the investor deposit additional money or securities into the trading account or close positions within a certain period of time in order to bring the account up to the required minimum margin. If the investor fails to do so, the broker closes that investor's short positions so that the margin again meets the minimum margin requirement.

As already explained above, in some cases the margin does not necessarily have to be deposited in cash but may also consist of collateral of a different nature, such as equity or bond portfolios (which are, however, not accepted to their full value, but only up to a certain percentage).

The required minimum margin always consists of two components:

$$\begin{aligned} \textit{required margin amount} &= \textit{current closing price of the option} \\ &\quad + \textit{additional safety buffer} \end{aligned}$$

The required value of the safety buffer varies from exchange to exchange and from product to product. In some cases, the required margin is also not simply the sum of the required margins for each of the products but may depend on the combination of options. For example, the long butterfly spread (see Fig. 2.46) generally requires a lower margin than would have to be deposited in total if the spread's two short positions were simply added.

Moreover, in some cases brokers tend to require a higher margin from the investor than would be required by the exchange itself. For these reasons, we will provide only some elementary information below about margin requirements for CBOE-traded options on the S&P500 index. (These are the options that we will use in most of the following examples, and the margin rules we present here can serve as an approximate reference for margin requirements in general.)

Full details of the margin requirements for CBOE-traded options as well as examples and sample calculations can be found in the CBOE Margin Manual (see https://www.cboe.com/learncenter/pdf/margin2-00.pdf).

For our purposes, the most important **margin requirements for short positions in CBOE-traded S&P500 options** are as follows:

1. Required **margin for one short put position** (we refer to that as an "uncovered position" or "naked position")
   The **current market value of the put option plus the greater of the following two values**:
   (a) 10% of strike price K
   (b) 15% of the current price S of the S&P500 less max $(0, S - K)$

2. Required **margin for one short call position** (“uncovered position” or “naked position”)
   The **current market price of the call option plus the greater of the following two values**:
   (a) 10% of strike price K
   (b) 15% of the current price S of the S&P500 less max $(0, K - S)$
   The same margin requirements apply to stock options traded on the CBOE (e.g. the option considered for the XXX stock in Sect. 2.12), except that in item (b), 15% is replaced by 20%.
   Here is an example:

Example 2.2 The current 11 October 2021 values of the S&P500 is 4415 points.

- The put option on the S&P500 with expiration 17 March 2022 and strike price $K = 4350$.
  - The current closing price of the option is 194.60
  - 10% of the strike price is 435
  - 15% of the current value is 662.25
  - $S - K = 65$ and therefore max $(0, S - K) = 65$
  - As $662.25 - 65 = 597.25 > 435$ rule (b) applies
  - Therefore: margin $= 194.60 + 597.25 = 791.85$\$ per option, i.e. 79.185\$ per contract
- The call option on the S&P500 with expiration 17 March 2021 and strike price $K = 5000$
  - The current closing price of the option is 10.80
  - 10% of the strike price is 500
  - 15% of the current value is 662.25
  - $K - S = 650$ and therefore max $(0, K - S) = 650$
  - As $662.25 - 650 = 12.25 < 500$ rule (a) applies
  - Therefore: margin $= 10.80 + 500 = 510.80$\$ per option, i.e. 51,080\$ per contract

3. Required **margin for one short call position and one short put position** on the S&P500 (both being “uncovered positions” or “naked positions”)
   For each of the two short positions, we need to calculate the additional amounts Z(1) and Z(2), respectively, resulting from rules (a) and (b) in items (1) and (2) and then take the higher Z of the two amounts. The **total margin** then is:

current market value of the call option plus current market value of the put option plus Z

Example 2.3 If we hold the two short positions from Example 2.2, the margin per option of the combination is:
Margin $= 194.60 + 10.80 + \max(597.25, 500) = 194.60 + 10.80 + 597.25 = 802.65$\$, i.e. 80.265\$, per contract of the combination.

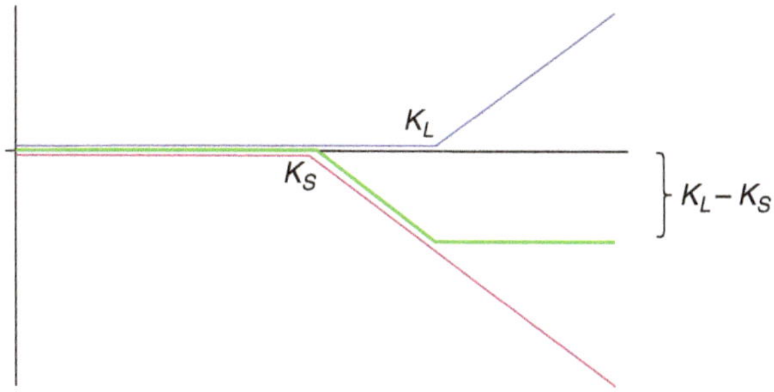

Fig. 2.62 Call spread

Margin requirements can sometimes be drastically reduced if uncovered short positions are hedged by adequate long positions:

4. Required **margin for one short call position and one long call position** on the S&P500, where the **long position's strike K_L** is higher than the **short position's strike K_S** and the long position's time to expiration is at least as long as the short position's time to expiration. The payoff function of this combination has the shape shown in Fig. 2.62 (green line). The maximum payoff to be paid by the holder of the combination is $K_L - K_S$.
   Accordingly, the margin requirement is bounded below by this maximum payable amount. Written as a formula:

$$\textbf{Total Margin} = \textbf{min}\,(K_L - K_S,\ \textbf{naked position margin})$$

5. Required **margin for one short put position and one long put position** on the S&P500, where the **long position's strike K_L** is lower than the **short position's strike K_S** and the long position's time to expiration is at least as long as the short position's time to expiration. The payoff function of this combination has the shape shown in Figure 2.63 (green line). The maximum payoff to be paid by the holder of the combination is $K_S - K_L$.
   Accordingly, the margin requirement is bounded below by this maximum payable amount. Written as a formula:

$$\textbf{Total Margin} = \textbf{min}\,(K_S - K_L,\ \textbf{naked position margin})$$

Example 2.4 We return to the put option in Example 2.2 with strike 4350 at an index value of 4415. For an uncovered short position in this option, we had to provide a margin of 51,080 dollars per contract.

If we combine this short position with a long position in a put with the same expiration and—for example—strike 4200, we only need a margin of 4350−4200 = 150$ per combination, i.e. 15,000$ per combination contract.

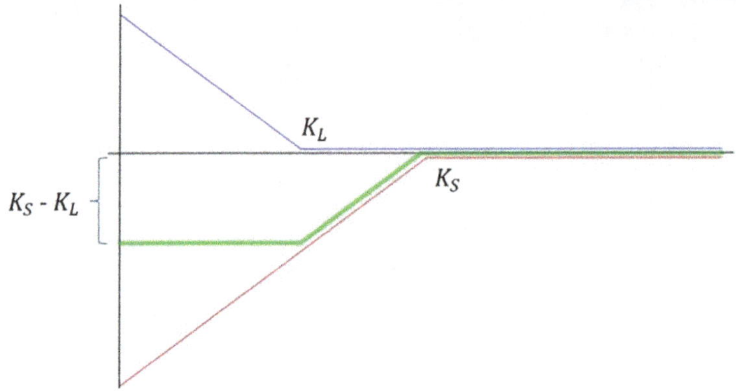

Fig. 2.63 Put spread

For this hedge of the naked short position, we obviously have to pay the price for the long position, which was approximately 154.66$ when we entered it. The disadvantage of the combination is therefore the lower premium income of approximately 194.60 – 154.60 = 40$ instead of 194.60$, yet it has the advantage of the lower margin requirement and the limitation of possible losses even in the event of an extreme decline in the S&P500.

Many brokers do not allow their clients to take uncovered short positions. In such cases, a short position can only be ordered and entered into in combination with a hedging long position. That long position can then only be closed if the short position is either closed at the same time or before that.

In the event of extreme rapid changes, uncovered short positions may even require the investor to remargin.

Example 2.5 Suppose we hold an uncovered short position in the put option with the strike price of 4350 of the previous example. Our deposited margin is 79,185$. This is covered by 90,000$ in our trading account. If, due to exceptional events, the index drops to—for example—3000 points within a very short period of time and in such a way that we have no way of closing the short position in time, and if the index does not rebound to more than 3000 points until expiration, we will have to cover a loss of at least 135,000$ in any case, even though we only have 90,000$ in our trading account. Thus, by entering into uncovered short positions, one also enters into a potential remargining commitment.

The trading platform Interactive Brokers allows investors to take naked positions. Clicking on "Account" in the IB Trader Workstation's top menu opens a window with the account information of the relevant trading account (see Fig. 2.64).

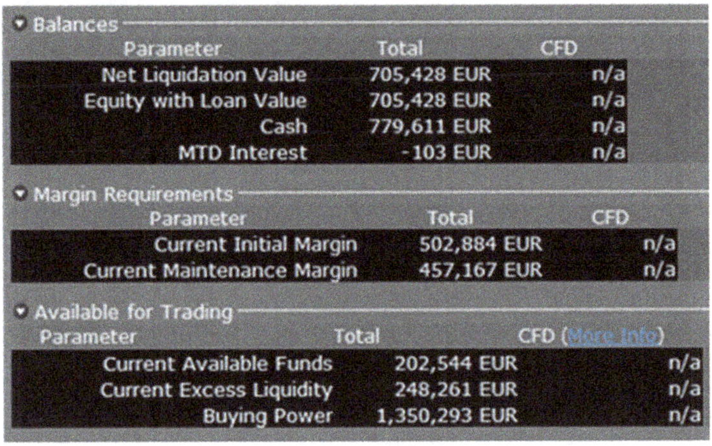

Fig. 2.64 Account information in the IB Trader Workstation

The second section that you see above is labelled "Margin Requirements", and the second line in that section states that the currently required margin for this account is 457,167 euros.

2.16 CBOE-Traded Options on the S&P500 Index, Market-Maker System, Settlement of SPX Options

In our options examples so far, we dealt almost exclusively with options on the S&P500 index (SPX) traded on the CBOE (Chicago Board Options Exchange) (hereinafter referred to as SPX options). We will continue to do so in most of the following. The SPX options market is one of the world's largest and most liquid exchange-traded options markets.

It is not the aim nor the task of this book to present a panoply of other options markets. Rather, we want to impart insights into the processes and fundamentals of options trading, using an important example, and will therefore limit our discussions mostly to options on the SPX. Trading in other index or stock options is basically analogous. Of course, before you launch yourself into trading a particular product type, you need to be clear on that product's specific trading rules (almost all of which are accessible on the websites of those products).

Below is some important information specifically for SPX options.

- SPX options are European style
- Trading hours in CET are Monday to Friday from 15:30 to 22:15
- Some brokers (including Interactive Brokers) allow options trading from 9:00 am CET!

- There are essentially three different types of SPX options:
  - The classic SPX index options expiring on the third Friday of each month
  - The SPX EOM (end of month) options expiring on the last trading day of each month
  - The SPXW (SPX weekly) options which expire on Monday, Wednesday, and Friday of each week
- A tight market-maker system guarantees bid and ask prices for each option contract for a certain minimum quota of contracts at a given maximum distance between bid and ask prices. Detailed information can be found on the CBOE website:
  https://www.cboe.org/general-info/liquidity-provider-info/market-maker-mm-program-information

It is essential to clarify one thing: until now, we have always referred somewhat vaguely to expiration dates, giving a specific date as the day on which an option expires. Yet it is highly relevant to define at exactly what time what price of the underlying asset is used to determine the payoff on the expiration date. Again, there are many different regulations that vary from option type to option type, and from exchange to exchange, so that when you intend to trade in a particular product, you will need to read up specifically on the relevant information in that particular product's specifications. We will again restrict ourselves to explaining the rules and regulations of CBOE-traded options on the S&P500 index. You will see that even those come with different rules.

The following applies to **SPX Index Options**:

Usually, these options are offered with expiration on the third Friday of the next 4-6 months, as well as with time to expiration of up to about 2 years for the months of March, June, September, and December. The options can always be traded until the close of trading on the third Thursday. Their settlement value is calculated as the (weighted) average of the prices of the first trade for each individual stock in the S&P500 on the third Friday. This settlement price can sometimes differ considerably from the opening price of the S&P500 on the third Friday! Usually, the settlement value is also not immediately set at the opening of the exchange, since it cannot be calculated until each of the 500 S&P500 stocks has been traded at least once.

Settlement values are published, for example, on
http://www.cboe.com/data/historical-options-data/index-settlement-values

In Bloomberg, the settlement value can be retrieved at
SPXSETindex

The following applies to **SPX EOM options**.

These options are usually offered with expiration on the last trading day of each month for up to 12 months. They always expire at the close of trading on the last trading day of each month. Their settlement value is calculated as the (weighted) average of the prices of the last trade for each individual stock in the S&P500 on the last trading day of the month.

Settlement values are published, for example, on

http://www.cboe.com/data/historical-options-data/index-settlement-values/end-of-month-settlement-values

The following applies to **SPXW options**.

These options are usually offered with expiration on Monday, Wednesday, and Friday of each month for up to 6 weeks. They always expire at the close of trading on the expiration day. Their settlement value is calculated as the (weighted) average of the prices of the last trade for each individual stock in the S&P500 on the expiration day.

Settlement values are published, for example, on

http://www.cboe.com/data/historical-options-data/index-settlement-values/weeklys-settlement-values

Further details and specifications of S&P500 options can be found on the following pages:

http://www.cboe.com/products/stock-index-options-spx-rut-msci-ftse/s-p-500-index-options/s-p-500-options-with-a-m-settlement-spx/spx-options-specs

http://www.cboe.com/products/stock-index-options-spx-rut-msci-ftse/s-p-500-index-options/end-of-month-spx-options

http://www.cboe.com/products/stock-index-options-spx-rut-msci-ftse/s-p-500-index-options/spx-weeklys-options-spxw

2.17 Futures, Basic Characteristics, Trading, Margin

Now that we have dealt with options extensively, we can be more brief in explaining and describing the basic features of **futures**.

Futures are—mostly exchange-traded—derivatives and thus derive their value from another product, i.e. an underlying asset, and, like options, they expire at a certain date in the future.

A significant difference in the parameters as compared to options is that the **purchase price of a futures contract is always zero**!

Another important difference in the parameters as compared to options is that the **strike price of a futures contract** is a **parameter that varies over time**.

The **fixed parameters of a futures contract** are:

- the underlying asset (an index, a stock, a commodity, a bond, etc.)
- the expiration date T (we often also denote the time from now to expiration by T)
- the price $= 0$ of the futures contract

The **variable parameter** of a futures contract is:

- the strike price (which is not fixed but changes constantly over time depending on supply and demand on the exchange)

We can also take **long positions** or **short positions** in futures. This is also sometimes referred to as “buying” or “selling” futures.

By taking a **long position** in a futures contract on an underlying asset A with expiration date T and strike price K (“buying a futures contract”), we enter into an agreement with the counterparty (“seller of the futures contract”) **to buy one unit of the underlying asset A at price K at time T**.

By taking a **short position** in a futures contract on an underlying A with expiration date T and strike price K (“selling a futures contract”), we enter into an agreement with the counterparty (“buyer of the futures contract”) **to sell one unit of the underlying asset A at price K at time T**.

Both parties to the agreement enter into an irrevocable obligation. Both parties must therefore deposit a margin to prove their ability to meet that obligation.

The agreement is made at the price of 0. At the moment of the agreement, therefore, no premium flows from one position to the other.

In the overwhelming majority of cases, there is no actual process of buying and selling the underlying asset, instead a cash settlement of the profit is carried out.

The **profit (= payoff) of the long position** at expiration is $\boldsymbol{S(T) - K}$.
The **profit (= payoff) of the short position** at expiration is $\boldsymbol{K - S(T)}$.
Here again, $S(T)$ denotes the price of the underlying asset at time T.

The following fact is important, however:

The profit or loss resulting from a futures position is settled continuously (i.e. at the respective daily settlement value at the close of trading) in the margin account (see the following Example 2.6).

Futures can be traded on the exchange at the price of 0 at any time until expiration, but at the new strike price K. Existing positions can therefore be closed again at any time until expiration. This is possible because of the continuous settlement in the margin account that we mentioned above (see Example 2.6).

In the following example, we are going to describe the trade of a futures contract on the S&P500 index via Interactive Brokers and the associated continuous settlement in the margin account.

More specifically, we are going to trade Mini S&P500 futures instead of S&P500 futures. Both products are traded on the CME (Chicago Mercantile Exchange). Both products have the same terms with the only significant difference that a mini-futures contract refers to 50 times the S&P500, while a standard futures contract has a multiplier of 250. The detailed specifications of Mini S&P500 futures can be found on the following page:
http://www.cmegroup.com/trading/equity-index/us-index/e-mini-sandp500_contract_specifications.html
The CME group, which is comprised of the CME (Chicago Mercantile Exchange), CBOT (Chicago Board of Trade), and NYMEX (New York Mercantile Exchange), is the world’s largest market place for futures.

An overview of the wide range of futures on various underlying assets (including futures on the price movements of the Bitcoin cryptocurrency or weather derivatives based on the development of temperatures or rainfall in a certain location) can be found at

http://www.cmegroup.com/trading/products.

Example 2.6 If we want to trade Mini S&P500 futures via Interactive Brokers, but don't find the product in our Trader Workstation, we can, for example, open the address mentioned above

http://www.cmegroup.com/trading/products

search for US indices in the Product Group menu and will immediately find (first line) Mini S&P500 futures (officially named "E-mini S&P500 futures").

If we now enter the corresponding abbreviation "ES" in the left column on a page of the IB Trader Workstation, a range of financial products that are related to the abbreviation ES will be displayed, from which we can select the mini-future with the desired expiration date. Default expiration dates for mini S&P500 futures are usually the third Friday next March, June, September, and December. In Fig. 2.65, we selected two mini-futures that expire on 17 December 2021, 18 March 2022, and 17 June 2022.

The final "settlement value" $S(T)$ that determines the payoff at expiration is calculated in the same way as the settlement value for the SPX options, i.e. as the average price of the first trade of each individual component of the S&P500 at the expiration date.

However, as we mentioned earlier, profits and losses of futures contracts are calculated on a daily basis (at the daily settlement value).

How exactly these daily settlement prices are determined (put simply, the average of the strike prices at which **the futures contract** was last traded on the respective trading day) can be found, for example, on the following CME page:
https://www.cmegroup.com/confluence/display/EPICSANDBOX/Standard+and+Poors+500+Futures#StandardandPoors500Futures-FinalSettlement.1
(This page also describes the procedures for calculating the final and daily settlement prices in certain special cases, for example, if the third Friday falls on a public holiday or if a component of the S&P500 was not traded on the expiration day, etc.)

Our focus is now on trading in an E-Mini futures contract expiring on 17 December 2021. If we right-click on the abbreviation of the product (blue arrow in Fig. 2.66) and select the contract description option in the menu that opens, we see the information as shown in Fig. 2.67.

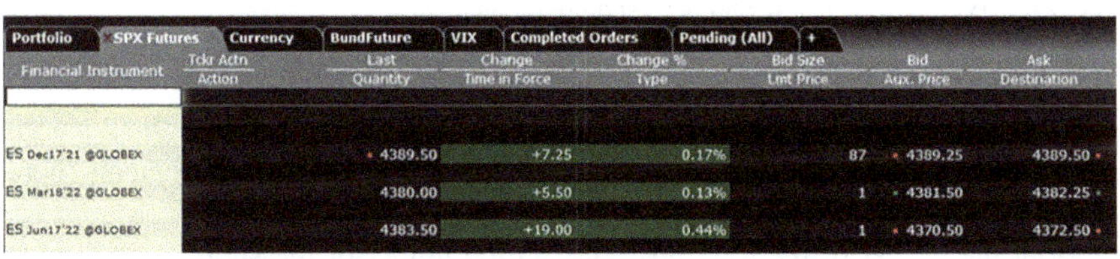

Fig. 2.65 ES page (mini S&P500 futures) in IB

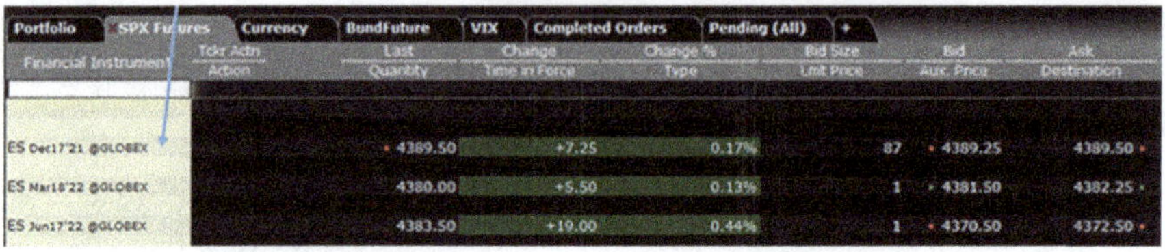

Fig. 2.66 ES page (Mini S&P500 futures expiring December 2021) in IB

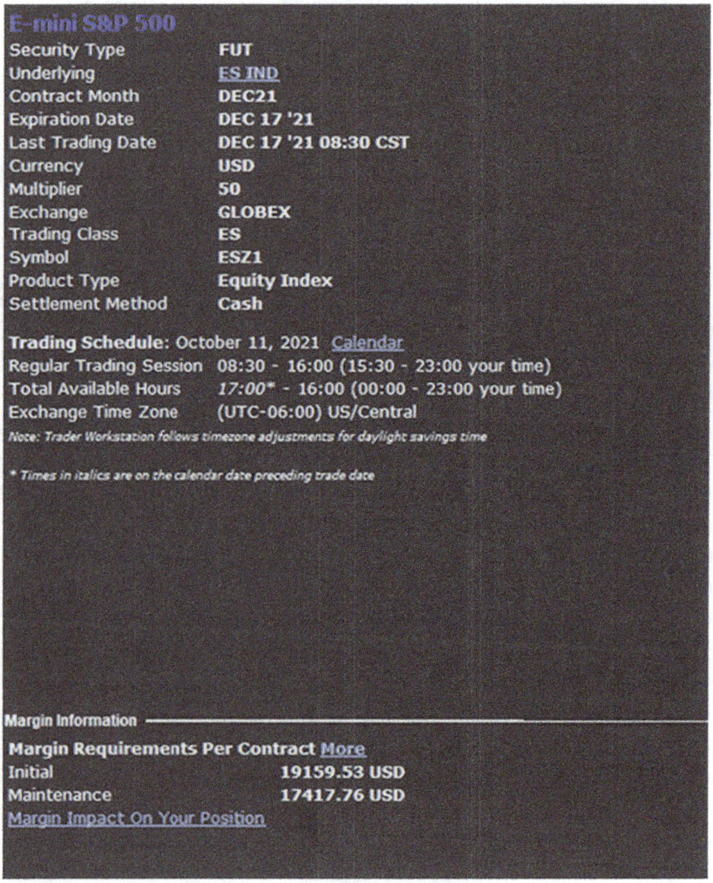

Fig. 2.67 Description of an E-Mini S&P500 futures contract

Of particular interest in the contract description is the information about margin requirements. In the last two lines, we see the broker's margin requirements (here: IB) to the investor.

Both the holder of the long position and the holder of the short position must provide this margin amount.

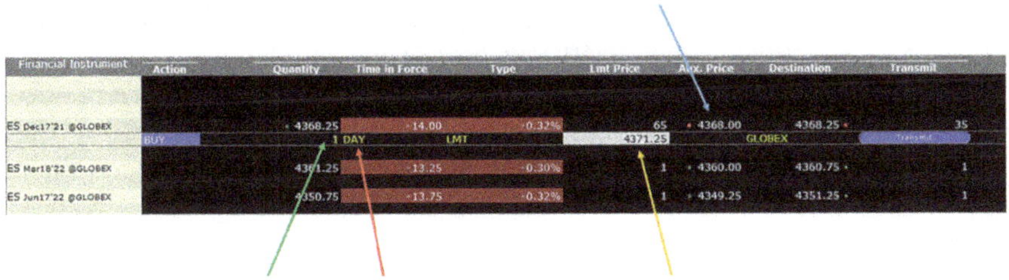

Fig. 2.68 Trading in ES: E-Mini S&P500 futures contract

Margin settings can vary based on account type, portfolio positions, and other factors. In the current example, the margin is approximately equal to current strike price of the futures contract × contract size × 8% USD.

We will have a closer look at the above-mentioned daily settlement and continuous valuation method further below.

We now want to short a mini-futures contract on the S&P500. To do this, we press the Bid button in the IB Trader Workstation (blue arrow in Fig. 2.68).

The bid and ask price (4370.50//4370.75) as well as the adjacent values 79 ("bid size", to the left of the bid price) and 35 ("ask size", to the right of the ask price) tell us as follows: at the moment, it is definitely possible to go short on up to 79 futures contracts with strike price 4370.75 or go long on up to 35 futures contracts with strike price 4370.50.

The order line below opens on which we can again select: the quantity of contracts (green arrow), the time in force (TIF) (red arrow), and the limit price (orange arrow). To transmit the order, we click "Transmit".

So, let us assume that we have taken a short position in a Mini S&P500 futures contract with strike 4370.50 and expiration date 17 December 2021. The price—as always for any futures contract—was 0$!

Let us also assume that we have USD 30,000 in our trading account to invest at our discretion. Of these 30,000 dollars, an amount of 19,159.53 dollars is tied up as margin for intraday trades.

Daily settlement:

Suppose the strike price of our futures contract changes to 4360 by the end of the business day.

The futures position is then settled as follows:

Upon taking the short futures position, we assumed the right and the obligation to sell the S&P500 at the price of 4370.50.

The product we now hold, however, only guarantees us the sale at a price of 4360. This means that we need to be compensated at exactly 10.50 dollars per future, i.e. $50 \times 10.50 = 525$ dollars per contract. A total of 525 dollars are credited to our account, increasing its balance to 30,525 dollars.

Suppose the strike price of our futures contract changes to 4380 by the end of the **next** business day. The futures position is then settled at the end of the next business day as follows:

Upon taking the short futures position, we secured the right and the obligation to sell the S&P500 at the price of 4360, which was valid until last night. The product we are now holding, however, now even guarantees us to sell at a price of 4380. This means that we need to pay exactly 20 dollars per future, i.e. $50 \times 20 = 1000$ dollars per contract. An amount of 1000 dollars are debited to our account, reducing its balance to 29,525 dollars.

Closing:

At some point during trading hours on the next following day, the strike price of the futures contract has changed to approximately 4350. We decide to **close** the contract. Before the position is actually closed, our trading account needs to be settled first. The procedure is the same as for the daily final settlement:

The product I now hold only guarantees us the sale at a price of 4350. This means that we need to be compensated at exactly 30 dollars per future, i.e. $50 \times 30 = 1500$ dollars per contract. Consequently, 1500 dollars are credited to our account, increasing its balance to 31,025 dollars. Then the futures contract is closed—at the price of 0, of course.

The result is a profit of $1025 = 50 \times (4370.50 - 4350)$ dollars, which has already been automatically credited to my account through the continuous settlement process.

It is important to keep this continuous settlement process in mind when trading in futures, as the fact that futures are always traded at a price of 0—even when the strike price changes—can easily cause confusion!

This procedure is also consistent with the way the payoff is settled if the futures position is not closed before expiration.

Let us assume, for example, that by the close of trading on 16 December 2021, the strike price of the futures contract has fallen to 4300. After settlement, the account balance would be $30,000 + 50 \times (4370.50 - 4300) = 33{,}525$ dollars.

Assume further that the futures settlement price $S(T)$ at trade opening on 17 December 2021 is 4310. The final settlement would then result in another $50 \times 10 = 500$ dollars being debited, bringing the final account balance to $33{,}025 = 30{,}000 + 50 \times (4370.50 - 4310) = 30{,}000 + 50 \times (K - S(T))$.

2.18 Long and Short Trades of Underlying Assets with Futures

As already noted and as can be seen from the CME Group's product overview, futures contracts are available for a variety of underlying instruments (which in some cases (e.g. weather derivatives) do not even have to be tradable financial products or commodities).

We are going to start this chapter by stating the following and explaining the statement in detail later.

The strike price of a futures contract is almost always close to the current price of the underlying asset, i.e. $K \approx S(0)$.

This applies all the more the closer the futures contract is to expiration.

This also means that with futures, you can go both long and short on underlying assets to achieve almost exactly the same results as with direct trades.

In absolute numbers, the profit made with a **long** position in a futures contract on an underlying A, $(S(T) - K)$, therefore **matches** almost exactly the **profit** you would have made if you had traded the underlying A itself $(S(T) - S(0))$.

(However, the investment budget that you need to trade a futures contract (i.e. essentially the amount of the margin) is much lower than the sum $S(0)$ that you would need to trade the underlying A directly.)

In absolute numbers, the profit made with a **short** position in a futures contract on an underlying A, **matches** almost exactly the **loss** you would have made if you had traded the underlying A itself.

Using long and short futures, a variety of financial products, commodities, metals, etc., can thus approximately be traded both long and short with little investment!

Let us look at another example, where we are going to trade another interesting underlying asset via the trading platform of Interactive Brokers:

Example 2.7 In this example, we want use futures to trade the underlying asset "gold". We start by opening the overview of the CME Group's product range at http://www.cmegroup.com/trading/products/ and begin searching in the product group titled "Metals". The gold futures we are looking for are already displayed in the second line. The corresponding abbreviation (tick symbol) is "GC".

Clicking on the product name "Gold Futures" provides us with the detailed product specifications of the gold futures. An excerpt of this is shown in Fig. 2.69. In particular, we see (in line 1) that the size of a futures contract is 100 troy ounces. Internationally, the price of gold is usually quoted in dollars per troy ounce (one troy ounce equals 31.034768 grams).

Contract Unit	100 troy ounces
Price Quotation	U.S. Dollars and Cents per ounce
Minimum Price Fluctuation	$ 0.10 per troy ounce
Product Code	CME Globex: GC
Listed Contracts	Trading is conducted for delivery during the current calendar month, the next two calendar months; any February, April, August, and October falling within a 23-month period; and any June and December falling within a 72-month period beginning with the current month.
Settlement Method	Deliverable

Fig. 2.69 CME Gold Futures specification (excerpt)

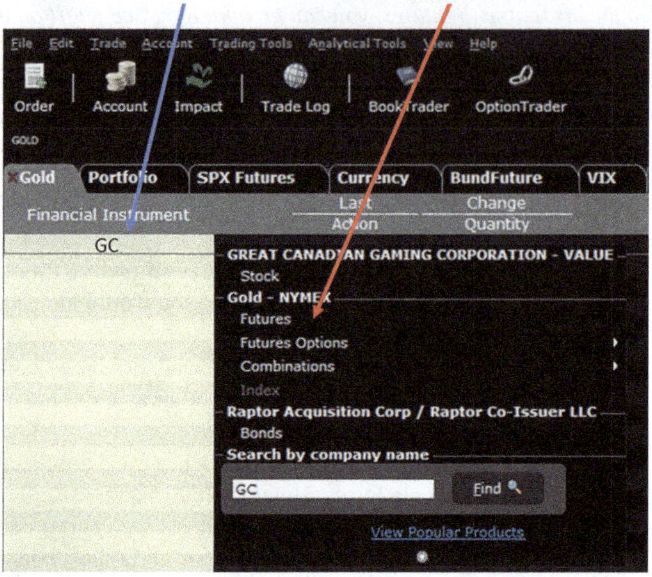

Fig. 2.70 Gold Futures in the IB Trader Workstation

It should also be noted (see "Deliverable" in the last line in Fig. 2.69,) that profits are not settled in cash at expiration but that the purchase and sale require actual physical delivery.

However, this rarely happens. In general, practically all futures contracts are closed before expiration. The details of how to conduct such a trade are explained in further specifications.

In order to trade the gold futures contract via IB, we open the IB Trader Workstation and enter the symbol "GC" in a line in the first column (blue arrow in Fig. 2.70) either on "the Portfolio page" or on another subpage and confirm the entry. A window opens in which you can select "Futures" (red arrow in Fig. 2.70). This opens another window (Fig. 2.71) in which you can select the expiration date. We choose the contract with expiration date 26 November 2021. The corresponding line opens in the IB Trader Workstation (Fig. 2.72), where we can again comfortably place our order as usual. The quotes are 1759.10//1759.30.
A look at the contract information (right-click on the product name "GC Nov26'21 @ NYMEX" and select "contract information") provides, among other things, information about the required margin, which amounts to approximately 5.5% of the strike price.

Finally, in Fig. 2.73 we see the gold price movements in 2021. At the time of selecting the above images and related futures contract data, the gold price was 1759.21 dollars. This confirms our statement that the strike price of the futures contract is indeed close to the current price of the underlying asset.

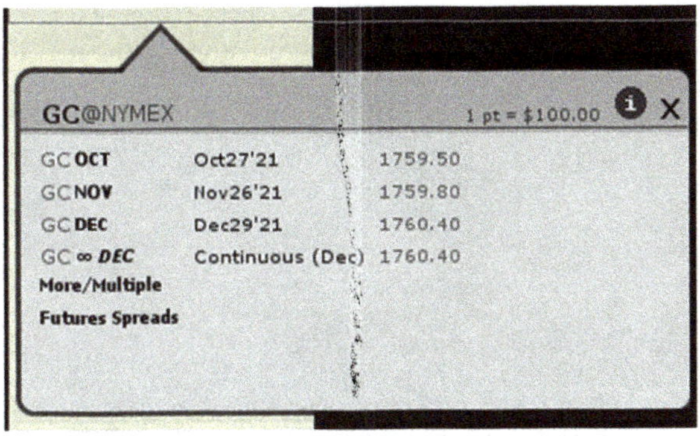

Fig. 2.71 Selecting the gold futures expiration date

Fig. 2.72 Gold futures trade in the IB

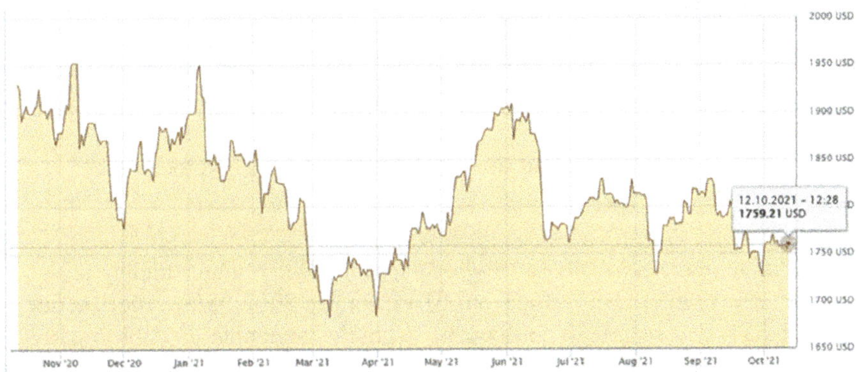

Fig. 2.73 Gold price movements in 2021

Take, for example, today on 12 October 2021 an investment in a short position in a gold futures contract (anticipating a decline in the gold price to around 1700 over the next 2 months). With a contract size of 100 troy ounces, a strike price of 1759, and a margin of 6%, this investment requires a margin of $6 \times 1759 = 10{,}554$ dollars. If the gold price actually falls from 1759 to 1700 over the next 2 months, as hoped, we can expect the strike price of the futures contract to also decline by approximately 59 points.

The result would then be a profit of approximately 5900 dollars, which, on a margin of 10,554 dollars, is a profit of approximately 55.9%.

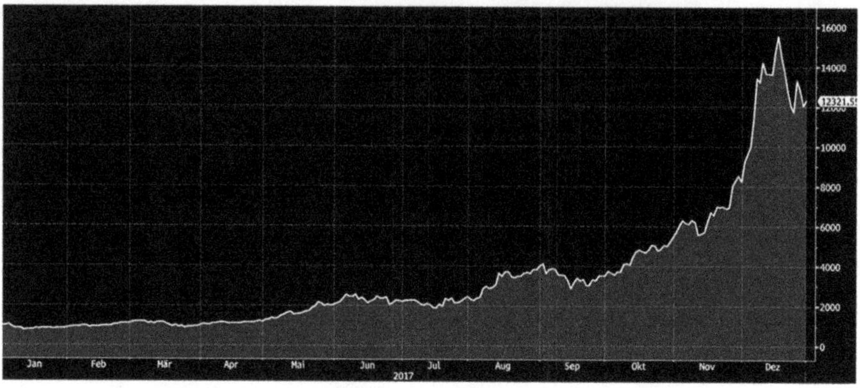

Fig. 2.74 Bitcoin price chart 2017 (source: Bloomberg)

Other interesting examples of futures that can be traded on the CME Group's markets via IB (listed below with their ticker symbols) include:

SI	Silver futures
CL	Crude oil futures (long and short trades in the oil price)
EC	Euro FX futures (long and short trades in the euro/dollar exchange rate)
RY	Euro/Japanese yen futures (long and short trades in the euro/yen exchange rate)
BTC	Bitcoin futures (long and short trades in the Bitcoin cryptocurrency

It is quite interesting to observe what happened when futures trading in the Bitcoin cryptocurrency started on 10 December 2017. While the Bitcoin price had risen seemingly unstoppably from the beginning of 2017 to early December (up to around 15,000 euros; see Fig. 2.74 and 2.75), this hike slowed down markedly with the opening of the futures market (at least for the first few weeks). With the introduction of futures trading, it became possible to speculate on falling bitcoin prices and thus make money on falling bitcoin prices. In October 2021, the price was in the range of 40,000–50,000 euros.

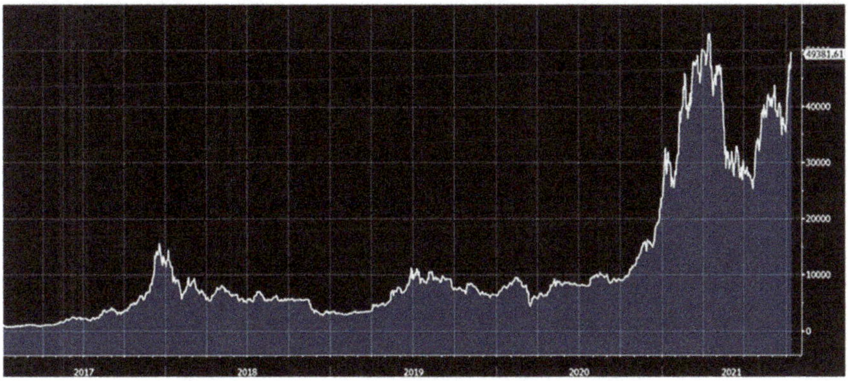

Fig. 2.75 Bitcoin price movements 2017 to October 2021 (source: Bloomberg)

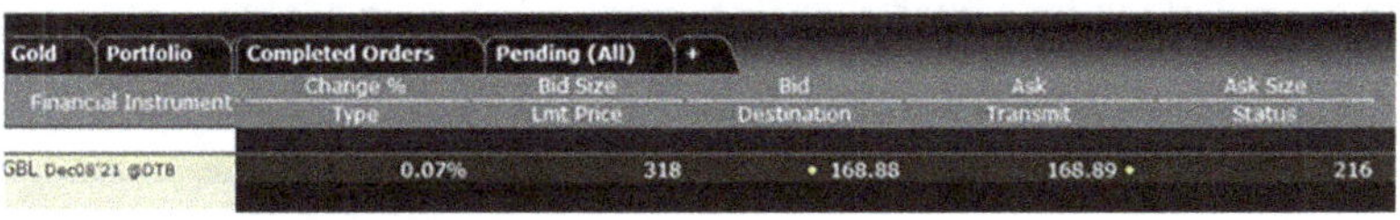

Fig. 2.76 Trading the GBL with expiration date 8 December 2021 via IB

2.19 The "Euro-Bund Future"

A special example of futures is the "Euro-Bund Future", or "Federal Government Bond Liability" (FGBL), as it is also known in English. As it is a very efficient tool for trading in euro interest rates, we are going to describe it in more detail below.

Simply put, it is a futures contract on the risk-free 10-year euro interest rate.

It can be traded on EUREX via IB using the ticker symbol GBL (see Fig. 2.76). A slightly more precise (but still not exact) description is the following.

The Euro-Bund Futures is a futures contract on a 10-year German government bond ("Bund" for short), with a coupon of 6%.

The reason why this description cannot be exact is, of course, that it will not always be possible to get a German government bond at any moment in time with exactly that coupon and exactly 10 years' time to maturity. The actual underlying of this futures contract is therefore a fictitious bond whose price is calculated as the average of the prices of German government bonds with a maturity of approximately 10 years (8.5–10.5 years) and a coupon of 6%, adjusted by certain conversion factors.

The exact specifications of the Euro-Bund Future can be found on the EUREX website at

https:///www.eurexchange.com/exchange-en/products/int/fix/government-bonds/Euro-Bund-Futures-137298

or directly in the Contract Information within the IB Trader Workstation (see Fig. 2.77).

One Euro-Bund Futures contract refers to a nominal value of 100,000 euros. A long position in a Euro-Bund Futures contract expiring on 8 December 2021 is currently (see Fig. 2.76) available at a strike price of 168.89 euros.

With a futures contract, you thus acquire the right and obligation to buy, on 8 December 2021, a (fictitious) German government bond with a nominal value of 100,000 euros, a term of 10 years, and a coupon of 6% at the price of EUR 168.89.

The margin to be deposited for this investment is only EUR 3341 (see Fig. 2.77).

If you decide to keep the futures contract until expiration, then, since the ideal bond type does not normally exist, you have to buy one of these bonds from a certain selection of EUREX-listed German government bonds, based on a certain conversion factor, which is also determined by EUREX.

Now, if the ideal market interest rates for 10-year investments were to rise over the next few months, the price of a 10-year government bond and thus the strike price of the Euro-Bund Ffutures contract would fall (e.g. to 167.00).

As a result of our long futures position, we would sustain a loss (in our example: loss $=$ 168.890 $-$ 167.000 $=$ 1890 euros. Given the margin of 3341 euros, this means a loss of 56.6%.)

Conversely, of course:

If the ideal market interest rates for 10-year investments were to fall over the next few months, the price of a 10-year government bond and thus the strike price of the Euro-Bund Futures contract would rise (e.g. to 170.00). As a result of our long futures position, we would make a profit (in our example: profit $=$ 170,000 $-$ 168,890 $=$ 1110 euros. Given the margin of 3341 euros, this means a profit of 33.2%.)

Below we will examine the approximate (!) relationship between the change in 10-year market interest rates and the profits or losses in a Euro-Bund Futures contract in a more detailed manner:

First, we assume that the yield of the ideal 10-year German government bond ("Bund") with a coupon of 6% matches the 10-year euro-market interest rate one-to-one. This means that an increase in the 10-year euro market interest rate by one basis point ($=$ 0.01%) increases the yield of the Bund by one basis point and vice versa. Second, we assume that a change in the Bund's price implies an equally large change in the strike price of the Euro-Bund Futures.

The formula for calculating the yield r of a bond (see Sect. 1.9) gives the following relationship between the yield r of the ideal Bund and its price A

(continued)

(we interpret $A = A(r)$ as a function of the yield r (stated as a percentage below)):

$$A(r) = \frac{100+6}{\left(1+\frac{r}{100}\right)^{10}} + \frac{6}{\left(1+\frac{r}{100}\right)^{9}} + \ldots + \frac{6}{\left(1+\frac{r}{100}\right)}$$

The derivative of A with respect to r is then given by

$$A'(r) = \frac{1}{100}\left(-10 \cdot \frac{106}{\left(1+\frac{r}{100}\right)^{11}} - 9 \cdot \frac{6}{\left(1+\frac{r}{100}\right)^{10}} - \ldots - 1 \cdot \frac{6}{\left(1+\frac{r}{100}\right)^{2}}\right)$$

A change in the yield r by one basis point, i.e. from r% to $(r+0.01)\%$, thus causes a change in the price of the Bund approximately from $A(r)$ to $A(r+0.01) \approx A(r) + A'(r) \cdot 0.01$.

Figure 2.78 shows the graph of the function $A(r)$ (for r ranging from -2% to 10%), i.e. the dependence of the ideal Bund's price on the yield. For example, a yield of 0% implies a price of 160, and a yield of 10% implies a price of approximately 75.

In Fig. 2.79 we plotted the function $A'(r) \cdot 0.01$. For a given yield r, the values of that function indicate the approximate change in the price of the Bund (and thus the strike price of the Bund Futures) if the yield increases by 0.01%.

For example, with a change in the yield from 0% to 0.01%, the price of the Bund (and thus the strike price of the Bund Futures) decreases by approximately 0.135.

So, if we hold a long position in a Euro-Bund Futures contract—assuming the simplified conditions given above—and the 10-year market interest rate goes from 0% to 0.01%, we will sustain a loss of approximately 0.135% of the nominal amount of 100,000, i.e. a loss of approximately 135 euros (which corresponds to 4.76% based on the margin of 2838 euros).

With a change in the yield from 10% to 9.99%, the price of the Bund (and thus the strike price of the Euro-Bund Futures contract) increases by approximately 0.05. A long position in a Euro-Bund Futures contract—assuming the simplified conditions given above—thus yields a profit of approximately 0.05% from the nominal amount of 100,000, i.e. a profit of approximately 50 euros (corresponding to 1.76% based on the margin of 2838 euros) when the 10-year market interest rate decreases from 10% to 9.99%.

Description/Name	Euro Bund (10 Year Bond) (GBL@)
Description/Name	Euro Bund (10 Year Bond)
Symbol	GBL
Exchange	DTB
Contract Type	Futures
Country/Region	Germany
Closing Price	168.78
Currency	Euro (EUR)
PRIIPS KID	PRIIPS KID link
Conid	475676590
Futures Type	Fixed Income
First Notice Date?	08/12/2021
First Position Date?	08/12/2021
Last Trading Date	08/12/2021
Expiration Date	08/12/2021
Multiplier	1000
Intraday Initial Margin	3,341
Intraday Maintenance Margin	2,673

Fig. 2.77 Euro-Bund Future specs (IB Trader Workstation)

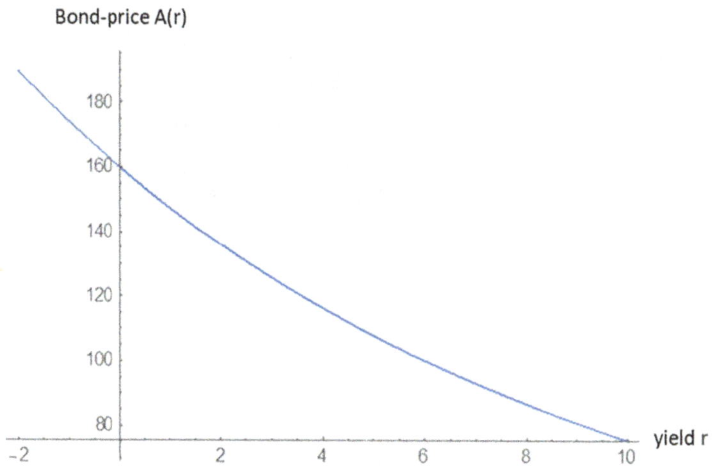

Fig. 2.78 Dependence of the ideal Bund's price A on the yield r

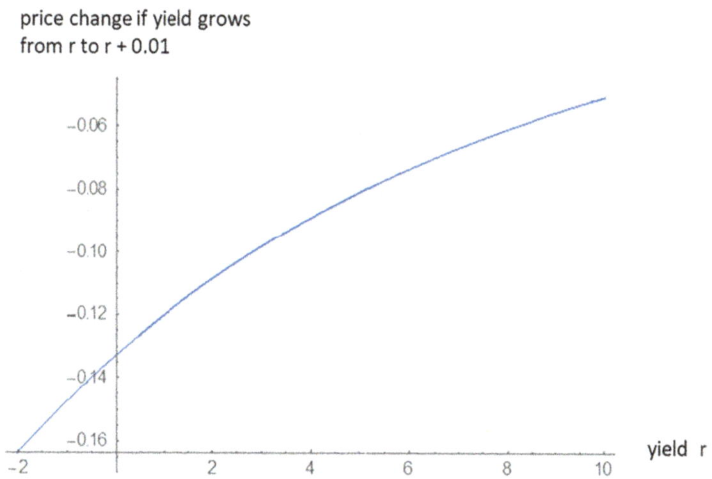

Fig. 2.79 Change in bond price for a given yield and yield increase by one basis point

2.20 More Comments on Futures Contracts (Rolling, Futures Options, Forwards)

Rolling Futures Contracts

Futures contracts usually expire within a maximum of 1 year. A Mini S&P500 futures contract, for example, is usually available with expiration within the next 3 months in the March, June, September, and December cycles. Of these futures contracts, it is usually only the ones with the nearest and second nearest expiration

Financial Instrument	Bid Size Lmt Price	Bid Aux. Price	Ask Destination	Ask Size Transmit
ES Dec17'21 @GLOBEX	57	4355.75	4356.00	29
ES Mar18'22 @GLOBEX	2	4347.75	4348.75	2
ES Jun17'22 @GLOBEX	2	4336.50	4339.75	1

Fig. 2.80 Quotes for Mini S&P500 futures on 12 January 2018

date that are highly liquid. However, this is not much of a limitation in the pursuit of long-term trading strategies in futures contracts. If a futures position (long or short) approaches its expiration date, you can roll over that position to a longer-dated but otherwise identical position. To do this, you close the existing soon-to-expire contract (cost = 0) and open the same contract with a longer time to expiration (again cost = 0). The only difference when switching to the new contract is that the strike price of that new contract usually deviates (slightly) from the strike price of the old contract. Let us take a look at an example where we "roll" a Mini S&P500 futures contract.

Example 2.8 Figure 2.80 shows the quotes on 12 October 2021 for the Mini S&P500 futures contracts expiring in December 2021, March 2022, and June 2022 (on the third Friday of the month in each case).

The S&P500 index stood at 4364 points at the time of quotation.

It is evident from the data for the bid size and the ask size (for the quotes given) and from the bid/ask spreads (0.25 for the December contract, 1.00 for the March contract, 3.25 for the June contract) that the December futures contract is clearly the most liquid of the three contracts. The strike price of the December contract is closest to the spot price of the underlying asset (the current price of the S&P500) of 4364. This, of course, represents a certain difference between the new contract and the old contract (assuming, e.g. that a Long March contract is rolled over to a Long June contract). However, this difference plays virtually no role for shorter trading horizons and only a minor role for longer trading horizons.

Short trading horizon:

For example, the "Change" and "Change in %" columns in Fig. 2.80 show the change in the strike prices of the three futures contracts during business hours on 12 October 2021. The change is virtually the same for the first two contracts, both in absolute terms (4.75 to 5.00) and in percentage terms (0.11%), and corresponds essentially to the change in the price of the S&P500 on that day.

Long trading horizon:

What is the difference if we switch from the December to the March contract? To answer that question, we will hypothetically assume that we could have traded the December contract until March. Let us also assume, for the time being, that the S&P500 will move to 4500 points by the third Friday in March 2021. Both the December and March contracts will then also simulate the increase of the strike price

Standard and E-mini S&P 500 Index Options on Futures Contract Specifications

	E-mini S&P 500 Index options	S&P 500 Index options
Ticker Symbols	**QTLY:** ES **EOM:** EW **WEEKLY:** EW1, EW2, EW3, EW4 **WED:** E1C, E2C, E3C, E4C, E5C	**QTLY:** SP **EOM:** EV **WEEKLY:** EV1, EV2, EV3, EV4 **WED:** S1C, S2C, S3C, S4C, S5C
Contract Size	One E-mini S&P 500 futures contract	One S&P 500 futures contract
Underlying Index	SPX	SPX
Minimum Price Fluctuation (Tick Size)	**Full:** 0.25 index points = $12.50 for premium > 5.00 **Cab:** 0.05 = $2.50 **Reduced Tick:** 0.05 = $2.50 for premium < or = 5.00	**Full:** 0.10 index points = $25.00 for premium > 5.00 **Cab:** 0.05 index points = $12.50[1] **Reduced Tick:** 0.05 = $12.50 for premium < of = 5.00
Trading Hours	**GLBX:** Monday - Friday 5:00 p.m. previous day - 4:00 p.m.; trading halt from 3:15 p.m. - 3:30 p.m.	**OO:** Monday - Friday 8:30 a.m. - 3:15 p.m. **GLBX:** Monday - Friday 5:00 p.m. previous day - 8:15 a.m.; trading halt, reopen 3:30 p.m. - 4:00 p.m.
Contract Months	**QTLY:** Four months in the March Quarterly Cycle (Mar, Jun, Sep, Dec) **EOM:** Six consecutive calendar months **WEEKLY:** Week 3 options on three nearest non-quarterly months, and four nearest weeks of Week 1, 2 and 4 **WED:** Two nearest Wednesdays	**QTLY, OO:** Eight months in the March Quarterly Cycle **QTLY, GLBX:** One month in the March Quarterly Cycle **EOM:** Six consecutive calendar months **WEEKLY:** Week 3 options on three nearest non-quarterly months, and four nearest weeks of Week 1, 2 and 4 **WED:** Two nearest Wednesdays
Last Trading Day	**QTLY:** 8:30 a.m. on the third Friday of the contract month **EOM:** 3:00 p.m. on the last business day of the month **WEEKLY:** 3:00 p.m. on the last business day of the week (usually Friday; 12:00 noon on shortened trading day) **WED:** 3:00 p.m. on the expiration Wednesday of the week.	**QTLY, OO:** 3:15 p.m. on the Thursday prior to the third Friday of the contract month **QTLY, GLBX:** 8:15 a.m. on the third Friday of the contract month **EOM, OO:** 3:00 p.m. on the last business day of the month **EOM, GLBX:** 8:15 a.m. on the last business day of the month **WEEKLY:** 3:00 p.m. on the last business day of the week (usually Friday; 12:00 noon on shortened trading day) **WED:** 3:00 p.m. on the expiration Wednesday of the week.
Price Limits	Halted when futures is locked limit overnight or experiencing circuit breaker event	
Strike Listing	25-point intervals within +/- 50% previous day's settlement price of the underlying futures 10-point intervals within +/- 20% previous day's settlement price of the underlying futures Once the contract becomes the second nearest contract, 5-point intervals within ﬀl 10% previous day's settlement price of the underlying futures will be available	
Exercise Procedure	At Expiration: All in-the-money **(ITM)** options on the last day of trading are exercised automatically as follows: **ITM QTRLY:** In the absence of contrary instructions delivered to Clearing by 7:00 p.m. on the expiration day, exercised into expiring cash-settled futures. **ITM EOM/WEEKLY:** A 3:00 p.m. fixing price based on the weighted average trading price of E-mini S&P 500 futures in the last 30 seconds of trading on expiration day (2:59:30 p.m.-3:00:00 p.m.) will be used to determine which options are ITM options auto-exercised and contrarian instructions not accepted.	
Block Trade Eligibility	No	Yes, minimum 250 contracts.

All times are listed in Central Time.

1 For Quarterly S&P 500 Index options only.

Fig. 2.81 Specifications of options on Mini S&P500 futures and S&P500 futures

to 4500 points by the third Friday in March. With the December contract, we would make a profit of $4500-4356 = 144\$$ per future (i.e. $50 \times 144 = 7200\$$ per contract), and with the March contract we would make a profit of $4500 - 4348.75 = 151.25\$$ per future (i.e. $50 \times 151.25 = 7562.50\$$ per contract). If we relate these hypothetical gains to the necessary margin of around 20,000$ (see Fig. 2.67), we would have made a profit of 36% with the December contract and a profit of 37.8% with the March contract.

Futures Options

Many futures (including futures on the S&P500 or the Euro-Bund Future) are themselves used as underlying assets for options.

Figure 2.81, for example, lists the specifications of options on Mini S&P500 Futures contracts (left column) and of options on S&P500 futures contracts. It is very interesting and worth noting that the trading hours for options on Mini S&P500 futures contracts are identical with the trading hours for the Mini Futures. This

Fig. 2.82 IB OptionTrader, options on E-mini S&P500 futures contracts

means that, unlike to the options taken on the S&P500 index directly, these options can be traded virtually around the clock.

Some brokers (including Interactive Brokers) also allow traditional SPX options to be traded over a longer period of time than the SPX stocks themselves, so that trades can be made from 9:00 to 22:15 CET.

However, liquidity is significantly lower during off-exchange hours , i.e. from 9:00 to 15:30.

Figure 2.82 shows a screenshot taken from the IB OptionTrader with quotes for the options expiring on 1 November 2021 on the Mini S&P500 futures contract expiring in March 2022.

The product specifications for Euro-Bund Futures options expiring in January 2022 on the Euro-Bund Future expiring March 2022 are shown in Fig. 2.83.

Forwards

A forward contract has exactly the same parameters and properties as a futures contract. Just like a futures contract, a forward is based on an underlying instrument, has a specific expiration date, is entered into at a price of 0 between an investor who takes the long position and another investor who takes a short position, and has a variable ("negotiable") strike price at which the holder of the long position buys a unit of the underlying asset from the holder of the short position at the expiration date.

A forward contract differs from a futures contract in that futures have standardized terms and are traded on an exchange, while forwards are often individually agreed arrangements that are traded over the counter (OTC), i.e. directly between investors or between an investor and a financial intermediary. It is often not possible to trade the forward contract before the end of the contract (the transaction is entered

	Underlying Information
Description/Name	Euro Bund (10 Year Bond) (GBL@)
	Contact Information
Description/Name	Euro Bund (10 Year Bond)
Symbol	GBL
Exchange	DTB
Contract Type	Futures Options
Country/Region	Germany
Closing Price	0.4
Currency	Euro (EUR)
PRIIPS KID	PRIIPS KID link
	Contact Identifiers
Conid	517303378
	Option Features
Expiration Date	23/12/2021
Last Trading Date	23/12/2021
Strike	166.5
Right	Put
Multiplier	1000
Exercise Style	American
	Eurex Germany (DTB) Top
Local Name	P OGBL JAN 22 16650
Local Class	OGBL
Settlement Method	Delivery
Exchange Website	http://www.eurexchange.com

Trading Hours	Sun	00:00-00:00
	Mon	08:00-17:15
	Tue	08:00-17:15
	Wed	08:00-17:15
	Thu	08:00-17:15
	Fri	08:00-17:15
	Sat	00:00-00:00

Fig. 2.83 Specifications of Bund Futures options

into solely with regard to the expiration date) and also in contrast to futures, there is usually no daily settlement.

Forwards are often entered into directly between companies to hedge risks. To illustrate this, let us look at a fictitious (and greatly simplified) example:

Airline A knows in January that it will have to stock up on fuel again in June (we assume oil for the sake of simplicity). Given the (fictitious!) oil price movements over the past year (250 trading days; see Fig. 2.84), the airline is very worried about the price it might have to pay in June (see forecast, Fig. 2.85).

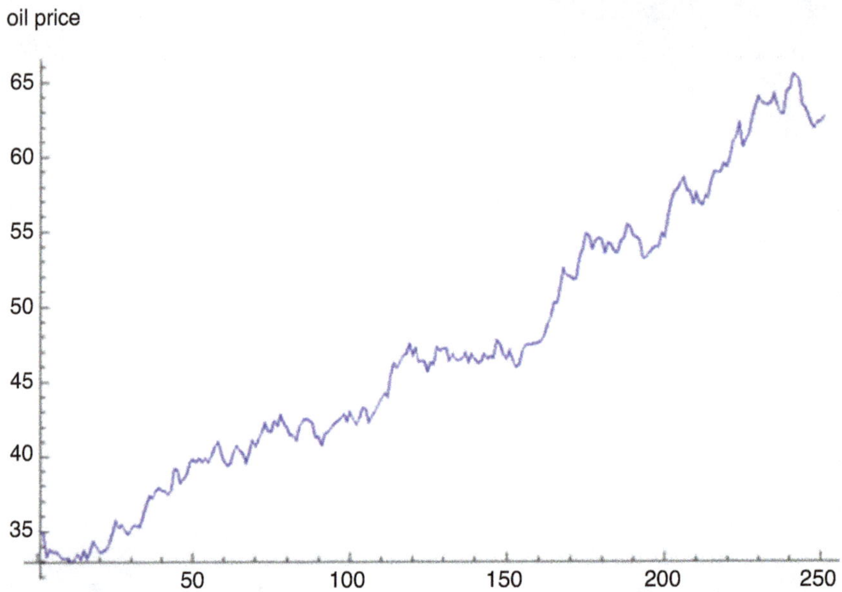

Fig. 2.84 Fictitious oil price chart for the past business year

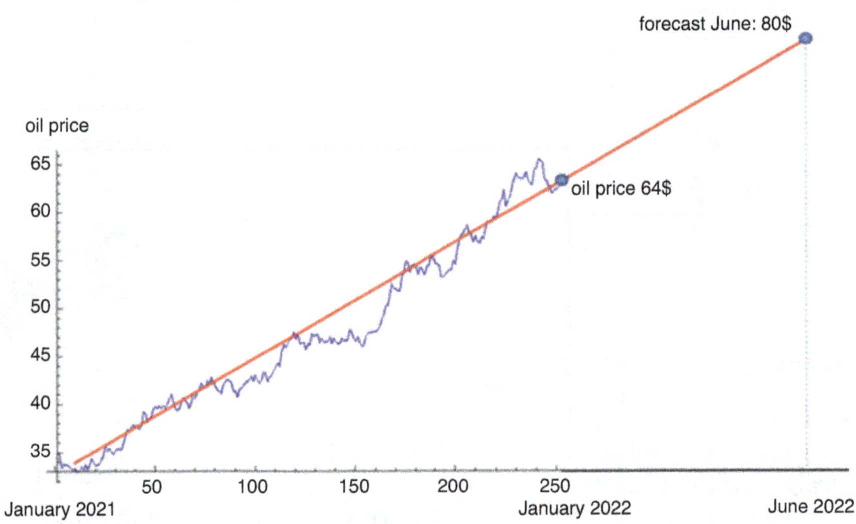

Fig. 2.85 Fictitious oil price movements in the past business year (January 2021 to January 2022) and forecast until June 2022

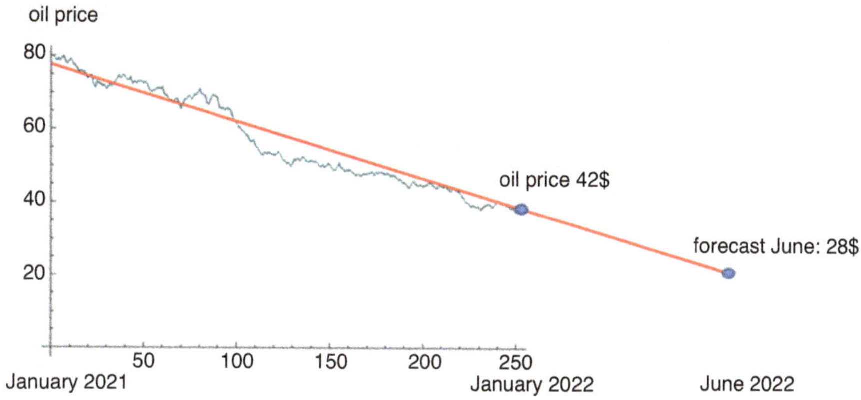

Fig. 2.86 Another fictitious oil price chart for the past business year (January 2021 to January 2022) and forecast until June 2022

Airline A therefore reaches out to an oil company B with the intention of entering into a forward contract on the amount of oil required for June 2022 (meaning: long position for airline A and short position for oil company B). The forward contract is to be agreed at a price of 0. It is of the essence for both companies that the strike price be negotiated to mutual satisfaction. Airline A will of course push for the lowest possible strike price, while oil company B will aim for the highest possible strike price. Intuitively, an agreement on a strike price close to the forecast of 80 dollars would appear plausible.

A completely different agreement on the strike price would presumably be reached (perhaps close to the new forecast of 28 dollars) if last year's price had evolved in the other direction (as shown in Fig. 2.86).

The question as to what extent these purely intuitive considerations actually hold (from a financial mathematical point of view and in reality) will be discussed in detail in the next chapter (among many other topics).

References

1. Marc Chesney, Remo Crameri, and Loriano Mancini. "Detecting abnormal trading activities in option markets". In: Journal of Empirical Finance 33 (2015), pp. 263–275.
2. Michael Khouw and Mark Guthner. The Options Edge - An Intuitive Approach to Generating Consistent Profits for the Novice to the Experienced Practitioner. New York:Wiley Trading Series, 2016. ISBN: 978-1-119-21241-6.
3. Michael Mullaney. The Complete Guide to Option Strategies - Advanced and Basic Strategies on Stocks, ETFs, Indexes, and Stock Index Futures. New York: Wiley Trading, 2009. ISBN: 978-0-470-24375-6.205

Basics of Derivative Valuation

3

Abstract

We define frictionless markets and clarify what we mean by a "fair value" of a financial product in a frictionless market. Then we state and discuss the fundamental axiom in quantitative finance, the "no-arbitrage principle". We give first applications of this NA principle and thereby derive the put-call-parity-equation and the formula for the fair price of futures. Finally, we provide the first steps towards the valuation of options: We define binomial stock-models, we give the formulas for the fair value of derivatives in such binomial models, and we show how hedging of derivatives in a binomial model is carried out.

Keywords

Frictionless markets · The no-arbitrage principle · Fair price of futures · Put-call-parity-equation · Luck and skill in games · The binomial stock-model · Derivative-valuation in a binomial stock-model · Hedging in a binomial model

One of the central themes in modern financial mathematics is the valuation of derivatives, the key question being: Does a particular financial product have a "fair price"—in the strict sense of that term (to be defined in more detail below)? And if so: How do you calculate that price and what are the consequences if the product's price on the market differs from the theoretically fair price?

This is the question that will largely be at the centre of this book.

In this chapter, we are going to take first steps towards clarifying and finding answers to this question. And we will do so using only basic –elementary– mathematical tools. The reason for using elementary methods is that we want to **provide an easy and generally comprehensible way of deriving the classic Black-Scholes formula** for the valuation of options in a Wiener process (the

G. Larcher, *The Art of Quantitative Finance Vol.1*, Springer Texts in Business and Economics, https://doi.org/10.1007/978-3-031-23873-4_3

"Wiener stock price model") and help you gain a basic understanding of this famous formula.

But before we do that we need to clarify our working premises. The most important premise that we need to be clear on is that we operate in **frictionless markets** where the **"no-arbitrage principle"** applies. However, since this means that we move in an ideal, purely theoretical environment, we will always refer back to real world and relate the **theoretical environment** to the **realities of the relevant financial market**, and we will test to what extent our theoretical results are applicable in reality.

Over the course of this chapter, we will already need and provide some basic techniques of time series analysis and probability theory. We will get to know and simulate the classic model for stock price modelling –the **Wiener stock price model**– and we will start considering possible arbitrage strategies to exploit potential price inconsistencies in the derivatives markets. For illustrative purposes, we will again concentrate on the S&P500 options market and on trading in this market via Interactivebrokers.

In this chapter we will not yet need the much more sophisticated mathematical techniques of stochastic analysis. These are tools that we will deal with much later in this book, in a crash-course style and heuristic manner. Due to their sophistication, they allow us to dive deeply into the theory of derivative valuation.

3.1 Frictionless Markets and the No-arbitrage Principle

From now on, we will proceed on two different levels:

On the one hand, we are going to define **frictionless markets**. These are **financial markets under ideal conditions**, where we are going to apply mathematics and derive theoretical results. In doing so, we will always be aware that the results we obtain in exactly that form are only valid in these ideal –non-existent– markets.

For this reason, we will always come back to real-world financial markets and ask ourselves to what extent those theoretical results are also valid and thus relevant when trading under real market conditions.

So, when we define the (mostly unrealistic) conditions of a frictionless market in the following, it is important to understand that we, just like you, our reader, are acutely aware that our assumptions differ from reality. Yet we need them in order to ensure mathematical efficiency. Rest assured, however, that we will repeatedly compare the theoretical results with reality and will be highly critical in doing so and that they will prove to be highly relevant to real-world situations, with certain adaptations.

That said, let us now define the prerequisites for a frictionless market:

(a) We assume ideal and identical risk-free interest rates for both investment and borrowing purposes

We already discussed ideal interest rates in an earlier chapter. So, we are going to assume that the same interest rates apply for both investors and borrowers and that they are generally known and accessible. Furthermore, we proceed on the theoretical assumption that these interest rates are not exposed to any credit risk, meaning that all interest payment and loan repayment agreements will be honoured without fail.

Bonds in particular are assumed to not be exposed to any default risk.

(We will be dealing with the problem of possible bond default in later chapters on credit risk management, however.)

Furthermore, we assume that any investment of cash in the course of a trading strategy is always made at either the risk-free spot interest rate (shortest-term interest rate) if the length of the investment period has not been unambiguously defined in advance or at interest rate $f_{0,T}$ if an investment (a borrowing) is planned in advance for a set period from now until time T.

(b) We assume arbitrarily fine subdivisions of financial products

In real financial markets, we can only trade whole-number units of a product or contract, of course. In a frictionless market, however, we assume that products can be arbitrarily divided. This means that in this theoretical market, we can trade arbitrarily small fractions of financial products. We therefore accept, unprotestingly, that a calculation may be based on an assumed trade of, say, 3.57 options on the S&P500 index (although in reality, it is only possible to trade in multiples of 100 units). Of course, when applying those theoretical calculations to real trades, we will take into account the limitation that we can only transact whole-number contracts.

(c) We assume the possibility of unlimited short selling

"Short selling" is defined as selling a product without owning that product. De facto, it means you trade this product directly or indirectly such that you profit one-to-one from falling prices of that product and incur losses one-to-one with rising prices of that product.

Short selling can sometimes be carried out directly (as described below); in many cases, however, traders engage in indirect short selling, for example, via certificates (such as the S&P500 short certificate in Sect. 1.18), through a short position in a futures contract on the product as its underlying asset or through a suitable combination of a call and a put option (see Sect. 2.13) on the product as its underlying.

Direct short selling requires a financial institution that offers interested investors the possibility of shorting a chosen financial product. Schematically (see Figs. 3.1 and 3.2), the process of short selling a financial product is as follows:

An investor A wants to short sell a product, i.e. sell a product she doesn't own. One possible reason for investor A's desire to do so is that A is convinced that the price of product X will decline (at least in the short term). A turns to financial institution B, which offers the possibility of short selling product X. Such a possibility is occasionally offered by institutions that wish to hold a larger quantity

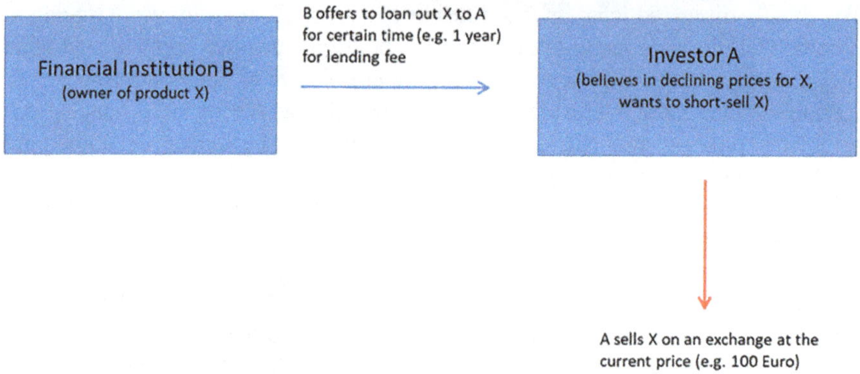

Fig. 3.1 Short selling, opening a short position

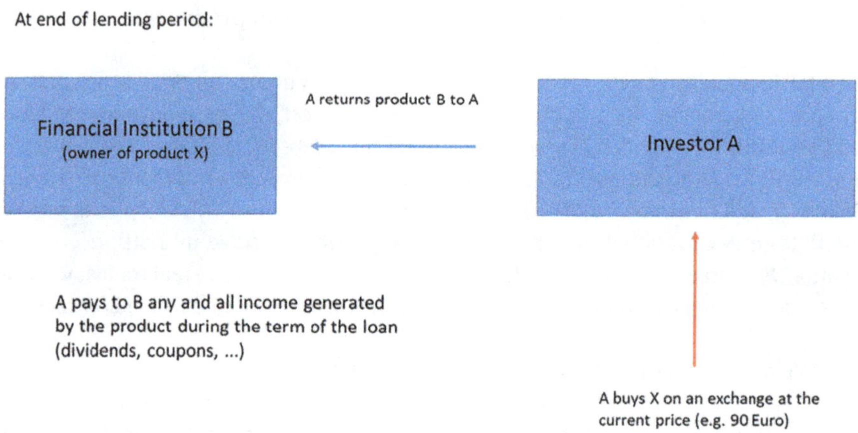

Fig. 3.2 Short selling, closing a short position

of product X for a longer period of time (perhaps because they expect prices of that product to rise in the long term). As a way to procure additional income, the institution offers to loan out product X for a lending fee. Such a loan is agreed for a certain period of time (e.g. 1 year) at a certain fee (e.g. 1 EUR per unit) between A and B. One condition that B stipulates, however, is that A pays to B any and all income generated by the product during the term of the loan. Such income would stem from dividends in the case of stock shares or coupons in the case of bonds. B understandably does not want to forgo any income from product X as a result of lending it to A.

Investor A can now sell product X on the stock exchange at its currently prevailing price (e.g. 100 EUR) to a buyer C. At the end of the loan period, A has to buy back product X on the stock exchange (from a seller D) at the then current price (e.g. 90 EUR) and return it to B. A can also buy it back at an earlier point, of course, if deemed preferable.

For the entire duration of the short selling agreement, buyer C (and not A) will receive all income from product X, yet A is the one who has to make the payments to B in the amount of that income (e.g. 2 EUR).

In addition, A has to deposit collateral with B as a guarantee until product X is safely returned to B at the end of the loan period.

All in all, investor A in our example would have made a profit of $100 - 90 - 2 - 1 = 7$ EUR per unit of product X through this short selling activity.

So, in a frictionless market, we assume the possibility of unlimited short selling for any product. Of course, for specific real-world applications (of our theoretical results), we need to examine in detail how any necessary short selling strategy can actually be carried out.

(d) We do not take into account transaction costs, fees, margins, and bid/ask spreads

In a frictionless market, we assume that there will be no transaction costs. We further assume that no other fees will apply (e.g. no short selling fees). We also do not consider margin requirements. And finally, we assume that a product can always be bought and sold at the same price, meaning that no additional trading losses occur due to bid/ask spreads. Specifically, in a frictionless market, we assume the possibility of executing trades at midpoint between bid and ask quotes.

Again: In practical, real-world applications, we will of course work with transaction costs and realistic buy and sell prices.

(e) In a frictionless market, the no-arbitrage principle applies

The "no-arbitrage principle (NA principle)" is basically something like the fundamental axiom of financial mathematics.

Put simply, the **NA principle** means that:

It is not possible in the financial markets to achieve a safe and risk-free profit in a certain period of time without investing any money.

The NA principle is sometimes also paraphrased with the saying "There is no such thing as a free lunch".

We will apply the NA principle mainly in the following form, or based on the following conclusion from the NA principle.

Theorem 3.1 *In a frictionless market: Be $[0, T]$ a time period and be P_1 and P_2 any two portfolios of financial products. For a $t \in [0, T]$, we denote the value of the two portfolios at time t by $P_1(t)$ and $P_2(t)$, respectively.* ***If it is known at a time $t \in [0, T]$ that $P_1(T) = P_2(T)$ will definitely hold at time T, then $P_1(t) = P_2(t)$ must also hold.***

Proof If $P_1(t) \neq P_2(t)$ at time $t \in [0, T]$. For example, $P_1(t) > P_2(t)$. Then (in a frictionless market!) we could sell (short sell) the portfolio P_1 at the price $P_1(t)$ at time t and use the proceeds from that sale to simultaneously buy the portfolio P_2 at the price $P_2(t)$. This gives us a positive difference $P_1(t) - P_2(t)$ in cash, which through interest accrued at the risk-free rate until time T develops to $(P_1(t) - P_2(t)) \cdot e^{r(T-t)}$.

At time T, we can then sell the portfolio P_2 at the price $P_2(T) = P_1(T)$ and use the proceeds to buy the portfolio P_1 and thus close the short position. We end up with a positive risk-free profit of $(P_1(t) - P_2(t)) \cdot e^{r(T-t)}$ without having invested any of our own money, which is counter to the no-arbitrage principle.

Observations:

- We will get to know and use the no-arbitrage principle in a mathematically even more precise version in the chapter on stochastic financial mathematics.
- We already applied the NA principle in an informal version in an earlier section, when we discussed the process of discounting future payments to arrive at the fair price of a bond.
- We will sometimes use the conclusion derived from the NA principle in the following variant:

  If it is known at a time $t \in [0, T]$ that $P_1(T) < P_2(T)$ will definitely hold at time T, then $P_1(t) < P_2(t)$ must also hold. The proof of this version is analogous to the proof of Theorem 3.1.

  And the result is correct, of course, even if the two “$<$” are replaced by “$\leq$”.
- The attentive reader, having critically reviewed Theorem 3.1 and carefully reading the section on short selling, might raise the following objection:

  If we short the portfolio P_1, then payments (coupons, dividends, ...) may have to be made to the lender of the shorted products although the amount of such payments are not known in advance. Where was this taken into account in the proof?

  This is an excellent objection, as it makes us aware that the condition “It is known at time t that $P_1(T) = P_2(T)$ will hold” really has to apply in the strict sense. So, it is only when we know for certain that “$P_1(T) = P_2(T)$” will hold, considering all payments incurred up to then (coupons, dividends, etc.), that the conclusion “$P_1(t) = P_2(t)$” is permissible. In analysing derivatives, this fact will often force us to take into account whether or not payments (or costs) will be incurred over the life of the derivative due to the underlying asset.
- A common –serious!– error in formulating the result of Theorem 3.1, which is often made by students in financial mathematics exams, is the following: The formulation

  “If the equation $P_1(T) = P_2(T)$ is true at time T, then $P_1(t) = P_2(t)$ must be true for any point in time $t \in [0, T]$”.

  is generally incorrect, of course! For this result to be valid, it is important to note:

  It must be known already at time t that $P_1(T) = P_2(T)$ will hold, only then can one conclude that $P_1(t) = P_2(t)$!

So, a frictionless market requires the applicability of the no-arbitrage principle. What is the reality of the financial markets in terms of arbitrage opportunities? To answer this question, we first have to make some distinctions.

There are almost constantly situations in which the conclusion from Theorem 3.1 does not hold, i.e. situations in which it is known that the equation $P_1(T) = P_2(T)$ will hold at a certain later point in time T, but in which the equation $P_1(t) = P_2(t)$ does not necessarily hold for all earlier points in time $t \in [0, T]$.

We will get to know and discuss such examples in the next few sections. The decisive question then is:

Are these –frequently occurring– "price inconsistencies" large enough to actually run a risk-free profit-making strategy without the use of own funds in real markets (taking into account transaction costs, bid/ask spreads, short selling opportunities, different interest rates for loans and investments, provision of margin, etc.)?

In other words:

Arbitrage opportunities in the ideal, strict sense occur practically continuously. Relatively rarely, however, do the prices differ so massively from the ideal, arbitrage-free prices that the differential can be used for running concrete, pure arbitrage strategies. The prerequisites for exploiting such arbitrage opportunities are access to favourable trading terms (e.g. low to no transaction costs) and a very fast response to price inconsistencies, as they usually occur only for a short time. This is why there are so-called arbitrageurs in the financial markets, whose job it is to identify arbitrage opportunities and act on them instantly (mostly on behalf of large financial institutions). Due to their immediate reaction, consisting in most cases in trading two portfolios that are either too expensive (P_1) or too cheap (P_2) in relation to each other –meaning that they sell P_1 and simultaneously buy P_2– the price of P_2 goes up (due to increasing demand) and the price of P_1 falls (due to increasing supply). And so the price inconsistency is mostly neutralized within a very short time.

The search for and exploitation of arbitrage opportunities thus consists of two stages: First, we look for price inconsistencies in a frictionless market. Once such a price inconsistency has been discovered, we have to analyse it to determine whether the inconsistency is large enough to actually be exploited.

In real settings, we will occasionally find opportunities where arbitrage in the strict sense may not be possible, but where making a profit is still highly probable or where a safe profit can be made if we invest a certain amount of money.

Searching for and identifying arbitrage opportunities in real settings is a very subtle task and is very rarely anything like the banal standard example of arbitrage cited in many textbooks (namely, *Short selling a stock A on a stock exchange X on which the stock is traded at a higher price than on the stock exchange Y, and buying stock A on stock exchange Y at a lower price*).

3.2 Application of the NA Principle, Put Call Parity Equation

In this section we will carry out our first (non-trivial) application of the NA principle.

For that purpose, let us use, for example, today's (13 October 2021) quotes for CBOE S&P500 options expiring on 17 December 2021, using Fig. 3.3. The list gives us a choice of strike prices ranging from 4270 to 4425. At the time of quotation, the S&P500 stood at 4358.50 points and the E-Mini Future on the S&P500 (ES) expiring on 17 December 2021 at 4350.00 points.

Of course, what we have noticed before, and what we notice again in Fig. 3.3, and what would seem only logical, is that the quotes for call options decrease as the strike price increases, while the quotes for put options rise as the strike price increases.

It is "logical" for the following reason:

Be $K_1 < K_2$.

The right to buy an underlying asset at time T at the price K_1 is obviously more valuable than the right to buy the underlying at that time at the higher price K_2.

CALLS			PUTS	
LAST	BID x ASK	STRIKE	LAST	BID x ASK
c173.65	179.20 x 180.00	4270	c103.65	99.20 x 99.90
c170.10	175.60 x 176.40	4275	c105.10	100.60 x 101.30
c166.55	172.00 x 172.80	4280	c106.60	102.00 x 102.70
c163.05	168.50 x 169.20	4285	c108.05	103.40 x 104.10
c159.60	164.90 x 165.70	4290	c109.60	104.90 x 105.60
c156.10	161.40 x 162.20	4295	c111.10	106.30 x 107.00
c152.70	157.90 x 158.70	4300	c112.70	107.80 x 108.50
c149.30	154.50 x 155.20	4305	c114.30	109.40 x 110.10
c145.90	151.00 x 151.70	4310	c115.90	110.90 x 111.60
c142.50	147.60 x 148.30	4315	c117.50	112.50 x 113.20
c139.15	144.20 x 144.90	4320	c119.15	114.10 x 114.80
c135.85	140.80 x 141.50	4325	c120.85	115.70 x 116.40
c132.55	137.50 x 138.20	4330	c122.55	117.30 x 118.10
c129.30	134.20 x 134.90	4335	c124.30	119.00 x 119.70
c126.10	130.90 x 131.60	4340	c126.10	120.70 x 121.40
c122.90	127.60 x 128.30	4345	c127.90	122.50 x 123.20
c119.75	124.40 x 125.10	4350	c129.75	124.20 x 124.90
c116.60	121.20 x 121.90	4355	c131.60	126.00 x 126.70
c113.50	118.00 x 118.70	4360	c133.50	127.80 x 128.60
c110.45	114.90 x 115.60	4365	c135.45	129.70 x 130.40
c107.40	111.80 x 112.50	4370	c137.40	131.60 x 132.30
c104.40	108.70 x 109.40	4375	c139.40	133.50 x 134.20
c101.40	105.60 x 106.30	4380	c141.40	135.40 x 136.20
c98.50	102.60 x 103.30	4385	c143.50	137.40 x 138.20
c95.60	99.70 x 100.40	4390	c145.60	139.50 x 140.20
c92.75	96.80 x 97.40	4395	c147.75	141.50 x 142.30
c89.90	93.90 x 94.60	4400	c149.90	143.60 x 144.40
c87.15	91.00 x 91.70	4405	c152.15	145.80 x 146.50
c84.40	88.20 x 88.90	4410	c154.40	148.00 x 148.70
c81.75	85.50 x 86.10	4415	c156.75	150.20 x 151.00
c79.10	82.70 x 83.40	4420	c159.10	152.50 x 153.20
c76.50	80.10 x 80.70	4425	c161.50	154.80 x 155.60

Fig. 3.3 Quotes for S&P500 options on 13 October 2021

And:

The right to sell an underlying asset at time T at a price K_1 is obviously less valuable than the right to sell the underlying at that time at the higher price K_2.

We will see later that, when it comes to financial mathematics, relying purely on logical arguments may sometimes be a slippery slope and that seemingly logical arguments can occasionally lead to incorrect results. We should therefore get into the habit of working strictly with NA arguments, even in such simple cases as this one. We will do exactly that in the following, purely for exercise purposes, as it is still a trivial application of the NA principle:

Let us take the following example: At the current point in time (time 0), we have a portfolio P_1 consisting of a call option with strike K_1 and a portfolio P_2 consisting of a call option on the same underlying with the same expiry T and a strike K_2 that is greater than K_1.

At time T, the value of the call options and thus the value of the respective portfolio consists precisely of the payoff we receive from these call options.

So we have:

$P_1(T) = \max(0, S(T) - K_1)$ *and* $P_2(T) = \max(0, S(T) - K_2)$.

Where $S(T)$ denotes the value of the underlying at time T.

As K_2 is greater than K_1, it follows that $P_2(T) \leq P_1(T)$.

Therefore, due to the no-arbitrage principle, the relation $P_2(t) \leq P_1(t)$ holds, and thus the price of the option with strike K_2 has to be less than or equal to the price of the option with strike K_1 for each point in time $t \in [t, T]$.

For the put options, we argue completely analogously.

As said, this was a quite logical relationship that has to exist between option prices.

But we now ask ourselves a much more subtle question:

To that end, let us take a line, i.e. a strike price, from the table in Fig. 3.3, for example, strike 4325 (Fig. 3.4).

The quotes for this call option are 140.80 // 141.50; the quotes for the put option are 115.70 // 116.40.

And we ask ourselves: Do these values come about purely as a result of supply and demand, or is there a certain necessary dependence here between the quotes of a call and a put option with the same underlying asset, the same expiry date, and the same strike price?

In more general terms, we can phrase our problem as follows:

We have a time period $[0, T]$ and an underlying asset S, whose price at time t is denoted by $S(t)$. Furthermore, we consider a call option and a put option on the underlying asset S with expiry date T and strike price K. $C(t)$ denotes the price of

Bid x Ask Call	Strike	Put Bid x Ask
140.80 x 141.50	4325	115.70 x 116.40

Fig. 3.4 S&P500 options with strike 4325

the call option, and $P(t)$ denotes the price of the put option at time t. Is there any correlation between the two option prices $C(t)$ and $P(t)$, or can these two prices take certain values independently of one another? We begin our analysis with an important **preliminary note**.

In principle, underlying assets for derivatives are instruments that can be traded directly or indirectly (e.g. equities, indices, commodities, foreign currencies, bonds, etc.), but they can also be non-tradable "products" such as weather events. In the following, we assume that our underlying assets are tradable products.

Furthermore, the underlying assets of derivatives may or may not generate some kind of (positive or negative) income over their life. The amounts of any such income may or may not be explicitly known in advance. If the amounts are not known in advance, then maybe reasonably reliable estimates are available, or maybe not.

Let us look at some examples of possible underlying assets in this context.

Bonds:

Over its life, a bond yields income that is known in advance, namely, the coupons.

Foreign currencies:

A foreign currency yields income that is known in advance, namely, the interest that accrues at the foreign currency's interest rate for the period in question.

Stocks:

A stock yields income that is not known in advance, namely, the dividend—in case a dividend is paid out while the investor is holding that stock. While it is not possible to project the exact amount of a dividend, an approximate value can be estimated based on historical data. If we are analysing a short-term derivative on a stock shortly after a dividend has been paid out, it can be assumed that no dividend will be distributed on the underlying stock during that short-term period.

Stock index:

In the case of a stock index, we need to establish whether we are dealing with a price index or a performance-based index and in what form that index can be traded. If the index can only be traded through a weighted trade in the constituent stocks of that index, dividends accrue through the stocks held. If the index is a performance-based index, any accruing dividends are already included in the price of the underlying to which the derivative relates, and no further payments will be made.

However, if the index is a price index (meaning that the derivative only relates to changes in the price without dividends), dividends will also accrue. When valuing a derivative, the expected amount of dividends needs to be estimated and taken into account. Dividend estimates for a broad index are generally more reliable than those for an individual stock, as they represent an average value of many stocks, which is normally more stable than a single individual value. Figure 3.5 illustrates the dividend yields of the S&P500 index since 1871.

Commodities:

Buying and owning commodities can cause warehousing costs, which can usually be estimated quite accurately or even agreed on exactly.

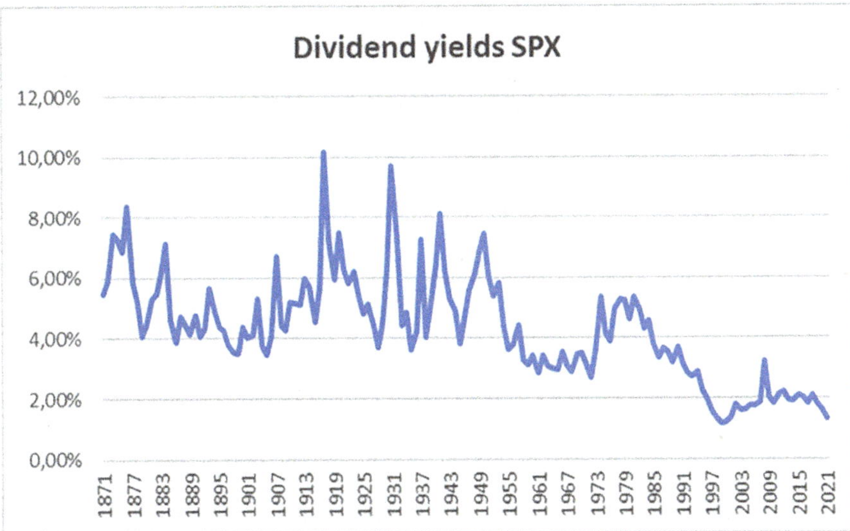

Fig. 3.5 Average dividend yields of S&P500 stocks, 1871 to 2021

In the following, we will first deal with the case of an underlying asset that does not make any payments or incur any cost over the life of the derivative.

This specific example refers to the S&P500 index, which is a price index. So dividends would have to be taken into account if we made an investment in the S&P500 by investing directly in the constituent stocks of the S&P500. For the time being, however, we assume that we can invest directly in the price index (e.g. through a certificate), so that we do not have to take dividends into account. The extent to which potential price inconsistencies can be exploited specifically when it comes to derivatives on the S&P500 will be dealt with in detail in a separate chapter.

To answer the question of a possible correlation between the prices of call and put options with the same strike price, we will again consider two different portfolios. In this case, to avoid confusion between the price P of the put option and the values of the two portfolios, we will not denote the two portfolios by P_1 and P_2 as we have done so far, but by F_1 and F_2, as an exception.

Let the first **portfolio** $\mathbf{F_1}$ consist of the following at **time 0**:
1 unit of the above long call option,
cash in the amount of $K \cdot e^{-r \cdot T}$

Let the second **portfolio** $\mathbf{F_2}$ consist of the following at **time 0**:
1 unit of the above long put option,
1 unit of the underlying

Here r is the risk-free interest rate for the time period $[0, T]$, i.e. $r = f_{0,T}$, and here the cash amount in portfolio F_1 is stated in the same currency in which the underlying asset and the options are quoted (i.e. USD in the case of S&P500 options). The interest rate used must, of course, also be for the currency in question.

Since the value of an option at expiration T is exactly equal to the payoff of that option, and since cash amounts are always automatically assumed to be invested at the risk-free interest rate, the values $F_1(T)$ and $F_2(T)$ of the two portfolios at time T are as follows:

$$\begin{aligned}\boldsymbol{F_1(T)} &= \max(S(T) - K, 0) + \left(K \cdot e^{-r \cdot T}\right) \cdot e^{r \cdot T} = \\ &= \max(S(T) - K, 0) + K \boldsymbol{= \max(S(T), K)} \\ \boldsymbol{F_2(T)} &= \max(K - S(T), 0) + S(T) = \\ &\boldsymbol{= \max(K, S(T))}\end{aligned}$$

We see: Regardless of how the underlying asset S is going to perform, the value of the two portfolios F_1 and F_2 at time T will always be equal, namely, $\max(S(T), K)$! While we do not yet know, at time 0, what value the two portfolios are going to have (since $S(T)$ is not yet known), we do know that the value of the two portfolios is going to be the same. Given the NA principle, we conclude that $F_1(0) = F_2(0)$.

As $F_1(0) = C(0) + K \cdot e^{-r \cdot T}$ and $F_2(0) = P(0) + S(0)$, the relation $C(0) + Ke \cdot^{-r \cdot T} = P(0) + S(0)$ follows.

This equation between the price of a put option, a call option, and an underlying asset is referred to as the put-call parity equation. The above arguments hold, of course, for any point in time $t \in [0, T]$ and not only for time 0. We summarize our findings in Theorem 3.2:

Theorem 3.2 (Put-call Parity Equation, Dividend-free Underlying)
Let $[0, T]$ be a time interval, and for $t \in [0, T]$, let $C(t)$ be the price of a call option with the underlying asset S (which can be traded in a way such that during the time interval $[0, T]$, no payments and no costs are incurred due to the underlying), with expiry T and strike K, and let $P(t)$ be the price of a put option with the underlying S, expiry T and strike K. The following then holds

$$C(t) + K \cdot e^{-r \cdot (T-t)} = P(t) + S(t)$$

Here, $S(t)$ denotes the price of the underlying asset at time t and r is the risk-free interest rate $r = f_{t,T}$.

In this context, it is important to be clear on the following: The result is completely independent of any assumptions about the kind of modelling of the underlying's price development.

We say: The **put-call parity equation is a model-independent result.**

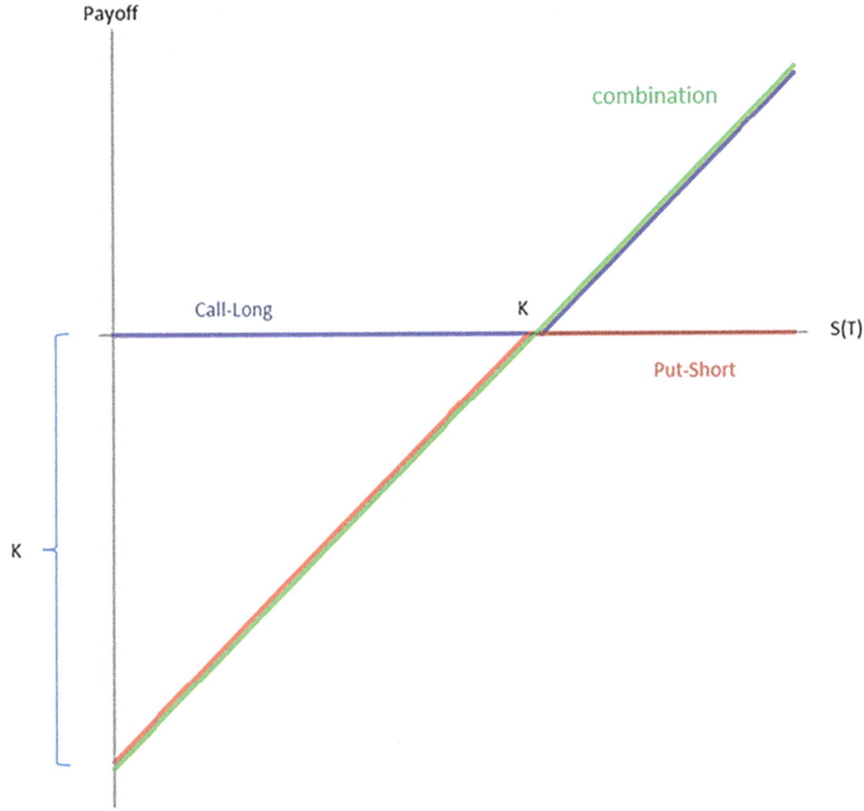

Fig. 3.6 Combination of a long call with a short put results in underlying minus K cash

Incidentally, we already obtained this result for time T in a slightly different way earlier in this book, namely, in Sect. 2.13, where we visualized it graphically, in Fig. 2.55. We republish the chart here in Fig. 3.6.

Figure 3.6 depicts the payoff (at time T) from a long call (blue) with a short put (red), which gives exactly the payoff of the underlying asset (green) less K. At time T, therefore, we have $C(T) - P(T) = S(T) - K$, which corresponds exactly to the put-call parity at time T.

This theorem, this put-call parity equation, holds in a strict sense. If this equation is not satisfied, then arbitrage is possible (in a frictionless market!)! Yet **how can arbitrage take place** (in a frictionless market!) in cases where the put-call parity does not hold?

If, for example, the left side of the equation were larger than the right side (we can limit ourselves to $t = 0$ below), i.e. if $C(0) + K \cdot e^{-r \cdot T} > P(0) + S(0)$, an arbitrage profit could be realized in the following way:

We short sell one call and take out a loan in the amount of $K \cdot e^{-r \cdot T}$ until time T at interest rate r. With this cash amount of $C(0) + K \cdot e^{-r \cdot T}$, we buy one put and one underlying S at a total price of $P(0) + S(0)$. We end up with a positive difference of $\left(C(0) + K \cdot e^{-r \cdot T}\right) - (P(0) + S(0))$ in cash. This positive amount yields interest until time T, changing our positive amount to $\left(\left(C(0) + K \cdot e^{-r \cdot T}\right) - (P(0) + S(0))\right) \cdot e^{r \cdot T}$. At time T, we sell the underlying at the price $S(T)$ and receive a payoff of $\max(K - S(T), 0)$ through the put option.

In total, we get $S(T) + \max(K - S(T), 0) = \max(K, S(T)) = K + \max(S(T) - K, 0)$.

With this money, we can pay back our loan in the amount of K and pay the payoff of our short call position in the amount of $\max(S(T) - K, 0)$. We therefore end up with a risk-free positive profit of $\left(\left(C(0) + K \cdot e^{-r \cdot T}\right) - (P(0) + S(0))\right) \cdot e^{r \cdot T}$ without having invested any of our own money.

If the right side of the equation were larger than the left side, that is, if $C(0) + K \cdot e^{-r \cdot T} < P(0) + S(0)$, then –as can be easily understood– we simply proceed the other way around.

We short sell one put option and one underlying, which gives us $P(0) + S(0)$. With this money, we buy one call option at $C(0)$ and invest the amount of $K \cdot e^{-r \cdot T}$ at the risk-free interest rate. This gives us the positive difference $(P(0) + S(0)) - \left(C(0) + K \cdot e^{-r \cdot T}\right)$. At time T we can then settle all our payables (resulting from the short put and the short underlying) with our income from the investment and the call option, again ending up with the risk-free positive profit of $\left((P(0) + S(0)) - \left(C(0) + K \cdot e^{-r \cdot T}\right)\right) \cdot e^{r \cdot T}$ without having invested any own funds.

Let us now check to what extent our example in Fig. 3.3 actually satisfies the put-call parity. Remember that for purposes of this example, we operate in a frictionless market. This means that, instead of using the bid/ask quotes, we use the midpoint prices between bid and ask as our option prices. (NB: The actual option prices, that is, the prices at which the options were last traded, are not suitable in this case, as it is highly likely that both transactions were executed at different times and at different prices of the underlying!) The mid-prices between 140.80 // 141.50 and 115.70 // 116.40 are 141.15 ($= C(0)$) resp. 116.05 ($= P(0)$).

From 13 October 2021 until expiration of the options on 17 December 2021, we are looking at a time-to-expiration period of practically 2 months exactly, that is, $\frac{1}{6}$ of a year. The risk-free interest rate for 2-month USD investments on 13 October 2021 was approximately + 0.05%. The price of the S&P500 was 2,785.96.

The put-call parity equation in our example thus has the form

$$C(0) + K \cdot e^{-rT} = P(0) + S(0) \Leftrightarrow 141.15 + 4,325 \cdot e^{-0.0005 \cdot \frac{1}{6}} =$$

$$= 116.05 + 4,358.50 \Leftrightarrow 4,465.79 = 4,474.55$$

We recognize a discrepancy in the amount of 8.76 dollars between the values on both sides of the equation. While this is not a huge difference, it is definitely noticeable.

But remember: That is assuming a frictionless market!

Now we will first draw some further simple conclusions from the put-call parity equation in the following sub-section and then use another application of the NA principle to draw conclusions as to the strike price of futures in the section following that one.

3.3 Simple Conclusions from the Put-call Parity Equation

In the last section we already came to the conclusion that the **price of call options** (with the same underlying and same expiration date) **decreases as the strike price increases** and that the **price of put options increases as the strike prices increases.**

An option always has a price greater than or equal to zero. With an option you always buy a right that you can choose to exercise or not exercise. The value of such a right cannot be negative.

In particular, the payoff of a call option is $\max(0, S(T) - K)$, thus always greater than or equal to zero, and the payoff of a put option is $\max(0, K - S(T))$, thus always greater than or equal to zero; it follows therefore that the value of such an option that delivers such a payoff is always greater than or equal to zero.

The value of a put option with strike 0 on an underlying whose price will always be greater than or equal to zero is equal to zero. The right to sell the underlying at the price $K = 0$ has no value. This means $\boldsymbol{K = 0 \rightarrow P(t) = 0}$ **for all t.**

The value of a call option with strike infinite ($K = \infty$) is equal to zero. The right to buy the underlying at the price $K = \infty$ has no value. This means $\boldsymbol{K = \infty \rightarrow C(t) = 0}$ **for all t.**

We can therefore conclude:

When the strike price K goes to 0, the prices $P(t)$ of a put option with strike K decrease monotonically to 0.

(For small strikes K, i.e. strikes far out of the money in the case of a put option, the prices of a put option are very small, close to zero.)

When the strike price K goes to ∞, the prices $C(t)$ of a call option with strike K decrease monotonically to 0.

(For large strikes K, i.e. strikes far out of the money in the case of a call option, the prices of a call option are very small, close to zero.)

Let us now return to the put-call parity equation at an arbitrary point in time $t \in [0, T]$:

$$C(t) + K \cdot e^{-r \cdot (T-t)} = P(t) + S(t)$$

We first look at the case of a very small strike price K. We know that in this case – given our observations above– $P(t)$ is very small, i.e. very close to 0. Consequently –due to the put-call parity– the price of the call option is very close to $S(t) - K \cdot e^{-r \cdot (T-t)}$. In the extreme case that $K = 0$, meaning $P(t) = 0$, $C(t) = S(t)$ even holds for all t.

If the time to expiration $(T - t)$ is small, or the interest rate r is close to zero –as is the case at the time of writing– then $e^{-r \cdot (T-t)}$ is very close to 1. In our example at the end of Sect. 3.2, for instance, we had the value $e^{-0.0005 \cdot \frac{1}{6}} = 0.999917$. So in this case, the price of the call option $C(t)$ is very close to $S(t) - K$, i.e. to the price of the underlying minus the fixed constant K.

With a small strike K, a short time to expiration, or an interest rate close to 0, the price behaviour of a call option is therefore very similar to the price movements of the underlying (minus a fixed value). This means that we can trade the underlying –at least approximately so– by taking a long or a short position in a call option with a very small strike and a short time to expiration.

In analogy with that we can thus argue that for a put option with a large strike K and a short term to expiration, the value $P(t)$ of that put option will be close to $K - S(t)$.

Let us look at that situation specifically, using the quotes for options on the S&P500 on 13 October 2021 (see Figs. 3.7, 3.8, and 3.9) to verify the above argumentation.

At the moment of the quotes shown in Figs. 3.7 and 3.9, the S&P500 stood at around 4342 points and the December futures at around 4333 points.

For strikes of 4910 and higher, the call option quotes are under a dollar. For strikes of 2300 and lower, the put option quotes are under a dollar.

Let us take the call option with the lowest strike $K = 1600$ listed in Fig. 3.7. The quotes are 2724.30 // 2733.20. This is a very wide bid/ask spread. Proceeding again from our assumption of a frictionless market, we are going to use the mid-value of 2728.75. The approximate value for the price of this call option is $S(t) - K = 4342 - 1600 = 2747$. The difference between the mid-price and the approximate value is 13 dollars (around 0.3% of the underlying asset's value). At expiration, the price will most likely be exactly $S(T) - 1600$. It is only in the very unlikely event that the S&P500 were to drop to below 1600 points by 17 December 2021 that this wouldn't hold (in that case $C(T) = 0$).

The difference between the price $C(t)$ of the call option and $S(t) - K$ tends to decrease over time. This means that the current difference of 13 dollars will most likely not increase (significantly) until the option expires. So, if we use this call option to simulate long or short trades in the underlying asset, the error margin would be limited to around 0.3% at most.

Finally, let us consider the put option with the highest strike $K = 5900$ listed in Fig. 3.9. The quotes are 1562.90 // 1571.80. This too is quite a wide bid/ask spread. Proceeding again from our assumption of a frictionless market, we are going to use the mid-value of 1567.35. The approximate value for the price of this put option is $K - S(t) = 5900 - 4342 = 1558$. The difference between the mid-price and the approximate value is thus around 9 dollars (around 0.21% of the underlying asset's value).

Let us now return once more to the (dividend-free version) of the put-call parity:

$$C(t) + K \cdot e^{-r \cdot (T-t)} = P(t) + S(t)$$

If the interest rate is $r > 0$:
Then $e^{-r \cdot (T-t)} < 1$, and thus $C(t) + K > C(t) + K \cdot e^{-r \cdot (T-t)} = P(t) + S(t)$, i.e. $C(t) > P(t) + S(t) - K > S(t) - K$, and thus $C(t) > S(t) - K$.

Fig. 3.7 S&P500 options, quotes on 13 October 2021, low strikes, calls

Calls		
DEC 16 \`21 SPX 100 64 DAYS	DEC 17 \`21 SPXW 100 65 DAYS	
BID x ASK		Strike
2724.30 x 2733.20		1600
2698.90 x 2708.20		1625
2673.90 x 2683.20		1650
2649.20 x 2658.40		1675
2624.40 x 2633.30		1700
2599.20 x 2608.40		1725
2574.10 x 2583.40		1750
2549.30 x 2558.60		1775
2524.40 x 2533.30		1800
2499.20 x 2508.50		1825
2474.20 x 2483.50		1850
2449.40 x 2458.70		1875
2424.70 x 2433.60		1900
2399.50 x 2408.80		1925
2374.40 x 2383.70		1950
2349.60 x 2358.80		1975
2324.80 x 2333.70		2000
2299.70 x 2308.90		2025
2274.80 x 2284.10		2050
2249.80 x 2259.00		2075
2224.90 x 2233.80		2100
2200.00 x 2209.20		2125
2175.00 x 2184.30		2150
2150.10 x 2159.30		2175
2125.10 x 2134.00		2200
2100.20 x 2109.50		2225
2075.30 x 2084.50		2250
2050.40 x 2059.60		2275
2025.40 x 2034.30		2300
2000.50 x 2009.60		2325
1975.60 x 1984.80		2350
1950.60 x 1959.80		2375

Fig. 3.8 S&P500 options, quotes on 13 October 2021, low strikes, puts

Puts	
Strike	BID x ASK
1600	0.15 x 0.25
1625	0.15 x 0.30
1650	0.20 x 0.30
1675	0.20 x 0.30
1700	0.20 x 0.30
1725	0.25 x 0.35
1750	0.25 x 0.35
1775	0.25 x 0.35
1800	0.30 x 0.40
1825	0.30 x 0.40
1850	0.30 x 0.40
1875	0.35 x 0.45
1900	0.35 x 0.45
1925	0.40 x 0.50
1950	0.40 x 0.50
1975	0.40 x 0.55
2000	0.45 x 0.55
2025	0.45 x 0.60
2050	0.50 x 0.60
2075	0.55 x 0.65
2100	0.55 x 0.70
2125	0.60 x 0.70
2150	0.65 x 0.75
2175	0.65 x 0.80
2200	0.70 x 0.85
2225	0.75 x 0.85
2250	0.80 x 0.90
2275	0.85 x 0.95
2300	0.90 x 1.00
2325	0.95 x 1.05
2350	1.00 x 1.15
2375	1.05 x 1.20

Calls			Puts	
BID x ASK	STRIKE		STRIKE	BID x ASK
1.10 x 1.25	4890		4890	554.80 x 564.00
1.05 x 1.15	4900		4900	564.90 x 573.70
0.95 x 1.10	4910		4910	574.60 x 583.80
0.90 x 1.00	4920		4920	584.50 x 593.70
0.85 x 0.95	4925		4925	589.50 x 598.70
0.80 x 0.95	4930		4930	594.40 x 603.70
0.75 x 0.90	4940		4940	604.40 x 613.60
0.70 x 0.85	4950		4950	614.30 x 623.50
0.65 x 0.80	4960		4960	624.20 x 633.50
0.60 x 0.75	4970		4970	634.20 x 643.40
0.60 x 0.70	4975		4975	639.20 x 648.40
0.55 x 0.70	4980		4980	644.10 x 653.40
0.55 x 0.65	4990		4990	654.10 x 663.30
0.50 x 0.60	5000		5000	664.20 x 673.10
0.40 x 0.55	5025		5025	688.90 x 698.20
0.35 x 0.45	5050		5050	713.80 x 723.10
0.30 x 0.40	5075		5075	738.80 x 748.00
0.25 x 0.40	5100		5100	763.90 x 772.80
0.25 x 0.35	5125		5125	788.70 x 797.90
0.20 x 0.30	5150		5150	813.60 x 822.80
0.20 x 0.30	5175		5175	838.50 x 847.80
0.15 x 0.25	5200		5200	863.70 x 872.60
0.15 x 0.25	5250		5250	913.40 x 922.70
0.10 x 0.20	5300		5300	963.50 x 972.40
0.10 x 0.20	5350		5350	1013.30 x 1022.50
0.05 x 0.20	5400		5400	1063.40 x 1072.30
0.05 x 0.15	5500		5500	1163.30 x 1172.20
x 0.15	5600		5600	1263.20 x 1272.10
x 0.10	5700		5700	1363.10 x 1372.00
x 0.10	5800		5800	1463.00 x 1471.90
x 0.10	5900		5900	1562.90 x 1571.80

Fig. 3.9 S&P500 options, quotes on 13 October 2021, expiration 17 December 2021, high strikes

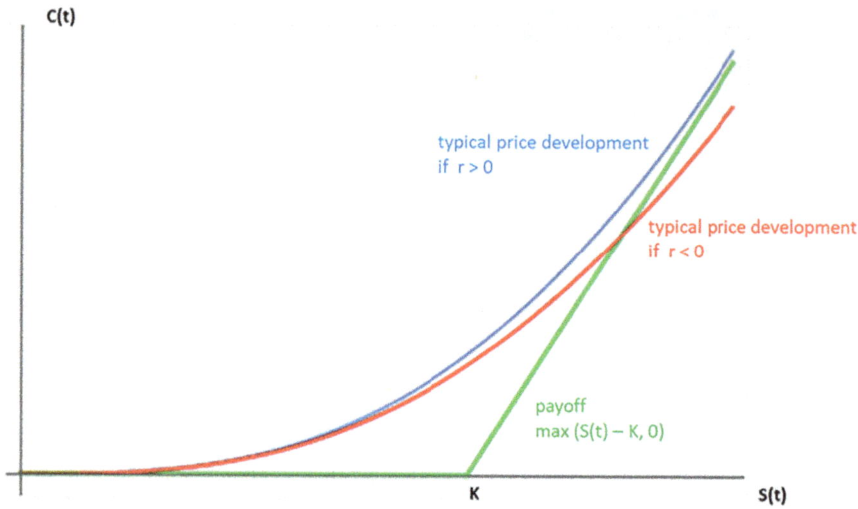

Fig. 3.10 Approximate price curve $C(t)$ of a call option for a given t and K in relation to $S(t)$

And because $C(t)$ is always greater than 0, it follows that
$\boldsymbol{C(t) > \max(0, S(t) - K)}$ always **holds.**

If the interest rate is $r < 0$:
Then $e^{-r\cdot(T-t)} > 1$, and thus $P(t) + S(t) = C(t) + K \cdot e^{-r\cdot(T-t)} > C(t) + K$, i.e. $P(t) > C(t) + K - S(t) > K - S(t)$, and thus $P(t) > K - S(t)$.
And because $P(t)$ is always greater than 0, it follows that
$\boldsymbol{P(t) > \max(0, K - S(t))}$ always **holds.**

We will deal specifically and in detail with the price trajectories of call and put options at a later point. But based on the above observations alone, price curves of call and put options (on dividend-free underlying assets) are approximately as shown in Figs. 3.10 and 3.11. These charts illustrate the plausible approximate price curve of a call option and a put option for a given point in time t and a given strike K in relation to the underlying asset's price $S(t)$. The blue curve plots the case where $r > 0$ and the red curve where $r < 0$.

3.4 Another Application of the NA Principle: The "Fair" Strike Price of a Futures Contract (on a Dividend-free/Cost-free Underlying Asset)

Figure 3.12 shows the quotes for futures on the S&P500 index on 14 October 2021 for the three expiration dates 17 December 2021, 18 March 2022, and 17 June 2022. The price of the S&P500 at the moment of these quotations was 4,387.24 points.

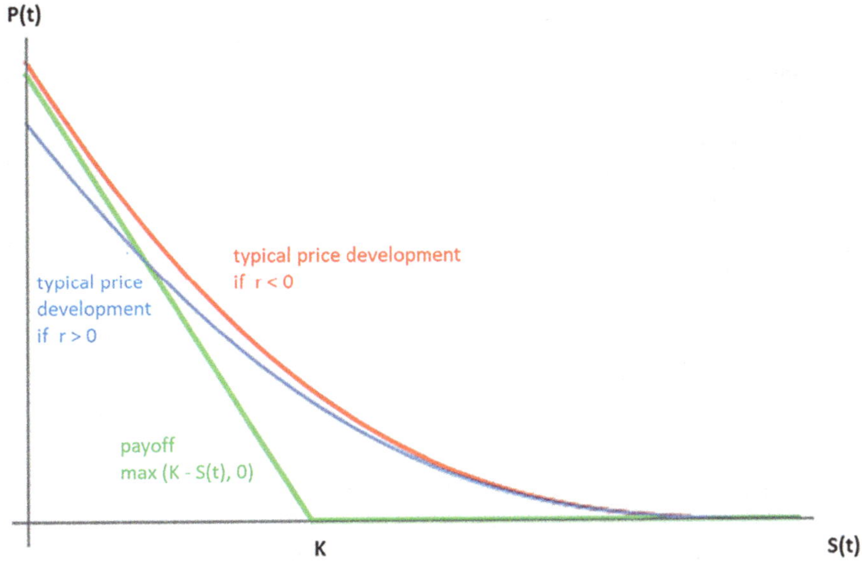

Fig. 3.11 Approximate price curve $P(t)$ of a put option for a given t and K in relation to $S(t)$

Financial Instrument	Bid Size Lmt Price	Bid Aux. Price	Ask Destination	Ask Size Transmit
ES Dec17'21 @GLOBEX	50	• 4378.50	4378.75 •	23
ES Mar18'22 @GLOBEX	2	• 4370.50	4371.50 •	2
ES Jun17'22 @GLOBEX	1	• 4359.75	4362.50 •	1

Fig. 3.12 Futures on the S&P500, 14 October 2021

Financial Instrument	Bid Size	Bid	Ask	Ask Size
DAX INDEX				
DAX Dec17'21 @DTB	10	• 15305.00	15307.00 •	2
DAX Mar18'22 @DTB	25	• 15283.00	15297.00 •	25
DAX Jun17'22 @DTB				

Fig. 3.13 Futures on the DAX, 14 October 2021

Figure 3.13 shows the quotes for futures on the DAX on 14 October 2021 also for the three expiration dates 17 December 2021, 18 March 2022, and 17 June 2022. The price of the DAX at the moment of these quotations was 15,324.96 points.

The put-call parity equation has given us a fair price **relation** between two derivatives but not an explicit fair price of any one instrument. In the following we want to derive the **explicit fair strike price of a futures contract**.

Just like the put-call parity, this fair strike price will be model-independent, yet to begin with, we are again going to **assume that the respective underlying asset does not imply any payments nor any costs during the derivative's time to expiration.**

Our **question** therefore is:

We have a futures contract on an underlying asset S with expiration T. We are at time t of the time interval $[0, T]$. The price of the futures contract is 0 (as always). Is there a fair (arbitrage-free) strike price K to be had for that futures contract and if so, what is that price?

Remember that we already asked this question when we discussed the "airline/oil company" example at the end of Sect. 2.20. We surmised, or at least suggested, that the strike price of a futures or a forward contract would likely depend on how the two parties estimated the future development of the underlying asset's price (i.e. the oil price) (see Figs. 2.86 and 2.85). Surprisingly enough, this is not the case—as we will see in a moment:

For that purpose, let us again look at two portfolios P_1 and P_2.

Let the first **portfolio P1** consist of the following at **time 0**:
1 unit of a long futures contract,
cash in the amount of $K \cdot e^{-r \cdot t}$

Let the second **portfolio P2** consist of the following at **time 0**:
1 unit of the underlying asset

Here r is the risk-free interest rate for the time period $[0, T]$, i.e. $r = f_{0,T}$, and the cash amount in portfolio P_1 is given in the same currency in which the underlying asset and the futures contract are quoted (i.e. USD in the case of S&P500 futures).

Since the value of a futures contract at expiration T is exactly equal to the payoff of that contract, and since cash amounts are automatically always assumed to have been invested at the risk-free interest rate, the values $P_1(T)$ and $P_2(T)$ of the two portfolios at time T are as follows:

$$\boldsymbol{P_1(T)} = S(T) - K + \left(K \cdot e^{-rT}\right) \cdot e^{r \cdot T} = \boldsymbol{S(T)}$$

$$\boldsymbol{P_2(T) = S(T)}$$

We see: Regardless of how the underlying asset S is going to perform, the value of the two portfolios P_1 and P_2 will always be equal, namely, $S(T)$! While we do not yet know, at time 0, what value the two portfolios are going to have (since $S(T)$ is not yet known), we do know for certain that the value of the two portfolios is going to be the same.

Given the NA principle, we conclude that $P_1(0) = P_2(0)$.

Bearing in mind that the price of a futures contract is always 0, it follows that

$$0 + K \cdot e^{-rT} = P_1(0) = P_2(0) = S(0)$$

and this equation $K \cdot e^{-rT} = S(0)$ gives us the **fair strike price K of a futures contract**

$$\boldsymbol{K = S(0) \cdot e^{r \cdot T}}.$$

Again, that relation holds for any point in time t over the futures contract's life, of course. We summarize this in a theorem.

Theorem 3.3 *Let F be a futures contract with expiration T on an underlying S without any payments or cost until that contract's expiration. The fair strike price K of the futures contract at time $t \in [0, T]$ then is*

$$\boldsymbol{K = S(t) \cdot e^{r \cdot (T-t)}}.$$

r being the risk-free interest rate for the time period $[t, T]$.

So, contrary to our speculations at the end of Sect. 2.20, the fair strike price is (at least in theory) completely independent of estimates as to the future development of the underlying asset's price!

Example 3.4 Figures 3.12 and 3.13 show the quotes for futures contracts on the S&P500 and on the DAX for three specific dates. Our result as to the fair strike price of a futures contract can only be applied to the DAX futures, as the DAX (contrary to the S&P500) is already a performance-based index. Based on Theorem 3.3 we now calculate the theoretical fair strike prices for the DAX futures shown in Fig. 3.13 and compare them with the actual bid/ask prices given in Fig. 3.13. We carry this out only for the first two dates, since there are no current quotes for the June futures.

The time periods to expiration for the two futures contracts as of 14 October 2021 were around 2 and 5 months, respectively. The Euribor interest rates that are relevant and available for purposes of our calculations are the 1-month Euribor (−0.568%), the 3-month Euribor (−0.552%), and the 6-month Euribor (−0.523%). From those, we linearly interpolate the interest rates for the time periods that we need, as follows:

$$r_1 = -0.359\% \text{ for computing } K_1 \text{ for 2 months}$$
$$r_2 = -0.303\% \text{ for computing } K_2 \text{ for 5 months}$$

Using the formula $K = S(0) \cdot e^{r \cdot T}$ and the then current value of the DAX of $S(0) = 15{,}324.96$ points, we get the following theoretical fair strike prices:

$$K_1 = 15{,}310.70$$
$$K_2 = 15{,}291$$

We compare with the actual bid-ask midpoint prices, namely,

$$15{,}306 \text{ for } K_1$$
$$15{,}290 \text{ for } K_2$$

There is a certain difference in the case of K_1. So theoretically, in a frictionless market, there ought to be an opportunity for an arbitrage strategy, as follows:

- We go long on the December futures contract with strike 15,306
- We sell one DAX unit at the price of $S(0) = 15{,}324.96$

These transactions would be executable at no cost.

At expiration T in December, the DAX unit is bought back (as agreed in the futures contract) at a price of 15,306. We end up with an arbitrage profit of $15{,}324.96 - 15{,}306 = 18.96$ EUR.

Let us now analyse –based on the result in the above example– to what extent the theoretical arbitrage profit in that example could be achieved under real trading conditions. Two components of the strategy described above in particular need to be discussed in more detail:

- What kind of transaction to consider for selling the DAX
- Margin requirements

The shortening of the DAX could be carried out by buying one DAX-short-index certificate. However, this needs money, at about in the height of the current DAX price of 15,325 EUR. Also for margin we need money, say, at about 10% of the strike, i.e. about 1,530 EUR. So altogether we would need about 17,000 EUR. We could try to take out a loan at low lending rates. (Taking out a loan at negative lending rates is still not possible at the current time (October 2021).) It is easy to check: As long as the lending rates are below 0.67%, the credit costs are below the 18.96 EUR. But we also have to take into account transaction costs and further cost for taking the credit. Further, buying the DAX-short-index via an index certificate is possible in principle, but the price trajectory of such certificates often does not align exactly with that for the short index.

3.5 Valuation of Futures for Underlying Assets with Payouts or Costs

In the previous sections, we determined the put-call parity equation and the fair price of futures for options and of futures on underlying assets that did not carry payments or costs over their life. In the following, we will attempt to obtain similar results for underlying assets that do come with payments or costs. For simplicity, we will refer to such payments or costs collectively as "dividends" in the following. (These "dividends" can be positive or negative.) In order to arrive at a result, we have to assume that, upon entering into the derivative (at time 0), we already know the amount of the underlying asset's payments or costs to be paid over the derivative's life, discounted to time 0, or that we can at least state a very precise estimate of that amount.

The amount of these discounted payments can either be given (or estimated) as an absolute value Z (e.g. coupons of a bond, warehousing costs per unit of a commodity, etc.), or it can be expressed as a continuously compounded percentage return d of the underlying asset's price (e.g. average dividend yield of an index of 2% p.a. relative to the respective price of the index). This information can then be interpreted as follows: From any point in time t to any later point in time u, the price of the underlying asset no longer moves from $S(t)$ to $S(u)$, but from $S(t)$ to $e^{d\cdot(u-t)} \cdot S(u)$. A negative Z or a negative d means that there are costs arising from the underlying asset. And remember that any payments Z must already have been discounted to time 0!

We encourage the reader to attempt to determine the fair strike price of the futures contract in both settings, analogous to our procedure in the previous section, before reading the solution given below.

Derivation of the fair strike price for underlying assets with discounted payouts or costs in the amount *of* Z euros:

We are going to set up two portfolios again:

Let the first **portfolio P1** consist of the following at **time 0**:
1 unit of a long futures contract,
cash in the amount of $K \cdot e^{-r\cdot t} + Z$

Let the second **portfolio P2** consist of the following at **time 0**:
1 unit of the underlying asset

Here r is the risk-free interest rate for the time period $[0, T]$, i.e. $r = f_{0,T}$, and the cash amount in portfolio P_1 is given in the same currency in which the underlying asset and the futures contract are quoted (i.e. USD in the case of S&P500 futures).

Since the value of a futures contract at expiration T is exactly equal to the payoff of that contract and cash amounts are automatically always assumed to have been invested at the risk-free interest rate, and since the payments made over the life of the futures contract have a value of Z at time 0 (i.e. a value of $Z \cdot e^{r\cdot T}$ at time T), the values $P_1(T)$ and $P_2(T)$ of the two portfolios at time T are as follows:

$$P_1(T) = S(T) - K + \left(K \cdot e^{-rT} + Z\right) \cdot e^{r\cdot T} = \mathbf{S(T) + Z \cdot e^{r\cdot T}}$$

$$\mathbf{P_2(T) = S(T) + Z \cdot e^{r\cdot T}}$$

We see: Regardless of how the underlying S is going to perform, the value of the two portfolios P_1 and P_2 will always be equal, $S(T) + Z \cdot e^{rT}$!

While we do not yet know, at time 0, what value the two portfolios are going to have (since $S(T)$ is not yet known), we do know that the value of the two portfolios is going to be the same.

Given the NA principle, we conclude that $P_1(0) = P_2(0)$.

Bearing in mind that the price of a futures contract is always 0, it follows that:

$$0 + K \cdot e^{-rT} + Z = P_1(0) = P_2(0) = S(0)$$

and this equation $K \cdot e^{-rT} + Z = S(0)$ gives us the **fair strike price K of a futures contract with discounted payments/costs in the amount of Z euros as follows**:

$$\boldsymbol{K = (S(0) - Z) \cdot e^{r \cdot T}}.$$

If the amount of the discounted payments is given as an average rate of return d, then our two comparative portfolios are as follows:

Let the first **portfolio P1** consist of the following at **time 0**:
$e^{d \cdot T}$ long futures,
cash in the amount of $K \cdot e^{(d-r) \cdot T}$

Let the second **portfolio P2** consist of the following at **time 0**:
1 unit of the underlying asset
The value of the two portfolios at time T then is

$$\boldsymbol{P_1(T)} = e^{d \cdot T} \cdot (S(T) - K) + K \cdot e^{(d-r) \cdot T} \cdot e^{r \cdot T} = \boldsymbol{e^{d \cdot T} \cdot S(T)}$$

$$\boldsymbol{P_2(T) = e^{d \cdot T} \cdot S(T)}$$

And based on the NA principle, we again conclude that $P_1(0) = P_2(0)$, which means that

$$K \cdot e^{(d-r) \cdot T} = S(0)$$

and thus

$$\boldsymbol{K = e^{(r-d) \cdot T} \cdot S(0)}$$

Again, that relation holds for any point in time t over the futures contract's life, of course. We summarize both results in a theorem.

Theorem 3.5 *Let F be a futures contract with expiration T on an underlying asset S with discounted payments or costs in the amount of Z euros during the period $[t, T]$. The fair strike price K of the futures contract at time $t \in [0, T]$ then is*

$$K = (S(t) - Z) \cdot e^{r \cdot (T-t)}.$$

r being the risk-free interest rate for the time period $[t, T]$.

Theorem 3.6 *Let F be a futures contract with expiration T on an underlying S with an average continuously compounded return d over the life of the contract. The fair strike price K of the futures contract at time $t \in [0, T]$ then is*

$$K = S(t) \cdot e^{(r-d) \cdot (T-t)}.$$

r being the risk-free interest rate for the time period $[t, T]$.

Example 3.7 As an example, we will determine the fair strike prices of the futures listed in Fig. 3.12 on the S&P500 index on 14 October 2021. We bear in mind that the S&P500 is a price index. An investor who holds the S&P500 by having bought the corresponding stocks is therefore faced with the price movements of those stocks, reflected precisely by the S&P500, and additionally receives the dividends paid by the stocks held.

However, unlike the valuation of DAX futures, we now have to work with risk-free US interest rates. We use the relevant available USD Libor rates from October 2021:

2-month LIBOR	0.10%
3-month LIBOR	0.13%
6-month LIBOR	0.16%
12-month LIBOR	0.26%

The time to expiration for the futures under consideration are 2, 5, and 8 months. We interpolate the interest values for 5 and 8 months linearly from the 3-, 6-, and 12-month LIBOR rates and get the values we are going to work with:

$$f_{0,\frac{2}{12}} = 0.10\%$$
$$f_{0,\frac{5}{12}} = 0.15\%$$
$$f_{0,\frac{8}{12}} = 0.19\%$$

An average dividend yield of the S&P500 stocks from the perspective of October 2021 can be assumed to be $d = 1.5\%$ (see also Fig. 3.5).

Using the formula $K = S(0) \cdot e^{(r-d) \cdot T}$ we then get (with the S&P500 standing at 4387 points on 14 October 2021):

$$K_1 = 4387 \cdot e^{(0.001-0.015) \cdot \left(\frac{2}{12}\right)} = 4377$$
$$K_2 = 4387 \cdot e^{(0.0015-0.015) \cdot \left(\frac{5}{12}\right)} = 4362$$
$$K_3 = 4,387 \cdot e^{(0.0019-0.015) \cdot \left(\frac{8}{12}\right)} = 4349$$

The actual futures quotes on 14 October 2021 were as follows:

Quote for March futures: 4378.50 // 4378.75
Quote for June futures: 4370.50 // 4371.50
Quote for September futures: 4359.75 // 4362.50

In conclusion, a **comment** on the valuation (i.e. fair strike price determination) of forwards:

Due to the fact that futures are settled on a daily basis and profits/losses are credited/debited to the investor's account on a daily basis, whereas profits/losses in the case of forwards are calculated and settled at the end of the agreement only, there is (generally) a slight difference in their valuation. However, we will not discuss this detail any further at this point.

3.6 The Put-call Parity Equation for Underlying Assets with Payouts or Costs

In the following, we want to find the version of the put-call parity equation for an underlying asset with a continuously compounded return d. (We will also present the version for the case of discounted payments in the amount of Z, but will leave the proof to the reader).

To that end, we will follow the proof of the put-call parity in Sect. 3.2 and simply make the necessary adjustments:

Let the first **portfolio $\mathbf{F_1}$** consist of the following at **time 0**:
1 unit of the above call option long,
cash in the amount of $K \cdot e^{-r \cdot T}$

Let the second **portfolio $\mathbf{F_2}$** consist of the following at **time 0**:
1 unit of the above put option long,
$e^{-d \cdot T}$ units of the underlying asset

It follows:

$$\begin{aligned}\boldsymbol{F_1(T)} &= \max(S(T)-K,0)+\left(K\cdot e^{-rT}\right)\cdot e^{r\cdot T} = \\ &= \max(S(T)-K,0)+K = \\ &= \mathbf{max}\,(\boldsymbol{S(T)},\boldsymbol{K}) \\ \boldsymbol{F_2(T)} &= \max(K-S(T),0)+\left(\mathbf{e}^{-\mathbf{d}\cdot\mathbf{T}}\cdot S(T)\right)\cdot \mathbf{e}^{\mathbf{d}\cdot\mathbf{T}} = \\ &= \mathbf{max}\,(\boldsymbol{K},\boldsymbol{S(T)})\end{aligned}$$

Given the NA principle, we conclude that $F_1(0) = F_2(0)$, thus

$$C(0) + K \cdot e^{-r\cdot T} = P(0) + \mathbf{e}^{-\mathbf{d}\cdot\mathbf{T}} \cdot S(0).$$

Again, the above arguments hold for any point in time $t \in [0, T]$, of course, and not only for time 0. To summarize:

Theorem 3.8 (Put-call Parity Equation: Underlying Asset with Yield) *Let* $[0, T]$ *be a time interval, and for* $t \in [0, T]$, *let* $C(t)$ *be the price of a call option on the underlying asset* S *with yield* d, *expiration date* T, *and strike price* K, *and let* $P(t)$ *be the price of a put option with the underlying asset* S, *expiration date* T, *and strike price* K. *The following then holds*

$$C(t) + K \cdot e^{-r(T-t)} = P(t) + e^{-d\cdot(T-t)} \cdot S(t)$$

Here $S(t)$ *denotes the price of the underlying asset at time* t *and* r *is the risk-free interest rate* $r = f_{t,T}$.

The result in the case of an underlying asset with fixed discounted payments Z can be obtained in a very similar way.

Theorem 3.9 (Put-call Parity Equation, Underlying Asset with Payments/Costs) *Let* $[0, T]$ *be a time interval, and for* $t \in [0, T]$, *let* $C(t)$ *be the price of a call option with the underlying asset* S *incurring discounted payments in the amount of* Z *during that time interval, with expiration date* T *and strike price* K, *and let* $P(t)$ *be the price of a put option with the underlying asset* S, *expiration date* T, *and strike price* K. *The following then holds*

$$C(t) + K \cdot e^{-r\cdot(T-t)} = P(t) + S(t) - Z$$

Where $S(t)$ *denotes the price of the underlying asset at time* t *and* r *is the risk-free interest rate* $r = f_{t,T}$.

In Sect. 3.2, we examined the two options on the S&P500 with regard to put-call parity. However, at that point, we did not take into account that the S&P500 is a price index and dividends would therefore have to be taken into account.

We will now examine the two options the correct way.

On 13 October 2021, we had a call and a put with strike 4325 and expiration date 17 December 2021. The mid-prices for the call option, quoted at 140.80 // 141.50, and for the put option, quoted at 115.70 // 116.40, are 141.15 ($= C(0)$) and 116.05 ($= P(0)$), respectively. From 13 October 2021 until expiration of the options on 17 December 2021, we are looking at a time-to-expiration period of practically 2 months exactly, that is, $\frac{1}{6}$ of a year. The risk-free interest rate for 2-month USD investments on 13 October 2021 was approximately $+0.05\%$. The price of the S&P500 was 4358.50.

As the **dividend yield** for the S&P500, we will again assume (as we did in our valuation of the futures contract in the previous section) the value $d = 1.50\%$.

The correct form of the put-call parity equation in our example is thus

$$C(0) + K \cdot e^{-rT} = P(0) + e^{-d \cdot T} \cdot S(0) \Leftrightarrow 141.15 + 4.325 \cdot e^{-0.0005 \cdot \frac{1}{6}} =$$

$$= 116.05 + e^{-0.015 \cdot \frac{1}{6}} \cdot 4,358.50 \Leftrightarrow 4,465.79 = 4,463.67$$

In the earlier (dividend-free) version, we had 4,474.55 on the right, so the difference is now somewhat smaller.

In Sect. 3.3, we drew some conclusions from the put-call parity. These conclusions also need to be adjusted for underlying assets carrying payouts/yields, of course. Let us take a closer look at one of these conclusions.

For underlying assets with yield d, it follows directly from the put-call parity that

$$C(0) > e^{-d \cdot T} \cdot S(0) - K \cdot e^{-rT}$$

and

$$P(0) > K \cdot e^{-rT} - e^{-d \cdot T} \cdot S(0)$$

and thus:

If the interest rate $r > 0$

Then $C(0) > e^{-d \cdot T} \cdot S(0) - K$. And because $C(0)$ is always greater than 0, it follows that $\boldsymbol{C(0) > \max\left(0, e^{-d \cdot T} \cdot S(0) - K\right)}$ always holds. If $\boldsymbol{d}$ is **negative**, i.e. if the underlying asset causes underlying costs, we can conclude that:
$\boldsymbol{C(0) > \max(0, S(0) - K)}$.

If the interest rate $r < 0$

Then $P(0) > K - e^{-d \cdot T} \cdot S(0)$.

And because $P(0)$ is always greater than 0, it follows that $\boldsymbol{P(0) > \max(0, K - e^{-d \cdot T} \cdot S(0))}$ always holds. If $\boldsymbol{d}$ is **positive**, i.e. if the underlying asset generates a positive yield, we can conclude that:
$\boldsymbol{P(0) > \max(0, K - S(0))}$.

Let us now verify these results using two options on the S&P500, which we already looked at briefly in Sect. 3.3 (see also quotation Figs. 3.7 and 3.9):
Price of S&P500 on 13 October 2021: $S = 4342$ points

Call option on the S&P500, expiration date 17 December 2021, strike $K = 1600$
Quotes on 13 October 2021: 2724.30 // 2733.20 (mid-price 2728.75)
rough approximation (call with low strike): $S - K = 2742$

Put option on the S&P500, expiration date 17 December 2021, strike $K = 5900$
Quotes on 13 October 2021: 1562.90 // 1571.80 (mid-price 1567.35)
rough approximation (put with high strike): $K - S = 1558$

Assumed dividend yield $d = 1.5\%$
2-month USD LIBOR $r = 0.05\%$

A more accurate approximation for the call option is obtained through the inequality:

$$C(0) > e^{-d \cdot T} \cdot S(0) - K \cdot e^{-r \cdot T} = e^{-0.015 \cdot \frac{1}{6}} \cdot 4342 - 1600 \cdot e^{-0.005 \cdot \frac{1}{6}} = 2732.49$$

A more accurate approximation for the put option is obtained through the inequality:

$$P(0) > K \cdot e^{-r \cdot T} - e^{-d \cdot T} \cdot S(0) = 5900 \cdot e^{-0.005 \cdot \frac{1}{6}} - e^{-0.015 \cdot \frac{1}{6}} \cdot 4,342 = 1563.93$$

Both approximations are relatively close to the mid-prices of these two options.

3.7 Basics of Derivative Valuation and Pricing Models

One of the most important tasks of modern financial mathematics consists of valuing derivatives. Derivatives occur in a variety of forms, and in a wide variety of contexts on the financial markets, and the question as to the "fair" price of what are frequently very complex products is therefore often a fundamental one. In addition, derivatives can often occur in a hidden, not immediately recognized form. For example, everyday arrangements such as a termination right or an interest rate cap on a loan are both derivatives.

Volume III Chapter 3 of this book discusses a number of real case studies based on the author's work as a court-appointed expert and on project work, and most of them come down to valuations of derivatives.

To recall:

The question as to "the fair price" of a derivative, or to be more precise (as we will see), of "**a** fair price" of a derivative, is not about finding a reasonable estimate of a financial product's price, but all about finding a clear answer to the following question: *Which derivative prices allow arbitrage opportunities –at least under ideal conditions– and which prices don't?*

Prices that allow arbitrage opportunities are what we call price inconsistencies. The only prices that we will refer to as fair prices of a derivative are those where no arbitrage opportunities can occur.

So far, we have looked at two situations where we applied the no-arbitrage principle to determine fair prices, or fair price relationships: when we derived the put-call parity equation and when we determined the fair strike price of futures contracts. In this two cases, we were able to establish a well-defined fair relationship between the prices of call and put options with the same strike price and the same expiration date and a well-defined fair strike price of a futures contract.

In both cases, the results were completely independent of any assumptions as to the modelling of the underlying asset. The only requirement was that we know the amount of future payments or costs incurred from the underlying.

That situation, however, is the exception!

In order to determine fair prices of a derivative, it is usually necessary to make assumptions about a model for the underlying asset's development and, in addition, to have reliable estimates of parameters occurring in that model.

In most cases, the result of our analyses is not going to be "the fair price of a derivative", but rather "the fair price of the derivative on the assumption that the underlying asset develops according to a particular model".

There may be cases therefore, where there is a fair price or a whole range of fair prices for a derivative in one model and where the very same derivative has another fair price or a whole range of other fair prices in another model.

The "assumption of a model for the underlying asset's development" has nothing to do with a forecast of a financial product's future prices. All the models that we are dealing with are stochastic models: in other words, models with probability parameters.

We want to illustrate the process of choosing a stochastic model for price developments using a well-known example:

You see a sequence of integers between 1 and 6. For example,

5, 2, 1, 6, 6, 3, 5, 3, 6, 2, 4, 1, 5, 2, 2, 1, 4, 1, 6, 2, 4, 5, 3, 5, 4, 3, 1, 4, 1, 3, 6, 5, 4, 4, 5, 1, 6, 1, 1, 2, 4, 3, 2, 6, 5, 1, 5, 3, 4, 3, 1, 4, 2, 6, 6, 5, 6, 2, 6, 1, 4, 2, 3, 1, 5, 2, 1, 6, 4, 3, 3, ...

You note that each of the six numbers occurs approximately equally frequently and that the order of these numbers appears to be random, at least at first glance. One **possible model** that this number sequence might follow could be a roll with a fair dice, that is, a dice where each number is likely to occur at a probability of $\frac{1}{6}$ and where the outcome of a roll is independent of the outcome of the preceding rolls. This is an example of a **stochastic model**. However, choosing a model does not give us any control whatsoever over a forecast regarding the outcome of the next roll of the dice; we can only calculate a probability for the next score.

A closer analysis of the number sequence reveals that a particular pair of two consecutive numbers –1, 6– occurs more frequently than other pairs:

5, 2, **1, 6**, 6, 3, 5, 3, 6, 2, 4, 1, 5, 2, 2, 1, 4, **1, 6**, 2, 4, 5, 3, 5, 4, 3, 1, 4, 1, 3, 6, 5, 4, 4, 5, **1, 6**, 1, 1, 2, 4, 3, 2, 6, 5, 1, 5, 3, 4, 3, 1, 4, 2, 6, 6, 5, 6, 2, 6, 1, 4, 2, 3, 1, 5, 2, **1, 6**, 4, 3, 3, ...

whereas the 4, 6 pair does not occur at all.

Another possible stochastic model for the sequence of numbers, which is perhaps more representative of reality, might be as follows:

"Each number occurs with the same probability $\frac{1}{6}$. However, there is one (and only one!) relationship between consecutive numbers, which is:

The probability that a number 1 is followed by a number 6 is $\frac{1}{3}$ (instead of $\frac{1}{6}$), and all other numbers have the same probability to occur ($\frac{2}{15}$). And:

The probability that a number 4 is followed by a number 6 is 0 (instead of $\frac{1}{6}$), and all other numbers have the same probability to occur ($\frac{1}{5}$)".

Of course, what we have done here is a far too superficial analysis of a far too short sample of our random sequence of integers to arrive at a reasonably consistent and reliable model. The analysis and discussion and "generation" of randomness is in fact a very complex and fascinating topic, and we will get back to it from time to time in the course of this book (especially in the sections on Monte Carlo and quasi-Monte Carlo methods). By the way, a classic work on the analysis of (pseudo-) random sequences, which is also easy to read for non-specialists, is Chapter III in the second volume of *The Art of Computer Programming* by Donald Knuth [1].

For our purposes of financial mathematics, it will also be essential that we choose a price model for the underlying that is designed such that the model itself doesn't allow any arbitrage. We are going to illustrate what we mean by this in the next section, using the so-called one-step binomial model.

In the next sections, we will go through a process of several steps, starting from the simplest possible model (the above-mentioned one-step binomial model), to finally arrive at the so-called "Wiener model", which, since it reflects reality relatively well, is the most commonly used stock pricing model, and we will use the respective models to calculate valuations of arbitrary derivatives.

Our first major objective will be to derive the fair price formulas for call and put options in a Wiener model, i.e. the classic Black-Scholes formulas.

As our earlier considerations on the put-call parity equation and on the fair strike price of futures suggest, the fair price of a derivative will again depend on whether or not the underlying asset makes payments (or causes costs) over the life of the derivative. This is how we are going to proceed in the following.

For the time being, we are going to derive all our results for dividend-free underlying assets only. In a final wrap-up section, we will then discuss the case of underlying assets with payments/costs, for all models together.

3.8 The One-step Binomial Model and Derivative Valuation in the One-step Binomial Model, Part I

We are going to start this section with a very simple game, which we will call **Game A**

Game A is organized by a game master, Mr. W.

Ms. S. is Mr. W.'s opponent.

Mr. W. offers Ms. S. a **lottery ticket D** *for her to buy.*

Ms. S. buys the ticket at a certain price.

Mr. W. then takes a fair (!) dice and rolls it.

If the score is a 6 (which has a probability of happening of $\frac{1}{6}$), Ms. S. wins 100 EUR. If Mr. W. rolls one of the lower five numbers (the likelihood of which is $\frac{5}{6}$), the ticket expires worthless. We schematically illustrate the game in Fig. 3.14:

The game is actually extremely simple.

We therefore ask ourselves just one ***question***: "What is a reasonable price for that ticket?" *What price will Ms. S. be willing to pay? What price will Mr. W. charge for the ticket?*

The obvious approach would be the following:

We know that a reasonable price for the ticket is one that is equal to the average expected payoff E that the holder of the ticket can win (in other words, the expected value of the payoff).

That expected payoff value is calculated as follows:

$$\mathbf{E} = \frac{1}{6} \cdot 100 + \frac{5}{6} \cdot 0 = \mathbf{16.66}\ \text{EUR}$$

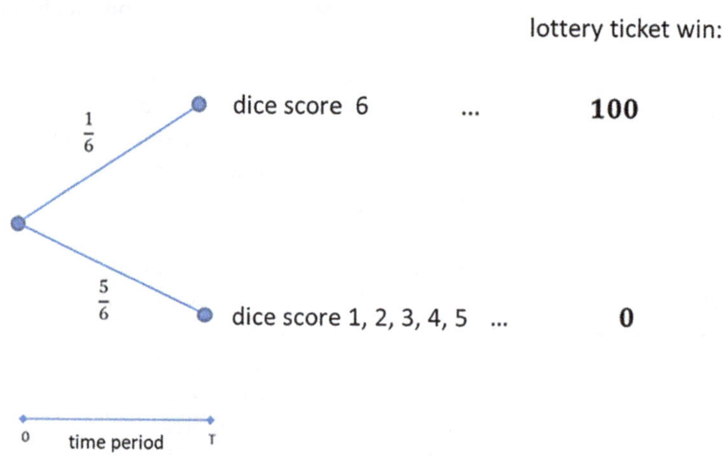

Fig. 3.14 Game of dice

(On average, you will receive a payoff of 100 EUR in one sixth of all cases, while in five sixths of the cases on average you will get nothing.)

The reasonable price of the ticket is therefore 16.66 EUR.

However, this reasonable price is not a fair price in our strict financial-mathematical sense. If, for example, Ms. S. gets the ticket at the price of 10 EUR, that would certainly be a very attractive deal for her, but she would not have any guarantee of making a safe and risk-free profit, meaning there would be no arbitrage opportunity!

And one more thing to note: It is not entirely correct to put the reasonable price at 16.66 EUR: While Ms. S. has to pay the price of the ticket now (at time 0), she will not receive any potential payoff (100 EUR) until a later point in time (let us call it T), after Mr. W. has thrown the dice. If a risk-free (continuously compounded) interest rate r applies for the time period $[0, T]$, then the 100 EUR amount that she might potentially win presently only has the discounted value $100 \cdot e^{-rT}$.

The correct reasonable price of the ticket is therefore

$$\textbf{reasonable price} = \frac{1}{6} \cdot 100 \cdot e^{-r \cdot T} + \frac{5}{6} \cdot 0 = \mathbf{16.66 \cdot e^{-rT}} \text{ EUR.}$$

The difference between E and the reasonable price in reality only matters if the time period $[0, T]$ is not too short, of course.

As you will see below, the **one-step binomial model** (see Fig. 3.15) which will be introduced in the following has nothing to do with the evolution of real prices of

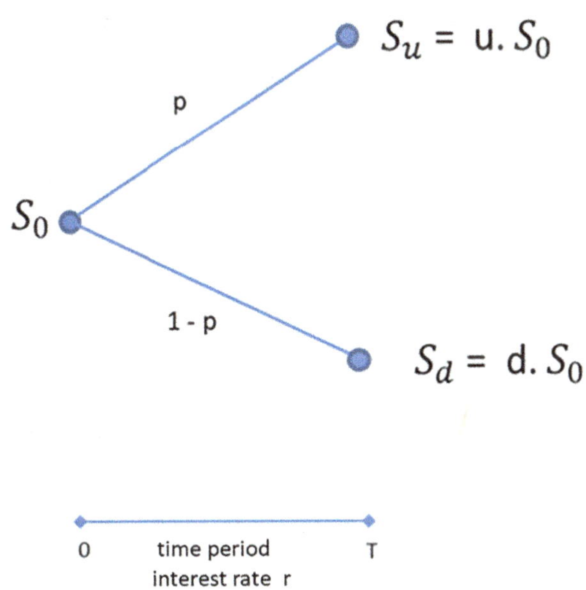

Fig. 3.15 One-step binomial model

financial products. Nevertheless, this model is very important and we will deal with it very closely for two reasons:

For one, this example will provide a very accurate illustration of the basic ideas when it comes to the valuation of derivatives. Secondly, the results that we are going to achieve in this chapter represent the first essential step on our way to the classic Black-Scholes formula.

The model could come from a simple financial strategy board game (such as a kind of stock-market monopoly game). For the sake of simplicity, going forward, we will always be referring to the development of a **stock price** that we want to model (even though it could just as well be any other financial product, an index, a currency, a commodity, etc.):

- Our time range consists of only one single time-step from time 0 (now) to time T. It is only at time 0 and at time T that can we trade the stock (the underlying).
- The current stock price is S_0 EUR.
- Until time T, the stock price can only take two possible values: $S_u = u.S_0$ (the probability of that to happen is p) or $S_d = d.S_0$ (the probability of which is $1 - p$).

  In this context $S_0 > 0, d < u$ and $0 < p < 1$ are arbitrary real numbers.
- In the time range $[0, T]$, interest accrues continuously at the rate r.

An example of this is shown in Fig. 3.16.

Next we will look at a derivative on this underlying in the one-step binomial model. The derivative is well defined by the underlying asset to which it relates (our stock), by its expiration date T, and by the payoff to be paid by the holder of the short position to the holder of the long position at time T based on the underlying asset's development. We illustrate the underlying/derivative model in Fig. 3.17.

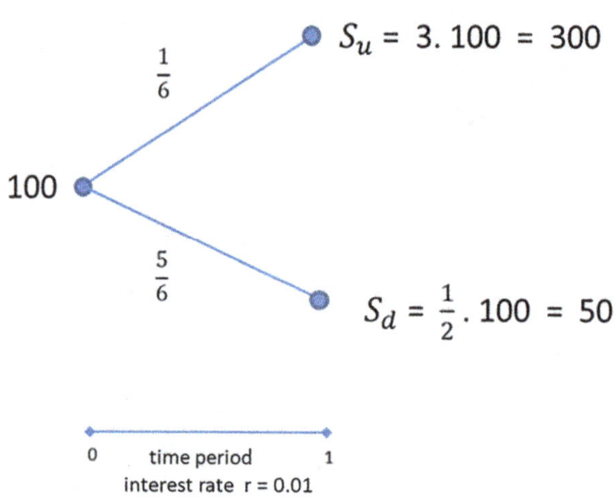

Fig. 3.16 One-step binomial model, numerical example

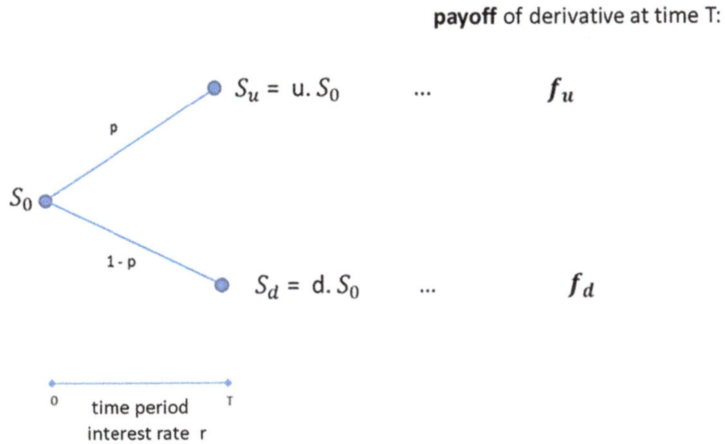

Fig. 3.17 Derivative on an underlying in the one-step binomial model

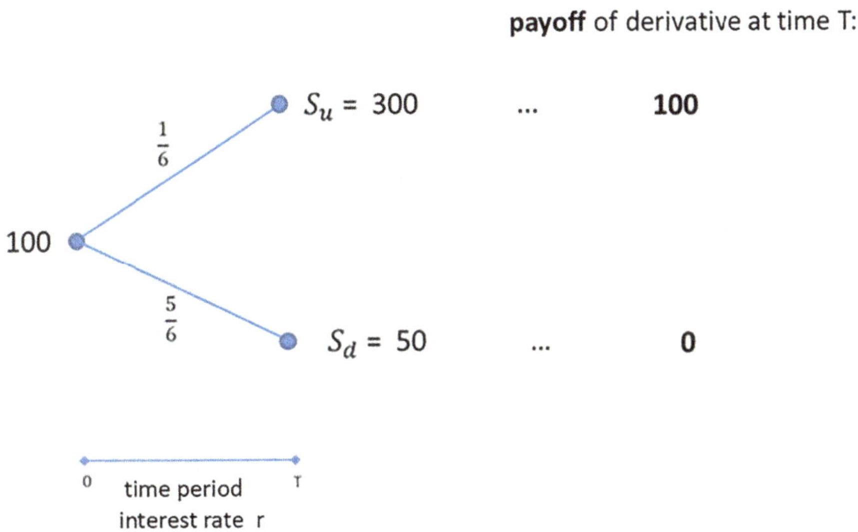

Fig. 3.18 Derivative on an underlying in the one-step binomial model, numerical example

In the above illustration, we denoted the **payoffs of the derivative** by f_u and f_d, respectively.

f_u is the payoff that the holder of the short position has to pay to the holder of the long position if the underlying moves up to the value S_u.

f_d is the payoff that the holder of the short position has to pay to the holder of the long position if the underlying moves down to the value S_d. We illustrate a numerical example in Fig. 3.18, where we chose $f_u = 100$ and $f_d = 0$.

Thus, if the stock takes the value $S_u(= 300)$, the holder of the long position in the derivative will receive EUR 100 from the holder of the short position in the derivative. If the stock takes the value $S_d(= 50)$, the derivative expires worthless.

On closer scrutiny, we see that this numerical example is actually a call option on the underlying with strike $K = 200$.

Our **question** of course is: **What is the fair price of this derivative at time 0 ?**

The answer to this question would seem to be **very simple**.

After all, we have exactly the same situation here as in the dice game we described in the preamble to this section. In both cases, when purchasing the ticket in the dice game or purchasing the derivative in our "stock-market game", we have an equal likelihood of $\frac{1}{6}$ to win an amount of 100 EUR and an equal likelihood of $\frac{5}{6}$ that the ticket or the derivative will expire worthless.

Consequently, we think that we can draw the same conclusions as to the price of the derivative in the "stock-market game" as we did for the price of the ticket in the dice game: So, for the time being, we think that the derivative does not have a "fair price" in the strict sense of financial mathematics ("the no-arbitrage theory"), but only "a reasonable price" in the informal sense, that is, a price that equals the discounted expected payoff, like in the dice game.

This would mean:

The reasonable price for the derivative shown in Fig. 3.17 –denoted by f_0– is

$$f_0 = e^{-r \cdot T} \cdot (p \cdot f_u + (1-p) \cdot f_d) \tag{3.1}$$

And for our numerical example shown in Fig. 3.18 (again, we neglect the interest rate r, assuming that it is very close to zero anyway and that the time period T is very short), this would mean the reasonable price of the derivative in our example is equal to the reasonable price of the ticket in the dice game, i.e.

$$f_0 = \frac{1}{6} \cdot 100 + \frac{5}{6} \cdot 0 = 16.66 \text{ EUR}$$

Suppose the participants in the financial market would follow this line of reasoning; in that case, the quotes on an exchange where this derivative is traded would usually be around those 16.66 EUR, so maybe bid/ask quotes of 16.50 // 17.00 would indeed be common. Let's be pessimistic and assume an even bigger spread, for example, 16.00 // 18.00.

Would you be willing to trade this derivative at one of these quotes, i.e. sell it at 16 EUR or buy it at 18 EUR?

Most of you will presumably say: No way!

And so you may be surprised when I say that I would definitely want to buy the derivative at a price of 18 EUR. I would even try to get as many units of that derivative as possible at the price of 18 EUR.

So, I would buy, for example, five units of the derivative at the price of 18 EUR, for a total of 90 EUR.

At the same time, however, I would do the following: I would go short on two units of the underlying. The price of the underlying is currently 100 EUR. On shorting it, I receive 200 EUR. As a result, I do not need any funds on my own for purchasing the derivative; instead I actually hold an amount of 110 EUR after these transactions.

To summarize:

My portfolio consists of five long positions in the derivative, two short positions in the underlying, and 110 EUR in cash.

Now we wait and see what happens until time T.

Two different situations can occur:

Situation 1: The stock price increases to 300 EUR.

Through the long positions in the derivative, we receive $5 \times 100 = 500$ EUR. At the same time, we buy two units of the underlying at $2 \times 300 = 600$ EUR, thus closing our short position in the underlying. We take a profit of $110+500-600 = 10$ EUR.

Situation 2: The stock price decreases to 50 EUR.

Through the long positions in the derivative we receive $5 \times 0 = 0$ EUR. At the same time, we buy 2 units of the underlying at $2 \times 50 = 100$ EUR, thus closing our short position in the underlying. We take a profit of $110 + 0 - 100 = 10$ EUR.

This means although it would seem that the derivative we bought was overpriced (we paid 18 EUR instead of the reasonable price of 16.66 EUR), we made a safe risk-free profit –i.e. an **arbitrage profit**– of 10 EUR without investing any of our own money. The price of 18 EUR (and all the more so any price below 18 EUR and especially 16.66 EUR) cannot be a fair price for the derivative!

What happened here?

3.9 The One-step Binomial Model and Derivative Valuation in the One-step Binomial Model, Part II

The **beauty of financial mathematics** is that as has already become obvious through this very simple example of a derivative valuation, the situation in financial mathematics is so much more complex and thus more beautiful and challenging than a banal profit expectation calculation. The financial markets –even their simplest models– with their intricate interplay between basic underlying products and derivative products, form complex dynamic systems, in which many different players continuously react to one another using a wide variety of interdependent products. And that is precisely why the most sophisticated mathematical techniques play a prominent role in modern financial mathematics.

So, what happened in our example in the previous section?

Due to the possibility of combining the derivative with positions in the underlying asset, the reasonable price of the derivative appears to be significantly above the mere average payoff that can be achieved through the derivative alone.

And we also observed something else: It is possible to play the derivative/underlying game more or less skilfully. Our strategy of making a safe profit through a suitable combination of derivative and underlying without the use of money is certainly preferable to the naíve strategy where you use money to buy just the derivative and be stuck waiting for a potential profit or loss.

While we have only briefly touched upon the topic of the "degree of skill and chance in a game" here, we will discuss it in more detail in a section further below.

Before we do that, however, we want to ask once again the question as to the fair price of the derivative in our specific case or, more generally, of any derivative in the one-step binomial model: Does such a fair price even exist and what would it have to look like if it existed?

Let us answer the second question first: What would the fair price look like if it existed? We'll go straight to the general case.

For that purpose, let us again look at two portfolios P_1 and P_2:

Portfolio P_1 consists of:
1 unit of the derivative

Portfolio P_2 consists of:
x units of the underlying asset,
y EUR

We are now going to proceed as follows: We determine the two unknowns x and y such that the two portfolio values $P_1(T)$ and $P_2(T)$ are the same upon expiration in any of the two possible situations.

Situation 1: The stock price takes the value $S_u = u.S_0$.
It follows that $P_1(T) = f_u$ and $P_2(T) = x.u.S_0 + e^{r \cdot T} \cdot y$
Situation 2: The stock price takes the value $S_d = d.S_0$.
It follows that $P_1(T) = f_d$ and $P_2(T) = x.d.S_0 + e^{r \cdot T} \cdot y$
And so we have to satisfy the following two equations:

$$
\begin{aligned}
f_u &= x.u.S_0 + e^{r \cdot T} \cdot y \\
f_d &= x.d.S_0 + e^{r \cdot T} \cdot y
\end{aligned} \tag{3.2}
$$

We compute the two unknowns x and y from these two equations (which we can determine unambiguously when $u.S_0 \neq d.S_0$, which is precisely the case, given that we required that $S_0 > 0$ and $d < u$).

A simple calculation yields the two solutions:

$$
\begin{aligned}
x &= \frac{f_u - f_d}{S_0.(u-d)} \\
y &= e^{-rT} \cdot \frac{u.f_d - d.f_u}{u-d}
\end{aligned} \tag{3.3}
$$

So, if we create the portfolio P_2 with this x and this y, then the equation $P_1(T) = P_2(T)$ holds in any case, and it follows from the no-arbitrage principle that $P_1(0) = P_2(0)$.

We get

$$f_0 = P_1(0) = P_2(0) = x.S_0 + y = \frac{f_u - f_d}{S_0.(u-d)} \cdot S_0 + e^{-rT} \cdot \frac{u.f_d - d.f_u}{u-d}$$

We rearrange the right side of this equation somewhat (by isolating $e^{-r \cdot T}$ and arranging for f_u and f_d), which gives us –as can be easily verified– the following form:

$$f_0 = e^{-rT} \cdot \left(\frac{e^{r \cdot T} - d}{u-d} \cdot f_u + \left(1 - \frac{e^{r \cdot T} - d}{u-d}\right) \cdot f_d \right) \tag{3.4}$$

This formula is extremely important and will accompany us very intensely for a while. We will therefore restate this formula in the following very intuitive form:

$$f_0 = e^{-rT} \cdot (p' \cdot f_u + (1-p') \cdot f_d)$$
$$\text{where } p' = \frac{e^{r \cdot T} - d}{u-d} \tag{3.5}$$

And this formula, in this form, is something you should memorize, because it contains some very surprising insights (but also questions that we will address in due course)! The formula tells us **if a fair price for the derivative exists, then it is well-defined and it is calculated as specified in Formula** (3.5).

Before we discuss these insights and questions in more detail, let us just quickly apply the formula to our above numerical example:

Example 3.10 The parameters we need to calculate the value f_0 for our numerical example in Fig. 3.18 are:

$S_0 = 100$
$u = 3$
$d = \frac{1}{2}$
$f_u = 100$
$f_d = 0$
$r = 0$ (as said before, we will ignore the interest rate for purposes of this example)

And I think that if it hasn't been clear so far, it now becomes eminently clear that there is one parameter –an extremely essential one after all– which we do

not **need for calculating f_0, namely, the probability p for the stock price to go up in value!**

We get (bearing in mind that $e^0 = 1$) $p' = \frac{1-\frac{1}{2}}{3-\frac{1}{2}} = \frac{1}{5}$ and thus $f_0 = \frac{1}{5} \cdot 100 + \frac{4}{5} \cdot 0 = 20$.

We can conclude therefore: If there is a fair price for the derivative in our example, then that price is well-defined and has the value 20.

However, our result would not be complete until we prove that the f_0 calculated as per Formula (3.5) is indeed a price for the derivative where no arbitrage is possible (so far we have only shown: "If a fair price exists, then it must be as stated in Formula (3.5) ...").

We can do this in the following way: We show that if the derivative has the price f_0, then there are no arbitrage opportunities. Let us briefly recall what an arbitrage opportunity is. An arbitrage opportunity is a situation where you can trade the existing products –**derivative, underlying,** and **cash**– at no cost at time 0 and pocket a safe positive payoff at time T. If we denote the number of derivative units in this trade by a, the number of units of the underlying asset by b, and the cash amount in this arbitrage trade by c, we get

$$
\begin{aligned}
a \cdot f_0 + b \cdot S_0 + c = 0 \quad &\text{and} \\
a \cdot f_u + b \cdot u \cdot S_0 + c \cdot e^{r \cdot T} > 0 \quad &\text{and} \\
a \cdot f_d + b \cdot d \cdot S_0 + c \cdot e^{r \cdot T} > 0 &
\end{aligned} \tag{3.6}
$$

However, a simple calculation, which we are going to perform for the interested reader in the box below, reveals that such values a, b, and c cannot exist. So the value f_0 given in Formula (3.5) is indeed the well-defined fair price of the derivative. And in our numerical example, the derivative's well-defined fair price is indeed 20 EUR (and not the initially assumed 16.66 EUR).

We summarize this in a theorem.

Theorem 3.11 *For an underlying asset S in a one-step binomial model in the time range $[0, T]$ (without payments or costs) with the parameters S_0, u, d, p, and r (with $S_0 > 0$ and $d < u$), any derivative with payoffs f_u and f_d has a well-defined fair value f_0, which is given by*

$$f_0 = e^{-rT} \cdot \left(p' \cdot f_u + \left(1 - p'\right) \cdot f_d\right)$$

where

$$p' = \frac{e^{r \cdot T} - d}{u - d}.$$

It remains to be proven here that the equation/inequality system in Formula 3.6 cannot have a solution a, b, c. Conversely, suppose that the system

$$a \cdot f_0 + b \cdot S_0 + c = 0$$
$$a \cdot f_u + b \cdot u \cdot S_0 + c \cdot e^{r \cdot T} > 0$$
$$a \cdot f_d + b \cdot d \cdot S_0 + c \cdot e^{r \cdot T} > 0$$

does indeed have a solution a, b, c. Then $a \neq 0$ has to hold, as else the binomial model without a derivative is already not arbitrage-free.

Let us first consider the case $a > 0$ (the case $a < 0$ is handled analogously) and let us assume that $d > 0$ (the case $d < 0$ is also handled analogously with obvious adjustments and is left to the reader as an exercise).

Without any loss of generality, we can assume that $a = 1$ (otherwise we divide the three equations/inequalities by a and rename the parameters $\frac{b}{a}$ and $\frac{c}{a}$, respectively).

The system would then look like this:

$$f_0 + b \cdot S_0 + c = 0$$
$$f_u + b \cdot u \cdot S_0 + c \cdot e^{r \cdot T} > 0$$
$$f_d + b \cdot d \cdot S_0 + c \cdot e^{r \cdot T} > 0$$

If the two equations

$$f_u + b' \cdot u \cdot S_0 + c' \cdot e^{r \cdot T} = 0$$
$$f_d + b' \cdot d \cdot S_0 + c' \cdot e^{r \cdot T} = 0$$

are satisfied, then the b' corresponds to the $-x$ and the c' to the $-y$ in the above derivation of f_0. And from that automatically follows the equation

$$f_0 + b' \cdot S_0 + c' = 0$$

(As f_0 is exactly $x \cdot S_0 + y$.)

If b'' and c'' take other values, so that

$$f_u + b'' \cdot u \cdot S_0 + c'' \cdot e^{r \cdot T} > 0$$

$$f_d + b'' \cdot d \cdot S_0 + c'' \cdot e^{r \cdot T} > 0$$

holds
it follows that

$$f_d + b'' \cdot d \cdot S_0 + c'' \cdot e^{r \cdot T} > f_d + b' \cdot d \cdot S_0 + c' \cdot e^{r \cdot T}$$

(continued)

and therefore

$$\left(b'' - b'\right) \cdot S_0 > \left(c' - c''\right) \cdot \frac{e^{r \cdot T}}{d}.$$

Furthermore, we have

$$f_u + b'' \cdot u \cdot S_0 + c'' \cdot e^{r \cdot T} > f_u + b' \cdot u \cdot S_0 + c' \cdot e^{r \cdot T}$$

and therefore

$$(b'' - b') \cdot S_0 > (c' - c'') \cdot \frac{e^{r \cdot T}}{u}.$$

As we will see in the next chapter, $e^{r \cdot T}$ will always have to be greater than or equal to d and less than or equal to u; otherwise the binomial model would not be arbitrage-free. And from this we conclude that

$$\left(b'' - b'\right) \cdot S_0 > \left(c' - c''\right) \text{ thus } b'' \cdot S_0 > b' \cdot S_0 + c' - c''$$

and therefore

$$f_0 + b'' \cdot S_0 + c'' > f_0 + b' \cdot S_0 + c' - c'' + c'' = f_0 + b' \cdot S_0 + c' = 0.$$

So, to summarize:
If

$f_u + b'' \cdot u \cdot S_0 + c'' \cdot e^{r \cdot T} > 0$ holds and
$f_d + b'' \cdot d \cdot S_0 + c'' \cdot e^{r \cdot T} > 0$ holds,

then

$f_0 + b'' \cdot S_0 + c'' > 0$

must also hold, and that is a contradiction to the initial assumption and thus no arbitrage is possible.

3.10 Derivative Valuation in the One-step Binomial Model, Discussion of Outcomes

In the previous chapter we learned the following.

For any derivative D on an underlying asset that evolves according to the one-step binomial model (see Fig. 3.19), there is a well-defined fair price f_0, which is calculated as follows:

$$f_0 = e^{-rT} \cdot (p' \cdot f_u + (1 - p') \cdot f_d)$$

$$\text{where } p' = \frac{e^{r \cdot T} - d}{u - d}.$$

We will discuss this fact in various aspects below.

Independence from Parameter p

What seems particularly surprising is that the parameter p, i.e. the probability with which the underlying moves up or down, is not included in the formula for calculating f_0. The fair price of the derivative is independent of the likelihood with which the underlying asset is going to move up or down.

Especially in the case of a call option, this may seem surprising: Thinking in purely heuristic terms, one would immediately assume that a strong upward tendency in the price of an underlying would greatly increase the likelihood of

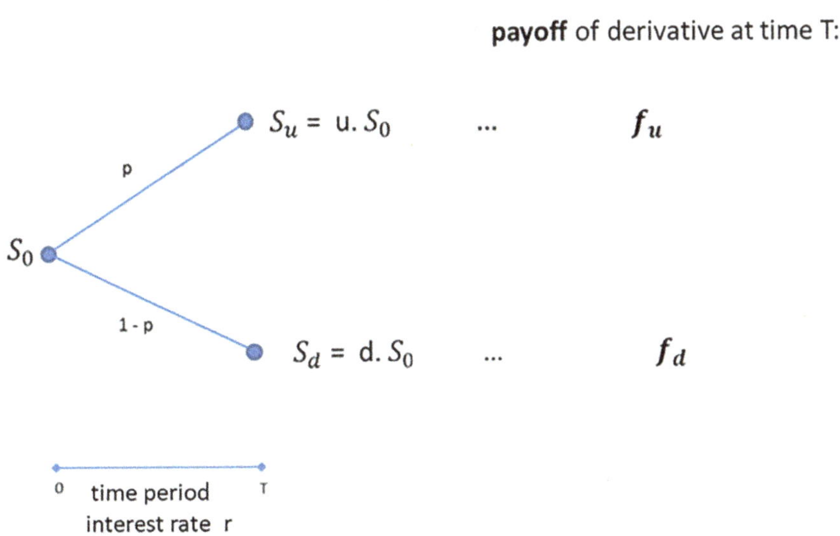

Fig. 3.19 Derivative on an underlying asset in the one-step binomial model

the call option generating a payoff and that this would then translate into a strong increase in the option's fair price.

However, that would be rather simplistic thinking, since it would only consider the option and not the possible combinations of option and underlying! Regardless of the formal mathematical calculation, this independence of the option price from the probability of the underlying asset's development can also be plausibilized in the following way: A strongly increasing likelihood of a strong price increase of the underlying increases the probability of making a profit with the call option. However, the strongly increasing probability also significantly increases the chances of making a profit when investing directly into the underlying (at the current price). This means that we don't necessarily need the call option in order to make that profit. And so, the value of the call option does not increase (significantly) compared to the current price of the underlying. A derivative should never be analysed in isolation from its underlying, but always in relation to it. (Provided of course that the underlying is actually a tradable asset. But this is an issue we will get back to later.)

Is it indeed true that, in the reality of the financial markets, the prices of derivatives are independent of the traders' assessment of the market situation? This is another issue that we will look into later (and based on more realistic stock price models).

The Role of the Parameters *u* and *d*

The parameters u and d essentially indicate how much the underlying asset can fluctuate. Put differently: They are a measure of the underlying asset's potential volatility. The greater u or the smaller d, the greater the possible fluctuation of the underlying asset's price until time T. The fair price of the derivative is fundamentally dependent on the values u and d, both of which are embedded in the definition of the new parameter p'.

Type of Dependency Between the Fair Price f_0 and the parameters *u* and *d*

How does the fair price f_0 of a derivative react to changes in the parameters u or d? We illustrate this question using our numerical example shown in Fig. 3.18, looking at two variants:

Variant 1:

In our example, let us first set the parameter u as a variable, the only prerequisite being that u be greater than d, i.e. greater than $\frac{1}{2}$. All the other parameters –including f_u– remain unchanged. For the fair price f_0, we then get the formula with the variable u, as follows:

$$f_0(u) = 100 \cdot p' = 100 \cdot \frac{\frac{1}{2}}{u - \frac{1}{2}} = \frac{50}{u - \frac{1}{2}}$$

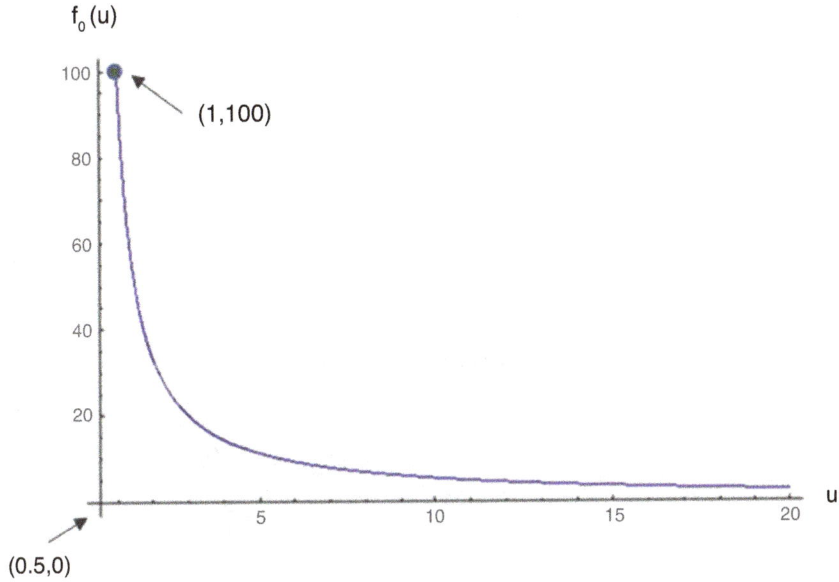

Fig. 3.20 Dependence of a derivative's fair price on parameter u, Numerical Example Variant 1

where we see that the fair price value f_0 decreases as u increases (see also Fig. 3.20). So, in this example, the fair price falls with increasing u and increasingly approaches 0.

It is also striking that the fair price $f_0(u)$ grows indefinitely when u approaches the value 0.5 (the origin in Fig. 3.20 is located at (0.5, 0)). How can we explain that the derivative's fair price increases to infinity for u going to 0.5? How can we explain, for example, that the derivative's price in the binomial model where $S_u = 51$ (i.e. $u = \frac{51}{100}$) takes the value $f_0 = \frac{50}{\left(u-\frac{1}{2}\right)} = 5000$, although the derivative's maximum payoff is 100? The model is shown in Fig. 3.21.

In fact, it cannot be explained, because it is incorrect: We will see further below that a binomial model must always be such that $d < e^{r \cdot T} < u$, otherwise the binomial model itself is already not arbitrage-free. In our specific example, $e^{r \cdot T} = 1$, so u can only take values greater than 1 (i.e. $u \cdot S_0 > 100$). For $u = 1$, we would get the fair price $f_0 = \frac{50}{u-\frac{1}{2}} = 100$. We see therefore that the fair price of the derivative is indeed always less than 100. At least here, our intuition proved to be right.

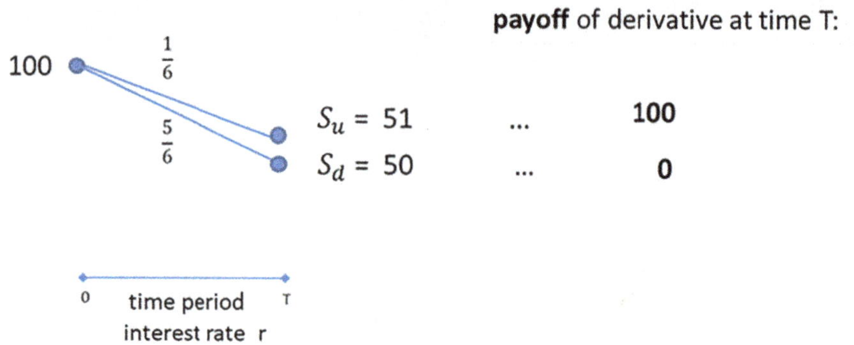

Fig. 3.21 How can a product with a maximum payoff of 100 EUR have a fair price of 5000 EUR?

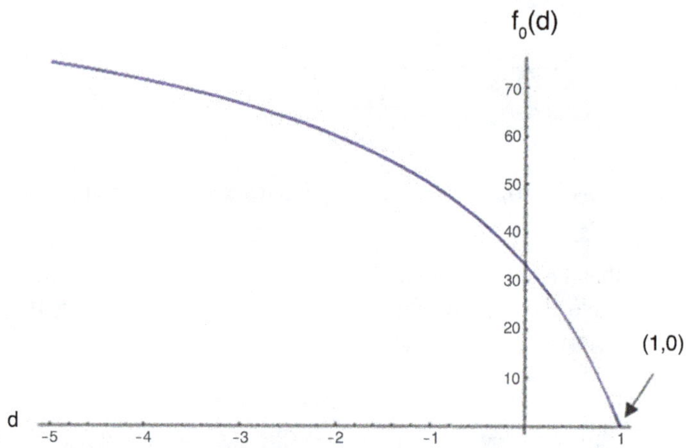

Fig. 3.22 Dependence of a derivative's fair price on parameter d, Numerical Example Variant 1

Let us now set the parameter d as a variable, the only prerequisite being that d be less than u, i.e. less than 3. However, we noted above that d must also be less than $e^{r \cdot T}$, which in our example has the value 1. All the other parameters –including f_u– remain unchanged. For the fair price f_0 we then get the formula with the variable d, as follows:

$$f_0(d) = 100 \cdot p' = 100 \cdot \frac{1-d}{3-d}$$

Differentiating $f_0(d)$ with respect to d gives us $f_0'(d) = -100 \cdot \frac{2}{(3-d)^2}$.

f_0' is therefore always negative (the number in the denominator is –since it is a square number– always positive!). f_0 is therefore strictly decreasing for increasing d (see also Fig. 3.22) or strictly increasing for decreasing d—where d can assume values less than 1 (and can even be negative). For $d = 1$, the derivative's fair price would be 0.

Variant 2:

In our example, let us again set the parameter u as a variable.

For $u = 3$ as set in the original numerical example, and $f_u = 100$, the derivative represents a call option with strike $K = 200$. If we vary u and leave $f_u = 100$ as fixed (as in Variant 1), we are still dealing with a call option, yet a call option whose strike price K changes as u varies. In Variant 2, we want to consider the case of a call option with a fixed strike $K = 200$ instead. We only look at $u > 2$ (so that $u \cdot S_0 > K$ is satisfied) and set $f_u = u \cdot 100 - 200\,(= u \cdot S_0 - K)$. Thus f_u will also vary. Yet, for any value that we choose for u, we get a call option with strike $K = 200$.

For the derivative's fair price f_0 depending on u, we then get

$$f_0(u) = (u \cdot 100 - 200) \cdot p' = (u \cdot 100 - 200) \cdot \frac{\frac{1}{2}}{u - \frac{1}{2}} = 100 \cdot \frac{u-2}{2u-1}.$$

If we differentiate $f_0(u)$ with respect to u, we get $f_0'(u) = \frac{3}{(2u-1)^2}$. $f_0'(u)$ is therefore always positive, and $f_0(u)$ is therefore strictly increasing in u (see also Fig. 3.23). For an ever-increasing u, f_0 increasingly approaches the value 50.

In the case of dependence on the variable d, nothing changes compared to Variant 1; the payoff f_d is still always equal to 0.

With regard to Variant 2, that is, in the case of a call option with fixed strike $K = 200$, we can therefore state as follows: The call option's fair price f_0 increases

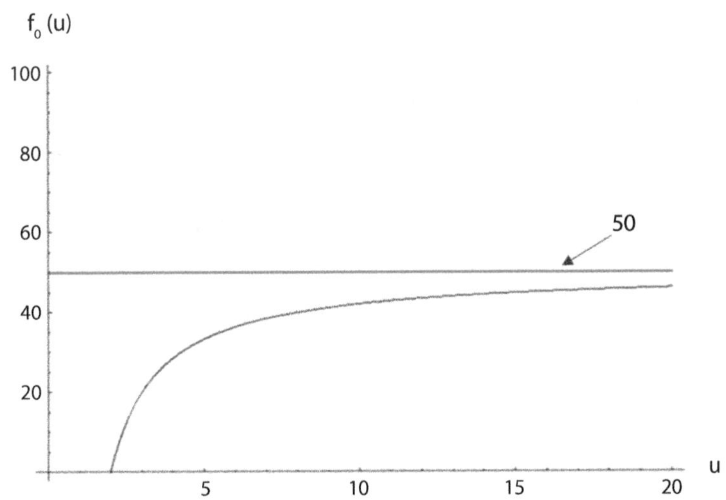

Fig. 3.23 Dependence of a derivative's fair price on parameter u, Numerical Example Variant 2

as u increases and d decreases. Meaning, *as the volatility of the underlying asset increases, the price of the call option increases.*

Variant 1 shows that it isn't generally always the case that the derivative price increases with increasing volatility.

The considerations we made for our numerical example above regarding the dependence of the fair price f_0 on the parameters u and d could of course also be made for the general case and discussed in detail. However, we will save that for when we deal with derivative pricing in more realistic models than the one-step binomial model.

Interpretation of a derivative's fair price as a discounted expected payoff with respect to an "artificial probability p'"

You probably noticed long ago the parallel between the correct formula for a derivative's fair price f_0 as given in Theorem 3.11

$$f_0 = e^{-rT} \cdot \left(p' \cdot f_u + \left(1 - p'\right) \cdot f_d\right)$$

and the initially (see Formula (3.1)) incorrect price assumption (discounted expected payoff)!

$$e^{-rT} \cdot (p \cdot f_u + (1 - p) \cdot f_d)$$

The correct fair price f_0 can thus also be seen as a discounted expected payoff. However, this discounted expected payoff MUST NOT be interpreted with respect to the REAL probability p in the binomial model, but in relation to the ARTIFICIAL probability p'.

This is a pivotal insight that will accompany us throughout the entire book—albeit in myriad variants and forms.

This "artificial probability" p' thus plays a highly essential (and at the same time somewhat mysterious) role, in that it replaces the real probability and makes it completely irrelevant when it comes to computing the fair price of a derivative.

So what is this artificial probability p', really?

The artificial probability p' and the absence of arbitrage in the one-step binomial model

We recall that p' has the form $p' = \frac{e^{r \cdot T} - d}{u - d}$. Above, we referred to p' as an (artificial) probability measure without specifying it any further. Yet stating a probability only makes sense if it has a value between 0 and 1. Does that hold for p'? Can we even refer to p' as a probability? In order to answer this, we want to verify under which conditions p' is greater than 0 and less than 1

$$\mathbf{0 < p'} \Leftrightarrow 0 < \frac{e^{r \cdot T} - d}{u - d} \Leftrightarrow 0 < e^{r \cdot T} - d \Leftrightarrow \mathbf{d < e^{r \cdot T}}$$

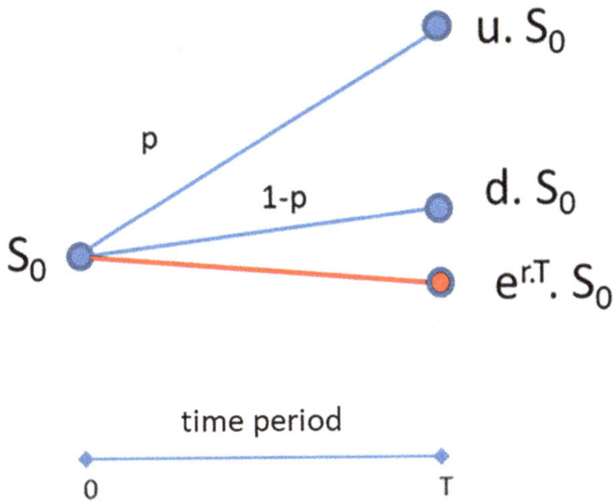

Fig. 3.24 Situation if $e^{r \cdot T} < d$

and

$$\boldsymbol{p' < 1} \;\Leftrightarrow\; \frac{e^{r \cdot T} - d}{u - d} < 1 \;\Leftrightarrow\; e^{r \cdot T} - d < u - d \;\Leftrightarrow\; \boldsymbol{e^{r \cdot T} < u}$$

p' can therefore be interpreted as a probability precisely when $\boldsymbol{d < e^{r \cdot T} < u}$ holds.

Note, however:

This inequality chain MUST ALWAYS be satisfied; otherwise the one-step binomial model itself would already not be arbitrage-free!

Because if $e^{r \cdot T} < d$, we would have the situation as shown in Fig. 3.24.

In this case, I could take out a loan, at time 0, in the amount of S_0 and until time T, and thus buy one share of stock at the price of S_0. By time T, that stock will have gone up in value to at least $d \cdot S_0$. I will then sell that share for at least $d \cdot S_0$ and repay the loan. The repayment cost is $e^{r \cdot T} \cdot S_0$. Given that $d \cdot S_0 > e^{r \cdot T} \cdot S_0$, I make a risk-free arbitrage profit of at least $d \cdot S_0 - e^{r \cdot T} \cdot S_0$ EUR. The model is therefore not arbitrage-free.

Analogously, we can show that there is also an arbitrage opportunity when $e^{r \cdot T} > u$. We leave the proof to the reader.

And –again, this is a simple exercise– it can be shown that if $d \leq e^{r \cdot T} \leq u$, then the one-step binomial model is indeed arbitrage-free (in the sense that we defined "arbitrage-free" above).

So the theorem is as follows.

Theorem 3.12 *The one-step binomial model is arbitrage-free precisely when the condition $d \leq e^{r \cdot T} \leq u$ is satisfied.*

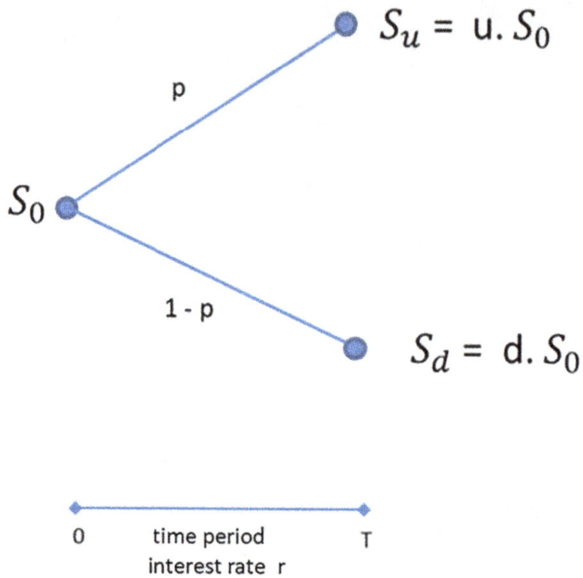

Fig. 3.25 The one-step binomial model revisited

What Is the Significance of the Artificial Probability p'?
Thus, if the one-step binomial model is arbitrage-free, i.e. if $d \leq e^{r \cdot T} \leq u$ is satisfied, we can indeed interpret p' as a probability measure, and so the fair price of any derivative in this model is precisely the discounted expected payoff of that derivative under the artificial probability p'.

Does the value p' have any intuitive meaning or is it really completely meaningless, given its artificialness?

To answer this question, let us take another look at the one-step binomial model itself (Fig. 3.25).

And we ask ourselves the following question:

What probability p would make the stock represented by the model a "fair game"?

What is that supposed to mean?

I buy one share of the stock at a price of S_0 EUR. The average payoff I can expect from that share is $p \cdot S_u + (1 - p) \cdot S_d$. In order for my "wager" S_0 to be justified (not too high for me and not too low from the counterparty's perspective), the expected payoff should correspond to my wager. More precisely:

The discounted payoff should correspond to my wager, since I am placing my bet today but will not pocket any profit before time T. This is why

$$S_0 = e^{-rT} \cdot (p \cdot S_u + (1-p) \cdot S_d) = e^{-rT} \cdot (p \cdot S_0 \cdot u + (1-p) \cdot S_0 \cdot d)$$

should apply.

A game in which at **any point in time the required wager is equal to the discounted expected payoff at any later point in time is called a "fair game" or a "martingale".**

So back to the question: *What probability p would make the stock represented by the model a "fair game"?*

To find the answer, all we need to do is calculate the value p from the equation

$$S_0 = e^{-rT} \cdot (p \cdot S_0 \cdot u + (1-p) \cdot S_0 \cdot d)$$

and rearrange it to get –perhaps not surprisingly– the following result:

$$p = \frac{e^{r \cdot T} - d}{u - d}$$

So we see:

The artificial probability p' is precisely the probability where trading the stock modelled by the one-step binomial model is a fair game (a martingale)!

p' is therefore referred to as the model's *"martingale measure"* or *"risk-neutral probability measure"*. ("With my bet, I risk exactly what I may expect to be my (discounted) payoff".)

And:

The fair price f_0 of any derivative in the one-step binomial model is the discounted expected payoff of the derivative under the risk-neutral probability measure in this model.

The replicating portfolio of a derivative

A very important piece of information is embedded in the very method that we used to derive the fair price of a derivative in the one-step binomial model. Let us recall: We constructed a portfolio (P_2) consisting of an underlying asset and cash (x units of the underlying, y cash) with exactly the same characteristics as the derivative. When we trade that portfolio, the effects will be exactly the same as if we trade the derivative. So the derivative was not actually needed here. Trading in the derivative could be simulated one-to-one by adequately trading the underlying asset and investing the requisite amount of cash. The values x and y in that setup were uniquely defined by two equations (see Formula (3.2)).

This well-defined portfolio of x units of the underlying and y amount of cash is called **the replicating portfolio** of the derivative D (which we will refer to in the following as "$RP(D)$"). The price of the $RP(D)$ is exactly the fair price f_0 of the derivative D.

If the replicating portfolio $RP(D)$ of the derivative is known, then it is also clear what the arbitrage strategy should look like, if the actual price of the derivative differs from f_0.

If the price of the derivative is greater than f_0, then we sell the derivative and buy the $RP(D)$; if the price of the derivative is less than f_0, we sell the $RP(D)$ and buy the derivative. In both cases, the arbitrage profit equals the absolute difference between the derivative's actual price and its fair price. In the numerical example in Sect. 3.8, we had initially assumed the reasonable price of the derivative to be 16.66 EUR, the expected payoff from the derivative. (Only later did we find that the actual fair price of this derivative is in fact 20 EUR.) Then we put on our arbitrageur's hat and pursued the following strategy (which may have appeared a bit like a magician's trick at first): "Buy five units of the derivative and sell two units of the underlying asset". The equivalent would have been: "Buy one unit of the derivative and sell two fifths of one underlying asset". To calculate the replicating portfolio of this derivative, we can use Formula 3.3 to get

$$x = \frac{100 - 0}{100 \cdot \left(3 - \frac{1}{2}\right)} = \frac{2}{5} \quad \text{und} \quad y = \frac{3 \cdot 0 - \frac{1}{2} \cdot 100}{3 - \frac{1}{2}} = -20$$

So in the arbitrage example in Sect. 3.8, we actually applied a specific method, using the replicating portfolio D, and not a magician's trick. The number of underlying units that we purchased was exactly equal to the number of underlying units in the replicating portfolio. (The cash amount was irrelevant in this example, as the assumed interest rate was $r = 0$.)

Hedging a Derivative Using the Replicating Portfolio

The replicating portfolio of a derivative is relevant for realizing arbitrage profit (later, in more realistic models, we are going to talk about "replicating trading strategies"), but it is even more important in another context.

Using replicating portfolios, we can eliminate risk arising from derivatives, a process we refer to as **hedging**. We illustrate below how hedging basically works (in the one-step binomial model), using the example of an OTC derivatives trade between an investor A and a bank B.

Bank B creates a derivative D on an underlying asset S with certain properties and offers this derivative for purchase to interested investors A.

Two questions arise for the bank:

- At what price should the derivative be offered on the market?
- What should the bank do about the risks it takes in selling the derivative?

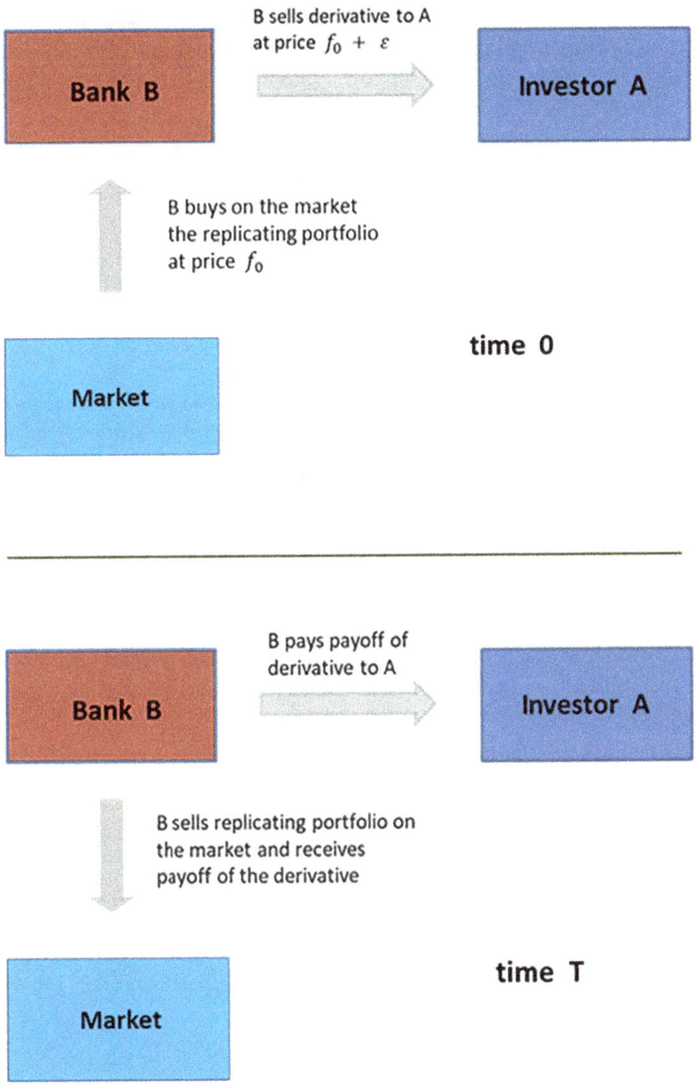

Fig. 3.26 Perfect hedging using the replicating portfolio

The bank will naturally proceed as follows (their approach is also illustrated graphically in Fig. 3.26):

The bank is unwilling to expose itself to risk by selling D. Therefore, it buys the replicating portfolio $RP(D)$ on the market at the same time that it sells D. As a result, the bank sells derivative D at the price $f_0 + \varepsilon$.

The price of the $RP(D)$ is f_0. The bank pockets the margin of ε EUR.

Since the derivative (short position from the bank's perspective) and the replicating portfolio (long position from the bank's perspective) completely cancel each other out at time T, the bank has fully eliminated any risk and secured the ε margin as a safe, risk-free profit.

In this example, the bank applied a **static, perfect hedging** strategy. "Static" in this context means that hedging was achieved through *one single* trade, and "perfect" means *complete* elimination of any risk. Such static, perfect hedging is rarely possible in the real world. Yet in our one-step binomial model (which is far removed from reality), and assuming a frictionless market, such static, perfect hedging is possible, as we have just seen.

To what extent can derivatives in the one-step binomial model be traded more or less "skilfully"?

We immediately have a quick answer to that: "We can proceed skillfully in that we wait for and identify discrepancies between a derivative's market price and its fair price (resulting from imbalances in supply and demand) and then immediately make a risk-free arbitrage profit by running appropriate strategies".

In principle, that is a correct answer.

But what if the entire system is arbitrage-free? If the one-step binomial model is arbitrage-free and the derivative is priced at the well-defined fair price, can the outcome still be influenced by a greater or lesser degree of skilfulness?

Let us go back to the very first game A that we dealt with in the preamble to Sect. 3.8. For illustration, we revisit it here in Fig. 3.27.

Again, we assume interest rate $r = 0$ and a very short time interval T between placing our bet and getting the payout.

The game is played with a fair dice. The ticket is bought at the reasonable price of 16.66 EUR.

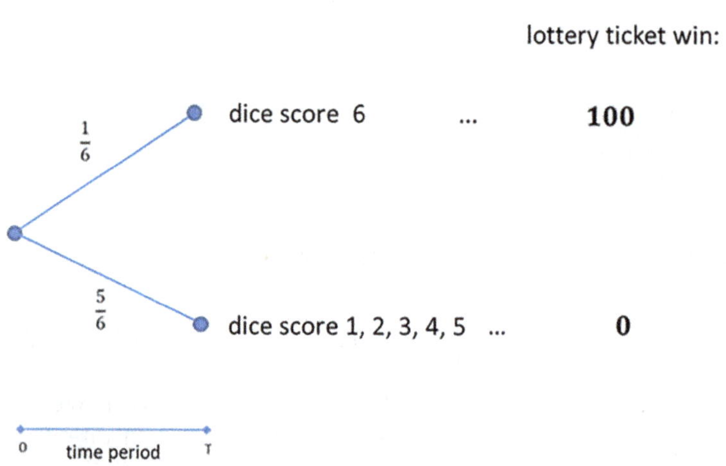

Fig. 3.27 The dice game revisited

You can approach this in whichever way you like, write down and analyse columns and strings of numbers, and try to recognize patterns and place your bets based on that—yet in the long average, your payoff in this game will be +/− 0 euros.

In the long run, even the best mathematician, the smartest hedge-fund manager, or incumbent world chess champion will not be able to prevail in any relevant way compared to a toddler placing haphazard bets.

The game is a pure game of chance that cannot be influenced by any amount of skill!

Let us now turn to our fairly priced derivative in the one-step binomial model. And don't expect much in terms of new insights here, just a comment.

Suppose you and the above-mentioned toddler are forced to invest 100 EUR each in a trade within this system. In total, there are three trading variants that you can choose from:

Trading Variant 1:
1 stock share long
Trading Variant 2:
2 stock shares long, 5 derivatives short
Trading Variant 3:
5 derivatives long

Each of these variants comes at a cost of 100 EUR. The toddler would always choose one of the variants at random. You on the other hand would calculate the expected payoff of each variant, find that the expected payoff is negative for variants 1 and 3 and equal to zero for variant 2, and would therefore always choose variant 2. In the long run, you would undoubtedly outperform the toddler's random play in this game.

It is therefore possible to "play the trading game" more or less skilfully (even if no arbitrage were ever to occur). This assertion is not about whether it is difficult or easy to play the game well or poorly, or whether your chances of winning are positive or negative; it simply states that there are better and worse strategies. And as soon as there are better and worse strategies, a game is no longer a pure game of chance.

The degree of skill that can influence the winning odds in a game is a very interesting question. It is also a highly relevant one for game providers, yet in many cases very difficult to answer, and mathematically demanding. But it is definitely worth a brief digression.

3.11 A Brief Excursus on the Degree of Luck and Skill in Games

The question of how much skill goes into games is not simply an academic one. It also has definite practical relevance—not only for players looking for superior playing strategies but also undoubtedly for the providers of the games themselves.

In many cases, for instance, such as on online gaming portals or in private casinos that pit players against the provider or against each other, games can only be run if they require a certain minimum degree of skill.

And, of course, people who work professionally on the financial markets will often find themselves being asked the same question: "Isn't your job just one big game of chance?" And they would be well advised to have a pithy response ready with which they can quickly convince their interrogator of the skill element involved (particularly when derivatives are in play).

The author of this book has written a series of expert opinions on the skill element in games, which have been used in the courts. These games include blackjack, two aces (a variant of blackjack), Eurolet 24 (a variant of roulette), Bejeweled, Hangman, Spider Solitaire, "Schnapsen" (an Austrian variant of Sixty-Six), and even the family favourite "Mensch ärgere dich nicht" (known as Parcheesi or Ludo in other countries).

The challenges of mathematics and game theory that this work posed, the people I met, and experiences I had along the way –both in a setting that often took some getting used to and in relevant court proceedings– were extremely fascinating and taxing, often highly enjoyable, and would fill an entire book by themselves. Maybe there will still be some time at the end of this chapter for a little anecdote on the topic.

As for now, however, how can we even begin to approach the task of determining the degree of skill required for a game? How do you define such a "degree of skill" in the first place? There is no one single universally valid answer to this fundamental question. Rather, there can be several different ones depending on your theoretical approach. For my expert opinions, I developed an approach that –in my view– is readily comprehensible and fairly easy to justify and that I would like to outline briefly here.

As I have already mentioned, there is currently no binding, generally accepted model or criterion for classifying games by the level of skill involved—at least as far as I am aware. In my view, the diagnostic criteria or skill coefficients that various observers have proposed furnish information that is neither adequate nor easily comprehensible.

I therefore based my investigations on a very clear, universally comprehensible model that, at the very least, is able to provide a solid lower bound for the skill element in a game.

(continued)

In order to present and justify this model, let us start by considering two games that can probably be regarded as extremes on the skill/chance scale: chess and tossing a fair coin.

We will also consider two different kinds of players of these two games and of others that will be analysed:

The first type of player is the "random chance player". By this, we mean a player who knows the rules of the game in question and is thus perfectly capable of playing it but who does so without deploying a particular strategy or making any other strategic considerations. The second type of player is the "skill player", i.e. someone with experience, who adopts a certain, solid playing strategy that need not be specified in any greater detail and that anyone could learn with some effort.

(In light of various court rulings made in the past, games are to be analysed under the assumption that they are being played by competent, ambitious moderately experienced players rather than elite professionals.)

Chess is undoubtedly a game that has no element of random chance whatsoever and thus is to be regarded as a pure game of skill. The "random chance player" as defined above would not win a single game of chess against the "skill player" as defined above. Even if such a victory were ever to happen, it would merely be down to an extremely unlikely piece of good fortune on the part of the "random chance player" (who just happened to make the right move at the right time). The probability of such a random event occurring is negligible, however. In other words, the "random chance player" will win 0% of their games; chess is thus 100% skill.

With the coin toss (each player puts up an equal stake and picks a side of the coin to bet on), both the "random chance player" and the "skill player" have a 50% chance of winning. Tossing a coin is thus 0% skill.

Let us now consider a game (which we shall call Combo(p)) that is a combination of chess and tossing a coin.

The game of Combo(p) consists of a random number generator that will produce the number 1 with a probability of p% (p being any number between 0 and 100) and the number 0 with a probability of 100% - p%.

If "1" is generated, the two players have to play a game of chess against each other. If "0" appears, they have to toss a coin. This game of Combo(p) is played repeatedly.

It is immediately plausible that the degree of skill involved in playing Combo(p) is p%. (The game of Combo(0) is only the coin toss, of course, and Combo(100) is playing only chess.)

(continued)

Let us now consider the average probability of the "skill player" winning compared with the average probability of the "random chance player" winning the game of Combo(p):

The "skill player" would win every game of chess and half the coin tosses, i.e.

$p\% + 0.5 \cdot (100\% - p\%) = 50\% + 0.5 \cdot p\%$ of the games.

The "random chance player" would lose all the games of chess but win half the coin tosses, i.e. he wins

$0.5 \cdot (100\% - p\%) = 50\% - 0.5 \cdot p\%$ of the games.

The difference between the average win probabilities of the "skill player" and the "random chance player" is thus exactly p%, which is exactly the degree of skill involved in the game.

It is on this observation that the author bases his analyses of any given game:

In each case, a lower estimate is made (in percent) for the difference between the average probability of a "skill player" winning and the average probability of a "random chance player" winning, and –based on the above observations for the game of Combo(p)– this difference is then used as the lower bound for the degree of skill involved in the game being analysed.

The trick –and the challenge– now lies in specifying an acceptable "good strategy" for each game and then calculating the probability of a player who has adopted this "good strategy" beating an opponent deploying a "random chance strategy". Although the outcome will naturally depend on the choice of "good strategy" in each case, it will always provide some perfectly serviceable and reliable lower estimates for the degree of skill involved in the games.

Let us end with a few results based on this method, which may prove highly interesting and, in some cases, perhaps surprising too. (We will not be saying anything about the individual games themselves, so the results will only really be meaningful to people already familiar with them.)

The game of **Spider Solitaire**:

- Playing with one suit only: degree of skill at least **71%**
- Playing with two different suits: degree of skill at least **78%**
- Playing with four different suits: degree of skill at least **82%**
- A two-player game of **Schnapsen (Sixty-Six)**:
- Degree of skill at least **60%**

In the game of Ludo or **Parcheesi** (those of you who have children or still remember your own childhood will be familiar with it), you can let your children win without having to cheat. In other words, there has to be a degree of skill involved.

(continued)

In a **two-player game**, the degree of skill is just a touch over **50%**.

The more players join in, however, the more it becomes a game of chance. With **four players**, the degree of skill is down to around **20%**.

One particular experience that the author had (and here comes the anecdote mentioned earlier) occurred when he was writing an expert opinion on a game called "**observation roulette**".

The main difference between standard roulette and observation roulette (of which there are several variants) lies in the properties of the wheel:

(a) Whereas, in standard roulette, the croupier sets the number ring in motion as soon as the ball is released, the ring used in observation roulette is fixed in place and does not move.
(b) The rim of a standard roulette wheel is relatively narrow and studded with diamond-shaped obstacles, while the rim of an observation roulette wheel is much wider and uniformly smooth. It also has three white concentric circles marked around it. Rather than presenting an obstacle, however, these are embedded smoothly into the rim.
(c) The numbers and colours are jumbled up in standard roulette, while they are arranged in order in observation roulette.

With observation roulette, the croupier has to release the ball in such a way that it completes at least three laps below the concave rim of the wheel before breaking away from the rim and continuing on for some more laps, eventually falling into the number ring. The ball's trajectory becomes increasingly elliptical the further it travels.

The main axes of these ellipses shift, spaced relatively evenly apart. The player may continue to place bets until the ball has reached the innermost of the three white concentric circles.

After some time spent observing, it becomes clear that the ball almost always falls into one of the numbered pockets when it is traveling on the shorter main axis of the ellipse, ending up in a pocket fairly close by.

Various experiments clearly demonstrated that, under ideal conditions (e.g. no vibrations affecting the wheel), a player can increase the probability of winning to as much as 70% by spending quite a while observing the ball's movements and betting on half the number ring each time. These experiments included asking a croupier at a private casino in the Netherlands, who had many years' experience of observation roulette, to play the game for a lengthy period of time and be observed in the process—and he too emerged with a win probability of nearly 70%.

When it was put to him that an experienced and highly focused player visiting the casino could naturally pocket significant winnings and put the casino in some jeopardy, and he was asked how he would handle this situation,

(continued)

the croupier smiled good-naturedly and replied, in a glorious, inimitable Dutch accent, with words to this effect:

"If I spot that someone's beginning to win on a regular basis, I've always got a few balls of different weights up my sleeve, which I switch without anyone noticing. And then it's not long before the situation starts looking completely different once again".

3.12 Fair Price of Derivatives in the Binomial Model on Underlying Assets with Payouts/Costs

As announced above, we are now going to derive (without commenting it much further) the fair price of a derivative D (on an underlying asset S in the one-step binomial model) which we dealt with in the last couple of sections, under the assumption that payments are made or costs incurred through the underlying asset over the term of the derivative.

Just as we did in Sects. 3.5 and 3.6, we will again assume that the amount of any such payments is given as an absolute value Z discounted to time 0 or as a continuously compounded return δ.

We will only discuss the case of a continuously compounded return δ for the underlying asset (leaving the case of a given discounted absolute payment Z to the reader).

In doing so, we assume that the dividends (payments, costs) are always immediately invested into the underlying asset. This means that between time 0 and time T, the underlying asset may go from S_0 to either $u \cdot e^{\delta \cdot T} \cdot S_0$ or to $d \cdot e^{\delta \cdot T} \cdot S_0$. However, this also means that we can use exactly the same method for valuing the derivative as we did for the dividend-free case, by simply replacing u from before by $u \cdot e^{\delta \cdot T}$ and the d from before by $d \cdot e^{\delta \cdot T}$. This gives us exactly the same formula for calculating the derivative's fair price f_0, namely,

$$f_0 = e^{-rT} \cdot \left(p' \cdot f_u + \left(1 - p'\right) \cdot f_d\right)$$

only now we have (because we replaced u and d by $u \cdot e^{\delta \cdot T}$ and $d \cdot e^{\delta \cdot T}$)

$$p' = \frac{e^{r \cdot T} - d \cdot e^{\delta \cdot T}}{u \cdot e^{\delta \cdot T} - d \cdot e^{\delta \cdot T}}$$

which we can simplify to

$$p' = \frac{e^{(r-\delta) \cdot T} - d}{u - d}$$

instead of the previous artificial probability $p' = \frac{e^{r \cdot T} - d}{u - d}$.

To summarize:

Theorem 3.13 *In a one-step binomial model in the time range* $[0, T]$ *with a continuously compounded payout/cost return* δ*, and with the parameters* S_0, u, d, p*, and* r *(where* $S_0 > 0$ *and* $d < u$*), any derivative with payoffs* f_u *and* f_d *has a well-defined fair value* f_0*, which is given by*

$$f_0 = e^{-rT} \cdot \left(p' \cdot f_u + \left(1 - p'\right) \cdot f_d\right)$$

where

$$p' = \frac{e^{(r-\delta)\cdot T} - d}{u - d}.$$

3.13 The Two-step Binomial Model

The one-step binomial model was extensively analysed and processed in the previous sections. Our main criticism of the model was, of course, that the model is not at all representative of real stock price developments.

Therefore, we complicate the model somewhat, with the intention of bringing it at least somewhat closer to reality: We move on to the two-step binomial model. (You need not worry, however, that we will then go on to the three-step model and the four-step model, etc.! Yet it is very useful to examine the two-step model in detail, as a way to illustrate and clarify some other basic principles.)

The two-step binomial model is the logical continuation of the one-step model, encompassing **two consecutive but mutually independent** steps. We illustrate the model schematically in Fig. 3.28.

A (European) derivative D on the underlying asset S is again given by the payoffs $\left(f_{u^2}, f_{ud}, f_{d^2}\right)$ from that derivative at time T.

Again we denote the derivative's fair price at time 0 by f_0. We now let f_u denote the derivative's fair price at time $dt \left(= \frac{T}{2}\right)$ if the stock has the value $u \cdot S_0$ at time dt and let f_d denote the derivative's fair price at time $dt \left(= \frac{T}{2}\right)$ if the stock has the value $d \cdot S_0$ at time dt.

As we can see, we can already calculate the values f_u and f_d. They are precisely the fair prices resulting from the two one-step models on the right hand-side of the two-step model (green ellipse for f_u and red ellipse for f_d in Fig. 3.29). By applying the one-step model method, we obtain (noting that the step size in the one-step model here is dt, not T)

$$f_u = e^{-rdt} \cdot \left(p' \cdot f_{u^2} + \left(1 - p'\right) \cdot f_{ud}\right)$$

and

$$f_d = e^{-rdt} \cdot \left(p' \cdot f_{ud} + \left(1 - p'\right) \cdot f_{d^2}\right) \tag{3.7}$$

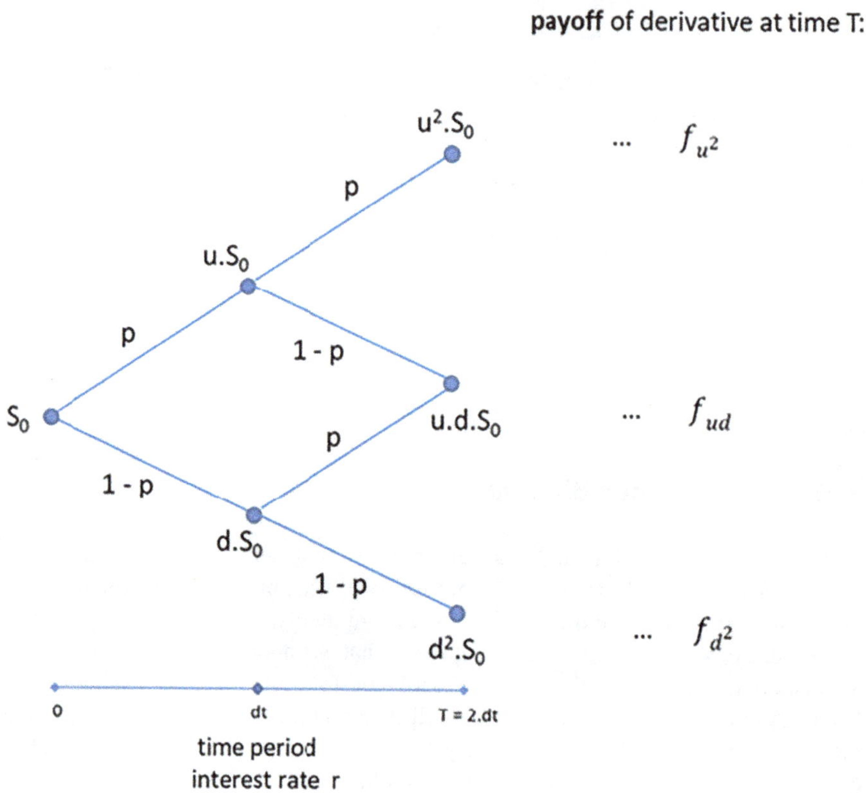

Fig. 3.28 Two-step binomial model

where

$$p' = \frac{e^{r \cdot dt} - d}{u - d}.$$

However, we are particularly interested in f_0, i.e. the derivative's fair value at time 0. The obvious approach would be to work in the yellow one-step model on the left hand-side of the two-step model.

If the underlying asset takes the value $u \cdot S_0$ in the first step, then our derivative has the (previously calculated) value (payoff ?) f_u; if the underlying asset takes the value $d \cdot S_0$ in the first step, then our derivative has the (previously calculated) value (payoff ?) f_d. Based on this, we can again calculate the fair value f_0 at time 0 using the one-step method:

$$f_0 = e^{-rdt} \cdot \left(p' \cdot f_u + \left(1 - p'\right) \cdot f_d\right) \tag{3.8}$$

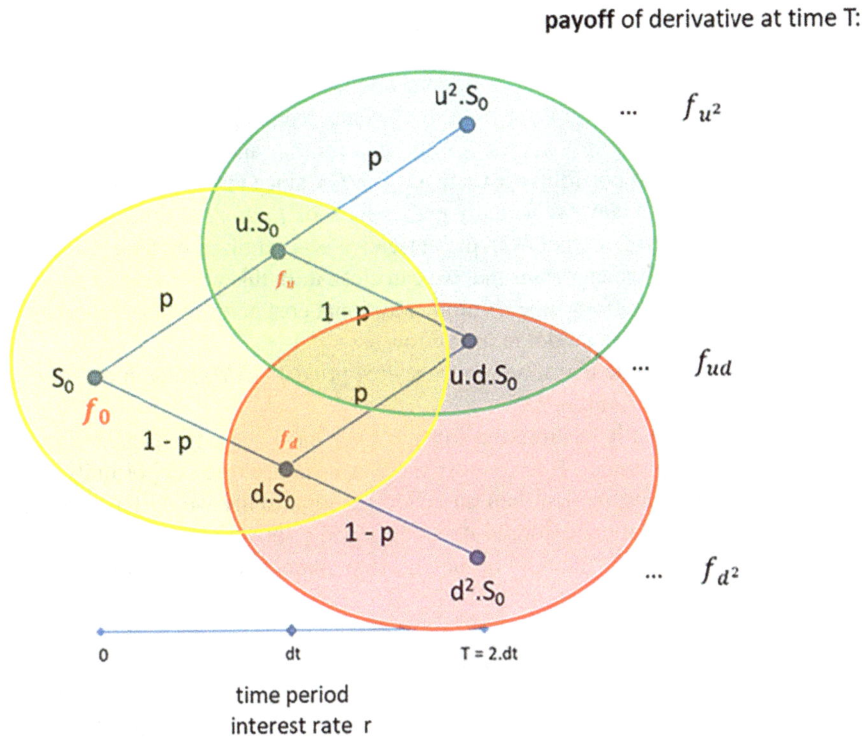

Fig. 3.29 Two-step binomial model as a combination of three one-step models

Now we plug the values for f_u and f_d from Formula 3.7 into this Formula 3.8, arrange the result by the values f_{u^2}, f_{ud} and f_{d^2}, observing that $e^{-r \cdot dt} \cdot e^{-r \cdot dt} = e^{-r \cdot T}$, and get

$$f_0 = e^{-rT} \cdot \left((p')^2 \cdot f_{u^2} + 2 \cdot p' \cdot (1 - p') \cdot f_{ud} + (1 - p')^2 \cdot f_{d^2} \right) \tag{3.9}$$

You probably noticed my reservations about using the one-step method again when I wrote "the obvious approach would be to ...". What is the difference between the one-step model in the yellow area compared to the green or red area? The difference is that the values f_{u^2}, f_{ud}, and f_{d^2} are fixed, **guaranteed payoffs** upon the derivative's maturity, while the values f_u and f_d only represent the derivative's **fair prices** at time dt. If those were the actual prices at time dt, then we could sell the derivative at time dt and actually get f_u and f_d as payoffs, and the argument for that approach would be correct. But the fact is that we have no guarantee that f_u and f_d are actually going to be the real prices at time dt.

Question: Can an argument still be made for the one-step model in the yellow area?

Answer: Yes, if we include the following additional argumentation:

If the price of the derivative at time dt deviates from f_u (or f_d), we short the replicating portfolio for the payoffs f_{u^2} and f_{ud} (or f_{ud} and f_{d^2}) at time dt. The price of the replicating portfolio is exactly f_u (or f_d), since that is precisely how f_u (f_d) is defined. In that way, we actually get a payoff of f_u or f_d at time dt.

At time T, the payoff we get from the derivative is neutralized by the price of the replicating portfolio, which means that we can close it. In this way, we have secured f_u or f_d as actual payoffs at time dt at no additional cost and can indeed make the argument for the one-step model in the yellow area.

So, we can again state that if an arbitrage-free price f_0 exists, then it must be of the form as shown in Formula 3.9.

Whether f_0 is actually an arbitrage-free price can easily be argued again with the help of the one-step model. It is easy to see that if an arbitrage opportunity were to exist in the two-step model, then an arbitrage opportunity would have to exist already in one of the three coloured one-step models shown above. However, we have already shown in the analysis of the one-step model that this is not the case.

For those of you interested in delving further into that last argument:

Let us assume that the price of the derivative is f_0 and that an arbitrage opportunity exists, i.e. a strategy which has the price 0 at time 0, will always have a value greater than or equal to 0 at time T and has a positive probability of taking a value greater than 0 at time T. We know from the one-step model that a strategy with the value 0 at the moment 0 will either have the value 0 in both possible cases at time dt or have a negative value in one of the two cases. In the first of these two cases, as we also know from the one-step model, the strategy must either have the value 0 at time T in each of the three possible cases or a negative value in at least one case. In the second of the two cases, we assume the situation in which the value of the strategy at time dt is negative (e.g. value $= -\varepsilon$). From the one-step model we know that, starting from that assumption, the final value has to be negative in at least one situation. (Rationale: If both final values were greater than or equal to zero, we would look at the same strategy at time dt including ε in cash. This new strategy has a value of 0 at time dt and definitely a positive value at time T. However, as we know, this is not possible in a one-step model.) No arbitrage is therefore possible in any of the cases.

Thus, we can summarize:

Theorem 3.14 *In a two-step binomial model in the time range* $[0, T]$ *(without payments or costs) with the parameters* S_0, u, d, p, *and* r *(where* $S_0 > 0$ *and*

$d < u$), any derivative with payoffs f_{u^2}, f_{ud}, and f_{d^2} has a well-defined fair value f_0, which is given by

$$f_0 = e^{-rT} \cdot \left(\left(p'\right)^2 \cdot f_{u^2} + 2 \cdot p' \cdot \left(1 - p'\right) \cdot f_{ud} + \left(1 - p'\right)^2 \cdot f_{d^2} \right)$$

where

$$p' = \frac{e^{r \cdot dt} - d}{u - d}.$$

3.14 Derivative Valuation in the Two-step Binomial Model, Discussion of Results

No Arbitrage in the Two-step Model

As in the case of the one-step model, it must again be argued that the two-step model (by itself) is arbitrage-free precisely when $d \leq e^{r \cdot dt} \leq u$. And that is exactly when p' is again a value between 0 and 1.

By the way, the two-step model drawing in Figs. 3.28 and 3.29 is geometrically incorrect, since the development, for example, from S_0 to $u \cdot S_0$ to $u^2 \cdot S_0$ is not linear, as suggested by the schematic representation, but exponential. Figure 3.30, for instance, is a geometrically exact depiction of how our numerical example from earlier sections was extended from the one-step model to the two-step model.

Nevertheless, in the following we will mostly stick to the illustrative schematic linear representation.

Interpretation of the Valuation Formula as a Discounted Expected Payoff

In analogy to the one-step model, could we also interpret the valuation formula in the two-step model (Theorem 3.14) as a discounted expected payoff under the probability p'?

Yes, we can indeed:

- There is **one** possible path for the underlying asset (going from $u \cdot S_0$ to $u^2 \cdot S_0$) that results in a derivative payoff of $\boldsymbol{f_{u^2}}$. The (artificial) probability that the underlying asset will take this path is $p' \cdot p' = \boldsymbol{(p')^2}$.

  (This is assuming that the two consecutive steps are mutually independent. Only then is the path the product of the two individual probabilities.)
- There are **two** possible paths that the underlying asset can take (one going from $u \cdot S_0$ to $u \cdot d \cdot S_0$ and one going from $d \cdot S_0$ to $u \cdot d \cdot S_0$), resulting in a derivative payoff of $\boldsymbol{f_{ud}}$. The (artificial) probability that the underlying asset will take the first path is $p' \cdot \left(1 - p'\right)$, while the (artificial) probability of it taking the second

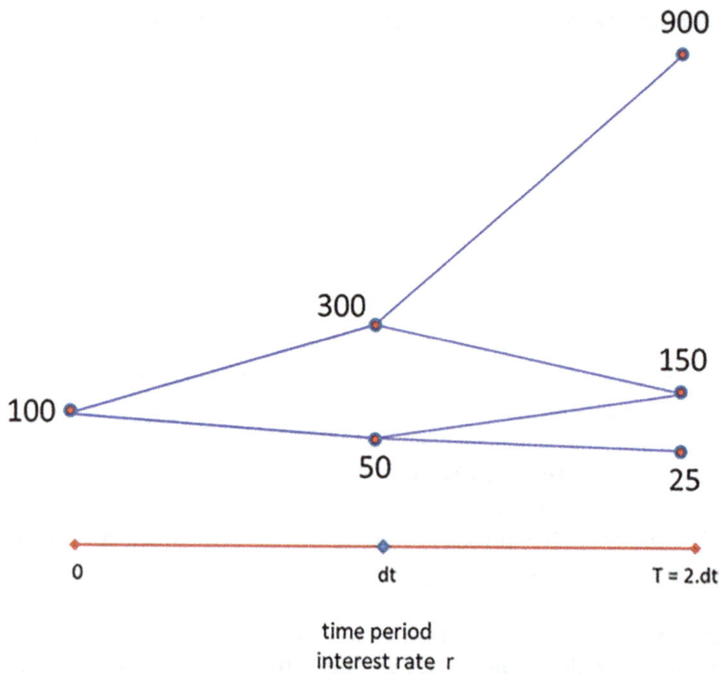

Fig. 3.30 Two-step model, specific numerical example, geometrically correct representation

path is $(1-p')\cdot p'$. In total therefore, the (artificial) probability that the payoff f_{ud} occurs is $\mathbf{2\cdot p'\cdot(1-p')}$.

- There is **one** possible path which the underlying asset could take (from $d\cdot S_0$ to $d^2\cdot S_0$) and which would result in a derivative payoff of $\boldsymbol{f_{d^2}}$. The (artificial) probability that the underlying asset will take this path is $(1-p')\cdot(1-p')=$ $\mathbf{(1-p')^2}$.

And so, under the artificial probability measure, the discounted expected payoff is indeed equal to the valuation formula

$$e^{-rT}\cdot\left((p')^2\cdot f_{u^2}+2\cdot p'\cdot(1-p')\cdot f_{ud}+(1-p')^2\cdot f_{d^2}\right).$$

p' as a Risk-neutral Measure in the Two-step Binomial Model

Using the probability p', the two-step binomial model also becomes a fair game. For this purpose, we need to verify whether the expected payoff discounted to time 0 from the underlying asset at time $2\cdot dt$ is equal to that underlying asset's current value S_0. (Actually, this would have to be carried out for each point in time in relation to each later point in time, but we already did so for each step in the one-step model.)

And in fact (when using the binomial formula $a^2 + 2 \cdot a \cdot b + b^2 = (a+b)^2$ for $a = u \cdot p'$ and $b = d \cdot (1 - p')$ and applying the above considerations about the individual paths in the two-step model and their probabilities), we get

discounted expected payoff =

- $e^{-rT} \cdot \left((p')^2 \cdot u^2 \cdot S_0 + 2 \cdot p' \cdot (1-p')\, u \cdot d \cdot S_0 + (1-p')^2 \cdot d^2 \cdot S_0 \right) =$
- $e^{-rT} \cdot S_0 \cdot (p' \cdot u + (1-p') \cdot d)^2,$

and because

$(p' \cdot u + (1-p') \cdot d)^2 = \left(\frac{e^{r \cdot dt} - d}{u-d} \cdot u + \left(1 - \frac{e^{r \cdot dt} - d}{u-d}\right) \cdot d \right)^2 = \left(e^{r \cdot dt}\right)^2 = e^{rT}$

the outcome for the discounted expected payoff is indeed

$e^{-rT} \cdot S_0 \cdot e^{rT} = S_0$, which is what we needed to show.

3.15 Hedging and Arbitrage in the Two-step Binomial Model

Let us revisit the situation illustrated in Fig. 3.31, where a derivative was traded OTC between a bank B and an investor A, with the bank hedging the transaction within a one-step binomial model.

We will now assume the same situation, yet in this case, the underlying asset of the derivative develops according to a two-step binomial model. In the two-step binomial model, we generally no longer have a replicating portfolio, i.e. a portfolio of x underlying units and y cash that is purchased at time 0 and the value of which at time T is equal to the derivative's payoff. This is because the two-step model has three different outcomes ($u^2 \cdot S_0$, $ud \cdot S_0$, and $d^2 \cdot S_0$) and thus generally three different payoffs f_{u^2}, f_{ud}, and f_{d^2}. Thus, three equations must be satisfied at time

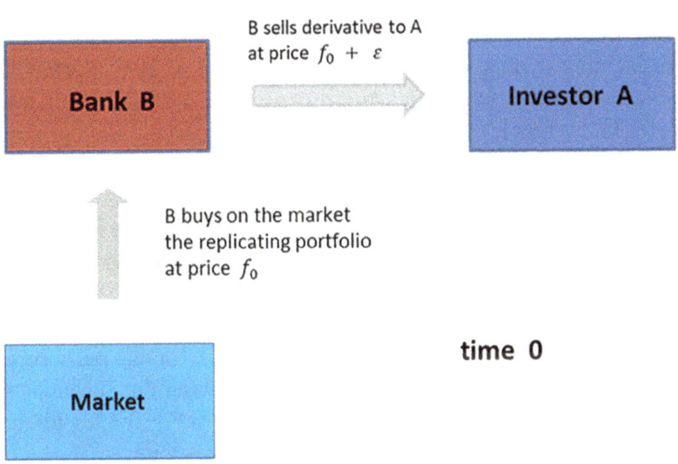

Fig. 3.31 Hedging process using a replicating portfolio in the one-step model

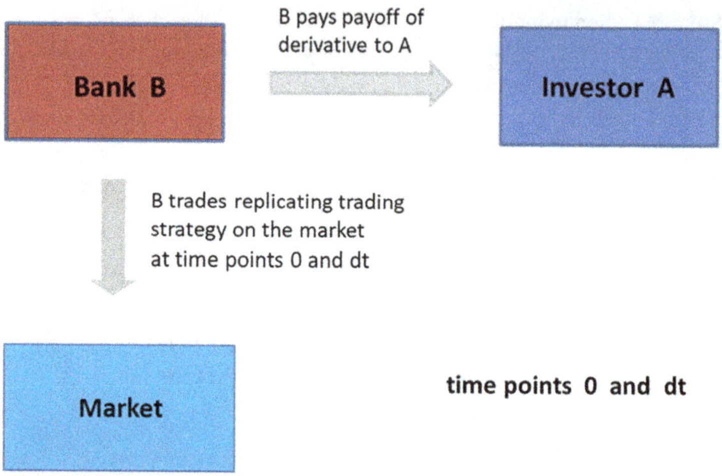

Fig. 3.32 Hedging process using a replicating trading strategy in the two-step model

T. However, we only have 2 variables to work with, x and y. Three equations in two variables generally have no solution.

So, while there is **no (static) replicating portfolio**, we do have what is called a **replicating trading strategy**, as follows:

At time 0, we create the replicating portfolio for the values f_u and f_d at time dt. The cost for this portfolio is f_0. This portfolio is sold at time dt. We take the proceeds from that sale, f_u (if S_0 takes the value $u \cdot S_0$), or f_d (if S_0 takes the value $d \cdot S_0$) to buy the replicating portfolio for f_{u^2}, f_{ud} in the first case, or f_{ud} and f_{d^2} in the second case, at time T. We can do so because the cost of these portfolios is precisely f_u and f_d, respectively.

This dynamic and no longer static trading strategy perfectly hedges the risk posed by the derivative. The bank can therefore eliminate its risk by using this dynamic replicating trading strategy (compare Fig. 3.32). Thus, perfect hedging is also possible in the two-step model, the process is just a bit more complex (and requires more know-how).

3.16 Numerical Example of How to Value and Hedge Derivatives and Execute Arbitrage Trades in a Two-step Binomial Model

We are now going to calculate a numerical example and, for that purpose, consider a call option D with strike $K = 200$ on the underlying asset that develops within the two-step model as shown in Fig. 3.30 and again below in Fig. 3.33, with the option's payoffs. We again choose $r = 0$ as the applicable interest rate.

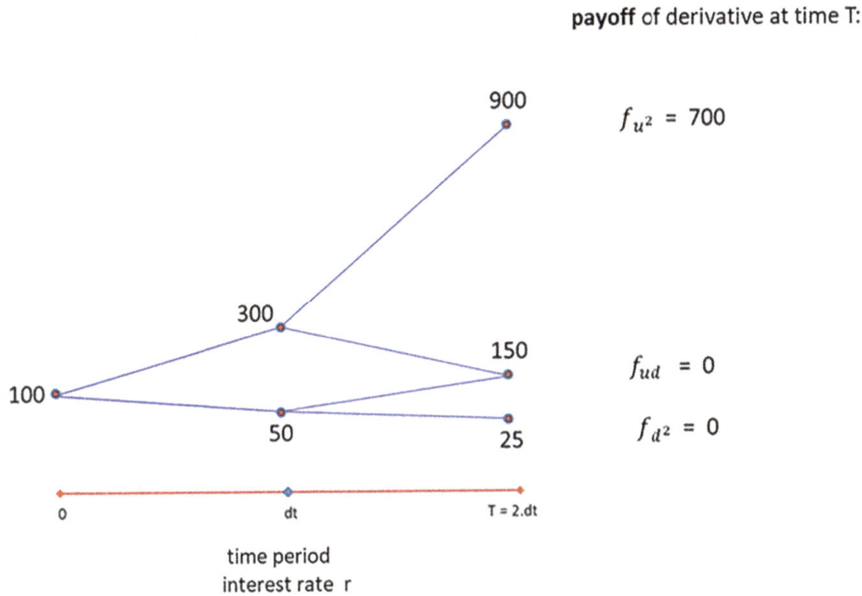

Fig. 3.33 Numerical example, call option with strike price 200 in the two-step model

u and d have the same values as in the numerical example of the one-step model –3 and $\frac{1}{2}$, respectively– and r is again equal to 0 (thus $e^{r \cdot dt} = e^{r \cdot T} = 1$), and therefore p' has the same value as before, namely, $p' = \frac{1}{5}$.

Therefore

$$f_0 = e^{-rT} \cdot \left((p')^2 \cdot f_{u^2} + 2 \cdot p' \cdot (1 - p') \cdot f_{ud} + (1 - p')^2 \cdot f_{d^2} \right) =$$

$$= \left(\frac{1}{5} \right)^2 \cdot 700 + 2 \cdot \frac{1}{5} \cdot \frac{4}{5} \cdot 0 + \left(\frac{4}{5} \right)^2 \cdot 0 = \mathbf{28}$$

However, in order to be able to implement the hedging strategy and, possibly, an arbitrage strategy, we need much more information. We need the values f_u and f_d as well as the form of the replicating portfolios in each of the one-step models.

We have (Formula 3.7)

$$f_u = e^{-rdt} \cdot \left(p' \cdot f_{u^2} + (1 - p') \cdot f_{ud} \right) = \frac{1}{5} \cdot 700 = \mathbf{140}$$

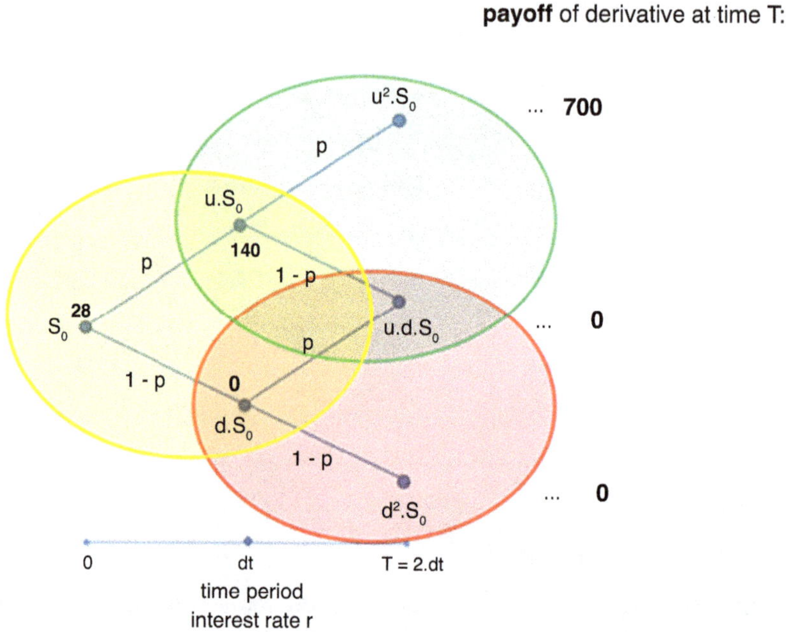

Fig. 3.34 Replicating in the two-step model

and

$$f_d = e^{-rdt} \cdot (p' \cdot f_{ud} + (1 - p') \cdot f_{d^2}) = \mathbf{0}$$

In the **red area** (Fig. 3.34), we simply replicate using the empty portfolio (i.e. $x_{ro} = 0, y_{ro} = 0$). Its cost is $f_d = 0$. For the **green area**, we calculate the replicating portfolio using Formula 3.3, adjusting the parameter notations to the green area, of course. Thus

$$x_{gr} = \frac{f_{u^2} - f_{ud}}{u \cdot S_0 \cdot (u - d)} = \frac{700 - 0}{300 \cdot \left(3 - \frac{1}{2}\right)} = \frac{14}{15}$$

$$y_{gr} = e^{-rdt} \cdot \frac{u \cdot f_{ud} - d \cdot f_{u^2}}{u - d} = \frac{0 - \frac{1}{2} \cdot 700}{\left(3 - \frac{1}{2}\right)} = -140$$

For the **yellow area**, we calculate the replicating portfolio directly from the Formulas 3.3 using the previously calculated values for $f_u = 140$ and $f_d = 0$:

$$x_{ye} = \frac{f_u - f_d}{S_0 \cdot (u - d)} = \frac{140 - 0}{100 \cdot \left(3 - \frac{1}{2}\right)} = \frac{14}{25}$$

$$y_{ye} = e^{-rT} \cdot \frac{u \cdot f_d - d \cdot f_u}{u - d} = \frac{0 - \frac{1}{2} \cdot 140}{\left(3 - \frac{1}{2}\right)} = -28$$

We now have all the information we need for the replicating strategy, and we move on to the **hedging** process. This is how it works:

Time 0:

The bank sells the derivative at $f_0 + \varepsilon = 28 + \varepsilon$ EUR.

The 28 EUR amount will be used for hedging. ε is the bank's profit. The bank takes out a loan in the amount of $-y_{ye} = 28$ EUR. This 28 EUR loan is used together with the $f_0 = 28$ EUR to purchase $x_{ye} = \frac{14}{25}$ units of the underlying asset. The cost of this purchase is in fact $\frac{14}{25} \cdot 100 = 56$ EUR.

Time dt if the underlying asset has taken the value $u \cdot S_0 = 300$

The bank closes the portfolio that it opened at time 0. This means:

The bank sells the $\frac{14}{25}$ units of the underlying asset and collects $\frac{14}{25} \cdot 300 = 168$ EUR. Of this amount, it uses 28 EUR to repay the loan and has $f_u = 140$ EUR left at its disposal. We are now in the green area (Fig. 3.34).

The bank takes out a loan in the amount of $-y_{gr} = 140$ EUR and uses that amount plus the $f_u = 140$ EUR to buy $x_{gr} = \frac{14}{15}$ units of the underlying asset. The cost of that purchase is in fact $\frac{14}{15} \cdot 300 = 280$ EUR.

Time dt if the underlying asset has taken the value $d \cdot S_0 = 50$

The bank closes the portfolio that it opened at time 0. This means:

The bank sells the $\frac{14}{25}$ units of the underlying asset and collects $\frac{14}{25} \cdot 50 = 28$ EUR. The bank uses this 28 EUR amount to repay the loan and has $f_d = 0$ EUR left at its disposal. We are now in the red area. Since $x_{re} = y_{re} = 0$, no further action is taken. Knowing that the derivative will definitely have a payoff of 0 at time T, no action is indeed required; there will be no payments flowing from the derivative nor from the hedging strategy at time T.

Time T if the underlying asset took the value $u \cdot S_0 = 50$ at dt:

In this case, the bank holds $\frac{14}{15}$ of the underlying asset and an outstanding loan amount of 140 EUR. The bank sells the $\frac{14}{15}$ of the underlying asset at $\frac{14}{15} \cdot 900 = 840$ EUR (if the underlying asset takes the value 900) or at $\frac{14}{15} \cdot 150 = 140$ EUR (if the underlying asset takes the value 150). The loan is paid back, leaving the bank a residual amount of 700 and 0 EUR, respectively, which it can use to pay the derivative's respective payoffs.

Thus, the derivative was perfectly hedged by this strategy!

Now it is also clear how the **arbitrage** strategy is to be executed in the event that the derivative's actual price D_0 deviates from its fair price f_0.

If $D_0 > f_0$:

Sell the derivative at D_0 *and execute the replicating strategy at the price of* f_0. *You retain* $D_0 - f_0 > 0$ *as risk-free profit without investing any capital.*

If $D_0 < f_0$:

Execute the opposite of the replicating strategy at the price of $-f_0$ *and buy the derivative at the price of* D_0. *You retain* $f_0 - D_0 > 0$ *as risk-free profit without investing any capital.*

3.17 Derivative Valuation in the N-step Binomial Model

The N-step binomial model is the logical continuation of the one-step and two-step models, encompassing N **consecutive yet mutually independent** steps. We present the model schematically (in a geometrically incorrect form) in Fig. 3.35 (together with the general payoff of a derivative) and, for illustration, also in a geometrically correct form, using a specific numerical example, in Fig. 3.36. For each point in all of the representations, the probability of an up move is p, and the probability of a down move is $1 - p$.

In the N-step binomial model, there are 2^N paths that the underlying can take. In the 9-step model shown in Figs. 3.36 and 3.37, for example, there are 512 such

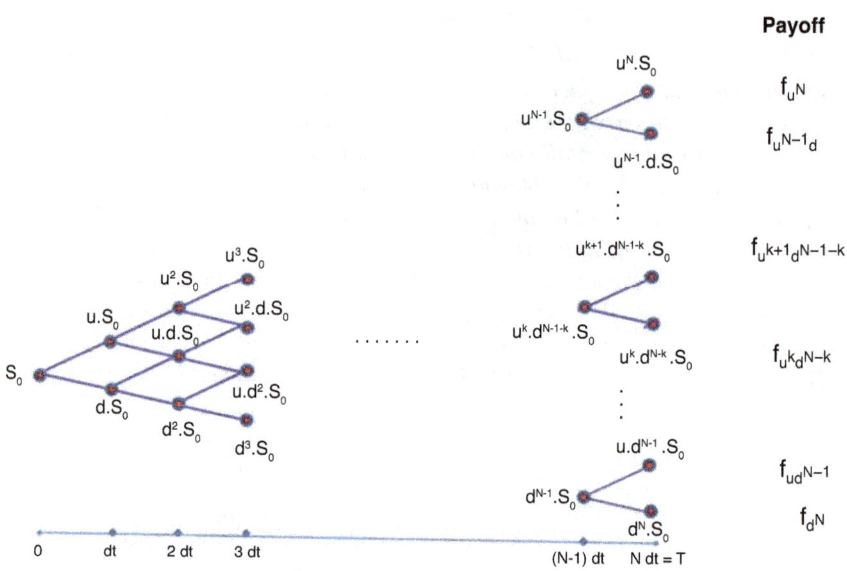

Fig. 3.35 N-step binomial model, schematic representation

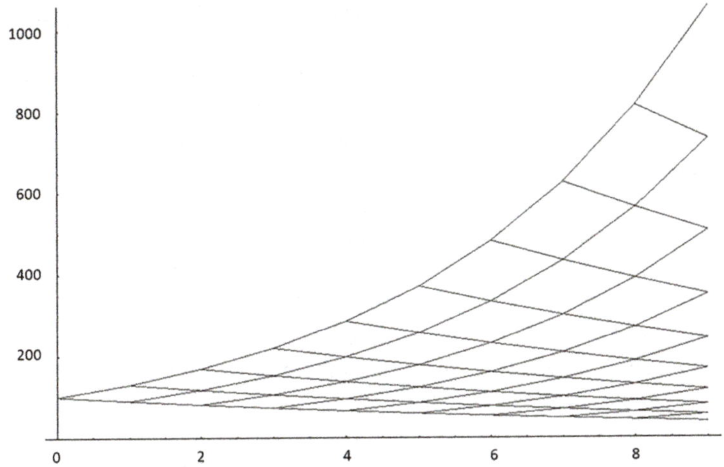

Fig. 3.36 Nine-step binomial model, numerical example, geometrically correct representation

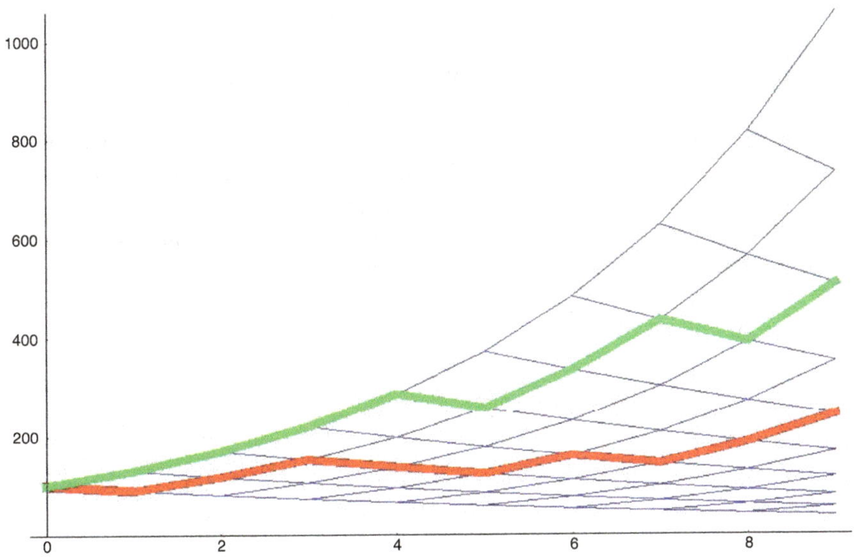

Fig. 3.37 Nine-step binomial model, numerical example, with two possible price paths

possible paths. For illustration, two of these paths have been coloured green and red in Fig. 3.37.

Given our preliminary work on the valuation of derivatives in the one-step and two-step models, we already have an idea as to what the fair price f_0 of the derivative shown in Fig. 3.35 is going to be in the N-step binomial model.

We **assume** that:

The derivative's fair price f_0 *is the discounted expected payoff from that derivative, calculated under the risk-neutral probability measure, i.e. the probability under which the N-step binomial model becomes a fair game. Just like in the one-step and two-step model, the risk-neutral probability is given by the probability* $p' = \frac{e^{r \cdot dt} - d}{u - d}$ *for each upward step at each of the model's nodes.*

Let us first consider what the valuation formula would look like if our assumption is correct. It would obviously be

$$f_0 = e^{-rT} \cdot \sum_{k=0}^{N} \left(f_{u^k d^{N-k}} \cdot P\left(\text{payoff } f_{u^k d^{N-k}}\right) \right).$$

where $P\left(\text{Payoff } f_{u^k d^{N-k}}\right)$ denotes the probability for the payoff $f_{u^k d^{N-k}}$ to occur.

What is the probability that the payoff $f_{u^k d^{N-k}}$ will occur?

Payoff $f_{u^k d^{N-k}}$ occurs when the underlying takes the value $u^k \cdot d^{N-k} \cdot S_0$. This in turn occurs precisely when the stock price moves up exactly k times and moves down exactly $N - k$ times during the N time steps. However, there are of course plenty of paths with exactly k up moves and exactly $N - k$ down moves (see, e.g. Fig. 3.38, which shows three example paths with the same endpoint, each with six upward movements and three downward movements).

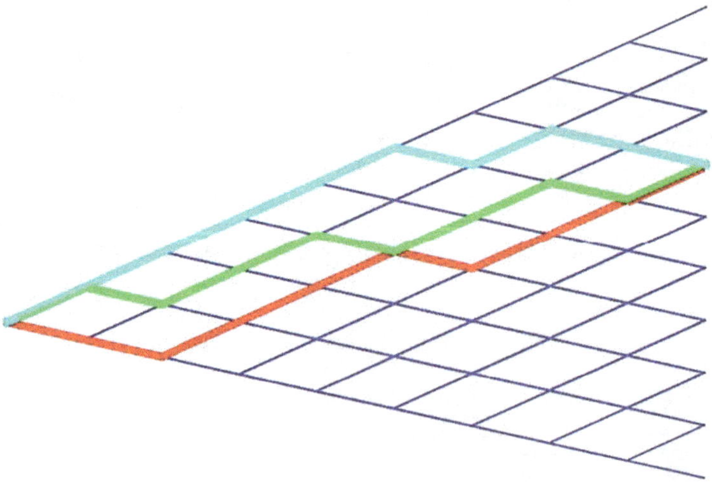

Fig. 3.38 Three price paths with the same endpoint in the nine-step model

Any given path out of these paths with endpoint $u^k \cdot d^{N-k} \cdot S_0$ has to make an upward movement at k given points in time and make a downward movement at the remaining $N-k$ points in time. Since the individual movements are independent of each other, the probability of this to happen is $(p')^k \cdot (1-p')^{N-k}$.

Now we only need to calculate how many paths we have that go from S_0 to $u^k \cdot d^{N-k} \cdot S_0$. Each of these paths is uniquely determined by the k points at which, out of the *total* N points in time, an upward movement is to take place. The number of possibilities to select k points in time from a total of N points in time is given by the binomial coefficient $\binom{N}{k} = \frac{N!}{k! \cdot (N-k)!}$.

Thus, the probability for the payoff $f_{u^k d^{N-k}}$ to occur is given by $\binom{N}{k} \cdot (p')^k \cdot (1-p')^{N-k}$, and so our assumption as to the form of the formula for f_0 is

$$f_0 = e^{-rT} \cdot \sum_{k=0}^{N} \left(f_{u^k d^{N-k}} \cdot \binom{N}{k} \cdot (p')^k \cdot (1-p')^{N-k} \right)$$

The assumption is indeed correct. We therefore phrase it as a theorem and will prove it in the following.

Theorem 3.15 *In an N-step binomial model in the time range* $[0, T]$ *(without payments or costs) with the parameters* $S_0, u, d, p,$ *and* r *(where* $S_0 > 0$ *and* $d < u$*), any derivative with payoffs* $f_{u^k d^{N-k}}$ *for* $k = 0, 1, \ldots, N$ *has a dynamic replicating portfolio and a well-defined fair value* f_0*, which is given by*

$$f_0 = e^{-rT} \cdot \sum_{k=0}^{N} \left(f_{u^k d^{N-k}} \cdot \binom{N}{k} \cdot (p')^k \cdot (1-p')^{N-k} \right) \tag{3.10}$$

where

$$p' = \frac{e^{r \cdot dt} - d}{u - d}$$

and p' *represents the risk-neutral measure in the N-step binomial model.*

f_0 *is therefore the discounted expected payoff under the risk-neutral measure.*

Proof The proof must consist of four steps:

We need to show that p' is indeed the risk-neutral probability in the N-step model; we need to show that the fair price f_0, if it exists, must have the form of Formula (3.10); we need to show that a dynamic replicating portfolio exists;

(continued)

and finally we need to prove that the value of f_0 from the Formula (3.10) does not allow any arbitrage.

We are going to prove this by induction on N, the number of steps.

Base case: All four steps of the proof have already been shown for the cases $N = 1$ and $N = 2$.

Induction assumption: We assume that all four steps of the proof have already been shown for the $1-, 2-, 3-, \ldots (N-1)$-step models.

Induction step: We will now prove that under the induction assumption (IA), all statements also hold in the N-step model.

In the following, we are going to use the following notation:

S_T = value of the underlying asset at time T,

D_T = payoff (= value) of the derivative at time T,

$E'(\ldots)$ = expected value of the argument under the risk-neutral probability (p').

$E'(\ldots \mid u \cdot S_0)$ = expected value of the argument under the risk-neutral probability (p') and under the assumption that the underlying asset has the value $u \cdot S_0$ at time dt.

$E'(\ldots \mid d \cdot S_0)$ = expected value of the argument under the risk-neutral probability (p') and under the assumption that the underlying asset has the value $d \cdot S_0$ at time dt. And we are going to refer to Fig. 3.39 in the following. We start by showing that p' is indeed the risk-neutral probability in the N-step model, i.e. that the following holds: $e^{-r \cdot T} \cdot E'(S_T) = S_0$.

The following calculations become easier to handle if we use the equivalent equation $E'(S_T) = S_0 \cdot e^{r \cdot T}$ below. From our (IA) we know that the corresponding relationship holds in all N-1-step models and in each one-step model, i.e. in the blue, in the red, and in the yellow model in Fig. 3.40.

This means:

$$E'(S_T \mid u \cdot S_0) = u \cdot S_0 \cdot e^{r \cdot dt \cdot (N-1)},$$

$$E'(S_T \mid d \cdot S_0) = d \cdot S_0 \cdot e^{r \cdot dt \cdot (N-1)}$$

and

$$p' \cdot u \cdot S_0 + (1 - p') \cdot d \cdot S_0 = S_0 \cdot e^{r \cdot dt}$$

Using the above three equations,

$$\begin{aligned} \mathbf{E'(S_T)} &= p' \cdot E'(S_T \mid u \cdot S_0) + (1 - p') \cdot E'(S_T \mid d \cdot S_0) = \\ &= p' \cdot u \cdot S_0 \cdot e^{r \cdot dt \cdot (N-1)} + (1 - p') \cdot d \cdot S_0 \cdot e^{r \cdot dt \cdot (N-1)} = \end{aligned}$$

(continued)

$$= \left(p' \cdot u \cdot S_0 + \left(1 - p'\right) \cdot d \cdot S_0\right) \cdot e^{r \cdot dt \cdot (N-1)} =$$
$$= S_0 \cdot e^{r \cdot dt} \cdot e^{r \cdot dt \cdot (N-1)} = \boldsymbol{S_0 \cdot e^{r \cdot T}}$$

which is what we needed to show.

Next, we are going to prove the formula for f_0, i.e. we will show that $f_0 = e^{-r \cdot T} \cdot E'(D_T)$.

According to (IA), this formula applies analogously again in the blue, in the red, and in the yellow model. Let f_u again denote the fair value of the derivative at time dt if the underlying asset is located at node $u \cdot S_0$ at that time, and let f_d denote the fair value of the derivative at time dt if the underlying asset is located in node $d \cdot S_0$ at that time; then the following holds:

$$f_u = e^{-rdt \cdot (N-1)} \cdot E'(D_T \mid u \cdot S_0) \text{ and}$$
$$f_d = e^{-rdt \cdot (N-1)} \cdot E'(D_T \mid d \cdot S_0),$$

and

$$f_0 = e^{-rdt} \cdot \left(p' \cdot f_u + \left(1 - p'\right) \cdot f_d\right)$$

With regard to the last equation, we have to add a short comment again (just like we did when we argued the proof in the two-step model, remember?):

f_u and f_d are not pre-guaranteed payoffs; they represent only the fair values of the derivative, so we cannot immediately assume that the third equation holds. Yet by our (IA), we know that in the blue and red models, there exist dynamic replicating portfolios for the derivative at the prices f_u and f_d. So, if we hold the derivative, we can indeed secure the amounts f_u and f_d at time dt by using these replicating portfolios, if we so wish (by selling the respective replicating portfolio at time dt).

Now, using the above three equations,

$$E'(D_T) = p' \cdot E'(D_T \mid u \cdot S_0) + \left(1 - p'\right) \cdot E'(D_T \mid d \cdot S_0) =$$
$$= e^{r \cdot dt \cdot (N-1)} \cdot \left(p' \cdot f_u + \left(1 - p'\right) \cdot f_d\right) =$$
$$= e^{r \cdot dt \cdot (N-1)} \cdot e^{r \cdot dt} \cdot f_0 = e^{r \cdot T} \cdot f_0,$$

we do indeed get

$$f_0 = e^{-rT} \cdot E'(D_T).$$

(continued)

The existence of a replicating portfolio immediately follows from the existence of the replicating portfolios at the prices f_u and f_d in the blue and the red model, respectively (according to (IA)), and from the existence of the replicating portfolio for f_u and f_d at the price f_0 in the yellow model (one-step model). So all that remains is to show that the price f_0 is indeed an arbitrage-free price. We do this by arguing the case analogously to the two-step model.

Let us assume that the price of the derivative is f_0 and that an arbitrage opportunity exists, i.e. a strategy which has the price 0 at time 0 and which will have a value greater than or equal to 0 at time T and a positive probability of taking a value greater than 0.

We know from the one-step model that a strategy with the value 0 at the moment 0 will either have the value 0 in both possible cases at time dt or have a negative value in one of the two cases.

In the first of these two cases, as we know, according to (IA) from the N-1-step model, the strategy must either have the value 0 at time T in each of the N-1 possible cases or a negative value in at least one case.

In the second of the two cases, we assume the situation in which the value of the strategy at time dt is negative (e.g. value $= -\varepsilon$). From the N-1-step model, we know according to (IA) that, starting from that assumption, the final value has to be negative in at least one situation. (Rationale: If all final values were greater than or equal to zero, we would consider the same strategy at time dt including ε in cash. This new strategy has a value of 0 at time dt and definitely a positive value at time T. However, as we know, this is not possible in an N-1-step model.) No arbitrage is therefore possible in any of the cases.

This successfully completes the proof of our theorem.

3.18 Comments on the Valuation of Derivatives in the N-step Binomial Model and an Example

Figure 3.40 depicts the network of paths in a binomial 40-step model with 4 randomly selected price paths that the underlying asset could take. Those paths already show a certain similarity to those that we are used to seeing for real stock price developments. We can therefore assume that the N-step model with a large N is more realistic than the one-step and the two-step model.

Our aim in the next paragraph is to develop a realistic stock price model, the so-called Wiener stock price model.

How **hedging –or, tantamount to that, the replicating strategy– works in the N-step binomial model** is clear both from the proof of Theorem 3.15 and from the

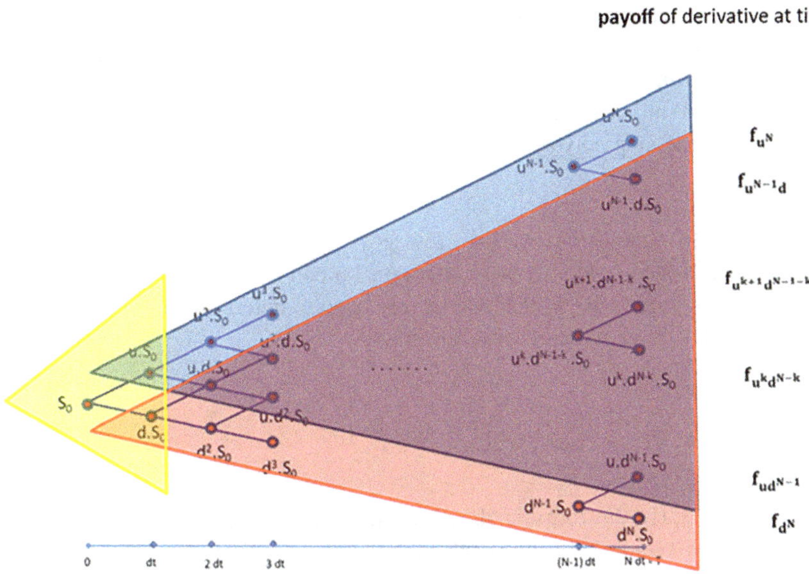

Fig. 3.39 The N-step model split into two N-1-step models and a one-step model

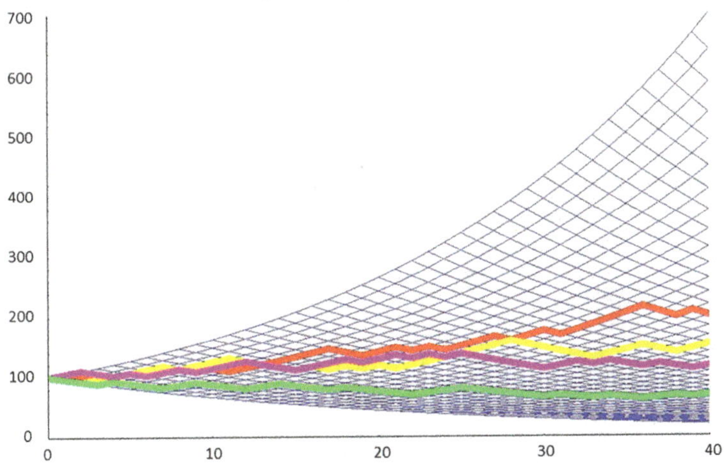

Fig. 3.40 40-step binomial model with 4 possible price paths

observations we made in the context of the two-step binomial model: At time 0, the replicating one-step portfolio is purchased for the values f_u and f_d at time dt. At any point in time $k \cdot dt$ and value of the underlying $S_0 \cdot u^l \cdot d^{k-l}$, the one-step portfolio entered into at time $(k-1) \cdot dt$ is closed out at the price of $f_{u^l d^{k-l}}$, and the replicating one-step portfolio for the values $f_{u^{l+1} d^{k-l}}$ and $f_{u^l d^{k+1-l}}$ is purchased at this price.

Example 3.16 In conclusion, we are now going to apply the general valuation formula in the N-step model to the valuation of a call option with strike K in the N-step binomial model. To do this, we simply need to determine the payoff values $f_{u^l d^{N-l}}$ and enter them in the general formula for

$$f_0 = e^{-rT} \cdot \sum_{l=0}^{N} \left(f_{u^l d^{N-l}} \cdot \binom{N}{l} \cdot (p')^l \cdot (1-p')^{N-l} \right)$$

of Theorem 3.15. (In the formula above, we denoted the summation index by “l” instead of “k” to avoid confusion between the summation index “k” and the strike price “K”).

We now have

$$f_{u^l d^{N-l}} = 0 \qquad \text{if } S_0 \cdot u^l \cdot d^{N-l} < K$$

and

$$f_{u^l d^{N-l}} = S_0 \cdot u^l \cdot d^{N-l} - K \qquad \text{if } S_0 \cdot u^l \cdot d^{N-l} > K.$$

In the formula for f_0, therefore, we only need to consider the indices l for which $S_0 \cdot u^l \cdot d^{N-l} > K$ is satisfied. Now

$$S_0 \cdot u^l \cdot d^{N-l} > K \Leftrightarrow$$

$$\left(\frac{u}{d}\right)^l > \frac{K}{S_0 \cdot d^N} \Leftrightarrow$$

$$l > \frac{\log\left(\frac{K}{S_0 \cdot d^N}\right)}{\log \frac{u}{d}}$$

In the last expression (and in all of the following), we use “log” to denote the natural logarithm, i.e. the logarithm to a base value of e.

We will now use the abbreviation “L” for the next larger integer to

$$\frac{\log\left(\frac{K}{S_0 \cdot d^N}\right)}{\log \frac{u}{d}}.$$

The fair price value of the call option with strike K in the binomial N-step model is therefore

$$f_0 = e^{-rT} \cdot \sum_{l=L}^{N} \left(S_0 \cdot u^l \cdot d^{N-l} - K \right) \cdot \binom{N}{l} \cdot (p')^l \cdot (1-p')^{N-l}.$$

3.19 Derivative Valuation in the N-step Binomial Model on Underlying Assets with Payouts or Costs

Here again, like in the one-step binomial model, we will only discuss the case of a continuously compounded return δ for the underlying asset and leave the case of a given discounted absolute payment Z to the reader.

We will again assume that the dividends (payments, costs) are always immediately invested into the underlying asset. This means that from any point in time $k \cdot dt$ until time $(k+1) \cdot dt$, the underlying asset can move from $S_{k \cdot dt}$ to either $u \cdot e^{\delta \cdot dt} \cdot S_{k \cdot dt}$ or $d \cdot e^{\delta \cdot dt} \cdot S_{k \cdot dt}$. However, this also means that we can use exactly the same method for valuing the derivative as we used for the case without dividends, by simply replacing the u from before by $u \cdot e^{\delta \cdot dt}$ and the d from before by $d \cdot e^{\delta \cdot T}$. We therefore get exactly the same formula for calculating the derivative's fair price f_0, namely,

$$f_0 = e^{-rT} \cdot \sum_{k=0}^{N} \left(f_{u^k d^{N-k}} \cdot \binom{N}{k} \cdot (p')^k \cdot (1-p')^{N-k} \right)$$

only now we have (because we replaced u and d by $u \cdot e^{\delta \cdot dt}$ and $d \cdot e^{\delta \cdot dt}$)

$$p' = \frac{e^{r \cdot dt} - d \cdot e^{\delta \cdot dt}}{u \cdot e^{\delta \cdot dt} - d \cdot e^{\delta \cdot dt}}$$

which we can simplify to

$$p' = \frac{e^{(r-\delta) \cdot dt} - d}{u - d}$$

instead of the previous artificial probability $p' = \frac{e^{r \cdot dt} - d}{u - d}$.

To summarize:

Theorem 3.17 *In an N-step binomial model in the time range* $[0, T]$ *with a continuously compounded payments or costs return* δ *and parameters* S_0, u, d, p *and* r *(where* $S_0 > 0$ *and* $d < u$*), any derivative with payoffs* $f_{u^k d^{N-k}}$ *for*

$k = 0, 1, \ldots, N$ *has a dynamic replicating portfolio and a well-defined fair value* f_0*, which is given by*

$$f_0 = e^{-rT} \cdot \sum_{k=0}^{N} \left(f_{u^k d^{N-k}} \cdot \binom{N}{k} \cdot (p')^k \cdot (1 - p')^{N-k} \right)$$

where

$$p' = \frac{e^{(r-\delta) \cdot dt} - d}{u - d}.$$

Reference

1. Donald Knuth. *The Art of Computer Programming*. Addison-Wesley Professional, 2011.

4 The Wiener Stock Price Model and the Basic Principles of Black-Scholes Theory

Abstract

We give basic tools for the statistical analysis of financial data. Especially we introduce the basic parameters trend, volatility, and correlation for stocks. Then we motivate and derive the Wiener stock price model (geometric Brownian motion). Thereby we discuss the goals and the meaning of mathematical modeling in general. We then show that—if parameters are chosen in a suitable way—the binomial N-step-model converges to the Wiener model. As a consequence we derive the Black-Scholes formula, i.e. the formula for the fair price of derivatives over an underlying which follows a Wiener model. We extensively discuss this pricing formula for plain vanilla call and put options and for strategies built by combinations of these options. We define the "Greeks" (i.e., the derivations of the fair prices with respect to different parameters), and we discuss the Greeks for various examples in detail. Finally we explain how hedging derivatives in a Wiener model is carried out.

Keywords

Trend · Volatility · Distribution of stock returns · Heavy-tail phenomenon · Correlations · Basic tools from statistics of financial markets · The role of mathematical modeling · The Wiener stock price model · Simulation of stock price developments · Brownian motion · Geometric Brownian motion · The Black Scholes formula · Hedging in the Wiener model · Discussion of pricing formulas for puts · Calls and option combinations · The Greeks · Delta/gamma hedging

The birth of modern financial mathematics can probably be said to have occurred when the famous Black-Scholes formulas for the valuation of stock options, developed by Fischer Black and Myron Scholes, were published in the early 1970s.

G. Larcher, *The Art of Quantitative Finance Vol.1*, Springer Texts in Business and Economics, https://doi.org/10.1007/978-3-031-23873-4_4

These formulas in their fundamental basic version is what we are going to focus on in this chapter. To do that, however, we first need a realistic stock price model in which to move and which to use in the valuation of derivatives. So far, our derivative valuations have been based on binomial models, whose real-world applicability is highly questionable.

In the following therefore, we are going to derive the so-called Wiener stock price model and verify to what extent it holds in real-world settings. Afterward, we will use that model for option pricing and, based on that, derive the basic Black-Scholes formulas in an elementary way. First, however, we need to discuss some basic tools from probability theory and key statistical concepts.

4.1 Basic Tools for Analysing Real Stock Prices: Trend, Volatility, Distribution of Returns, Skewness, and Kurtosis

The following observations are solely meant to serve **as an introduction to basic tools** which we are going to need below; they have nothing to do with professional time series analysis! For a more in-depth discussion of the analysis of financial data, you may find Schmid and Trede's introduction to financial market statistics [1] of interest, for example. We start by observing the price movements of a stock (or stock index) over a certain period of time. The observed stock prices $A_0, A_1, \ldots, A_N$ in this case are measured at (reasonably) regular intervals. They may be the closing prices of that stock on $N + 1$ consecutive trading days, or the opening prices of that stock on the first trading day of each month over a period of $N + 1$ months, or $N + 1$ price data at 10-second intervals—whatever we choose. ("Reasonably" here refers to the fact that while a unit of time, such as a month, may not be constant in size over time, it still represents a clearly defined, hence "reasonably" constant time interval.) Those "reasonably" regular consecutive intervals are referred to below as the **time unit** (trading day, month, 10 s, ...) We denote the time period from the first price determination A_0 to the last price determination A_N by $[0, T]$ and refer to this time period as the **time range**.

To illustrate the following arguments and for the computations that we are going to run as examples, we use the daily closing prices of the DAX stocks "Deutsche Lufthansa" and "SAP" as from 28 February 2017 to 28 February 2018 given in the Table 4.1 and Plot 4.1.

In the following we denote the LHA stock price by $A_0, A_1, \ldots, A_N$ and the SAP stock price by $B_0, B_1, \ldots, B_N$, where N is 256.

First, we calculate the daily returns based on the price movements of these two stocks. There are two types of returns, discrete-period returns and continuously compounded returns.

Table 4.1 Daily closing prices of various DAX stocks from 28 March 2017 to 13 March 2017

Date	LHA	SAP	LHA returns	SAP returns	LHA returns	SAP returns
28.02.2017	13.83	87.95	**Discrete**	**Discrete**	**Continuous**	**Continuous**
01.03.2017	14.06	89.37	0.01663051	0.01614561	0.016493741	0.016016584
02.03.2017	13.93	89.44	−0.00924609	0.00078325	−0.009289099	0.000782954
03.03.2017	13.9	89.28	−0.00251256	−0.00178894	−0.002155948	−0.001790511
06.03.2017	13.89	88.94	−0.00071968	−0.00380821	−0.000719683	−0.003815514
07.03.2017	13.73	89.16	−0.01152323	0.0024736	−0.011585937	0.002470523
08.03.2017	14.12	89.53	0.0287796	0.00414979	0.028009012	0.004141256
09.03.2017	14.49	89.5	0.02620397	−0.00033507	0.025866524	−0.000335139
10.03.2017	14.34	89.03	−0.01035197	−0.00525141	−0.010405921	−0.005265234
13.03.2017	14.32	89.16	−0.00174338	0.00146024	−0.001395674	0.001459117

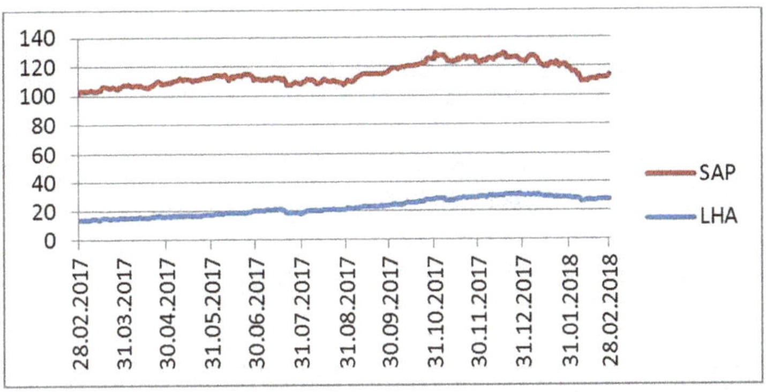

Fig. 4.1 Excerpted data/returns and charts for Deutsche Lufthansa and SAP 28 February 2017 to 28 February 2018

The **discrete return** a_i of the price movement from day i to day $i + 1$ (or in the general case, "from the i-th point in time to the $i + 1$-th point in time") is calculated as

$$a_i = \frac{A_{i+1} - A_i}{A_i} \qquad \text{(discrete return)}$$

a_i is therefore the relative price increase from time i to time $i + 1$.
a_i corresponds to the "discrete rate of return" that changes the value A_i to A_{i+1} (i.e. $A_{i+1} = A_i \cdot (1 + a_i)$).

For example: The discrete return for the LHA stock for the trading period from its closing price on 28 February 2017 to its closing price on 1 March 2017 is

$$a_0 = \frac{(14.06 - 13.83)}{13.83} = 0.01663\ldots$$

The **continuous return** a_i of the price movement from day i to day $i+1$ (or in the general case, "from the i-th time to the $i+1$-th time") is calculated as

$$a_i = \log\left(\frac{A_{i+1}}{A_i}\right) \qquad \text{(continuous return)}$$

where "log" denotes the natural logarithm.

a_i then corresponds to the "continuous return" that changes the value A_i to A_{i+1} (i.e. $A_{i+1} = A_i \cdot e^{a_i}$).

For short periods of time, the values of discrete and continuous returns usually differ only slightly.

For example: The continuous return for the LHA stock for the trading period from the closing price on 28 February 2017 to the closing price on 1 March 2017 is

$$a_0 = \log\left(\frac{14.06}{13.83}\right) = 0.01649\ldots$$

In the following, we are going to work with the continuous returns a_i for the LHA stock and b_i for the SAP stock, where i runs from 0 to $N-1 = 254$.

We calculate the mean over the returns a_i in the observed time range, which, for the LHA stock, gives us the value

$$\boldsymbol{\mu'_A = \frac{1}{N}\sum_{i=0}^{N-1} a_i} = 0.002713\ldots$$

μ'_A is the trend of the LHA stock per day in the time range from 28 February 2017 to 28 February 2018.

In general, we refer to $\boldsymbol{\mu'_A}$ as the (historical) **trend of stock *A* per unit of time in the time range [0, *T*]**.

For the SAP stock, we get the trend

$$\boldsymbol{\mu'_B = \frac{1}{N}\sum_{i=0}^{N-1} b_i} = -0.000079\ldots$$

The next metric we are going to look at is the standard deviation of the returns. Put simply, the standard deviation measures the **approximate (!) average deviation of**

each return from the mean of the returns, i.e. from the trend. More precisely, the standard deviation σ'_A measures the square root of the mean square deviation of the returns from the trend, i.e.

$$\sigma'_A = \sqrt{\frac{1}{N}\sum_{i=0}^{N-1}\left(a_i - \mu'_A\right)^2}$$

We refer to the standard deviation $\boldsymbol{\sigma'_A}$ as the (historical) **volatility of the stock *A* per time unit in the time range $[0, T]$**. Volatility is one of the most common measures of how strongly a stock price fluctuates.

In our example, we find that the volatility of the LHA stock per day in the time range from 28 February 2017 to 28 February 2018 is

$$\sigma'_A = 0.016548\ldots,$$

and for the volatility of the SAP stock per day in the time range from 28 February 2017 to 28 February 2018, we get

$$\sigma'_B = 0.009207\ldots.$$

To non-mathematicians, it is often not clear why "volatility" as a measure of fluctuation isn't calculated using the seemingly more intuitive average **distance** of returns from the trend.

To a mathematician, however, the correct definition of volatility is in fact the "clearer" term, namely, the conventional, geometrically illustrative **Euclidean distance** between two vectors (normalized by $\frac{1}{\sqrt{N}}$), i.e. between the vector of the actual returns and the constant vector in which all entries are constantly equal to the trend. The term "illustrative" is, of course, somewhat relative here, given that we are operating in an N-dimensional space. Yet in an N-dimensional space, the distance between two vectors is calculated in exactly the same way as in an—easily imagined—three-dimensional space:

The Euclidean distance between two three-dimensional vectors $\begin{pmatrix} a_1 \\ a_2 \\ a_3 \end{pmatrix}$ and $\begin{pmatrix} \mu \\ \mu \\ \mu \end{pmatrix}$ is given by the length of the difference vector $\begin{pmatrix} a_1 - \mu \\ a_2 - \mu \\ a_3 - \mu \end{pmatrix}$, i.e. by $\sqrt{(a_1 - \mu)^2 + (a_2 - \mu)^2 + (a_3 - \mu)^2}$.

(continued)

If we normalize this distance by multiplying $\frac{1}{\sqrt{3}}$, we get precisely the definition for σ'_A in the specific case of $N = 3$.
This type of distance is also much better to handle mathematically, since this concept of distance originates in an "inner product", meaning that we are operating within a so-called Hilbert space, in which much more sophisticated mathematical techniques can be applied.

We have now defined the "trend μ'_A of stock A in the time range $[0, T]$ per unit of time" and the "volatility σ'_A of stock A in the time range $[0, T]$ per unit of time".

Still, for better comparison, these variables are almost always "normalized", that is, **the trend and volatility variables are practically always normalized to "per annum" and are always given in this form** (unless explicitly stated otherwise).

This normalization is done as follows:

If the unit of time in the expression "trend μ'_A of stock A in the time range $[0, T]$" is given in years, i.e. if **unit of time = x years**, the **trend μ_A in the time range $[0, T]$ per annum** results from normalizing

$$\boldsymbol{\mu_A = \frac{1}{x} \cdot \mu'_A.}$$

(Later, we will occasionally denote the unit of time x by dt and set $T = N \cdot dt$. In this case, we would then have $\frac{1}{x} = \frac{1}{dt} = \frac{N}{T}$, and normalization of the trend would be $\mu_A = \frac{N}{T} \cdot \mu'_A$. meaning that in this case $\boldsymbol{T \cdot \mu_A = N \cdot \mu'_A}$.)

For example:

A year has 12 months. If the trend μ'_A in the time range $[0, T]$ was calculated from monthly data, meaning that the unit of time is "month", then the "time range $= \frac{1}{12}$ years", and we get the "trend μ_A per annum in the time range $[0, T]$" through

$$\mu_A = 12 \cdot \mu'_A.$$

Or, **for example**:

A year has approximately 255 trading days. If the trend μ'_A in the time range $[0, T]$—like in our example—was calculated from data on each trading day, meaning that the unit of time is "trading day", then the "time range $= \frac{1}{255}$ years", and we get the "trend μ_A per annum in the time range $[0, T]$" through

$$\mu_A = 255 \cdot \mu'_A.$$

In our example, this means:

For the LHA stock,

$$\mu_A = 255 \cdot \mu'_A = 255 \cdot 0.002713\ldots = 0.6917\ldots.$$

For the SAP stock,

$$\mu_B = 255 \cdot \mu'_B = 255 \cdot (-0.000079\ldots) = -0.0201\ldots.$$

The per annum values for the trend are often quoted in percent, i.e. $\mu_A = 69.17\%$ and $\mu_B = -2.01\%$.

The **mathematical rationale** for this type of trend normalization is as follows:
The time range that we are analysing is $[0, T]$, where T is given in years.
The period is x years. This means that $T = N \cdot x$.
The returns per x in the time range are $a_0, a_1, \ldots, a_{N-1}$ (and have an average value of μ'_A).
For each i, $A_{i+1} = A_i \cdot e^{a_i}$ and therefore

$$A_N = A_0 \cdot e^{a_0+a_1+\ldots+a_{N-1}} = A_0 \cdot e^{N\cdot\mu'_A} = A_0 \cdot e^{\frac{T}{x}\cdot\mu'_A}.$$

In the specific case that $T = 1$ year, or if we increase the interval $[0, T]$ to 1 year (assuming that the returns continue to behave approximately as in the interval $[0, T]$), that would mean $A_N \approx A_0 \cdot e^{\frac{1}{x}\cdot\mu'_A}$.
The continuous 1-year return μ_A, i.e. the value for which $A_N = A_0 \cdot e^{\mu_A}$, would thus satisfy $\boldsymbol{\mu_A \approx \frac{1}{x} \cdot \mu'_A}$.

Normalization of the standard deviation is done as follows:

If the unit of time in the expression "volatility σ_A of stock A in the time range $[0, T]$" is given in years, i.e. if **unit of time** = x **years**, the **volatility** $\boldsymbol{\sigma'_A}$ **in the time range** $\boldsymbol{[0, T]}$ **per annum** results from normalizing

$$\boldsymbol{\sigma_A = \frac{1}{\sqrt{x}} \cdot \sigma'_A}.$$

(As already mentioned above when we normalized the trend, we will occasionally denote the time unit x by dt and set $T = N \cdot dt$. In this case, we would then have $\frac{1}{x} = \frac{1}{dt} = \frac{N}{T}$, and the normalization of the volatility would be $\sigma_A = \frac{\sqrt{N}}{\sqrt{T}} \cdot \sigma'_A$, meaning that in this case $\boldsymbol{\sqrt{T} \cdot \sigma_A = \sqrt{N} \cdot \sigma'_A}$.)

For example:

A year has 12 months. If the volatility σ'_A in the time range $[0, T]$ was calculated from monthly data, meaning that the unit of time is "month", then the "time range = $\frac{1}{12}$ years", and we get the "volatility σ_A per annum in the time range $[0, T]$" through

$$\sigma_A = \sqrt{12} \cdot \sigma'_A.$$

Or **for example**:

A year has approximately 255 trading days. If the volatility σ'_A in the time range $[0, T]$—like in our numerical example—was calculated from data on each trading day, meaning that the unit of time is "trading day", then the "time range = $\frac{1}{255}$ years", and we get the "volatility σ_A per annum in the time range $[0, T]$" through

$$\sigma_A = \sqrt{255} \cdot \sigma'_A \approx 16 \cdot \sigma'_A.$$

In our example, this means:

For the LHA stock,

$$\sigma_A = \sqrt{255} \cdot \sigma'_A = 255 \cdot 0.016589 \ldots = 0.2649 \ldots.$$

For the SAP stock,

$$\sigma_B = \sqrt{255} \cdot \sigma'_B = 255 \cdot 0.00923 \ldots = 0.14747 \ldots.$$

The per annum values for volatility are also often quoted in percent, i.e. $\sigma_A = 26.49\%$ and $\sigma_B = 14.75\%$.

The **mathematical rationale** for this type of normalization of volatility is as follows:
As we noted above upon normalization of the trend, we have the relationship $A_N = A_0 \cdot e^{a_0+a_1+\ldots+a_{N-1}}$. If specifically $T = 1$, i.e. $N = \frac{1}{x}$, then the return per annum is given by the expression $Y := a_0 + a_1 + \ldots + a_{N-1}$. All the a_i are random variables with the mean value μ'_A and the standard deviation σ'_A, i.e. variance $\left(\sigma'_A\right)^2$. If we assume that the a_i are independent of one another, then the variance of the random variable Y (return per annum) is the sum of the variances of the a_i, i.e. $N \cdot \left(\sigma'_A\right)^2$.

(continued)

The standard deviation σ_A of the random variable Y (return per annum) is thus

$$\sigma_A = \sqrt{N \cdot \left(\sigma_A'\right)^2} = \sqrt{N} \cdot \sigma_A' = \frac{1}{\sqrt{x}} \cdot \sigma_A' .$$

With trend μ_A' and volatility σ_A', we now have values for the average magnitude and the average fluctuation range of the (continuous) returns a_i of a stock in the time interval $[0, T]$ per unit of time. If we were to present the values of the returns as a histogram, we would generally get a very typical picture, as can be seen in the example of the LHA and SAP stocks in Figs. 4.2 and 4.3.

By "typical" we mean a graphical representation reminiscent of a normal distribution. Judging by what these histograms look like, we might conclude that the stock returns behave like normally distributed random variables with the estimated trend μ_A' and estimated volatility σ_A', i.e. we might be led to assume that the density function of the returns takes the form

$$f_A(x) = \frac{1}{\sqrt{2\pi \left(\sigma_A'\right)^2}} \cdot e^{\frac{-\left(x-\mu_A'\right)^2}{2\cdot\left(\sigma_A'\right)^2}} .$$

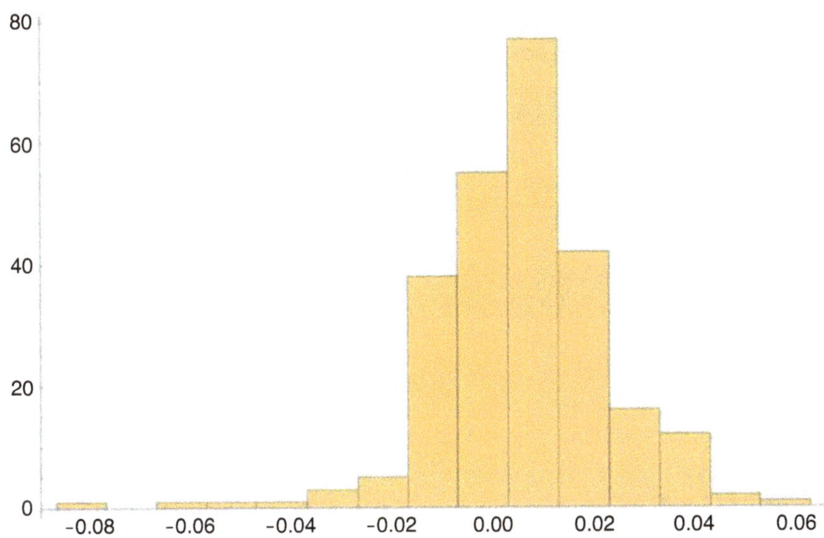

Fig. 4.2 Histogram of the distribution of LHA stock's daily returns from 28 February 2017 to 28 February 2018

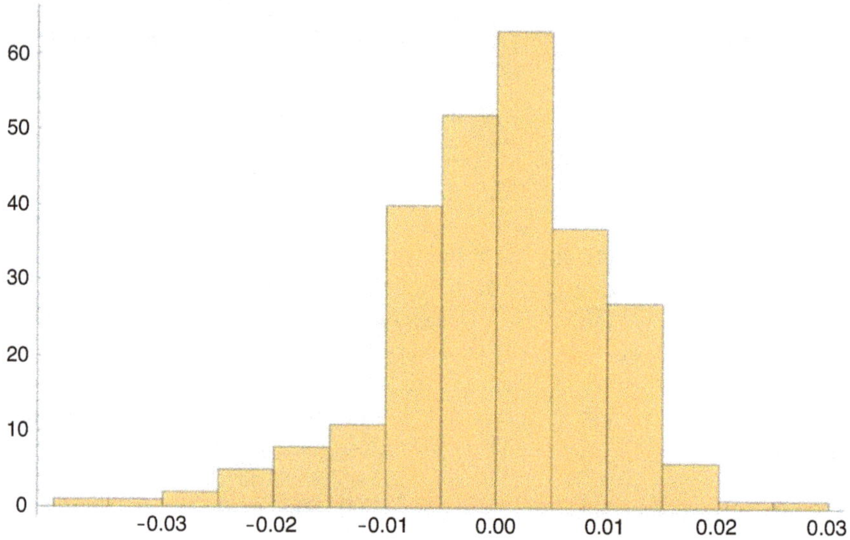

Fig. 4.3 Histogram of the distribution of SAP stock's daily returns from 28 February 2017 to 28 February 2018

In order to compare the histograms in Figs. 4.2 and 4.3 with the density function of the normal distribution, we first need to normalize the histograms, and we need to do that in a way so that the area of the histogram has the value 1 (as is the case for the area between the density function f_A and the x-axis). Right now, the area F of the columns ("bins") that form the respective histogram is obviously

$$F = \text{width of the histogram intervals} \times \text{number of returns considered}$$

If we divide the height of the histogram by this value F, then the area of the histogram has the value 1.

In the example of the LHA stock (Fig. 4.2), F has the value $F = 0.01 \times 256 = 2.56$, and in the example of the SAP stock (Fig. 4.3), F has the value $F = 0.005 \times 256 = 1.28$.

Figures 4.4 and 4.5 show the histograms after we have normalized them in this way, as well as the superimposed density functions of the normal distribution.

In principle, the comparison is reasonably satisfactory in both Figs. 4.4 and 4.5. The hypothesis that the returns of the stock prices show an "approximately normally distributed" behaviour is definitely tenable.

Let us look at some more visualizations (various additional examples of comparing empirical distribution functions with the associated normal distribution densities

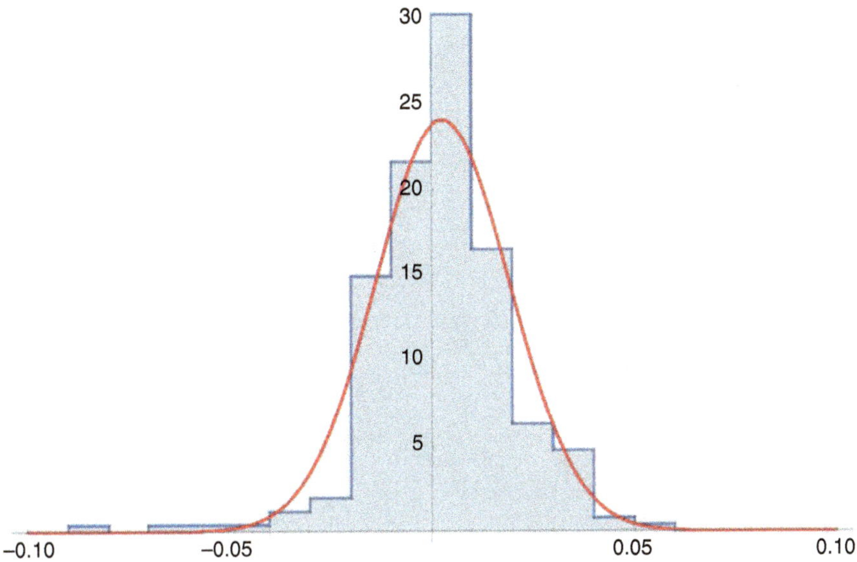

Fig. 4.4 Comparison of the empirical density function of LHA stock's daily returns with the density of the normal distribution

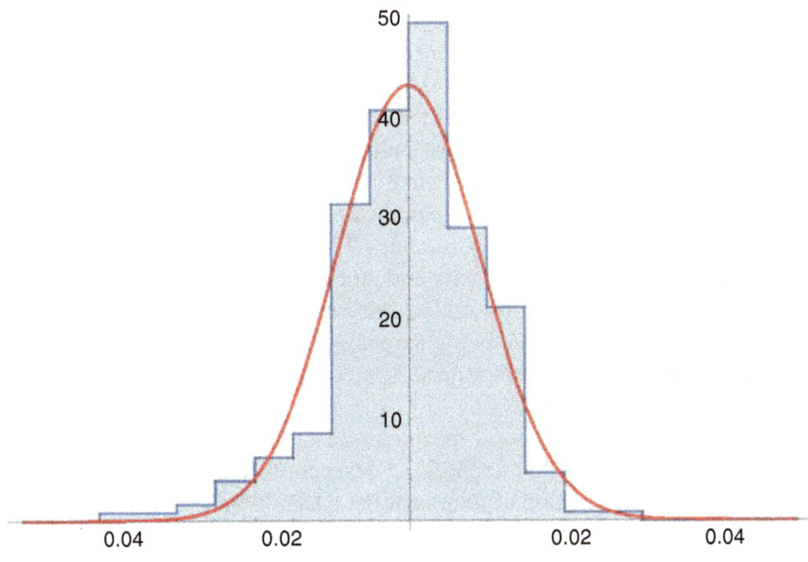

Fig. 4.5 Comparison of the empirical density function of SAP stock's daily returns with the density of the normal distribution

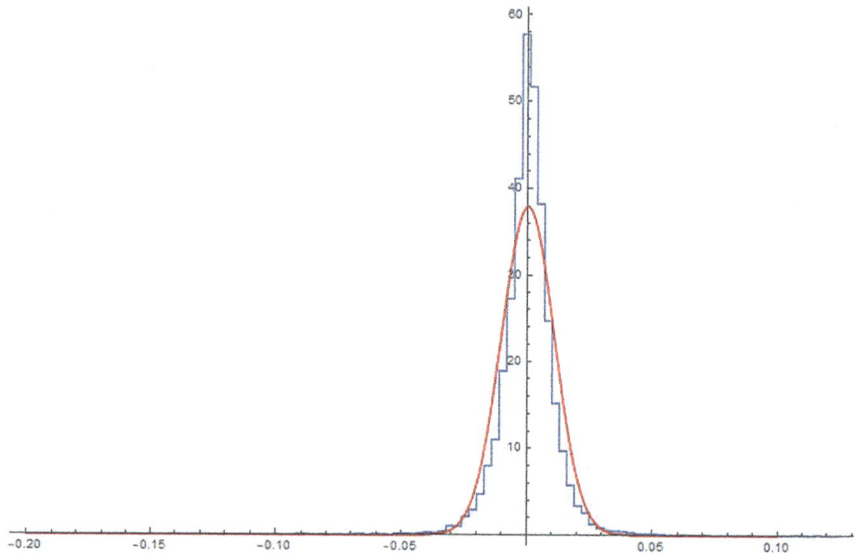

Fig. 4.6 Historical daily returns of S&P500 since 1970

can be found on the book's website and can also be created there; see: https://app.lsqf.org/book/normal-distributed-profits):

Figure 4.6 shows the historical distribution of the daily returns for the S&P500 index since 1970, compared to the density of the normal distribution function for the corresponding trend and volatility parameters.

Figure 4.7 shows the historical distribution of the daily returns for the Apple stock since 1980, compared to the density of the normal distribution function for the corresponding trend and volatility parameters.

In all of these Figs. 4.4, 4.5, 4.6, and 4.7 and also in the vast majority of other histograms of returns on stocks and stock indices, we immediately notice a strikingly systematic deviation of the historical distribution from the normal distribution: namely, a consistently higher peak of the historical distribution in the area of the trend and, on the other hand, somewhat lower values in the historical distribution's mean value range.

A second systematic deviation—meaning a deviation that can be observed quite frequently—is only noticeable upon closer inspection. Figures 4.8 and 4.9 show zooms of Fig. 4.6 (distribution of returns of the S&P500 index) for the tails of that distribution (from return values of 0.03 in Fig. 4.8 and up to return values of −0.03 in Fig. 4.9).

Figures 4.10 and 4.11 show zooms of Fig. 4.7 (distribution of Apple stock returns) for the tails of that distribution (from return values of 0.05 in Fig. 4.10 and up to return values of −0.05 in Fig. 4.11).

Here, as well as in the overwhelming number of further examples, we notice a heightened probability for strongly higher-than-average and strongly lower-than-

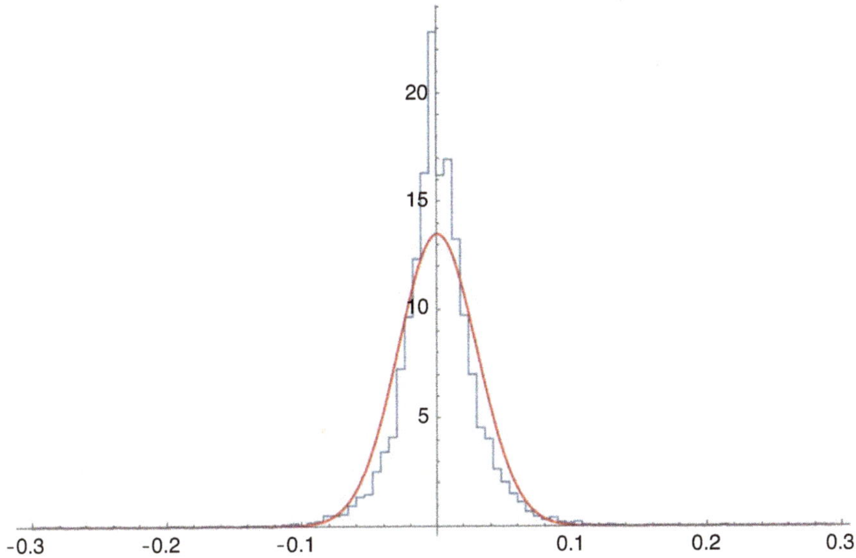

Fig. 4.7 Historical daily returns of Apple stock since 1980

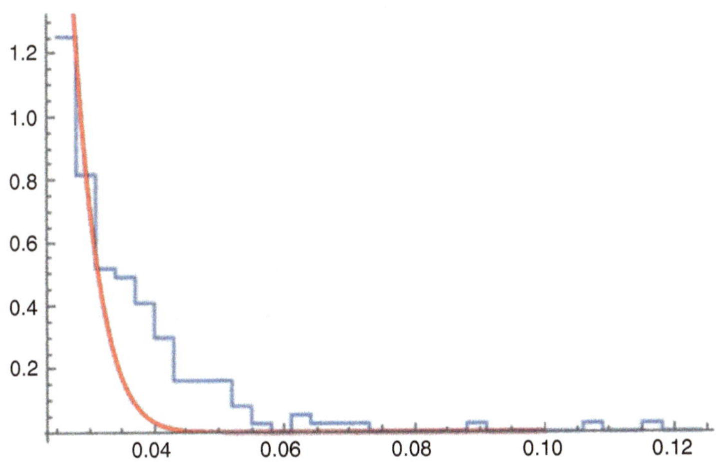

Fig. 4.8 Distribution of SPX returns, right tail

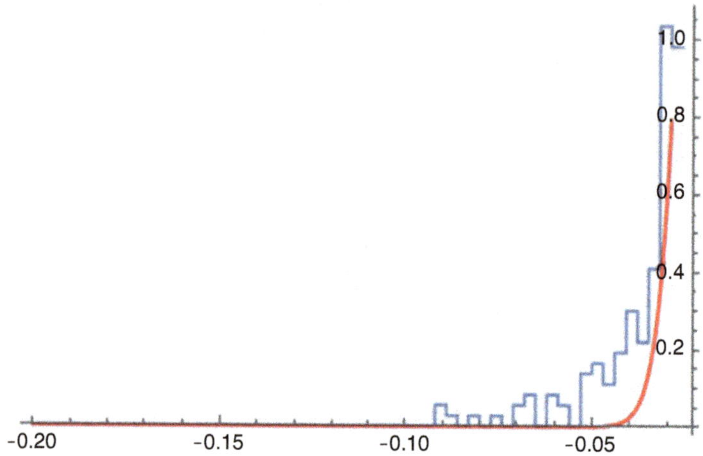

Fig. 4.9 Distribution of SPX returns, left tail

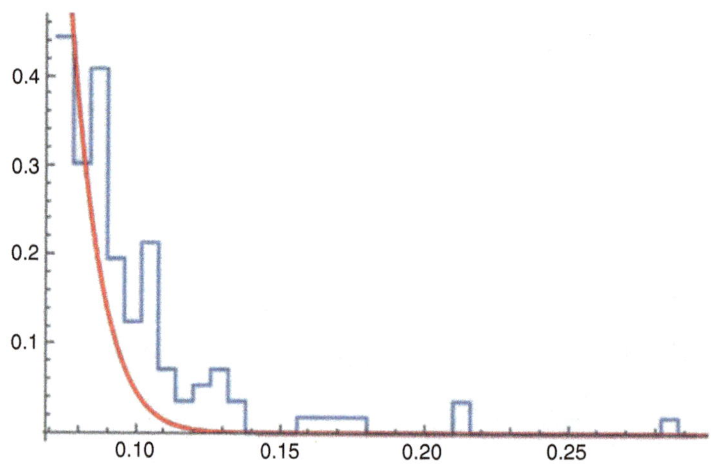

Fig. 4.10 Distribution of Apple returns, right tail

average returns in comparison to the normal distribution. This is referred to as the **fat-tail or heavy-tail phenomenon**. Extreme changes in returns occur much more frequently than would be predicted assuming a normal distribution of returns.

The fat-tail phenomenon can be illustrated even better by comparing the empirical and theoretical *distribution* functions instead of the empirical and theoretical *density* functions.

A distribution function F of a random variable Z indicates for each x the probability $F(x)$ that the value of that random variable Z is less than x. The following Figs. 4.12, 4.13, 4.14, 4.15, 4.16, and 4.17 show the empirical distribution

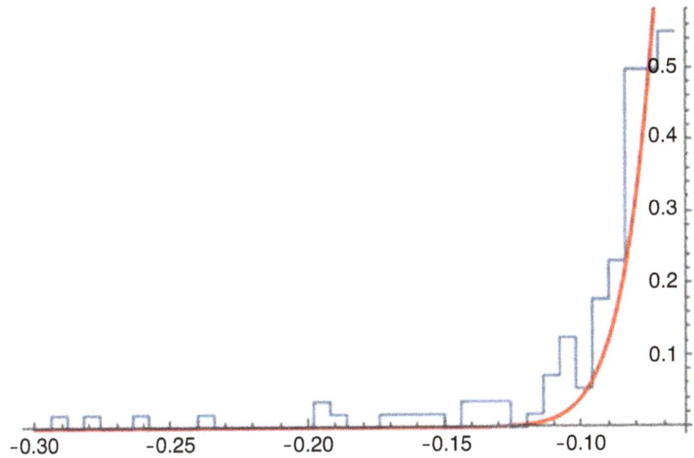

Fig. 4.11 Distribution of Apple returns, left tail

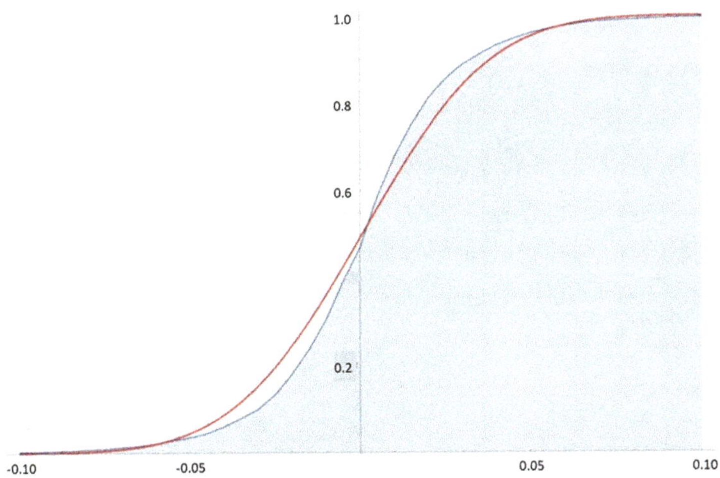

Fig. 4.12 Empirical distribution function (blue) of Apple stock returns vs. normal distribution function (red)

functions of the S&P500 index and the Apple stock in comparison with the theoretical distribution functions of the corresponding normal distributions, in each case with a section of the left and right tails of the figures.

In all of these representations, we can clearly see that the left tail of the empirical distribution function is always above the normal distribution, while the right tail is always below the normal distribution. This means that very low returns and very high returns are more likely to occur than a normal distribution would suggest.

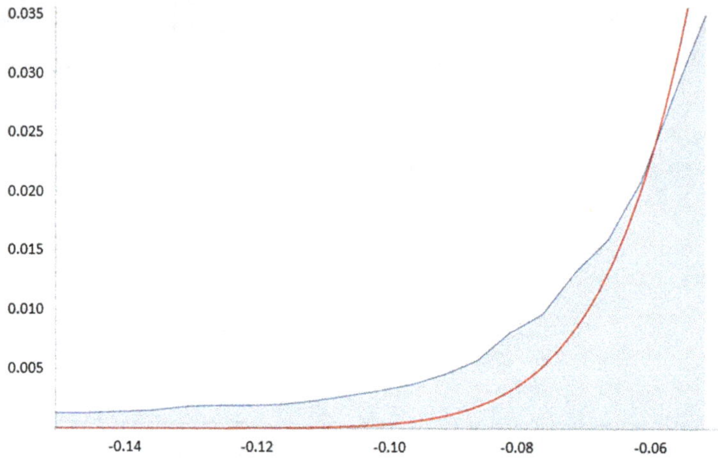

Fig. 4.13 Empirical distribution function (blue) of Apple stock returns vs. normal distribution function (red), left tail

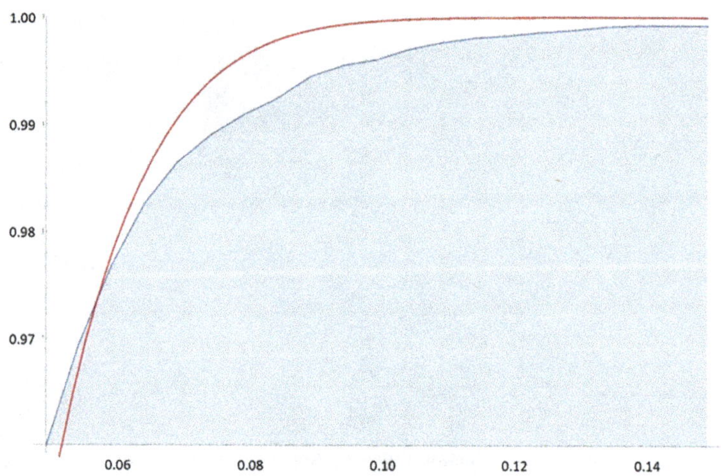

Fig. 4.14 Empirical distribution function (blue) of Apple stock returns vs. normal distribution function (red), right tail

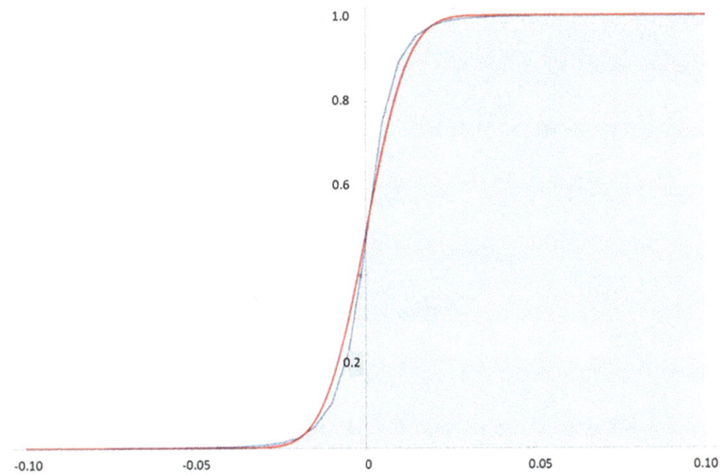

Fig. 4.15 Empirical distribution function (blue) of S&P500 returns vs. normal distribution function (red)

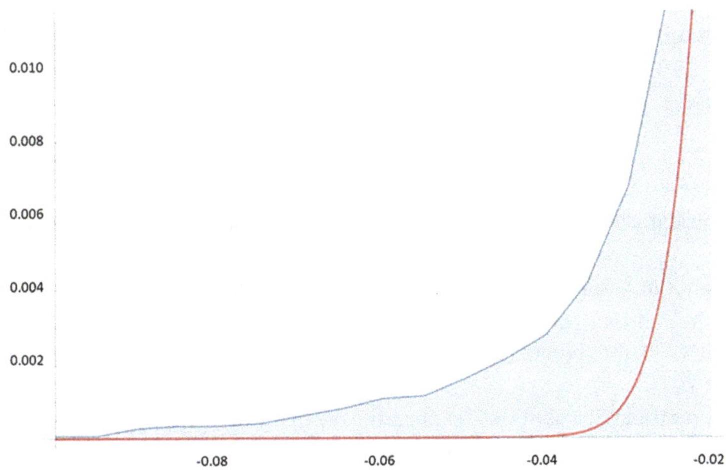

Fig. 4.16 Empirical distribution function (blue) of S&P500 returns vs. normal distribution function (red), left tail

So far, we have identified "narrower and higher middle parts" and "fat tails" as systematic deviations of most return distributions for stocks and stock indices.

Another systematic deviation is a slight "left skew" in empirical return distributions. This means the "peak" of the empirical density function is usually located slightly to the right of the peak of the normal distribution, and the empirical density function is more steeply inclined on the right side and less inclined (more drawn out) on the left. This type of deviation is often barely visible to the eye. However, it can be indexed by means of a further statistical measure, "the skewness" of distribution.

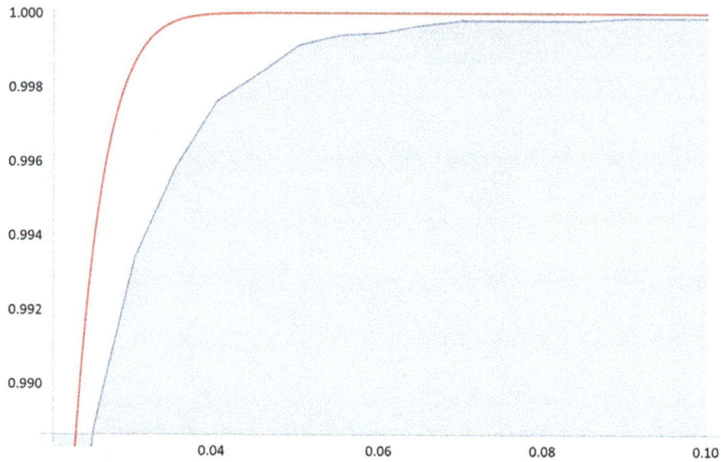

Fig. 4.17 Empirical distribution function (blue) of S&P500 returns vs. normal distribution function (red), right tail

The **empirical skewness** v'_A of the returns that we are analysing is calculated as

$$v'_A = \frac{1}{N} \sum_{i=0}^{N-1} \left(\frac{a_i - \mu'_A}{\sigma'_A} \right)^3 .$$

The empirical skewness is thus the **mean of the third powers of the normalized returns**.

If a random variable Z has an expected value of μ and a standard deviation of σ, then the random variable $Y = \frac{Z-\mu}{\sigma}$ has the expected value 0 and the standard deviation 1. In this context, we refer to Y as the **normalization of the random variable** Z.

A symmetrical distribution—such as the normal distribution—always has a skew equal to 0. A **positive skew** indicates a **right-skew distribution** (the peak of the distribution is located to the left of μ, and the left tail is steeper and shorter than the one on the right). A **negative skew** indicates a **left-skew distribution** (the peak of the distribution is located to the right of μ, and the right tail is steeper and shorter than the one on the left).

The skewness values we get in our above longer-term example data are:

$$\begin{array}{ll} \text{Apple} & -1.784\ldots \\ \text{S\&P500} & -1.018\ldots \end{array}$$

The slight left skew of empirical return distributions that we asserted above is thus confirmed in these two examples.

We would like to cite another measure used in distributions here, namely, the **kurtosis (= tailedness)** of a distribution, or the **excess** of a distribution. Kurtosis ω_A is calculated in the same way as skewness, except that kurtosis is the **mean value of the fourth powers of the normalized returns**, i.e.

$$\omega_A = \frac{1}{N} \sum_{i=0}^{N-1} \left(\frac{a_i - \mu'_A}{\sigma'_A} \right)^4.$$

The kurtosis of the normal distribution is 3.

The **excess γ compares the kurtosis of a random variable with the kurtosis of the normal distribution** and is therefore defined as

$$\gamma = \omega_A - 3.$$

The empirical distributions of stock price returns usually exhibit a distinctly positive excess. This points to a systematical presence of a narrower, steeper and higher middle part in the empirical distribution of stock price returns compared to the normal distribution (as can be clearly seen in Figs. 4.4 and 4.5 but above all in 4.6 and 4.7).

The values we get for the excess in our above longer-term sample data are:

Apple 43.836...
S&P500 24.599...

Let us briefly summarize the results so far with regard to a potential normal distribution of stock price returns:

Stock price returns generally exhibit an empirical distribution that is quite similar to a normal distribution. However, the empirical distributions typically have narrower, higher middle parts, and thicker ends toward the edges (fat tails, heavy tails) than the normal distribution; in particular we often see a negative skew of the distribution and a distinctly positive excess.

Yet, despite this observation, standard models for stock price movements are still based on the assumption of a normal distribution. This assumption is justifiable in many applications and is of great advantage for the mathematical treatment of many highly complex financial mathematical analyses. Of course there are mathematical models that allow a much better representation of the actual empirical distribution of stock price returns, but these models often have other serious disadvantages in terms of mathematical treatment. Yet this is something we will discuss in greater detail in subsequent chapters of this book.

The aim of this chapter is to derive *the* basic model for the simulation of stock price movements: the Wiener stock price model. And the first assumption underlying this model is precisely the assumption of normally distributed returns.

4.2 Basic Tools for Analysing Real Stock Prices: Covariances and Correlations

The second assumption underlying the Wiener model is the following:

Stock price returns are independent of past returns of that stock.

In other words: No relevant information on potential future returns of a stock can be derived from tracking and analysing past returns (e.g. those of the previous day, or of 2 days earlier, ...) of that stock.

Again a certain scepticism about this assumption is called for, of course, as we know from our results for the run analysis of the S&P500 index in Sect. 1.24. There, we found that, for example, 4 consecutive positive trading days are (empirically) more likely to be followed by a negative trading day (a negative return) than 4 consecutive negative trading days. Thus, this is a very strong indication of a certain interdependence of consecutive returns.

In order to analyse and quantify the strength of this dependence with greater accuracy, we need the terms **covariance** and **correlation** (and subsequently **auto-correlation**). Covariance and correlation are used to quantify dependencies between two time series.

As an example, we will again take the returns $a_0, a_1, \ldots, a_{N-1}$ of the Lufthansa stock (A) and the returns $b_0, b_1, \ldots, b_{N-1}$ of the SAP stock (B) (where $N = 255$ in our example). We defined the means of the returns (trends of stock A and B) as μ'_A and μ'_B and the standard deviations as σ'_A and σ'_B.

The empirical covariance in the time range $[0, T]$ per unit of time of these two return sequences is defined as

$$cov_{A,B} = \frac{1}{N}\sum_{k=0}^{N-1}\left(a_k - \mu'_A\right)\cdot\left(b_k - \mu'_B\right)$$

For the daily price data of SAP and LHA from 28 February 2017 to 28 February 2018, this gives us $cov_{A,B} = 0.000054$.

The covariance value does not generally tell us much, since it is dependent not only on the correlation of the two time series but also on the individual values and again especially on the unit of time.

If instead of the returns a_k, for example, we were to use the tenfold values $d_k = 10 \cdot a_k$, then the dependency structure between d_k and b_k would obviously be exactly the same as between a_k and b_k. However, the covariance between d_k and b_k is ten times the covariance between a_k and b_k.

So, in order to obtain a meaningful measure of the dependence, the covariance must first be adequately normalized (by the respective standard deviations). This gives us the **correlation $\rho_{A,B}$ between the returns of stock A and B** through

$$\boldsymbol{\rho_{A,B} = \frac{cov_{A,B}}{\sigma'_A \cdot \sigma'_B}}.$$

The correlation always has a value between −1 and 1. The reason for this can be easily demonstrated with a geometric argument (see box below). For the daily price data of SAP and LHA from 28 February 2017 to 28 February 2018, this gives us the value $\rho_{A,B} = 0.35416\ldots$.

The correlation is now no longer directly dependent on the time series values. For example, the correlation between d_k and b_k (see example above) has exactly the same value as the correlation between a_k and b_k, i.e. also $0.35416\ldots$.

A **clearly positive correlation indicates a positive dependence** between the movements of the two time series. The closer the correlation value is to 1, the stronger the indication that the dependence is positive.

A **clearly negative correlation indicates a negative dependence** between the movements of the two time series. The closer the correlation value is to −1, the stronger the indication that the dependence is negative.

A correlation close to 0 is an indication that the movements in the two time series are likely to be independent of one another.

At this point we need to explain and clarify some terms, especially with a view to preventing some of the mistakes that are often made in this context:

- **Caveat:** So far, we have always stated the values of means (trends), standard deviations (volatilities), covariances, correlations by the symbols of the respective stocks (stock prices, A, B, ...). However, these are **never the means, standard deviations, covariances, or correlations for the stock prices** but **always for the (continuous) returns** of those stock prices!
- A value that is qualified as being "clearly positive" or "clearly negative" is usually understood to mean a value greater than 0.2 or less than −0.2.
- While a value close to 0 (i.e. approximately in the range between −0.2 to 0.2) is an **indication** that the two trends are likely independent, it **does not necessarily imply** that they are actually independent!

The following simple example shows that the correlation can be very close to zero, despite the presence of a strong dependence between the two time series:
For the sequence $X = x_1, x_2, \ldots$, we choose an arbitrary sequence of 0 and 1, for example,

$$X: \; 0\;1\;0\;1\;1\;1\;0\;1\;0\;0\;0\;1\;1\;1\;0$$

We then define the sequence $Y = y_1, y_2, \ldots$ as follows:

- It also starts with 0.
- If x_{k+1} has the same value as x_k, then y_{k+1} also has the same value as y_k.

(continued)

- If x_{k+1} has a different value than x_k, then y_{k+1} increases by 1 relative to y_k.

In our example, this would mean

$$Y: \; 0\ 1\ 2\ 3\ 3\ 3\ 4\ 5\ 6\ 6\ 6\ 7\ 7\ 7\ 8$$

The Y sequence is thus completely defined by the X sequence, i.e. it is completely dependent on X. And yet there is only a very low correlation of -0.01508 between the two sequences.

- A misconception regarding correlations that we often hear is
A strong correlation (close to 1) between the returns of two stock prices A and B means that stock A usually increases when the other stock B also increases.

This interpretation is generally incorrect. To illustrate this, let us look at another simple example:

Stock price movement of A:

$$100, 101, 102, 106, 107, 111, 113, 117, 122, 127, 128, 129, 134, 140, 141$$

Stock price movement of B:

$$100, 94, 88, 87, 81, 80, 75, 73, 69, 68, 64, 60, 59, 57, 54$$

The movements of the two stocks are displayed in the following Fig. 4.18:

The correlation between the returns of these two stocks is 0.866 ... and thus actually close to 1, so there is indeed a strong dependence despite the fact that both stocks are clearly moving in different directions.

This shows that the statement (which we hear really often)

A strong positive correlation (close to 1) between the returns of two stock prices A and B means that stock A usually increases when the other stock B also increases, and vice-versa.

is *not* correct.

Much closer to the truth is the following modified—and much more complex—interpretation of a high correlation:

A strong positive correlation (close to 1) between the returns of two stock prices A and B means that stock A usually increases in relation to its average increase (trend) when the other stock B also increases in relation to its average increase (trend), and vice-versa.

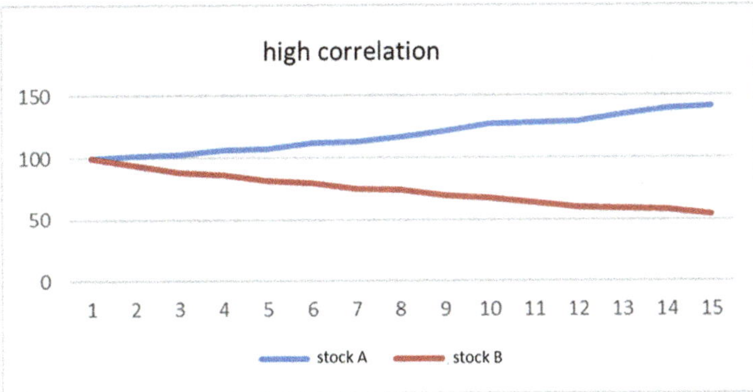

Fig. 4.18 Example of two (surprisingly) strongly positively correlated stock price movements

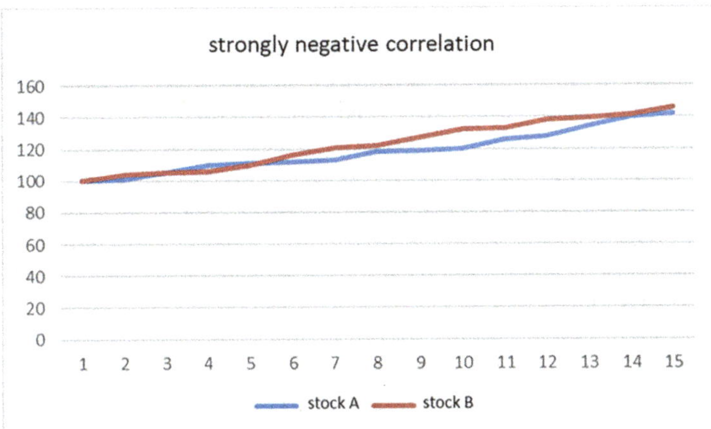

Fig. 4.19 Example of two (surprisingly) strongly negatively correlated stock price movements

In fact, in our example, A evidently exhibits a positive trend, while B evidently exhibits a negative trend. Both time series are strictly monotonic (A increasing, B decreasing). But whenever A rises sharply (e.g. from 102 to 106), B falls moderately (from 88 to 87), and whenever A rises moderately (e.g. from 100 to 101), B falls sharply (from 100 to 94).

The same holds true for the opposite case, of course:

A strong negative correlation (close to -1) between the returns of two stock prices A and B means that stock A usually increases in relation to its average increase (trend) when the other stock B falls in relation to its average increase (trend), and vice-versa. (see Fig. 4.19).

A look at the formula for calculating the covariance, or correlation, also confirms that this explanation of correlation is indeed correct:
After all, we have

$$\rho_{A,B} = \frac{cov_{A,B}}{\sigma_A \cdot \sigma_B}$$

and

$$cov_{A,B} = \frac{1}{N} \sum_{k=0}^{N-1} (a_k - \mu_A) \cdot (b_k - \mu_B)$$

The two standard deviations σ_A and σ_B in the correlation's denominator each depend on one single stock only. In order to obtain a high (positive) correlation, we therefore need the highest possible positive covariance. For this purpose, we need as many positive summands as possible in the above sum. Any summand in this sum, i.e. an expression $(a_k - \mu_A) \cdot (b_k - \mu_B)$, is positive if either both factors $(a_k - \mu_A)$ and $(b_k - \mu_B)$ are positive or if both factors are negative.
This means: In order to obtain a high positive correlation, it is necessary that both $a_k > \mu_A$ and $b_k > \mu_B$ or else both $a_k < \mu_A$ and $b_k < \mu_B$, for as many k as possible. So, what counts is not the absolute value of a_k and b_k but always the relation to μ_A and μ_B.
To mathematicians, the fact that the correlation

$$\begin{aligned} \rho_{A,B} &= \frac{cov_{A,B}}{\sigma_A \cdot \sigma_B} = \frac{\frac{1}{N}\sum_{k=0}^{N-1} (a_k - \mu_A) \cdot (b_k - \mu_B)}{\sigma_A \cdot \sigma_B} = \\ &= \frac{\frac{1}{N}\sum_{k=0}^{N-1} (a_k - \mu_A) \cdot (b_k - \mu_B)}{\sqrt{\frac{1}{N}\sum_{i=0}^{N-1} (a_i - \mu_A)^2} \cdot \sqrt{\frac{1}{N}\sum_{i=0}^{N-1} (b_i - \mu_B)^2}} = \\ &= \frac{\sum_{k=0}^{N-1} (a_k - \mu_A) \cdot (b_k - \mu_B)}{\sqrt{\sum_{i=0}^{N-1} (a_i - \mu_A)^2} \cdot \sqrt{\sum_{i=0}^{N-1} (b_i - \mu_B)^2}} \end{aligned} \tag{4.1}$$

(continued)

is always between -1 and 1 immediately follows from the Cauchy-Schwarz inequality. It is also easily understood when looking at a geometrical representation of the two N-dimensional vectors:

$$\begin{pmatrix} a_0 - \mu_A \\ a_1 - \mu_A \\ \vdots \\ a_{N-1} - \mu_A \end{pmatrix} \text{ and } \begin{pmatrix} b_0 - \mu_B \\ b_1 - \mu_B \\ \vdots \\ b_{N-1} - \mu_B \end{pmatrix}.$$

The numerator in the last expression of Formula (4.1) is the inner product of these two vectors. The value of the inner product of two vectors is, as we know, the Euclidean length of the first vector multiplied by the Euclidean length of the second vector's projection onto the first vector. This projection is of course shorter (or at most equal) in length to that of the second vector.
The denominator in the last expression of Formula (4.1) is precisely the product of the Euclidean lengths of the two vectors.
The absolute value of the correlation must therefore always be less than or equal to 1, meaning: The correlation must always lie between -1 and 1.

A good way to graphically illustrate strong positive or strong negative correlations between two time series is two-dimensional plots of those time series. Let us look at such a plot, for example, for the returns of our LHA and SAP stocks:

To do this, we bundle the returns $a_0, a_1, \ldots, a_{N-1}$ of the LHA stock and the returns $b_0, b_1, \ldots, b_{N-1}$ of the SAP stock into pairs of the form $(a_0, b_0), (a_1, b_1), \ldots,$

(a_{N-1}, b_{N-1}) and enter them into an a, b coordinate system (see Fig. 4.20).

We can see a certain alignment of the pairs of points along a straight line from "bottom left" to "top right" in Fig. 4.21.

This impression is reinforced when a binormal distribution (i.e. a contour plot of the binormal distribution) with the corresponding parameters of the two numerical sequences, i.e. with trends $\mu_A = 0.002713$, $\mu_B = -0.000079$, volatilities $\sigma_A = 0.016548$, $\sigma_B = 0.009207$, and correlation $\rho_{A,B} = 0.35416$, is placed over the image of the pairs.

This "alignment of the pairs of points along a straight line from bottom left to top right" reflects the clearly positive correlation of the two sequences. This alignment along a straight line from bottom left to top right generally becomes clearer the closer the correlation is to $+1$.

On the other hand, the closer the correlation is to -1, the more clearly the pairs of points are grouped from bottom right to top left.

The binormal distribution is a generalized representation of the normal distribution for two-dimensional random variables. Indeed, in our example above, the

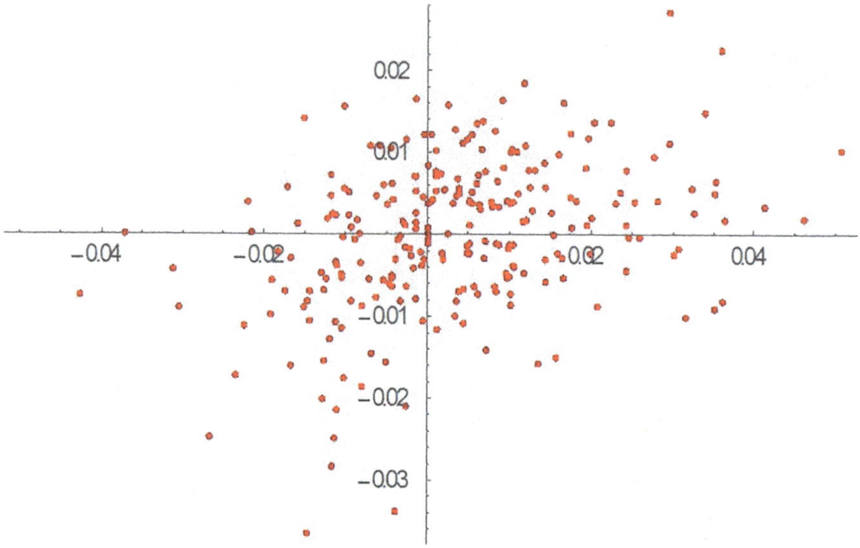

Fig. 4.20 Pairs of returns for the LHA and the SAP stock

pairs (a_k, b_k) form two-dimensional random variables (two-dimensional random vectors).

The density function of the two-dimensional binormal distribution for a sequence $(a_0, b_0), (a_1, b_1), \ldots, (a_{N-1}, b_{N-1})$ with trends μ_A, μ_B and volatilities σ_A, σ_B for each of the coordinate sequences and with correlation ρ is given by

$$f_{A,B}(x, y) = \frac{1}{2\pi\sigma_A\sigma_B\sqrt{1-\rho^2}} \times \exp\left(-\frac{1}{2\left(1-\rho^2\right)}\left[\frac{(x-\mu_A)^2}{{\sigma_A}^2} + \frac{(y-\mu_B)^2}{{\sigma_B}^2} - \frac{2\rho\,(x-\mu_A)\,(y-\mu_B)}{\sigma_A\sigma_B}\right]\right)$$

where $\exp(z) := e^z$ denotes the exponential function.

The density function of the binormal distribution, graphically represented as a two-dimensional area in three-dimensional space, has the shape of a more (correlation close to 0, Fig. 4.22) or less (correlation significantly different from 0, Fig. 4.23) symmetrical bell.

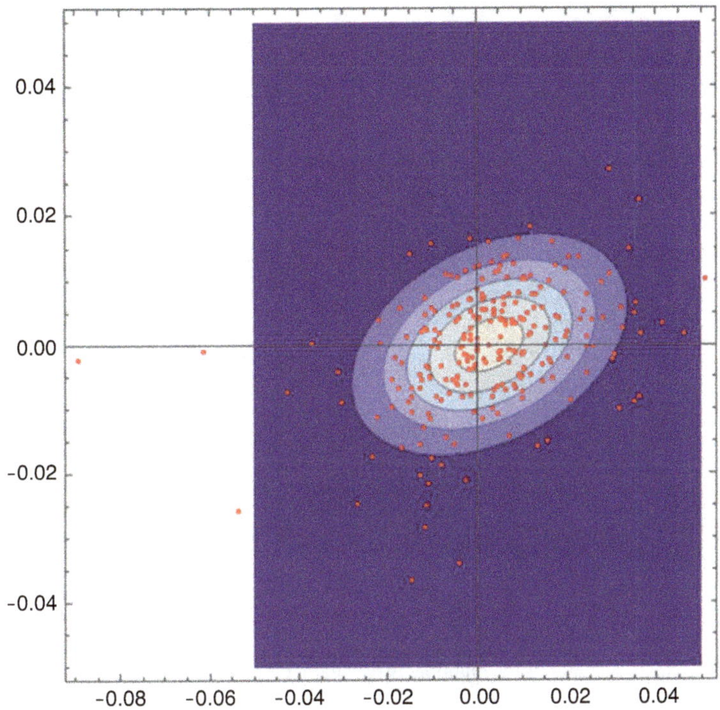

Fig. 4.21 Pairs of returns for the LHA and the SAP stock with a contour plot of a binormal distribution

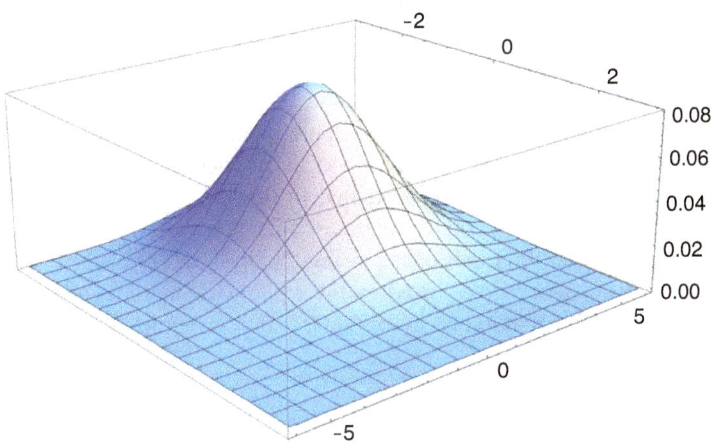

Fig. 4.22 Density function of the binormal distribution, correlation 0

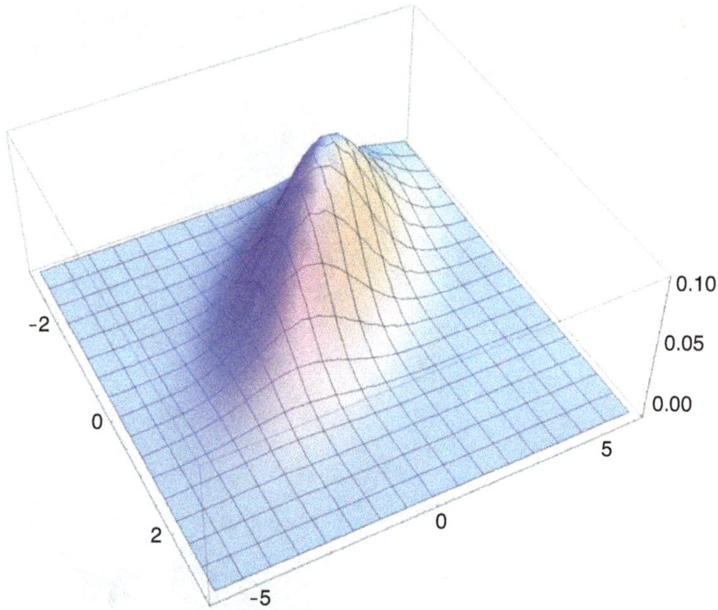

Fig. 4.23 Density function of the binormal distribution, correlation −0.7

The value of the covariance or correlation between the returns of different stocks is of essential relevance in various contexts (portfolio management, risk management). To give you an approximate idea of the typical covariance and correlation values between stocks from a specific stock universe, we are going to look at the stocks of the German stock index DAX below. Specifically, we are going to consider the daily closing rates from 3 consecutive years (18 May 2015 to 18 May 2016, 18 May 2016 to 18 May 2017, and 18 May 2017 to 18 May 2018).

For each of these 1-year periods, we have looked at the covariances or correlations between returns for two DAX stocks in each case. We are not listing all covariances and correlations determined in this way here but present only a small excerpt from the "covariance matrices" and "correlation matrices" of these DAX stocks for these three periods; see Figs. 4.24, 4.25, 4.26, 4.27, 4.28, 4.29, and 4.30. (The notation "E-05" used in the covariance matrices after a number means "division of this number by 100,000". For example: $8.8538E - 05 = 0.000088538$)

We observe the following:

- Almost all of these correlations are clearly positive.
- In all four correlation matrices, there are only two (just slightly) negative values. These are the correlations between Vonovia and Commerzbank in the 2016/2017 test period (correlation −0.049) and between Vonovia and Deutsche Bank in

	BMW	FME	LHA	CON	PSM	BEI	EOAN	HEI	DAI
BMW	8,8538E-05	4,688E-05	4,1029E-05	6,3051E-05	4,3674E-05	2,9406E-05	2,8941E-05	2,9586E-05	6,2213E-05
FME	4,688E-05	0,00014242	5,9975E-05	7,1818E-05	5,3435E-05	5,8461E-05	4,964E-05	5,1802E-05	5,2376E-05
LHA	4,1029E-05	5,9975E-05	0,00029755	6,0686E-05	4,0866E-05	1,6295E-05	2,6821E-05	3,8172E-05	4,1324E-05
CON	6,3051E-05	7,1818E-05	6,0686E-05	0,00013468	5,3944E-05	4,6909E-05	2,8628E-05	4,6909E-05	5,9849E-05
PSM	4,3674E-05	5,3435E-05	4,0866E-05	5,3944E-05	0,00034571	3,9224E-05	2,1769E-05	3,1751E-05	3,6455E-05
BEI	2,9406E-05	5,8461E-05	1,6295E-05	4,6909E-05	3,9224E-05	0,00010413	5,6043E-05	3,5162E-05	3,4189E-05
EOAN	2,8941E-05	4,964E-05	2,6821E-05	2,8628E-05	2,1769E-05	5,6043E-05	0,00021762	3,7716E-05	2,9392E-05
HEI	2,9586E-05	5,1802E-05	3,8172E-05	4,6909E-05	3,1751E-05	3,5162E-05	3,7716E-05	0,00012794	3,4206E-05
DAI	6,2213E-05	5,2376E-05	4,1324E-05	5,9849E-05	3,6455E-05	3,4189E-05	2,9392E-05	3,4206E-05	9,0192E-05

Fig. 4.24 Excerpt from the covariance matrix of DAX stocks in the period from 18 May 2017 to 18 May 2018

	BMW	FME	LHA	CON	PSM	BEI	EOAN	HEI	DAI
BMW	1	0,41747216	0,25278108	0,57739417	0,24963274	0,30624834	0,20849472	0,27798357	0,69619354
FME	0,41747216	1	0,29134288	0,51854241	0,2408123	0,48004079	0,28196059	0,38375316	0,46212116
LHA	0,25278108	0,29134288	1	0,30314803	0,12741647	0,0925726	0,10540192	0,19564335	0,25225672
CON	0,57739417	0,51854241	0,30314803	1	0,24999131	0,39609125	0,16721738	0,35735453	0,54301807
PSM	0,24963274	0,2408123	0,12741647	0,24999131	1	0,20672734	0,07936704	0,15097407	0,20644724
BEI	0,30624834	0,48004079	0,0925726	0,39609125	0,20672734	1	0,37228328	0,30462756	0,35277495
EOAN	0,20849472	0,28196059	0,10540192	0,16721738	0,07936704	0,37228328	1	0,22603608	0,20979382
HEI	0,27798357	0,38375316	0,19564335	0,35735453	0,15097407	0,30462756	0,22603608	1	0,31843558
DAI	0,69619354	0,46212116	0,25225672	0,54301807	0,20644724	0,35277495	0,20979382	0,31843558	1

Fig. 4.25 Excerpt from the correlation matrix of DAX stocks in the period from 18 May 2017 to 18 May 2018

	BMW	FME	LHA	CON	PSM	BEI	EOAN	HEI	DAI
BMW	0,00021782	5,9427E-05	0,00016529	0,00016344	0,0001108	3,9659E-05	0,00011963	0,00012292	0,00017433
FME	5,9427E-05	0,00015664	8,9948E-05	7,4828E-05	6,8705E-05	5,3763E-05	0,00010173	7,6403E-05	6,3468E-05
LHA	0,00016529	8,9948E-05	0,00045844	0,00014009	0,00011828	6,1499E-05	0,00014894	0,00014199	0,00014577
CON	0,00016344	7,4828E-05	0,00014009	0,00022338	0,0001117	4,8693E-05	0,00010508	0,00012034	0,00014375
PSM	0,0001108	6,8705E-05	0,00011828	0,0001117	0,00024461	4,9257E-05	0,00011257	0,00010461	0,00010232
BEI	3,9659E-05	5,3763E-05	6,1499E-05	4,8693E-05	4,9257E-05	9,5712E-05	7,3661E-05	4,7098E-05	3,8021E-05
EOAN	0,00011963	0,00010173	0,00014894	0,00010508	0,00011257	7,3661E-05	0,00034627	0,00010164	0,00011813
HEI	0,00012292	7,6403E-05	0,00014199	0,00012034	0,00010461	4,7098E-05	0,00010164	0,0002245	0,00011252
DAI	0,00017433	6,3468E-05	0,00014577	0,00014375	0,00010232	3,8021E-05	0,00011813	0,00011252	0,00018422

Fig. 4.26 Excerpt from the covariance matrix of DAX stocks in the period from 18 May 2016 to 18 May 2017

	BMW	FME	LHA	CON	PSM	BEI	EOAN	HEI	DAI
BMW	1	0,3217247	0,5230808	0,74094191	0,48002306	0,274669	0,43560348	0,55587795	0,87026848
FME	0,3217247	1	0,3356621	0,40002761	0,35099786	0,43908863	0,43681189	0,40743593	0,37362772
LHA	0,5230808	0,3356621	1	0,43775165	0,35319978	0,29359196	0,37382293	0,44259862	0,50161383
CON	0,74094191	0,40002761	0,43775165	1	0,47784791	0,33300878	0,37782763	0,53737736	0,70860537
PSM	0,48002306	0,35099786	0,35319978	0,47784791	1	0,32192292	0,38678448	0,44639266	0,48201261
BEI	0,274669	0,43908863	0,29359196	0,33300878	0,32192292	1	0,40461861	0,32130331	0,28633189
EOAN	0,43560348	0,43681189	0,37382293	0,37782763	0,38678448	0,40461861	1	0,36453605	0,46774004
HEI	0,55587795	0,40743593	0,44259862	0,53737736	0,44639266	0,32130331	0,36453605	1	0,55330432
DAI	0,87026848	0,37362772	0,50161383	0,70860537	0,48201261	0,28633189	0,46774004	0,55330432	1

Fig. 4.27 Excerpt from the correlation matrix of DAX stocks in the period from 18 May 2016 to 18 May 2017

	BMW	FME	LHA	CON	PSM	BEI	EOAN	HEI	DAI
BMW	0,00044817	0,00024624	0,00022834	0,00034545	0,00021201	0,00021014	0,00034448	0,00030725	0,00040368
FME	0,00024624	0,00032595	0,00019634	0,00021321	0,00020697	0,00020426	0,00024282	0,00022141	0,00024589
LHA	0,00022834	0,00019634	0,00041492	0,00018067	0,0001533	0,00014366	0,00023556	0,00018203	0,00020839
CON	0,00034545	0,00021321	0,00018067	0,00038651	0,0001933	0,00019424	0,00027898	0,00027289	0,00034297
PSM	0,00021201	0,00020697	0,0001533	0,0001933	0,00029464	0,00017879	0,00019056	0,00020892	0,00021301
BEI	0,00021014	0,00020426	0,00014366	0,00019424	0,00017879	0,00024236	0,0002136	0,00018919	0,000216
EOAN	0,00034448	0,00024282	0,00023556	0,00027898	0,00019056	0,0002136	0,00072606	0,00027993	0,00032473
HEI	0,00030725	0,00022141	0,00018203	0,00027289	0,00020892	0,00018919	0,00027993	0,00036066	0,00029147
DAI	0,00040368	0,00024589	0,00020839	0,00034297	0,00021301	0,000216	0,00032473	0,00029147	0,00042959

Fig. 4.28 Excerpt from the covariance matrix of DAX stocks in the period from 18 May 2015 to 18 May 2016

	BMW	FME	LHA	CON	PSM	BEI	EOAN	HEI	DAI
BMW	1	0,64426591	0,5295111	0,83001013	0,58341801	0,63760095	0,60388512	0,76423096	0,91999228
FME	0,64426591	1	0,53388452	0,6006812	0,66785643	0,72672531	0,49913943	0,64577787	0,65711881
LHA	0,5295111	0,53388452	1	0,45115001	0,438437	0,45302525	0,42917086	0,47056219	0,49358449
CON	0,83001013	0,6006812	0,45115001	1	0,57279295	0,63465293	0,52662349	0,73090496	0,84166757
PSM	0,58341801	0,66785643	0,438437	0,57279295	1	0,66904476	0,41199375	0,64088966	0,59871998
BEI	0,63760095	0,72672531	0,45302525	0,63465293	0,66904476	1	0,50920238	0,63992577	0,66940822
EOAN	0,60388512	0,49913943	0,42917086	0,52662349	0,41199375	0,50920238	1	0,54703455	0,58144905
HEI	0,76423096	0,64577787	0,47056219	0,73090496	0,64088966	0,63992577	0,54703455	1	0,7404983
DAI	0,91999228	0,65711881	0,49358449	0,84166757	0,59871998	0,66940822	0,58144905	0,7404983	1

Fig. 4.29 Excerpt from the correlation matrix of DAX stocks in the period from 18 May 2015 to 18 May 2016

	BMW	FME	LHA	CON	PSM	BEI	EOAN	HEI	DAI
BMW	1	0,51384003	0,46363853	0,76321229	0,44720907	0,483381	0,50040226	0,62686061	0,87879474
FME	0,51384003	1	0,40358904	0,5265647	0,44279579	0,60150621	0,43869772	0,52117542	0,54520509
LHA	0,46363853	0,40358904	1	0,4084341	0,30615936	0,30732577	0,33583286	0,39481976	0,43643112
CON	0,76321229	0,5265647	0,4084341	1	0,44187749	0,50446864	0,42109374	0,60305437	0,75467739
PSM	0,44720907	0,44279579	0,30615936	0,44187749	1	0,42669497	0,30306167	0,43527616	0,44440334
BEI	0,483381	0,60150621	0,30732577	0,50446864	0,42669497	1	0,45394133	0,48245941	0,5163194
EOAN	0,50040226	0,43869772	0,33583286	0,42109374	0,30306167	0,45394133	1	0,43574807	0,49594813
HEI	0,62686061	0,52117542	0,39481976	0,60305437	0,43527616	0,48245941	0,43574807	1	0,61796434
DAI	0,87879474	0,54520509	0,43643112	0,75467739	0,44440334	0,5163194	0,49594813	0,61796434	1

Fig. 4.30 Excerpt from the correlation matrix of DAX stocks in the period from 18 May 2015 to 18 May 2018

the 2016/2017 test period (correlation −0.064). (Vonovia is a German housing company, and—according to its website—"Europe's leading private residential real estate company".)

- Viewed over the entire 2015/2018 test period, the highest correlation between DAX stocks was between BMW and Daimler (correlation 0.879; see Fig. 4.30, bottom left). The lowest correlation was between Vonovia and Commerzbank (correlation 0.200).

	BMW	FME	LHA	CON	PSM	BEI	EOAN	HEI	DAI
DAX 15/16	0,88481038	0,78958574	0,58326456	0,81518974	0,69390309	0,77927174	0,68150525	0,810777	0,89676622
DAX 16/17	0,81142398	0,55019275	0,5793705	0,74630903	0,61679286	0,44529617	0,59387136	0,73399149	0,84410733
DAX 17/18	0,65592636	0,70996878	0,42076452	0,66538436	0,38330604	0,56274781	0,44044616	0,55100937	0,69876346
DAX 15/18	0,83532764	0,70856763	0,53149461	0,77276698	0,56917563	0,66180435	0,6219354	0,74776687	0,85787526

Fig. 4.31 Excerpt from the correlation matrix between DAX and DAX stocks in the four time periods

- The average correlation between the individual stocks in the DAX was

$$\begin{aligned} 2015/2016: &\quad 0.582 \\ 2016/2017: &\quad 0.491 \\ 2017/2018: &\quad 0.349 \\ \text{Total } 2015/2018: &\quad 0.486 \end{aligned}$$

  So, in the course of these 3 years, average correlations experienced a slight decline.

The correlation of the DAX stocks with the DAX itself is also interesting. We have listed an excerpt of these correlations in the Fig. 4.31 for the four analysed periods.

Here we see significantly higher correlations (of course) between the individual stocks and the DAX.

- The average correlations between the DAX and the individual stocks were:

$$\begin{aligned} 2015/2016: &\quad 0.766 \\ 2016/2017: &\quad 0.662 \\ 2017/2018: &\quad 0.601 \\ \text{Total } 2015/2018: &\quad 0.703 \end{aligned}$$

  Here, too, we notice a slight downward trend in the correlations.
- Looking at the entire test period, we see the highest correlation between DAX and the chemical group BASF (correlation: 0.874), and the lowest correlation between DAX and the utility company RWE (correlation: 0.489).

By way of conclusion, a final remark on covariance, which we will revisit in the section on portfolio selection:

The average covariance between the various DAX stocks in the test period from May 2015 to May 2018 was 0.000128. The average variance of the individual DAX stocks was 0.000288 (i.e. approximately 2.3 times that value).

4.3 Basic Tools for Analysing Real Stock Prices: Autocorrelations of Stock Returns

As we already noted above:

The aim of this chapter is to derive *the* basic model for simulating stock price movements: the Wiener stock price model.

The second assumption underlying the Wiener model, which we are going to focus on in this section, is

Returns of stock price movements are independent of past returns of that stock.

One way to address this assumption is to calculate and discuss **autocorrelations** of stock price returns.

To illustrate the concept of autocorrelation, we will again use the continuous returns of the LHA stock $a_0, a_1, \ldots, a_{N-1}$ in the period from 28 February 2017 to 28 February 2018. The obvious approach to obtaining an **indication** as to whether there is a connection between a_0 and a_1 or between a_1 and a_2 or between a_2 and a_3, etc. would be to form pairs first—(a_0, a_1), (a_1, a_2), (a_2, a_3) etc.—then enter these pairs in a two-dimensional plane, and, in a further step, calculate the correlations between the two component sequences, i.e. between $a_0, a_1, \ldots, a_{N-2}$ and $a_1, a_1, \ldots, a_{N-1}$. This is what we are going to do in the following.

Based on the Picture 4.32 of the return pairs, no obvious dependence is discernible at first glance. For the correlation between the two component sequences, i.e. between $a_0, a_1, \ldots, a_{N-2}$ and $a_1, a_2, \ldots, a_{N-1}$, we get a value—which we are going to denote by $\gamma(i, i+1)$—of $\gamma(i, i+1) = 0.0580\ldots$. We refer to this as

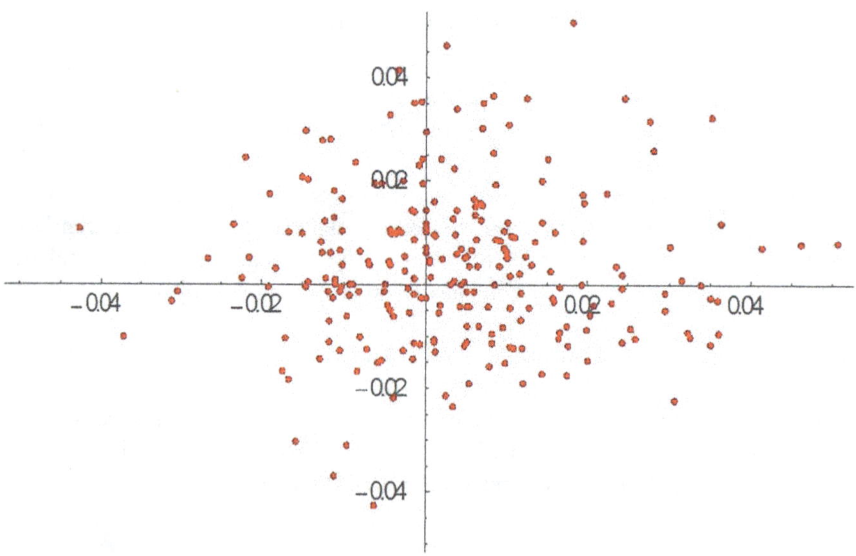

Fig. 4.32 Return pairs (a_i, a_{i+1}) of the LHA stock

	BMW	FME	LHA	CON	PSM	BEI	EOAN	HEI	DAI
1-day autocorrelation	0,10097207	-0,09426896	-0,01451135	-0,0108228	0,03840946	-0,07673136	-0,00301454	-0,00099889	0,07915681
2-day autocorrelation	0,07089343	-0,00914676	0,01421942	0,00139356	-0,0368657	-0,04597091	0,00550794	-0,0908187	0,06182592
3-day autocorrelation	0,06074273	0,04322176	-0,00581098	0,00138111	0,00957716	0,02798914	0,01984325	0,06336834	0,05048154
4-day autocorrelation	-0,11516735	-0,06712318	-0,04501662	-0,1035177	-0,00831764	-0,03212081	-0,01326222	-0,10539383	-0,1125652

Fig. 4.33 Excerpt from autocorrelations of DAX stocks in the period from 16 May 2015 to 16 May 2018

first-order autocorrelation. As we can see, the value is very close to 0 and thus indicates no dependence between consecutive returns

Similarly, we can zoom in on possible dependencies between a stock's return of 1 day and that of 2 days earlier by calculating the correlation $\gamma(i, i+2)$, i.e. the second-order autocorrelation, which is the correlation between $a_0, a_1, \ldots, a_{N-3}$ and $a_2, a_3, \ldots, a_{N-1}$. In our case this gives us $\gamma(i, i+2) = -0.0950\ldots$. Again, no obvious dependence can be recognized from the autocorrelation.

We can also generally calculate **p-th order autocorrelations $\gamma(i, i+p)$**, i.e. **correlations between $a_0, a_1, \ldots, a_{N-1-p}$ and $a_p, a_{p+1}, \ldots, a_{N-1}$**. As expected, the results in our case are again values close to 0.

For test purposes, we are also going to calculate first- to third-order autocorrelations for the DAX stocks in the period from 16 May 2015 to 16 May 2018 and present the results in the Fig. 4.33.

We see that, at least in our examples, calculating the autocorrelations gives us no indication of strong dependencies between consecutive returns. (The average 1-day autocorrelation of DAX stocks was 0.0134 during the test period. The strongest 1-day autocorrelations in the test period were observed for VW, at a value of 0.180, and BMW at a value of 0.101.)

This finding (non-significant autocorrelation) is also supported if we carry out analogous calculations on the time series of the Apple stock in the time range 12 December 1980 to 14 October 2021 and for the S&P500 index in the time range 4 March 1957 to 14 October 2021.

For the Apple stock, we get:

$$\gamma(i, i+1) = 0.0166\ldots$$
$$\gamma(i, i+2) = -0.0247\ldots$$
$$\gamma(i, i+3) = -0.0326\ldots$$

and for the S&P500-Index, we get:

$$\gamma(i, i+1) = -0.00689\ldots$$
$$\gamma(i, i+2) = -0.01744\ldots$$
$$\gamma(i, i+3) = -0.00315\ldots$$

These observations and examples of stock return autocorrelations are, of course, merely first impressions when it comes to dependencies between consecutive stock returns. A great deal of literature on financial market statistics addresses this question in great depth on the basis of statistical methods. For now, however, a general idea of it is all we want to provide at this point.

And autocorrelation is, of course, ultimately an insufficient measure of existing dependencies. While an autocorrelation that deviates significantly from 0 indicates the presence of dependencies, an autocorrelation close to 0 is in itself by no means proof of independence.

Finally, we are going to show a systematically recurring dependence (which is evidently not diagnosed by autocorrelation computations), using S&P500 daily data as an example.

The dependence that we are referring to here is the following:

Returns of a relatively high absolute value have an increased likelihood of being followed by a return of a relatively high absolute value.

We want to illustrate what we mean by this with an example. We examine all of the 16,267 returns of the S&P500 from 4 March 1957 to 14 October 2021. Of those returns, 767 have an absolute value greater than 0.02, and 15,500 have an absolute value less than or equal to 0.02. The empirical probability for a "high return" (greater than 0.02) to occur is therefore $\frac{767}{16,267} = 0.0471\ldots$, i.e. 4.71% in this time range.

Now let's just look at the 767 days with a high daily return and what the returns were on the immediately following days:

We note that out of these 767 following-day returns, 159 were high returns and 608 were low returns. This means:

The empirical probability that a high return (in absolute terms) is followed by a high return (in absolute terms) was $\frac{159}{767} = 0.2073\ldots$, i.e. 20.73%.

The relative frequency of a high return (in absolute terms) occurring after a high return is significantly higher (more than five times higher) than would otherwise be the case.

4.4 What Does Mathematical Modeling Mean? What Is a Stock Price Model?

The intention of this section is, among other things, to preclude a frequently encountered misunderstanding: If a financial mathematician talks about developing, programming, or using a certain stock price model for modeling and simulating stock prices (or prices of other financial products), this is often misunderstood to mean that the mathematician has a tool at hand for forecasting future stock prices. That is not the case, of course!

What is more: Forecasting future stock prices using purely mathematical models won't ever be the intention of any financial mathematician.

We want to explain the actual relationship between "the real world" and the mathematical model using a very simple example:

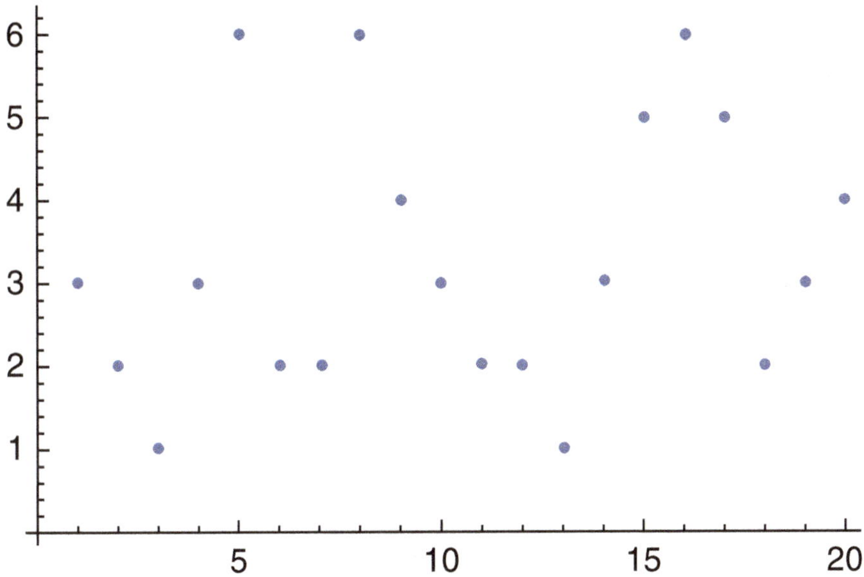

Fig. 4.34 Values of 20 consecutive rolls of a dice

We take a dice (any dice), roll it 20 times and add up the results successively. The results of each of those rolls could be something like this:

$$5, 2, 6, 6, 4, 3, 1, 3, 3, 4, 3, 6, 1, 4, 4, 6, 6, 2, 5, 1$$

The successively added values would then be

$$5, 7, 13, 19, 23, 26, 27, 30, 33, 37, 40, 46, 47, 51, 55, 61, 67, 69, 74, 75$$

The individual rolls and the values added up over time can then also be represented in the following Figs. 4.34 and 4.35:

If we repeat this procedure of rolling the dice 20 times several times over, the graph would look somewhat different each time, of course. For example, Fig. 4.36 shows the 10 paths that resulted when the 20 dice rolls were repeated 10 times.

While all of these paths differ from each other, all show a somewhat similar "typical" form.

Let us have another look at the one graph shown above in Fig. 4.35. In addition, let us pretend we do not know that this graph was created by adding up dice values.

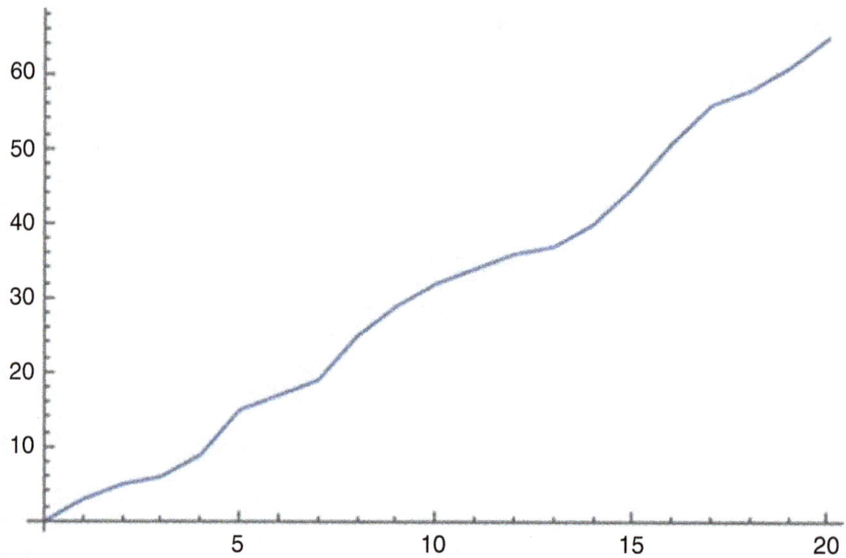

Fig. 4.35 Values of 20 consecutive rolls of a dice added up successively

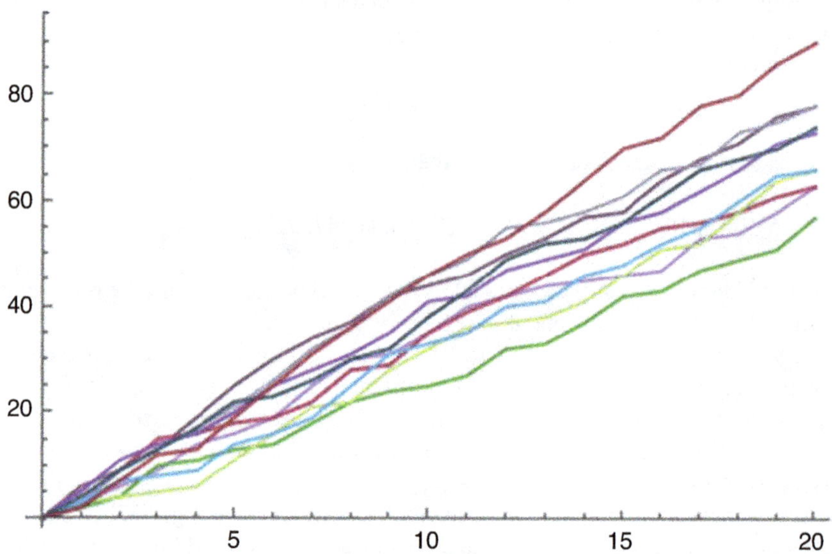

Fig. 4.36 10 sample paths resulting from adding up the values of 20 consecutive rolls of a dice

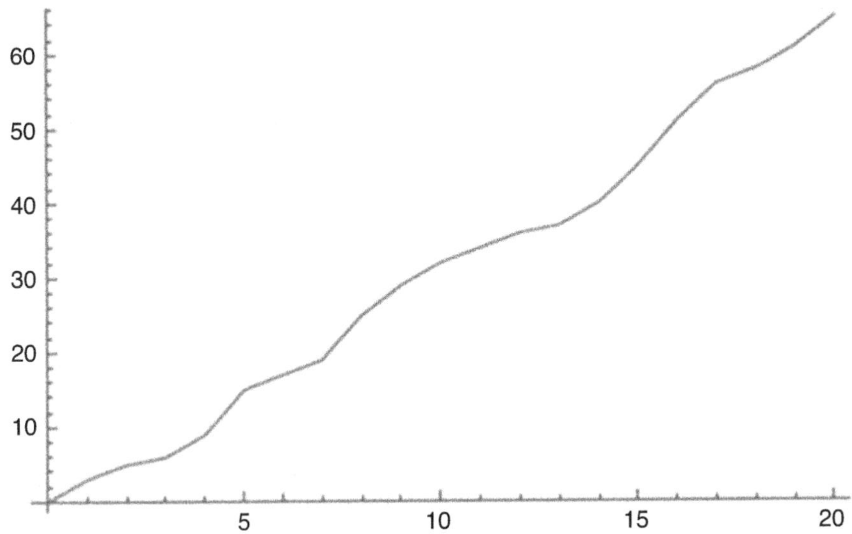

Fig. 4.37 Values of 20 consecutive rolls of a dice added up successively

If we then begin to analyse the values and the Graph 4.37, we might make the following **observations**:

- *The path is created by successively adding one of the numbers* $1, 2, \ldots, 6$.
- *On average, each of these numbers is likely to occur in an approximately equal frequency.*
- *There is no obvious relation between successive summands* $(1, 2, \ldots, 6)$.

Again: We assumed that we do not know how these values came about and so we do not know whether these three observations would actually hold in reality or withstand a long-term test. (For example, a 7 could suddenly appear as a summand in another path that might follow in the future. Or a relation between successive summands might very well exist, yet we did not discover it (perhaps because our analytical tools were not suited to this or because the test period was too short). Or it could turn out that, in the long run, summand 3 occurs slightly, yet significantly more often than summand 4, ...)

But based on the observations we have made so far, we may be inclined to **assume** a **mathematical model** for these paths. The mathematical model could look something like this:

Mathematical model:

- *We want to generate values* $X_1, X_2, X_3, \ldots, X_{20}$.
- *To do so, we start with the initial value* $X_0 = 0$.

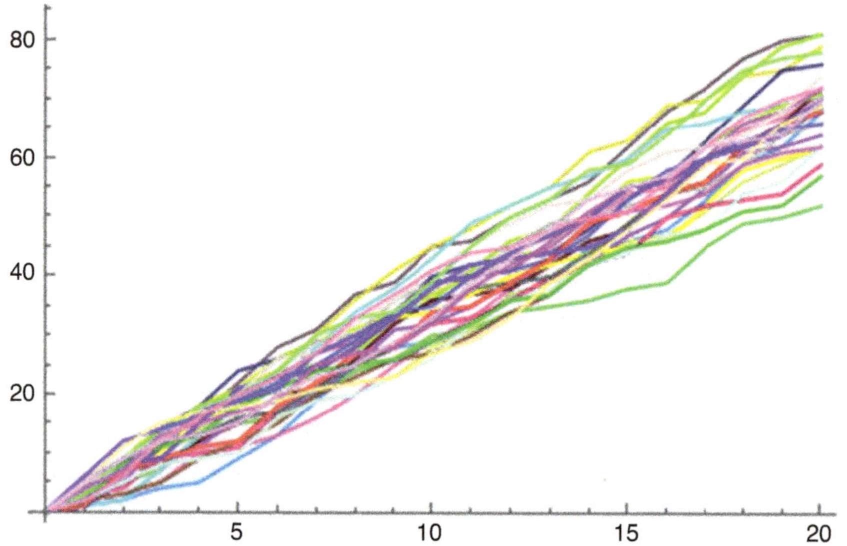

Fig. 4.38 30 realizations of paths using the mathematical model

- If the values $X_1, X_2, X_3, \ldots, X_{i-1}$ *have already been generated for some* $i = 1, \ldots, 20$, *then we need to generate* X_i *in the following way:*
  We generate an integer S_i *with an independent random number generator, where* S_i *is between 1 and 6 and where each of these numbers has a probability of occurring of* $\frac{1}{6}$. *Finally, we set* $X_i := X_{i-1} + S_i$.

This model delivers an exact rule for generating the number sequence X_1, X_2, $X_3, \ldots, X_{20}$. However, it still contains a (clearly defined) random component; it is a **stochastic model**, not a deterministic model!

Using this model and a computer, we can quickly generate any number of paths. Figure 4.38 shows 30 such **realizations** using this mathematical model.

The paths generated by the mathematical model present a very similar structure to the paths generated by the dice rolls in Fig. 4.36. The precisely defined mathematical (stochastic) model thus seems to be an apt tool for modeling the paths generated by a real-world process.

The great advantage of a mathematical model is:

It is precisely defined.

It allows us to perform **calculations** in it! For example, we can calculate:

What is the expected value $E(X_{20})$ *for* X_{20} *in this mathematical model?*

or:

What is the average deviation of the values X_{20} *(realized by the mathematical model) from the expected value* $E(X_{20})$?

or:

What is the probability that the values X_{20} (realized by the mathematical model) will lie in a range $[U, V]$?

Etc.:

...

Note that we can **accurately calculate** these values and probabilities in the model, not just **estimate** them!

But: While it is an exact mathematical (stochastic) model, it **does not allow us to predict** any specific future trajectories (think of the dice roll)! If we realize a specific trajectory using a mathematical (stochastic) model, it is impossible for us to say anything a priori about the actual form that this realization will take! A sequence of the form

$1, 2, 3, 4, 5, 6, 7, 8, 9, 10, 11, 12, 13, 14, 15, 16, 17, 18, 19, 20$
is just as likely to occur as a sequence of the form
$6, 12, 18, 24, 30, 36, 42, 48, 54, 60, 66, 72, 78, 84, 90, 96, 102, 108, 114, 120$
and this again is just as likely to occur as a sequence of the form
$5, 7, 13, 19, 23, 26, 27, 30, 33, 37, 40, 46, 47, 51, 55, 61, 67, 69, 74, 75.$

In other words:

The mathematical (stochastic) model does not help you to create specific forecasts of future developments, but you can use it to calculate a wide variety of key metrics and probabilities.

That is: If the mathematical model provides a "good" tool for modeling real-world processes, then (and only then) conclusions as to those real processes can be drawn (but no forecasts made) through calculations in that mathematical model.

Now, is our mathematical model (from the above example) a "good" modeling tool? This question is not easy to answer and depends on the information we have about the relevant real-world processes.

In our example, this means:

Let us assume that a **Person *A*** knows that the developments we observed in the real world are the result of rolling a dice and successively adding up the dice values. And let us further assume that a **Person *B*** merely had the number sequences (without any information as to how they had come about) and that this person *B* arrived at the three "observations" stated above simply by analysing the number sequence.

Both Person *A* and Person *B* have developed the same mathematical model as a tool to mathematically model the respective number sequences of the real world. While Person *A* can be quite sure, based on their background knowledge, that their model will reflect reality well (provided the dice is actually a fair dice), Person *B* will be much less sure. Person B's reasons for choosing this mathematical model are based purely on a few more or less solid observations (potential reservations have already been listed above).

If a third person M carries out calculations in the mathematical model and arrives at mathematical results (e.g. "the probability that the value X_{20} generated in the mathematical model lies in the interval $[U, V]$ is equal to P"), then both A and B can conclude that the probability for them to observe number sequences of the above form ending in the range of $[U, V]$ in the real world in the future is approximately P. However, Person A will have a higher assurance that this conclusion holds than Person B.

For the third person M ("M" being short for "Mathematician", perhaps), who uses the mathematical model for actual calculations, the possibility of drawing conclusions from those calculations as to reality may be completely irrelevant. M may be purely interested in performing calculations within the model and may have no interest as to whether or not or how well the model reflects reality.

"Conclusions to reality" and thus "the degree to which the real world is reflected by the mathematical model" are, however, of the essence to "Users" A and B, of course. Indeed, Users A and B are often interested **only** in the modeling quality and the conclusions as to reality and will not themselves perform any calculations in the mathematical model, leaving those to "a pure mathematician" instead.

To summarize: In all of this, it is important to be aware of the following terms and definitions:

- Certain observed real-world processes (in physics, finance, biology, medicine, ...) → Users A and B.
- Attempt to model these processes using a purely mathematical (stochastic) model → cooperation between Users A and B and Mathematician M.
- The modeling attempt is not a mathematically precise, provable process but rather an attempt that uses various (more or less accurately observed, but not provable) assumptions about possible regularities of real processes!
- The quality, accuracy, and suitability of the mathematical model for representations of the real world must be tested and be very high → cooperation between Users A and B and Mathematician M.
- Exact calculations in the mathematical model → Mathematician M.
- Drawing conclusions from the results of calculations in the exact mathematical model to real-world processes → Users A and B.

Sometimes it makes sense (or is simply fun because of the intellectual challenge) to delve into mathematical models (see Fig. 4.39) and perform calculations in them even if they reflect the real-world processes they are motivated by in a very incomplete and rudimentary manner.

Just look at the following two examples:

Example 4.1 In some cases, the real-world systems to be modeled may (still) be far too complex for a mathematical model to replicate them "exactly", or even just more or less so. Still, it may make sense to develop (greatly) simplified mathematical models that attempt to replicate some facets of the real system, as a way to gain at least a few insights into basic characteristics and fundamental processes.

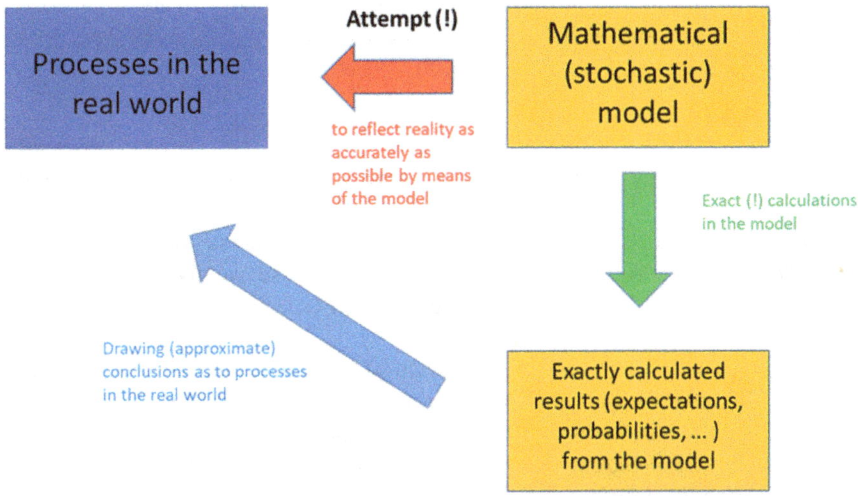

Fig. 4.39 Mathematical modeling

Example 4.2 Chess presumably has its original roots in—and was modeled on—real military battles. However, it is, of course, an extremely inadequate model for the actual events on any real battlefield. Anyone studying the game does not (therefore) do so out of a desire to use findings about strategies, positions, or openings from the chess game to draw conclusions as to real warfare but will do so completely decoupled from reality, purely out of intellectual, playful interest.

4.5 The Wiener Stock Price Model

Prepared by the observations on mathematical modeling in the previous section, we will now move on to our first attempt to describe the movements of stock prices by means of a mathematical model.

The assumptions that we want to make in this first attempt (and that we have classified as "approximately true" based on empirical observations) are as follows:

Basic Assumptions in the Wiener Stock Price Model

- The continuous returns a_i of a stock price (for equal time periods dt) are $N\left(\mu', \sigma'^2\right)$ distributed, i.e. normally distributed, with a certain expected value μ' (= estimated trend of the stock per time period dt) and a certain standard deviation σ' (= estimated volatility of the stock per time period dt).
- The continuous returns of a stock price in two disjoint (non-overlapping) time intervals are independent of one other.

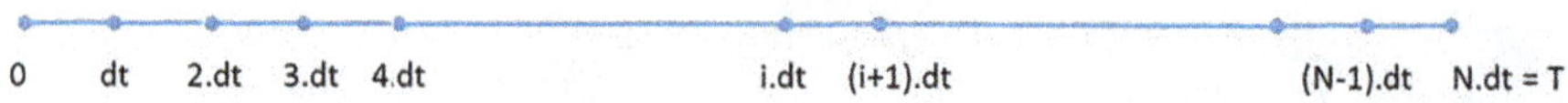

Fig. 4.40 "Discretization" of the time axis

Based on these key assumptions, we are now going to develop the Wiener model as follows:

We want to model a stock's price movements in the time interval $[0, T]$ where T is an arbitrary point of time in the future. For this purpose, we divide ("discretize") the time interval into N equal parts of the length dt (i.e. $T = N \cdot dt$). Bear in mind that all time data is given in years.

For any point in time t within the interval $[0, T]$, $S(t)$ denotes the stock price at time t. $S(0)$ is the stock price at time 0 (compare Fig. 4.40).

A_i denotes the continuous return of the stock price from time $i\cdot$ dt to time $(i+1)\cdot dt$. Based on the continuous return definition, we then get $S((i+1)\cdot dt) = S(i\cdot dt)\cdot e^{a_i}$ and so forth:

$$S((i+1)\cdot dt) = S(i\cdot dt)\cdot e^{a_i} = S((i-1)\cdot dt)\cdot e^{a_{i-1}+a_i} =$$
$$= \ldots = S(0)\cdot e^{a_0+a_1+\ldots+a_{i-2}+a_{i-1}+a_i}$$

Since according to the model's assumptions, a_i is $N\left(\mu', \sigma'^2\right)$ distributed, we can represent a_i in the following form:

$a_i = \mu' + \sigma' \cdot w_i$, where w_i is a random variable with a distribution of the form $\mathcal{N}(0,1)$ and where $w_0, w_1, \ldots, w_{N-1}$ are independent of one another.

Substituting this in the above equation, we get

$$S((i+1)\cdot dt) = S(0)\cdot e^{(i+1)\cdot\mu'+\sigma'\cdot(w_0+w_1+\ldots+w_{i-2}+w_{i-1}+w_i)}$$

Especially if $(i+1) = N$, i.e. $(i+1)\cdot dt = N\cdot dt = T$, we get

$$S(T) = S(0)\cdot e^{N\mu'+\sigma'\cdot(w_0+w_1+\ldots+w_{N-3}+w_{N-2}+w_{N-1})}.$$

The sum $w_0 + w_1 + \ldots + w_{N-3} + w_{N-2} + w_{N-1}$ of N independent (!) random variables with a distribution of the form $\mathcal{N}(0,1)$ is a random variable w' with a $\mathcal{N}(0,N)$ distribution.

w' can be written as $w' = \sqrt{N}\cdot w$ with a $\mathcal{N}(0,1)$-distributed random variable w. We then get

$$S(T) = S(0)\cdot e^{N\mu'+\sqrt{N}\sigma'\cdot w}$$

All that is left to do now is one last step. For that we go back to Sect. 4.1 and the relations $T\cdot\mu = N\cdot\mu'$ and $\sqrt{T}\cdot\sigma = \sqrt{N}\cdot\sigma'$ that we derived there, where μ denotes the trend per annum and σ denotes the volatility per annum. Substituting these relations, we get the Wiener stock price model $S(T) = S(0) \cdot e^{T\mu+\sqrt{T}\sigma \cdot w}$, where w denotes a $\mathcal{N}(0, 1)$-distributed random variable.

To summarize:

A stock (or any other financial product) moves according to the Wiener stock price model if, for any two points in time 0 and T, the following holds for the stock prices $S(0)$ and $S(T)$ at the points in time 0 and T :

$$\mathbf{S(T) = S(0) \cdot e^{T\mu+\sqrt{T}\sigma \cdot w}}$$

Here μ denotes the stock's trend per annum and σ denotes the stock's volatility per annum in the time range $[0, T]$ and w denotes a random variable with distribution $\mathcal{N}(0, 1)$.

Note This can of course be used to generate stock prices for any other time intervals $[T_1, T_2]$. The formula then looks like this:

$$S(T_2) = S(T_1) \cdot e^{(T_2-T_1)\mu+\sqrt{T_2-T_1}\sigma \cdot w}.$$

We will use this version further below for time intervals of the form $[i \cdot dt, (i+1) \cdot dt]$. As $T_2 - T_1 = dt$ we then get

$$S((i+1) \cdot dt) = S(i \cdot dt) \cdot e^{dt \cdot \mu+\sqrt{dt}\sigma \cdot w}.$$

This last equation can also be written (for later purposes) as follows:

$S((i+1) \cdot dt) = S(i \cdot dt) \cdot e^{dt\mu+\sigma w_i}$ with a random variable w_i that is now distributed in the form $\mathcal{N}(0, dt)$.

In continuation of this, we again get the representation $S(n \cdot dt) = S(0) \cdot e^{n \cdot dt \cdot \mu+\sigma \cdot (w_0+w_1+\ldots+w_{n-1})}$ for any n, with $\mathcal{N}(0, dt)$-distributed random variables w_i that are independent of one another.

What can this model help us with? It can **not** help us predict future stock prices. The model contains the normally distributed random variable w, which can take any value! But it can help us simulate possible (typical) price movements (this is an essential prerequisite for being able to carry out so-called Monte Carlo simulations later), and it enables us to make calculations in the model that then allow certain conclusions to be drawn about the real stock price movements. In the next two sections, we will therefore focus on the simulation of stock prices using the Wiener model, as well as on some basic calculations in the Wiener model and some key characteristics of the Wiener model.

4.6 Simulation of Stock Prices in the Wiener Model

We have a time range $[0, T]$, in which we want to simulate stock prices using the Wiener model. The simulation procedure described below can be carried out in many different variations with the relevant program on our website. See https://app.lsqf.org/book/wiener-model-from-prices.

Algorithm (Simulation of a Stock Price with the Wiener Model)

- First, we decide on a discretization length (period) dt for the time range. A new price is calculated for each $n.dt$. (We assume that $T = N \cdot dt$, i.e. that the length of the time range is an integral multiple of the period.)
- We choose an initial value $S(0)$ for the stock price.
- We choose a trend per annum μ and a volatility per annum σ.
- We use a random number generator to generate N standard normally distributed random numbers $w_0, w_1, \ldots, w_{N-1}$ that are independent of one another.
- Then the simulated stock prices $S(0), S(1 \cdot dt), S(2 \cdot dt), \ldots, S(N \cdot dt)$ are successively calculated using the formula

$$S((i + 1) \cdot dt) = S(i \cdot dt) \cdot e^{dt \cdot \mu + \sqrt{dt} \sigma \cdot w_i}$$

  (see end of the previous section).
- These prices can then be entered in a time/price diagram, connected, and thus illustrated in the form of a "price path".

Bear in mind, of course, that even if we choose the same parameters $T, dt, S(0), \mu$, and σ, each new execution of the simulation as described above will yield a different path, a different "realization of the model". Each realization will yield a possible (typical) path of a stock price based on these pre-set parameters $T, dt, S(0), \mu$ and σ.

Example 4.3 In the period from 18 May 2017 to 18 May 2018, the DAX had a trend $\mu = 0.021$ and a volatility $\sigma = 0.1275$ calculated on the basis of its daily closing rates.

The price $S(0)$ of the DAX on 18 May 2017 was 12,804 points.

We now want to carry out several simulations using the Wiener model and the same parameters and compare these realizations with the actual movements of the DAX during this period. For this purpose, we additionally set $T = 1$ and $dt = \frac{1}{255}$ and perform the simulation 20 times.

Figure 4.41 shows 20 simulations compared to the actual movements of the DAX (bolder, black graph). From a purely visual point of view, the simulations seem to produce a path that is "similar in type" to that of the DAX.

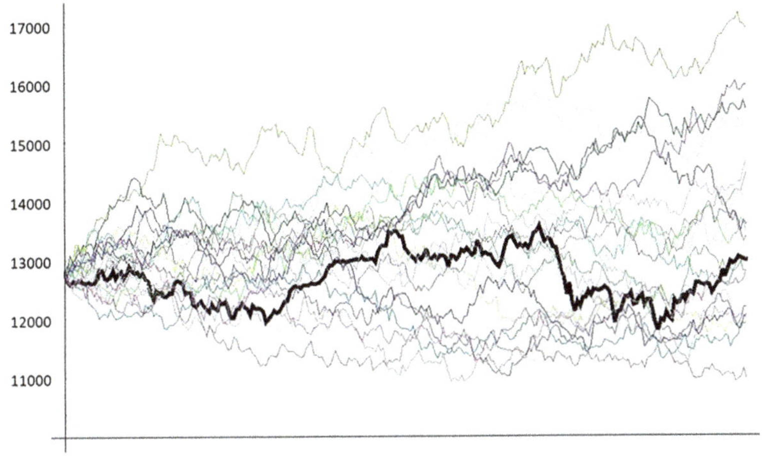

Fig. 4.41 20 simulations with DAX parameters from 18 May 2017 to 18 May 2018 compared to the DAX (black) in the same period

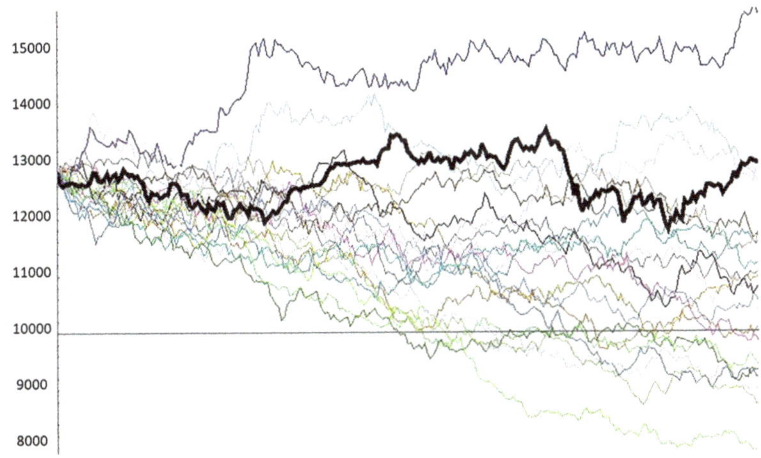

Fig. 4.42 20 simulations with DAX parameters from 18 May 2017 to 18 May 2018 with changed trend of -0.2 compared to the DAX (black) in the same period

The effect that changes in the two estimation parameters μ and σ have on the simulations is quite obvious:

On average (i.e. not necessarily in every single simulation!), an increase in μ leads to stronger price increases and an increase in σ leads to stronger price fluctuations. These two facts are illustrated in Figs. 4.42, 4.43, 4.44, 4.45, and 4.46:

In the last chart, volatility was set to 0. As a result, the Wiener model loses its randomness:

$$S(T) = S(0) \cdot e^{T\mu + \sqrt{T}\sigma \cdot w} = S(0) \cdot e^{T\mu}$$

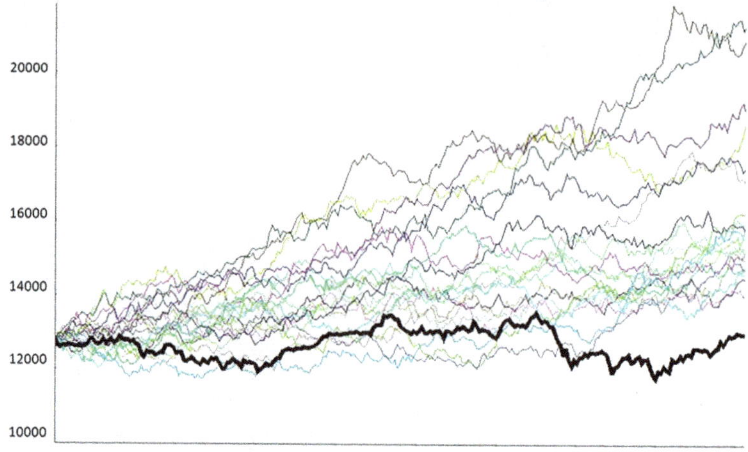

Fig. 4.43 20 simulations with DAX parameters from 18 May 2017 to 18 May 2018 with changed trend of $+0.2$ compared to the DAX (black) in the same period

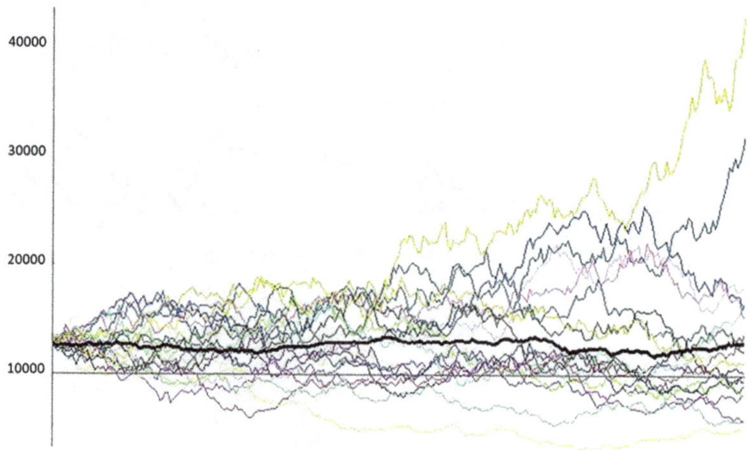

Fig. 4.44 20 simulations with DAX parameters from 18 May 2017 to 18 May 2018 with changed volatility of 0.5 compared to the DAX (black) in the same period

and corresponds to a (deterministic) continuously compounded interest rate μ on an investment $S(0)$ for the period $[0, T]$. Each simulation thus yields the same path (brown path in Fig. 4.46), which, incidentally, is not a straight line segment but a curve (in our example, with a low interest rate of 2.1% and a short maturity of 1 year) that grows slightly exponentially.

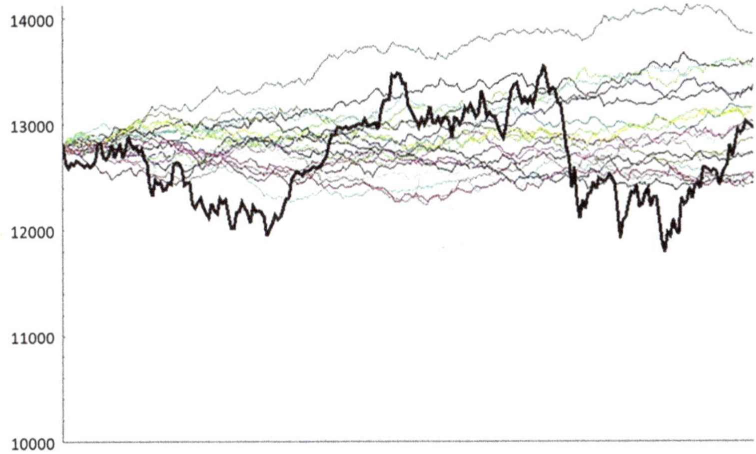

Fig. 4.45 20 simulations with DAX parameters from 18 May 2017 to 18 May 2018 with changed volatility of 0.03 compared to the DAX (black) in the same period

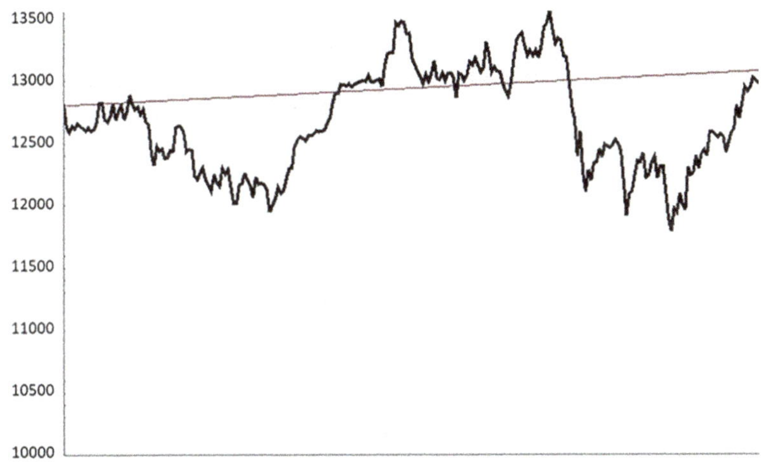

Fig. 4.46 20 "simulations" with DAX parameters from 18 May 2017 to 18 May 2018 with changed volatility of 0.00 compared to the DAX (black) in the same period

4.7 Simulation of Two Correlated Stock Prices

In some applications, two or more stocks (financial products) may have to be simulated simultaneously for the same period of time. We can of course proceed in exactly the same way as we did in the algorithm described above, by simulating two stocks independently of each other, with their respective parameters, and entering them into a common coordinate system for illustration:

Algorithm (Simulation of Two Independent Stock Prices Using the Wiener Model)

- First, we decide on a discretization length (period) dt for the time range. A new price is calculated for each new dt. (We assume that $T = N \cdot dt$, i.e. that the length of the time range is an integral multiple of the period.)
- We select initial values $S_1(0)$ and $S_2(0)$ for the stock prices S_1 and S_2.
- We select trends per annum μ_1 and μ_2 as well as volatilities per annum σ_1 and σ_2 for the stock prices S_1 and S_2.
- We use a random number generator to generate two sequences of N independent standard normally distributed random numbers $w_0, w_1, \ldots, w_{N-1}$ and $z_0, z_1, \ldots, z_{N-1}$.
- Then, the simulated stock prices $S_1(0), S_1(1 \cdot dt), S_1(2 \cdot dt), \ldots, S_1(N \cdot dt)$ are successively calculated using the formula

$$S_1((i+1) \cdot dt) = S_1(i \cdot dt) \cdot e^{dt \cdot \mu_1 + \sqrt{dt} \sigma_1 \cdot w_i}$$

  and $S_2(0), S_2(1 \cdot dt), S_2(2 \cdot dt), \ldots, S_2(N \cdot dt)$ are successively calculated using the formula

$$S_2((i+1) \cdot dt) = S_2(i \cdot dt) \cdot e^{dt \cdot \mu_2 + \sqrt{dt} \sigma_2 \cdot z_i}$$

- These prices can then be entered in a time/price diagram and connected to produce two "price paths".

If we proceeded in this way, we would indeed get paths of mutually independent stocks, especially paths with a return correlation very close to 0. Yet as we know, the assumption of mutually independent stock prices often does not hold in reality. The correlation between different stock prices (more precisely: between stock price returns!) is often significantly different from 0.

How then should we proceed if we want to simulate the prices of two (or more) stocks that are supposed to have (at least approximate) a given correlation?

Let us recall: The given correlation of two stocks is actually the correlation between the continuous returns of these two stocks (and not the correlation between the stock prices themselves)!

We first describe the suitable algorithm specifically for the case of two stocks and then generally for the case of any number of stocks (for which we need the concept of Cholesky factorization). The website of this book provides the program for simulating any number of stocks satisfying a given correlation structure, of course. See https://app.lsqf.org/book/correlated-wiener-model

Algorithm (Simulation of Two Stock Prices with a Given (Return) Correlation) ρ using the Wiener model):

- First, we decide on a discretization length (period) dt for the time range. A new price is calculated for each new dt. (We assume that $T = N \cdot dt$, i.e. that the length of the time range is an integral multiple of the period.)
- We select initial values $S_1(0)$ and $S_2(0)$ for the stock prices S_1 and S_2.
- We select trends per annum μ_1 and μ_2 as well as volatilities per annum σ_1 and σ_2 for the stock prices S_1 and S_2 and the (return) correlation ρ of the two stocks.
- We use a random number generator to generate two sequences of N independent standard normally distributed random numbers $w_0, w_1, \ldots, w_{N-1}$ and $z'_0, z'_1, \ldots, z'_{N-1}$.
- We set $z_i = \rho \cdot w_i + \sqrt{1-\rho^2} \cdot z'_i$ for all $i = 0, 1, \ldots, N-1$.
- Then, the simulated stock prices $S_1(0), S_1(1 \cdot dt), S_1(2 \cdot dt), \ldots, S_1(N \cdot dt)$ are successively calculated using the formula

$$S_1((i+1) \cdot dt) = S_1(i \cdot dt) \cdot e^{dt \cdot \mu_1 + \sqrt{dt}\sigma_1 \cdot w_i}$$

  and $S_2(0), S_2(1 \cdot dt), S_2(2 \cdot dt), \ldots, S_2(N \cdot dt)$ are successively calculated using the formula

$$S_2((i+1) \cdot dt) = S_2(i \cdot dt) \cdot e^{dt \cdot \mu_2 + \sqrt{dt}\sigma_2 \cdot z_i}$$

- These prices can then be entered in a time/price diagram and connected to produce two "price paths".

This algorithm for two dependent stocks represents only a special case of the algorithm for several stocks, which we are going to describe below. The only difference to the simulation of two independent stocks is that the second sequence of random numbers must first be suitably converted (from z'_i to z_i) before it can be used for the simulation).

The algorithm can also be used to simulate a new stock price S_2 with other—given—parameters μ_2 and σ_2 for a given stock price S_1 (for which a Wiener model is assumed), such that the (return) correlation between S_1 and S_2 is ρ:

- For this to happen, we need to calculate the random numbers $z_0, z_1, \ldots, z_{N-1}$, which are required for generating S_2, exactly as described in the algorithm, from the previously generated $z'_0, z'_1, \ldots, z'_{N-1}$ by means of the formula $z_i = \rho \cdot w_i + \sqrt{1-\rho^2} \cdot z'_i$ for all $i = 0, 1, \ldots, N-1$.
- Before that, however, we need the values $w_0, w_1, \ldots, w_{N-1}$, which have to be calculated from the given stock prices $S_1(0), S_1(1 \cdot dt), S_1(2 \cdot dt), \ldots,$

$S_1(N \cdot dt)$. Due to $S((i+1) \cdot dt) = S(i \cdot dt) \cdot e^{dt \cdot \mu + \sqrt{dt}\sigma \cdot w_i}$, this is done by means of the formula

$$w_i = \frac{\log \frac{S((i+1) \cdot dt)}{S(i \cdot dt)} - \mu \cdot dt}{\sigma \cdot \sqrt{dt}}.$$

Substituting this in the formula for calculating z_i, we then get

$$z_i = \rho \cdot \frac{\log \frac{S((i+1) \cdot dt)}{S(i \cdot dt)} - \mu \cdot dt}{\sigma \cdot \sqrt{dt}} + \sqrt{1 - \rho^2} \cdot z_i', \quad \text{for } i = 0, 1, \ldots, N-1$$

- We can then use this z_i to generate the stock price S_2 as usual.

This formulation too can be expanded to include any number of stocks (k stocks are already given, and additional l stocks are to be constructed in a way to obtain a given correlation structure between all stocks), and it is available as a program on this book's website.

Example 4.4 Simulate the movements of two stock prices S_1, S_2 based on the parameters

$T = 1$
$dt = \frac{1}{52}$

$\mu_1 = 0.2$
$\sigma_1 = 0.3$
$S_1(0) = 100$

$\mu_2 = -0.2$
$\sigma_2 = 0.3$
$S_2(0) = 100$

and with a high correlation $\rho = 0.8$

In Fig. 4.47, we see four sample simulations that were run for the above parameters using the specified algorithm and the program on our website.

Example 4.5 Simulate possible price movements (daily prices) of a stock over the course of a year, each having the same initial value, the same trend, and the same volatility as the daily prices of the DAX in the period 18 May 2017 to 18 May 2018 but a strong negative correlation with the DAX of -0.8.

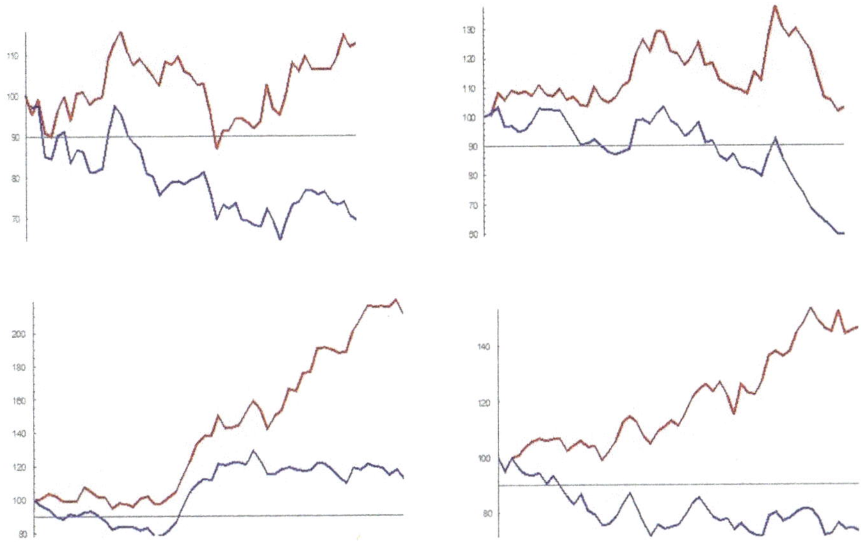

Fig. 4.47 Sample simulations of two price paths with correlation 0.8

To run this simulation, we can again use the DAX data that we used earlier for the same period. Specifically, we had the parameters

$T = 1$
$dt = \frac{1}{252}$

$\mu_1 = 0.021$
$\sigma_1 = 0.1275$
$S_1(0) = 12,804$

So we also have to choose

$\mu_2 = 0.021$
$\sigma_2 = 0.1275$
$S_2(0) = 12,804$
and correlation $\rho = -0.8$.

The program on our website will now run the desired simulations in analogy to the above algorithm and based on the respective daily rates of the DAX. Figure 4.48 again presents four examples of simulations (compared to the DAX (black)).

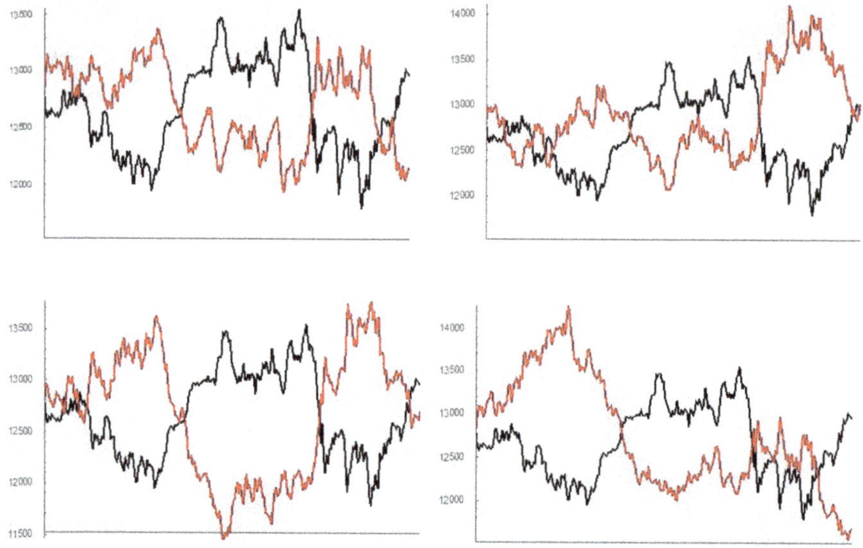

Fig. 4.48 Sample simulations of stock prices (red) with correlation -0.8 to the DAX (black) and otherwise the same parameters

4.8 Simulation of Several Correlated Stock Prices

As announced, we will now discuss the algorithm for simulating the prices of several stocks relative to a given return correlation matrix. For that purpose, we are going to generate m stock prices $S_1, S_2, \ldots, S_m$ such that the correlations between the returns of two stocks S_i and S_j each have a given value $\rho_{i,j}$.

It is obvious that this requirement cannot be satisfied for just any choice of $\rho_{i,j}$: For example, it is clearly not possible to generate $S_1, S_2,$ and S_3 in such a way that $\rho_{1,2} = \rho_{1,3} = 1$, and $\rho_{2,3} = -1$.

As preparatory work, we consolidate the correlations in an $m \times m$ correlation matrix $M = \left(\rho_{i,j}\right)_{i,j,=1,2,\ldots,m}$, where we set $\rho_{i,i} = 1$ for all i.

Provided that matrix M is positive definite, a Cholesky decomposition of the matrix M can be carried out for the matrix M using mathematical software. (A further prerequisite for performing a Cholesky decomposition is that the matrix M is symmetrical. This condition is definitely satisfied in that $\rho_{i,j} = \rho_{j,i}$.) This means that a lower $m \times m$ triangular matrix C can be found so that $C \cdot C^T = M$. We can then start the algorithm:

Algorithm (Simulation of m Stock Prices with a Given (Return) Correlation Matrix M Using the Wiener Model)

- First, we decide on a discretization length (period) dt for the time range. A new price is calculated for each $n.dt$. (We assume that $T = N \cdot dt$, i.e. that the length of the time range is an integral multiple of the period.)
- We select initial values $S_1(0), S_2(0), \ldots, S_m(0)$ for the stock prices $S_1, S_2, \ldots, S_m$.
- We select trends per annum $\mu_1, \mu_2, \ldots, \mu_m$ as well as volatilities per annum $\sigma_1, \sigma_2, \ldots, \sigma_m$ for the stock prices $S_1, S_2, \ldots, S_m$.
- Using a random number generator, we generate m sequences of N independent standard normally distributed random numbers $w^{(j)}{}_0, w^{(j)}{}_1, \ldots, w^{(j)}{}_{N-1}$ for $j = 1, 2, \ldots, m$.
- From the $w^{(j)}{}_0, w^{(j)}{}_1, \ldots, w^{(j)}{}_{N-1}$ for $j = 1, 2, \ldots, m$, we compute real numbers $z^{(j)}{}_0, z^{(j)}{}_1, \ldots, z^{(j)}{}_{N-1}$ for $j = 1, 2, \ldots, m$ in the following way:

  $$\text{For each } i, \begin{pmatrix} z_i{}^{(1)} \\ z_i{}^{(2)} \\ \vdots \\ z_i{}^{(m)} \end{pmatrix} = C \cdot \begin{pmatrix} w_i{}^{(1)} \\ w_i{}^{(2)} \\ \vdots \\ w_i{}^{(m)} \end{pmatrix}.$$

  Here C is the Cholesky decomposition matrix of M.
- Then the simulated stock prices $S_j(0), S_j(1 \cdot dt), S_j(2 \cdot dt), \ldots, S_j(N \cdot dt)$ for $j = 1, 2, \ldots, m$ are successively calculated using the formula

  $$S_j((i+1) \cdot dt) = S_j(i \cdot dt) \cdot e^{dt \cdot \mu_j + \sqrt{dt}\sigma_j \cdot z^{(j)}{}_i}$$

- These prices can then be entered in a time/price diagram and connected to produce m “price paths” for illustration.

Again, we are going to run an example, using the software on our website.

Example 4.6 Simulate sample paths for the three stocks S_1, S_2, and S_3 for the following parameters:

$T = 1$
$dt = \frac{1}{52}$

$\mu_1 = 0.2$
$\sigma_1 = 0.3$
$S_1(0) = 100$

$\mu_2 = -0.2$
$\sigma_2 = 0.3$
$S_2(0) = 120$

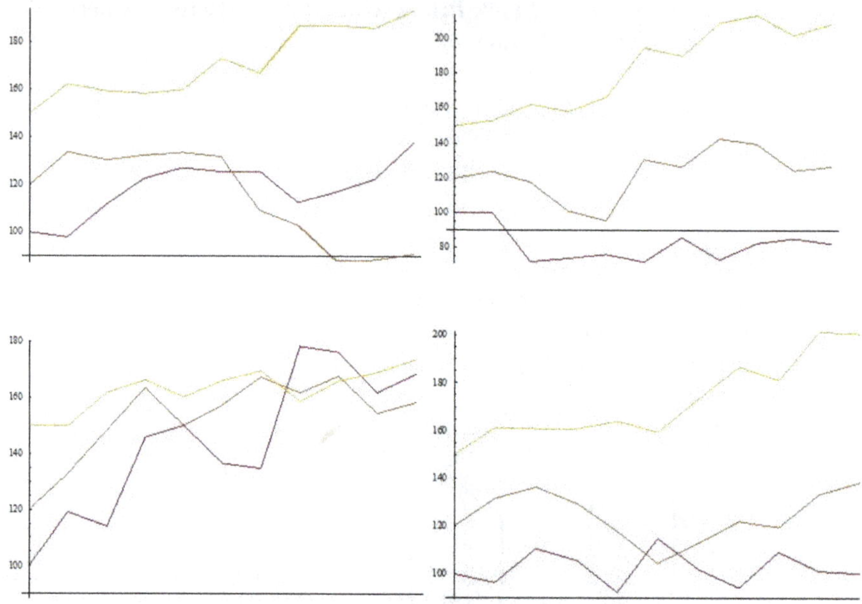

Fig. 4.49 Sample simulations of three stock prices with a given correlation matrix

$\mu_3 = 0$
$\sigma_3 = 0.5$
$S_3(0) = 150$

and for the correlation matrix $M = \begin{pmatrix} 1 & 0.5 & -0.7 \\ 0.5 & 1 & 0.2 \\ -0.7 & 0.2 & 1 \end{pmatrix}$.

Cholesky decomposition of M yields the Cholesky matrix

$$C = \begin{pmatrix} 1 & 0 & 0 \\ 0.5 & 0.866 & 0 \\ -0.7 & 0.635 & 0.327 \end{pmatrix}.$$

In Fig. 4.49 we see four realizations that we created using the software on our website.

Finally, we are going to specify a procedure by means of which we can construct a further stock price relative to k given prices so that the new stock price has given correlations with the given k stock prices. The program for this tool is available and can be used on our website!

This is therefore a generalization of the algorithm given at the end of the previous section, which generates a new price with a given correlation relative to *one* given stock price.

We proceed as follows:

We have the stock prices of k correlated stocks $S^{(1)}, \ldots, S^{(k)}$ at times $i \cdot dt$ for $i = 0, 1, \ldots, n$ and denote these by

$$S_0{}^{(1)}, \ldots, S_n{}^{(1)}$$
$$\ldots$$
$$S_0{}^{(k)}, \ldots, S_n{}^{(k)}$$

From those, we calculate the continuous returns

$$r_0{}^{(1)}, \ldots, r_{n-1}{}^{(1)}$$
$$\ldots$$
$$r_0{}^{(k)}, \ldots, r_{n-1}{}^{(k)}$$

and their trends $\mu^{(1)}, \ldots, \mu^{(k)}$, volatilities $\sigma^{(1)}, \ldots, \sigma^{(k)}$, and correlations ρ_{ij} for $i, j = 1, \ldots, k$. We define $\rho_{ii} := 1$.

We now want to generate a price path $S_0, S_1, \ldots, S_n$ with the given initial value S_0, the given trend μ, and the given volatility σ as well as the given correlations $\rho_{jk+1} = \rho_{k+1j}$ to the stock $S^{(j)}$.

We start the algorithm by normalizing the returns $r_i{}^{(j)}$:

$$\tilde{r}_i^{(j)} := \frac{r_i^{(j)} - \mu^{(j)} dt}{\sigma^{(j)} \sqrt{dt}}.$$

The $\tilde{r}_i^{(j)}$ are therefore (correlated) $\mathcal{N}(0, 1)$ distributed random variables.

We let $\widetilde{Y}_0, \widetilde{Y}_1, \ldots, \widetilde{Y}_{n-1}$ denote the $\mathcal{N}(0, 1)$ distributed random variables with a suitable correlation to $r_i{}^{(j)}$. These are the variables we are going to need to simulate the new stock price.

This means: Once we have created $\widetilde{Y}_0, \widetilde{Y}_1, \ldots, \widetilde{Y}_{n-1}$, we will proceed to simulate the path $S_0, S_1, \ldots, S_n$ by setting

$$S_{i+1} = S_i \cdot e^{\mu \cdot dt + \sigma \sqrt{dt} \cdot \widetilde{Y}_i}$$

To construct the $\widetilde{Y}_0, \widetilde{Y}_1, \ldots, \widetilde{Y}_{n-1}$, we let M denote the correlation matrix

$$M = \begin{pmatrix} \rho_{11} & \cdots & \rho_{1\,k+1} \\ \vdots & & \vdots \\ \rho_{k+1\,1} & \cdots & \rho_{k+1\,k+1} \end{pmatrix}.$$

Further let C be the lower left triangular matrix yielded by the Cholesky decomposition $M = C \cdot C^T$ and $\widetilde{C}$ the upper left $k \times k$ submatrix of C.

$\widetilde{C}$ is of course also a lower left triangular matrix.

(continued)

If we assume that, for a fixed i, the correlated return vector $\begin{pmatrix} \tilde{r}_i^{(1)} \\ \vdots \\ \tilde{r}_i^{(k)} \\ \widetilde{Y}_i \end{pmatrix}$ was generated by a vector $\begin{pmatrix} z^{(1)} \\ \vdots \\ z^{(k)} \\ x \end{pmatrix}$ of **uncorrelated** $\mathcal{N}(0, 1)$ distributed random variables and using the Cholesky matrix, i.e.

$$\begin{pmatrix} \tilde{r}_i^{(1)} \\ \vdots \\ \tilde{r}_i^{(k)} \\ \widetilde{Y}_i \end{pmatrix} = C \cdot \begin{pmatrix} z^{(1)} \\ \vdots \\ z^{(k)} \\ x \end{pmatrix} \tag{4.2}$$

then x can be generated completely independently as an independent $\mathcal{N}(0, 1)$ distributed random variable. $\widetilde{Y}_i$ then follows from the last equation in the above equation system (4.2):

$$\widetilde{Y}_i = c_{k+1\ 1} z^{(1)} + c_{k+1\ 2} z^{(2)} + \ldots + c_{k+1\ k} z^{(k)} + c_{k+1\ k+1} x$$

(if $C = \left(c_{ij}\right)_{i,j=1,\ldots,k+1}$).

For this, however, we also need the $z^{(1)}, z^{(2)}, \ldots, z^{(k)}$.

They can be obtained from the first k equations of the system (4.2):

$$\begin{pmatrix} z^{(1)} \\ \vdots \\ z^{(k)} \end{pmatrix} = \left(\widetilde{C}\right)^{-1} \cdot \begin{pmatrix} \tilde{r}_i^{(1)} \\ \vdots \\ \tilde{r}_i^{(k)} \end{pmatrix}$$

Here we made use of the fact that C and $\widetilde{C}$ are lower left triangular matrices.

Example 4.7 Below we see three prices (weekly prices over 1 year) of three different stocks.

100,112,115,117,117,124,121,116,111,120,119,116,113,114,121,128,133,139, 155,156,165,

159,172,177,169,177,189,198,191,198,211,209,209,222,228,208,209,214,222, 228,240,263,

245,221,245,235, 244,226,235,208,193,175,192

100,96,91,88,84,83,80,78,80,77,76,69,69,69,73,66,65,65,67,70,74,74,72,71,73, 77,72,77,

81,75,75,81,85,83,86,77,80,78,74,80,78,77,74,74,76,73,71,72,76,75,74,76,76

100,94,92,88,88,87,88,88,89,88,85,83,82,81,80,76,77,75,74,76,76,77,74,76,76, 77,73,77,

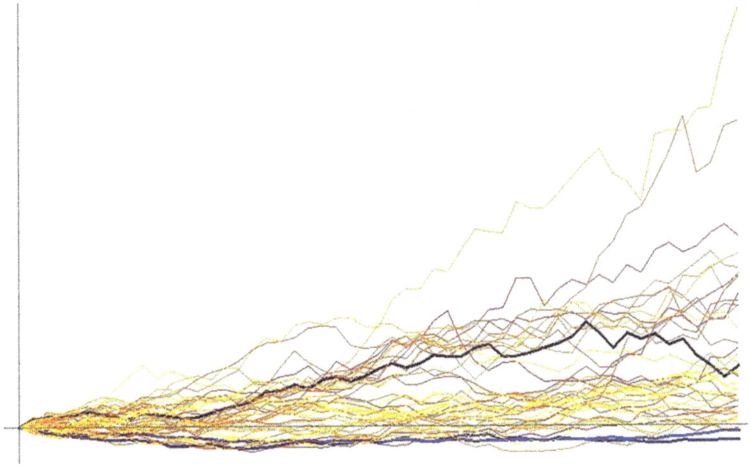

Fig. 4.50 40 simulated prices for given trend and volatility values and given correlations to the three given (blue) stock prices

78,75,74,77,77,75,76,75,77,76,75,77,78,77,79,80,77,79,77,81,82,84,87,91,89

Calculating the per annum trends of these three stocks, we get the values 0.65, −0.27, and −0.11, and for the per annum volatilities, we get 0.42, 0.31, and 0.19. The correlations are $\rho_{1,2} = 0.04$, $\rho_{1,3} = -0.58$, and $\rho_{2,3} = 0.56$.

We now want to simulate stock prices of the same length, with the same initial value of 100 and with trend $\mu = 0.4$, volatility $\sigma = 0.5$, and correlations $\rho_{1.4} = 0.1$, $\rho_{2.4} = 0.1$, and $\rho_{3.4} = 0.1$.

Using the simulation program on our website, we are going to generate 40 such realizations, which gives us the following Fig. 4.50. The three given stock prices are shown in shades of blue (the first price being the darkest and the third price the lightest). The 40 simulated prices are shown in yellow/red. Not all of the simulated prices satisfy the preset trend, volatility, and correlation values exactly, of course, being that they are randomly generated prices. It is only when several realizations are averaged that we obtain approximately these preset values!)

4.9 Simulation of a Wiener Model for Given Initial and Final Values: The Brownian Bridge

For various simulation tasks, it is necessary or at least desirable to simulate stock prices over a certain time range $[0, T]$, with an initial value $S(0)$ and a given trend μ and given volatility σ, using a Wiener model, yet in such a way that a fixed final value $S(T)$ is assumed for each simulation at time T.

Consider, for example, a simulation of the price path of a bond (we won't discuss here whether the Wiener model is in fact a suitable model for simulating bond prices) from the given initial offering price $S(0)$ to the final price, which is always 100.

We are only presenting a suitable methodology here, without analysing it further or demonstrating its functionality. Nor will we go into any further detail at this point as to why this methodology is called the "Brownian bridge". All this will be discussed in a subsequent chapter.

For the purposes of our simulation, we are going to represent $S(t)$ in the form $S(t) = S(0) \cdot e^{\mu \cdot t + \sigma \cdot X(t)}$ and simulate $X(t)$ so that the following conditions are satisfied:

- $X(0) = 0$
- $X(T) = \frac{\left(\log\left(\frac{S_T}{S_0}\right) - \mu \cdot T\right)}{\sigma}$ and thus $S(T) = S(0) \cdot e^{\mu T + \sigma X(T)}$
- For every $t \in [0, T]$ (for which a simulation point is generated) and all $0 \leq A < B$, the probability that $A < S(0) \cdot e^{\mu \cdot t + \sigma \cdot X(t)} < B$ holds is equal to the probability that $A < \hat{S}(t) < B$ holds for any Wiener model $\hat{S}$ with parameters μ and σ at time t provided that $\hat{S}(0) = S(0)$ and $\hat{S}(T) = S(T)$.

We are now going to present two possible construction types. The first one is much simpler than the second version. However, the second version has some advantages in complex simulations:

Variant 1

The interval $[0, T]$ is subdivided into N equal parts. We denote these subdivision points by $0 = t_0 < t_1 < t_2 < \ldots < t_{N-1} < t_N = T$.

We set $Y(0) = 0$, and for $k = 1, 2, \ldots, N$, we set $Y(t_k) = Y(t_{k-1}) + \sqrt{\frac{T}{N}} \cdot w_k$ where $w_1, w_2, \ldots, w_N$ are N independent standard normally distributed random numbers.

Finally, we set $X(t_k) = Y(t_k) - \frac{t_k}{T} \cdot (Y(T) - X(T))$.

Figure 4.51 presents two graphs with ten simulation paths, respectively, of a Wiener model on $[0, 1]$ with given values $S(0) = 100$ and $S(1) = 110$, once with the parameters $\mu = 0.1$ and $\sigma = 0.1$ and once with the parameters $\mu = 0.9$ and $\sigma = 0.9$. For subdivision of the time interval, we chose $N = 100$.

Variant 2

The interval $[0, T]$ is again subdivided into N equal parts, where N now needs to be a power of two, i.e. $N = 2^n$. The subdivision points of the interval are denoted by $0 = t_0 < t_1 < t_2 < \ldots < t_{2^n-1} < t_{2^n} = T$.

The values $X(0) = X(t_0)$ und $X(T) = X(t_{2^n})$ have already been set, as described above. In n steps we first simulate

$X\left(t_{2^{n-1}}\right)$
then
$X\left(t_{2^{n-2}}\right)$ and $X\left(t_{3.2^{n-2}}\right)$
then

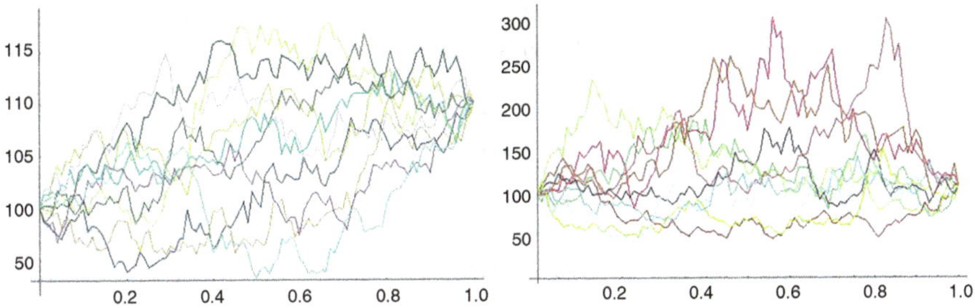

Fig. 4.51 Ten simulations, respectively, of the Wiener model from 100 to 110 ($\mu = 0.1$ and $\sigma = 0.1$ on the left, $\mu = 0.9$ and $\sigma = 0.9$ on the right)

$X\left(t_{2^{n-3}}\right)$ and $X\left(t_{3.2^{n-3}}\right)$ and $X\left(t_{5.2^{n-3}}\right)$ and $X\left(t_{7.2^{n-3}}\right)$
etc.
and finally
$X(t_1)$ and $X(t_3)$ and $X(t_5)$ and $X(t_7)$ and ... and $X(t_{2^n-3})$ and $X(t_{2^n-1})$

At the i-th step, a value $X\left(t_{k\cdot 2^{n-i}}\right)$ is constructed from the points $X\left(t_{(k-1)\cdot 2^{n-i}}\right)$ and $X\left(t_{(k+1)\cdot 2^{n-i}}\right)$ that were constructed in earlier steps.

For example $n = 4$:

$X(t_0)$ and $X(t_{16})$ are given.

Step 1:

Construction of $X(t_8)$ from $X(t_0)$ and $X(t_{16})$

Step 2:

Construction of $X(t_4)$ from $X(t_0)$ and $X(t_8)$ and
Construction of $X(t_{12})$ from $X(t_8)$ and $X(t_{16})$

Step 3:

Construction of $X(t_2)$ from $X(t_0)$ and $X(t_4)$
Construction of $X(t_6)$ from $X(t_4)$ and $X(t_8)$
Construction of $X(t_{10})$ from $X(t_8)$ and $X(t_{12})$
Construction of $X(t_{14})$ from $X(t_{12})$ and $X(t_{16})$

Step 4:

Construction of $X(t_1)$ from $X(t_0)$ and $X(t_2)$
Construction of $X(t_3)$ from $X(t_2)$ and $X(t_4)$
Construction of $X(t_5)$ from $X(t_4)$ and $X(t_6)$

Construction of $X(t_7)$ from $X(t_6)$ and $X(t_8)$
Construction of $X(t_9)$ from $X(t_8)$ and $X(t_{10})$
Construction of $X(t_{11})$ from $X(t_{10})$ and $X(t_{12})$
Construction of $X(t_{13})$ from $X(t_{12})$ and $X(t_{14})$
Construction of $X(t_{15})$ from $X(t_{14})$ and $X(t_{16})$

The construction formula for constructing the value $X\left(t_{k\cdot 2^{n-i}}\right)$ from the points $X\left(t_{(k-1)\cdot 2^{n-i}}\right)$ and $X\left(t_{(k+1)\cdot 2^{n-i}}\right)$ that we constructed in earlier steps is

$$X\left(t_{k\cdot 2^{n-i}}\right) = \frac{X\left(t_{(k-1)\cdot 2^{n-i}}\right) + X\left(t_{(k+1)\cdot 2^{n-i}}\right)}{2} + \sqrt{\frac{1}{2^{i+1}}} \cdot Z$$

with a standard normally distributed random variable Z (that is independent of all previously constructed X values).

4.10 Expectations, Variances, and Probability Distributions of Stock Prices in the Wiener Model

At the end of Sect. 4.4 we announced:

It (the Wiener model) can help us simulate possible (typical) price movements (this is an essential prerequisite for being able to carry out so-called Monte Carlo simulations later), and it enables us to make calculations in the model that then allow certain conclusions to be drawn about the real stock price movements.

In the previous three sections, we looked at simulations in the Wiener model. Now we are going to derive some basic characteristics of the model.

Modeling the price of a stock by means of a Wiener model provides that stock's price $S(T)$ at a future time T as a random variable:

$$S(T) = S(0) \cdot e^{T\mu + \sqrt{T}\sigma \cdot w}$$

As has often been pointed out, this modeling approach does not allow us to predict the future course of the price; what it allows us to do is calculate certain stochastic parameters and the probability distribution of a stock price that has been modeled in this way.

The Expected Value of $S(T)$

So the question we are now asking is what is the value $S(T)$ going to be on average?

An intuitive but incorrect argument would be to say: "The average value $E(w)$ of the random variable w is equal to 0, therefore the average value (expected value $E(S(T))$) of $S(T)$ is given by

$$E(S(T)) = S(0) \cdot e^{T\mu + \sqrt{T}\sigma \cdot E(w)} = S(0) \cdot e^{T\mu + \sqrt{T}\sigma \cdot 0} = S(0) \cdot e^{T\mu}."$$

Yet in fact, it is not possible to just lift the expected value in the exponents.

Why is that, and what can we actually expect to be the expected value of $S(T)$? The $\mathcal{N}(0,1)$ distributed random variable w assumes positive and negative values with, on average, equal frequency and margin of fluctuation. However, a positive deviation of w (e.g. $w = 1$) has a significantly stronger effect on the value of $e^{T\mu+\sqrt{T}\sigma\cdot w}$ than an equally large negative deviation of w (e.g. $w = -1$).

As an example, let us set $T = \mu = \sigma = 1$. In this case, $e^{T\mu+\sqrt{T}\sigma\cdot w}$ for $w = 0$, $w = -1$ and $w = 1$ has the following values:

$e^1 = 2.718\ldots$, $e^0 = 1$, and $e^2 = 7.389\ldots$

The distance between e^1 and e^0 is significantly smaller than the distance between e^1 and e^2. The average value of the three values is 3.702, i.e. greater than e^1.

So we will likely get an expected value $E\,(S(T))$ that is greater than the value $S(0)\cdot e^{T\mu}$ that was naively assumed above.

Indeed, calculating the expected value, we get

$$
\begin{aligned}
E\,(S\,(T)) &= \frac{1}{\sqrt{2\pi}}\int_{-\infty}^{\infty} S(0)\cdot e^{T\mu+\sigma\cdot\sqrt{T}w}\cdot e^{-\frac{w^2}{2}}\,dw = \\
&= S(0)\cdot\frac{1}{\sqrt{2\pi}}\int_{-\infty}^{\infty} e^{-\frac{\left(w-\sigma\sqrt{T}\right)^2}{2}+T\mu+\frac{\sigma^2 T}{2}}\,dw = \\
&= S(0)\cdot e^{T\left(\mu+\frac{\sigma^2}{2}\right)}\cdot\frac{1}{\sqrt{2\pi}}\int_{-\infty}^{\infty} e^{-\frac{\left(w-\sigma\sqrt{T}\right)^2}{2}}\,dw = \\
&= S(0)\cdot e^{T\left(\mu+\frac{\sigma^2}{2}\right)}\cdot\frac{1}{\sqrt{2\pi}}\int_{-\infty}^{\infty} e^{-\frac{y^2}{2}}\,dy = \\
&= S(0)\cdot e^{T\left(\mu+\frac{\sigma^2}{2}\right)}
\end{aligned}
$$

In fact, as anticipated, with

$\boldsymbol{E\,(S(T)) = S(0)\cdot e^{T\left(\mu+\frac{\sigma^2}{2}\right)}}$, we get an expected value greater than $S(0)\cdot e^{T\mu}$.

When T and μ are fixed, this expected value increases as the stock's volatility σ increases! This should be considered a highly essential finding!

On average, the greater the volatility of the stock (with the trend remaining unchanged), the more the stock's value can be expected to increase in the future!

And this is in fact the only argument to justify a risky investment in a stock as compared to an investment in a risk-free product. We illustrate this fact in Figs. 4.52 and 4.53:

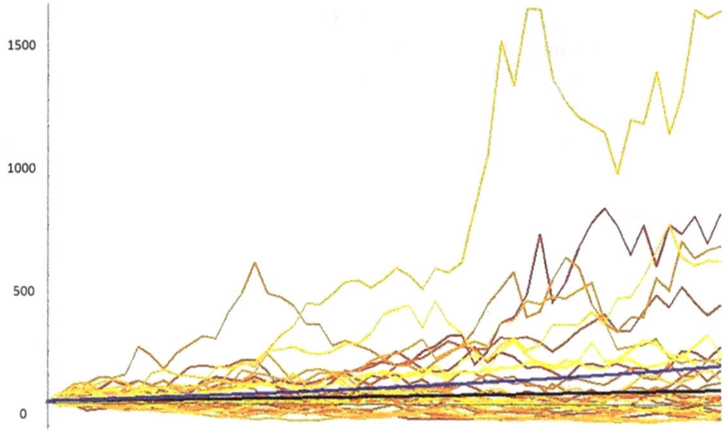

Fig. 4.52 30 simulated paths of the Wiener model with trend 0.4 and volatility 1.0

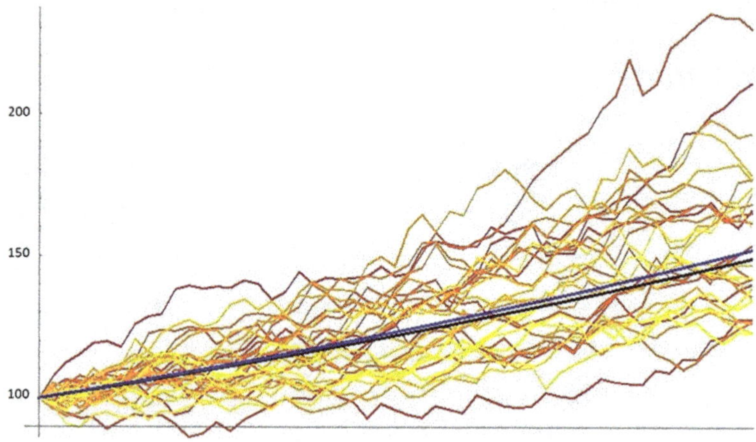

Fig. 4.53 30 simulated paths of the Wiener model with trend 0.4 and volatility 0.2

In each of the two Figs. 4.52 and 4.53, we simulated 30 paths of the Wiener model with initial value 100 and trend $\mu = 0.4$. For the volatility we chose $\sigma = 1.0$ in Fig. 4.52 and $\sigma = 0.2$ in Fig. 4.53.

In both Figs. 4.52 and 4.53, we plotted the path $S(0) \cdot e^{t\mu}$ of a risk-free product (balance in a savings account) over the period $t \in [0, T]$ with the same trend (black path) and the path of the expected value $S(0) \cdot e^{t\left(\mu+\frac{\sigma^2}{2}\right)}$ over the period $t \in [0, T]$ (blue, above).

We see the significantly stronger increase in the expected value compared to the risk-free investment in Fig. 4.52 (higher volatility) than in Fig. 4.53 (lower volatility).

In addition, we computed the actual average value at time $T = 1$ of the 30 simulations in both cases and compared it to the theoretical expected value $S(0) \cdot e^{T\left(\mu+\frac{\sigma^2}{2}\right)}$ and the "incorrect expected value" $S(0) \cdot e^{T\mu}$:

Variance and Standard Deviation of $S(T)$

We are now looking at the variance $V(S(T))$, the average square deviation of the value $S(T)$ in the Wiener model from its expected value $S(0) \cdot e^{T\left(\mu+\frac{\sigma^2}{2}\right)}$. For every random variable X, the relation $V(X) = E\left(X^2\right) - (E(X))^2$ always holds. It follows that

$$
\begin{aligned}
\boldsymbol{V(S(T))} &= E\left((S(T))^2\right) - (E(S(T)))^2 = \\
&= E\left(S^2(0) \cdot e^{2\mu T + 2\sigma\sqrt{T}w}\right) - \left(S(0) \cdot e^{T\left(\mu+\frac{\sigma^2}{2}\right)}\right)^2 = \\
&= S^2(0) \cdot e^{T\left(2\mu+\frac{(2\sigma)^2}{2}\right)} - S^2(0) \cdot e^{T\left(2\mu+\sigma^2\right)} = \\
&= \boldsymbol{S^2(0) \cdot \left(e^{T(2\mu+2\sigma^2)} - e^{T(2\mu+\sigma^2)}\right)} = \\
&= S^2(0) \cdot e^{T(2\mu+\sigma^2)} \cdot \left(e^{T\sigma^2} - 1\right)
\end{aligned}
$$

and for the standard deviation $\sigma(S(T))$ in the Wiener model, it follows therefore that

$$
\boldsymbol{\sigma(S(T)) = S(0) \cdot e^{T\left(\mu+\frac{\sigma^2}{2}\right)} \cdot \sqrt{e^{T\sigma^2} - 1}}
$$

We now add the standard deviation to the Table 4.2 that we created for our simulations and expected values and obtain the subsequent Table 4.3:

Probability Distribution of $S(T)$

Table 4.2 Comparison of the simulated, theoretical, and incorrect EV for different volatilities

	EV simulations	EV theoretical	EV incorrect
Vola = 0.2	153.16	152.20	149.18
Vola = 1.0	256.87	245.96	149.18

Table 4.3 Extension of Table 4.2 with the simulated and theoretical standard deviation

	EV simulations	EV theoretical	EV incorrect	Standard deviation simulations	Standard deviation theoretical
Vola = 0.2	153.16	152.20	149.18	25.94	30.75
Vola = 1.0	256.87	245.96	149.18	312.66	322.41

The random variable $S(T)$ **has a *log normal distribution***. The logarithm of $S(T)$ is normally distributed.

$S(T)$ **always has positive values** (provided, of course, that $S(0)$ is a positive initial value)!

For a given positive x, we can calculate the distribution function $F(x)$, which is $F(x) := W(S(T) < x)$, i.e. the probability that $S(T)$ assumes a value smaller than x, as follows:

$$\boldsymbol{F(x) := W(S(T) < x)} = W\left(S(0) \cdot e^{\mu T + \sigma\sqrt{T}w} < x\right) =$$

$$= W\left(w < \frac{\log \frac{x}{S(0)} - \mu T}{\sigma\sqrt{T}}\right) = \frac{\mathbf{1}}{\sqrt{2\pi}} \int_{-\infty}^{L(x)} e^{-\frac{u^2}{2}} du.$$

Here we use the abbreviation $L(x) := \frac{\log \frac{x}{S(0)} - \mu T}{\sigma\sqrt{T}}$ and once again the fact, of course, that w is standard normally distributed.

Example 4.8 We calculate the probability that $S(T)$ will assume a value that is less than its expected value $x = S(0) \cdot e^{\left(\mu + \frac{\sigma^2}{2}\right)T}$.

For this x we get $L(x) = \frac{\sigma\sqrt{T}}{2}$ and thus

$$W(S(T) < E(S(T))) = \int_{-\infty}^{\frac{\sigma\sqrt{T}}{2}} e^{-\frac{u^2}{2}} du.$$

This value is greater than $\frac{1}{2}$ and increases as the volatility σ increases.

The density $f(x)$ of the distribution of $S(T)$ is obtained by differentiating the distribution function $F(x)$ with respect to x:

$$\boldsymbol{f(x) = F'(x)} = \frac{1}{\sqrt{2\pi}} \cdot e^{-\frac{\left(L^2(x)\right)}{2}} \cdot L'(x) = \frac{\mathbf{1}}{\sqrt{2\pi T} \cdot \sigma x} \cdot e^{-\frac{\left(L^2(x)\right)}{2}}$$

For two arbitrary positive values A and B where $A < B$, the probability that $S(T)$ assumes a value between A and B is

$W(A < S(T) < B) = \int_{L(A)}^{L(B)} e^{-\frac{u^2}{2}} du$, where $L(x) := \frac{\log \frac{x}{S(0)} - \mu T}{\sigma\sqrt{T}}$ holds for any positive x.

We conclude this section with an illustrative sample simulation.

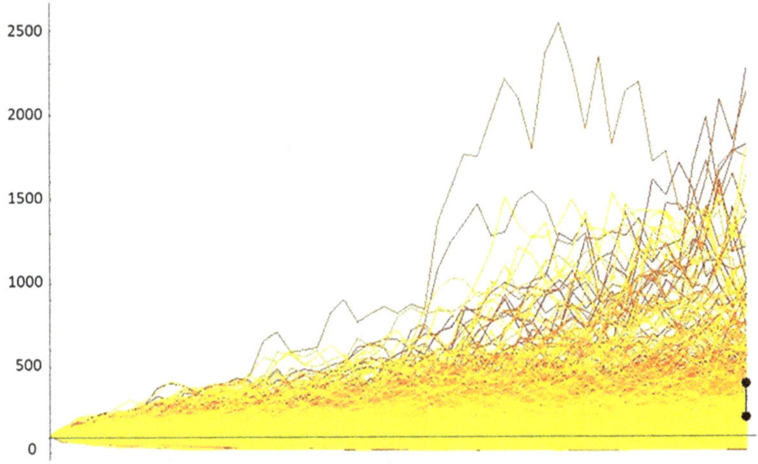

Fig. 4.54 1000 simulated paths of the Wiener model with trend 0.4 and volatility 1

Example 4.9 As in the previous examples, we use an initial value of 100, a time range $T = 1$, the trend $\mu = 0.4$, and volatility $\sigma = 1$. We want to compute the probability that $S(T)$ assumes a value between $A = 200$ and $B = 400$. Substituting, we get $L(B) = \log 4 - 0.4$ and $L(A) = \log 2 - 0.4$. Using Mathematica, for example, we get $W\,(A < S(T) < B) = \int_{L(A)}^{L(B)} e^{-\frac{u^2}{2}} du = 0.22271$.

We test the result by running 1000 simulations of $S(T)$ and counting how many times we get values for $S(T)$ between 200 and 400 (see Fig. 4.54).

In this particular simulation process, we obtained 231 values in the range between 200 and 400 (range marked in black), which, for this particular simulation, would correspond to an empirical probability of 0.231 (compared to the theoretical (rounded) probability of 0.223).

We now have a fairly good idea of the Wiener model. However, we are also aware of the reservations that must be made when using this model and what assumptions must be met in modeling stocks.

In the course of this book, we will get to know other models that cover some aspects of reality better than the Wiener model or are better suited for modeling other products (e.g. interest rates). Yet, as we will see, these models often have certain disadvantages compared to the Wiener model. We can thus say that, with certain reservations, the Wiener model holds up to reality, and it is therefore a model that is often used in real practice to simulate stock price movements. So, for now, we will continue to derive the basic principles of the classic Black-Scholes theory for the valuation of derivatives on underlyings following a Wiener-model.

4.11 Approximation of the Wiener Model Through Binomial Models: Preliminary Remarks

So far, our derivative valuations (in Chap. 3) have been based on binomial models, whose applicability in real practice is highly questionable. We will now show that, by selecting suitable parameters and a sufficiently large number of steps in binomial models, a Wiener model can be approximated to an arbitrarily high degree of accuracy by means of a binomial model.

And as a result, in a second step, we will be able to transfer the results of derivative pricing in binomial models to derivative pricing in the Wiener model. In this way, we are going to derive the results of the classic Black-Scholes theory and then of course use and illustrate them right away in specific applications.

4.12 Approximation of the Wiener Model Through Binomial Models: Preparation

Let us revisit the N-step binomial model once again, by looking at the schematic representation in Fig. 4.55 and the depiction of 4 sample paths in a 40-step binomial model in Fig. 4.56.

These sample paths already show a certain similarity to the realizations in a Wiener model, so the hypothesis that a Wiener model can be approximated to an arbitrarily high degree of accuracy by using a binomial model does not seem all that far-fetched. First, however, we need to define what exactly we mean when we say

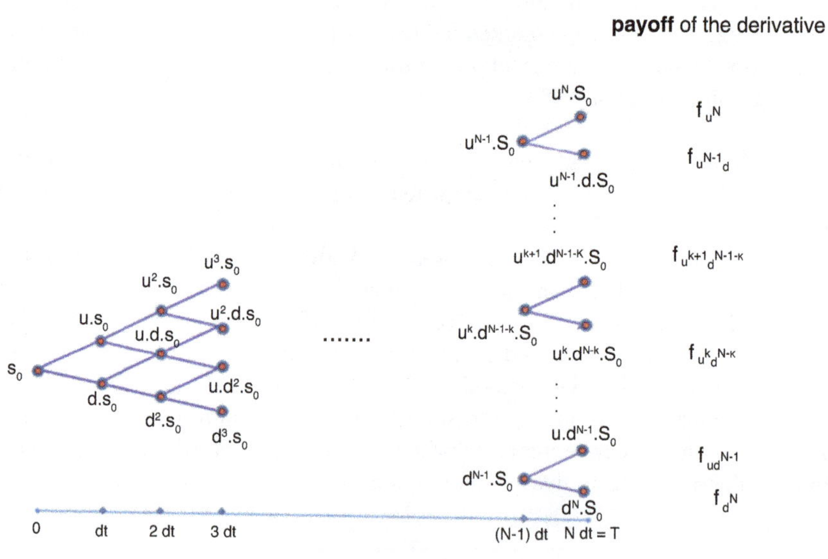

Fig. 4.55 N-step binomial model, schematic representation

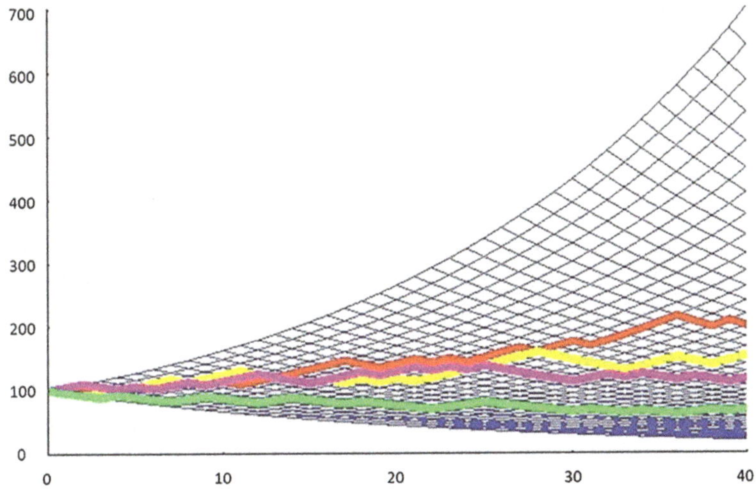

Fig. 4.56 40-step binomial model with 4 possible price paths

that one model "can be approximated to an arbitrarily high degree of accuracy" by means of another model.

- In both cases (Wiener model and N-step binomial model), we are looking at the time range $[0, T]$. We are currently at time 0.
- In both cases, we "start out" with an initial value $S(0)$.
- The "outcome" of the Wiener model at time T is a value $S(T) = S(0) \cdot e^{T\mu+\sqrt{T}\sigma \cdot w}$.
- The "outcome" of the N-step binomial model at time T is a value (we denote it by $BM(T)$) of the form $BM(T) = S(0) \cdot u^k \cdot d^{N-k}$.
- The parameters of the Wiener model are the trend μ and the volatility σ.
- The parameters of the N-step model are the number of steps N (and with that, the step length $dt = \frac{T}{N}$), the up-factor u, the down-factor d, and the probability p of an up-step occurring.
- Both $S(T)$ and $BM(T)$ are random variables (no deterministic values). Their particular values in a realization depend on which value the $\mathcal{N}(0, 1)$ distributed random variable w takes (in the Wiener model) or on which value the random variable k takes (which in the binomial model represents the number of up-moves).
- A random variable is uniquely defined by its probability distribution.

- We know the probability distribution of the random variable $S(T)$ from the previous section: $S(T)$ always takes positive values, and for arbitrary values A, B, where $0 < A < B$, the probability that $S(T)$ lies between A and B is
  $W(A < S(T) < B) = \int_{L(A)}^{L(B)} e^{-\frac{u^2}{2}} du$, where
  $L(x) := \frac{\log \frac{x}{S(0)} - \mu T}{\sigma \sqrt{T}}$. holds for any positive x.
- We now want to show: Let A and B be arbitrary values where $0 < A < B$. Then we can select the parameters u, d, and p for any N such that, for a growing N, the probability $W(A < BM(T) < B)$, i.e. the probability that $BM(T)$ lies between A and B, will converge to $W(A < S(T) < B) = \frac{1}{\sqrt{2\pi}} \int_{L(A)}^{L(B)} e^{-\frac{u^2}{2}} du$. "In the limit", $BM(T)$ then has the same distribution as $S(T)$.
- The suitability of the selected parameters u, d, and p depends, of course, on the parameters μ and σ and on the number N of steps (or its equivalent: the step length dt)!

This is the program and the clarification of our question.

We could now proceed as follows: *We could tell you in advance how to choose u, d and p so that, with this choice and for growing N, $W(A < BM(T) < B)$ will indeed converge to $W(A < S(T) < B)$.*

This is not what we are going to do, however. Instead, we want to enable you to come up with the suitable choice of parameters yourself. Before we do so, however, we will provide a technical tool that we are going to need in the following: It is a basic version of the "central limit theorem" from probability theory.

4.13 The Central Limit Theorem

This basic version of the central limit theorem that we are going to need states:

"Let $X^{(N)}$ be a $(1, -1)$ binomially distributed random variable with parameters N and p. The distribution of the (normalized) random variable $Y^{(N)} := \frac{X^{(N)} - N \cdot (2p-1)}{2\sqrt{N \cdot p \cdot (1-p)}}$ for $N \to \infty$ then converges to the standard normal distribution $\mathcal{N}(0, 1)$."

In the next section, we are going to use this result for the specific case where $p = \frac{1}{2}$. For this case we can formulate the result as follows:

Let $X^{(N)}$ be a $(1, -1)$ binomially distributed random variable with parameters N and $p = \frac{1}{2}$. We then have, $X^{(N)} = Y^{(N)} \cdot \sqrt{N}$, where $Y^{(N)}$ is a random variable whose distribution for $N \to \infty$ converges to the standard normal distribution $\mathcal{N}(0, 1)$.

We explain and illustrate the statement of the central limit theorem:

The $(1, -1)$ binomially distributed random variable $X^{(N)}$ with parameters N and p represents the sum $a_1 + a_2 + \ldots + a_{N-1} + a_N$ of N summands $a_1, a_2, \ldots, a_{N-1}, a_N$, which can independently assume the values 1 (with probability p) or -1 (with probability $1 - p$).

$X^{(N)}$ then is the sum of the N mutually independent random variables $a_1, a_2, \ldots, a_{N-1}, a_N$.

Each of these random variables a_i has the expected value
$E(a_i) = p \cdot 1 + (1 - p) \cdot (-1) = 2p - 1$.
And each of these random variables a_i therefore has the variance
$V(a_i) = p \cdot (1 - (2p - 1))^2 + (1 - p) \cdot (-1 - (2p - 1))^2 = 4p(1 - p)$.
The expected value $E\left(X^{(N)}\right)$ of $X^{(N)}$ is therefore equal to $N \cdot (2p-1)$, and the variance $V\left(X^{(N)}\right)$ (since the a_i are independent of one another) is equal to $4Np(1-p)$. For the standard deviation, $\sum\left(X^{(N)}\right)$ of $X^{(N)}$ obviously follows that $\sum\left(X^{(N)}\right) = 2\sqrt{Np(1 - p)}$.

Hence, $Y^{(N)}$ is precisely the normalized version of $X^{(N)}$ for every N, i.e. it has expectation 0 and variance 1.

$X^{(N)}$ can assume values between $-N$ and $+N$. Consequently, $Y^{(N)}$ can assume values between $-\sqrt{N} \cdot \sqrt{\frac{p}{1-p}}$ and $\sqrt{N} \cdot \sqrt{\frac{1-p}{p}}$, (which is easily seen by substituting $-N$ and N for $X^{(N)}$ in the definition of $Y^{(N)}$). The range of values for $Y^{(N)}$ thus continues to expand.

Now, if x is any arbitrary value between $-\sqrt{N}\sqrt{\frac{p}{1-p}}$ and $\sqrt{N} \cdot \sqrt{\frac{1-p}{p}}$, we can easily calculate the probability $W\left(Y^{(N)} < x\right)$, i.e. the probability that $Y^{(N)} < x$ (the distribution function $F^{(N)}(x)$ of $Y^{(N)}$):

For $Y^{(N)} = \frac{X^{(N)} - N\cdot(2p-1)}{2\sqrt{N\cdot p\cdot(1-p)}}$ to be less than x, it is necessary therefore that
$X^{(N)} < x \cdot 2\sqrt{N \cdot p \cdot (1 - p)} + N \cdot (2p - 1)$.

Thus, if $X^{(N)}$ is precisely the sum of k ones and of $N - k$ minus-ones, then $X^{(N)} = k - (N - k) = 2k - N$.

For $Y^{(N)} = \frac{X^{(N)} - N\cdot(2p-1)}{2\sqrt{N\cdot p\cdot(1-p)}}$ to be less than x, it is necessary therefore that
$X^{(N)} = 2k - N < x \cdot 2\sqrt{N \cdot p \cdot (1 - p)} + N \cdot (2p - 1)$, i. e.
$k < x \cdot \sqrt{N \cdot p \cdot (1 - p)} + N \cdot p$.

(continued)

The probability that $x^{(N)}$ is composed of exactly k ones and $N-k$ minus-ones is precisely $p^k \cdot (1-p)^{N-k} \cdot \binom{N}{k}$. This gives us the probability

$$W\left(Y^{(N)} < x\right) = \sum_{k=0}^{u(x)} p^k \cdot (1-p)^{N-k} \cdot \binom{N}{k},$$

where $u(x)$ is the largest integer that is less than $x \cdot \sqrt{Np(1-p)} + Np$.

This distribution function $F^{(N)}(x)$ of $Y^{(N)}$ can now be illustrated for various growing N and compared to the distribution function $F(x) = \frac{1}{\sqrt{2\pi}} \int_{-\infty}^{x} e^{-\frac{u^2}{2}} du$ of the $\mathcal{N}(0, 1)$ distribution.

For the comparative charts, we choose $p = \frac{1}{3}$, and for N we choose the values $N = 10, 20, 40$, and 200.

We can clearly see that with increasing N, the distribution function of the normalized binomial distribution is increasingly converging to the distribution function of the standard normal distribution.

Incidentally, the closer p is to the value $\frac{1}{2}$, the faster this convergence takes place. This can be seen, for example, in Fig. 4.58: This Fig. 4.58 compares the distribution functions for $N = 200$, i.e. as shown in the fourth chart of Fig. 4.57, but this time not for $p = \frac{1}{3}$ as in Fig. 4.57 but for $p = 0.01$ and $p = 0.99$, respectively. The convergence of the normalized binomial distribution to the standard normal distribution obviously takes much longer here than in the case where $p = \frac{1}{3}$ (Fig. 4.58).

Although we are not going to need this in the following, it should nevertheless be mentioned here that the **central limit theorem**, which is a key (hence, "central") result of probability theory, holds even under considerably more general conditions:

Whenever we have **any** mutually independent random variables $A_1, A_2, A_3, \ldots$ where all of them have the same yet arbitrary distribution (with finite expectation E and finite variance V), and if we then create the new random variables $X^{(N)} = A_1 + A_2 + \ldots + A_N$ and $Y^{(N)} = \frac{X^{(N)} - N \cdot E}{\sqrt{N \cdot V}}$, then the distribution of $Y^{(N)}$ (as explained above) for N to infinity converges to the standard normal distribution.

This statement holds even in the case of certain weak dependencies between the A_i. And the statement also holds if the distributions of the A_i, while not necessarily identical, are still such that none of the A_i overly dominate the other random variables.

This central result of probability theory also explains why normal distribution is of such great importance in so many areas of science, technology, and economics: Because of the interaction of many smaller, essentially

(continued)

independent random influences, many stochastic processes can be explained as the sum of many different random variables. And that is where, by way of the central limit theorem, the normal distribution comes into play.

A Simulation Experiment

Now, if the dominance of the normal distribution is indeed so overwhelming as has been said, how then is it possible that systematic deviations from the normal distribution occur where stock price returns are concerned? Remember: Noticeable "fat tails" and a narrower middle part in stock price return distributions!

Let us run a small experiment. This experiment is not intended to be a serious attempt at explaining systematic deviations from stock price returns, but it does, as you will see, yield quite an interesting result:

A continuous monthly return of a stock A is the sum of approximately 20 daily returns, an annual return the sum of 52 weekly returns, a daily return the sum of approximately 500 min returns, etc.

In the following, let us look at a continuous stock return R over a time interval $[0, 1]$ which results as the sum of 100 continuous stock returns $r_1, r_2, \ldots, r_{100}$, each over a smaller time interval dt. Thus

$$R = r_1 + r_2 + \ldots + r_{100}.$$

If all the r_i were independent of one another and $N(0, dt)$ distributed, then R would be standard normally distributed.

In the following we are going to modify this model somewhat:

We are now introducing certain dependencies into the system. The idea behind this is the following: If at any one time, an exceptionally strong positive or exceptionally strong negative return r_i occurs, the market (the stock market traders) often reacts very nervously. This nervousness can, for example, translate into heightened volatility for a brief period of time. Conversely, if a very quiet phase occurs, with returns close to zero, stock market traders often hold off on purchases and sales until a more active market promises new momentum and thus profit potential. In such cases, volatility will often be low.

We now want to integrate this observation into our system:

- Again we are going to model R by $R = r_1 + r_2 + \ldots + r_{100}$.
- The r_i remain normally distributed with mean $\mu = 0$.

(continued)

- If $r_i > \sqrt{dt}$ or $r_i < -\sqrt{dt}$, however, the volatility for r_{i+1} increases from dt to $5dt$.
- If, on the other hand, $-\sqrt{dt} < r_i < \sqrt{dt}$, then the volatility for r_{i+1} decreases from dt to $\frac{1}{2}dt$.

We have run this algorithm for calculating R 10,000 times. The mean of the resulting values $R^{(j)}$ is again 0, of course. Finally, as the standard deviation can deviate from 1, we normalize $R^{(j)}$ to $R'^{(j)}$ by dividing every $R^{(j)}$ by the standard deviation of the $R^{(j)}$. The mean of $R'^{(j)}$ then is 0 and the standard deviation 1.

If we plot the empirical distribution of the $R'^{(j)}$ (red curve in Fig. 4.59) and compare it to the distribution of the standard normal distribution (blue curve in Fig. 4.59), we typically get a result as shown in Fig. 4.59.

The beauty of this: The resulting picture is one that we recognize: Fat tails and a narrow middle . . .

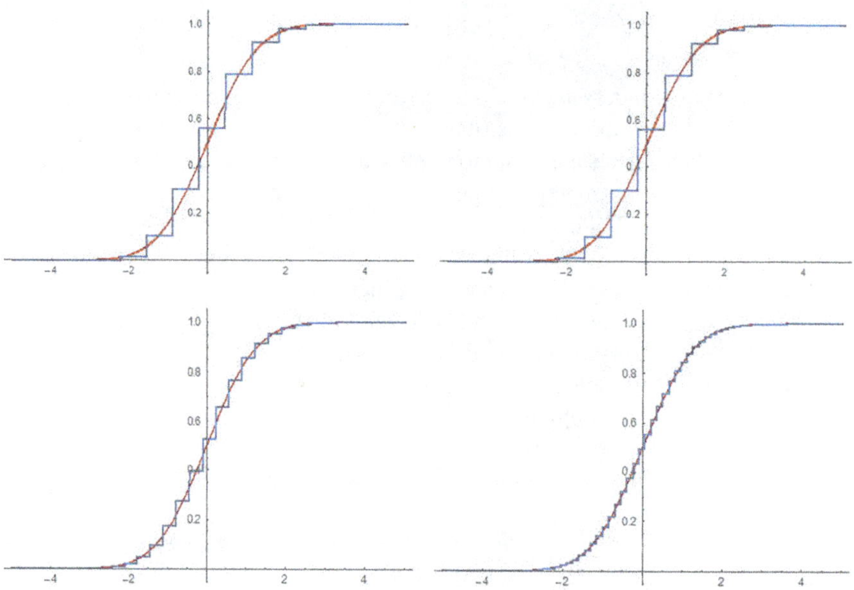

Fig. 4.57 Comparison of the distribution function of the normalized binomial distribution for $p = \frac{1}{3}$, $N = 10, 20, 40$, and 200 to the standard normal distribution

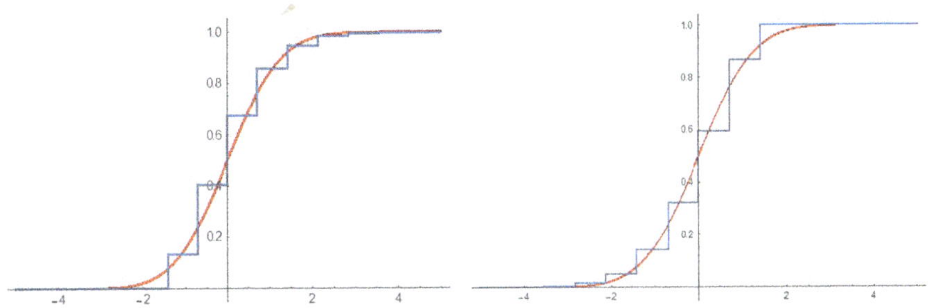

Fig. 4.58 Comparison of the distribution function of the normalized binomial distribution for $p = 0.01$, $p = 0.99$, and $N = 200$ with the standard normal distribution

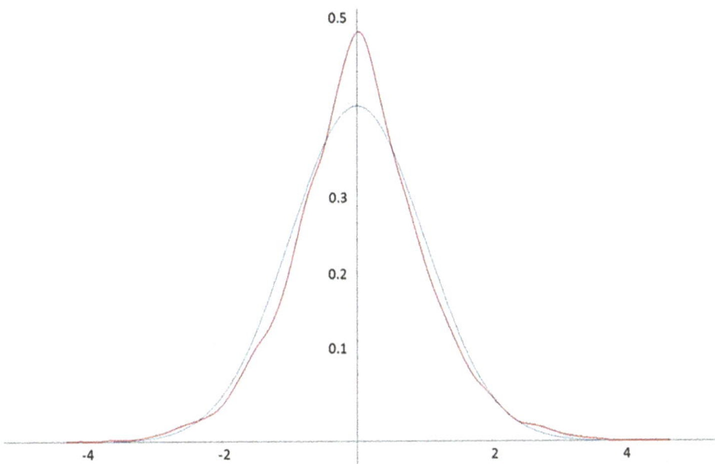

Fig. 4.59 Comparison of the empirical distribution function of the simulation experiment (red) to the standard normal distribution

4.14 Approximation of the Wiener Model Through Binomial Models: The Proof (Sketch of Proof)

We now want to prove the convergence of the N-step binomial model to the Wiener model. That is, we have to show that, for $N \to \infty$ and with a suitable choice of parameters u, d, and p, the probability

$W(A < BM(T) < B)$ converges to $W(A < S(T) < B)$.

(In the following, we will not deliver a strict formal proof but rather the clear idea of proof, so that a mathematically trained reader can easily carry out the exact proof correctly based on the idea of proof.)

Furthermore, as already mentioned, we do not want to reveal the correct values u, d, and p upfront but would like to try to find them ourselves instead.

We have

$$W(A < S(T) < B) =$$

$$W\left(A < S(0) \cdot e^{\mu T + \sigma\sqrt{T} \cdot w} < B\right) =$$

$$W\left(\log \frac{A}{S(0)} < \mu T + \sigma\sqrt{T} w < \log \frac{B}{S(0)}\right). \tag{4.3}$$

On the other hand,

$$W(A < BM(T) < B) =$$

$$W\left(A < S(0) \cdot u^K \cdot d^{N-K} < B\right) =$$

$$W\left(\log \frac{A}{S(0)} < K \cdot \log u + (N-K) \cdot \log d < \log \frac{B}{S(0)}\right) \tag{4.4}$$

We already see a certain similarity between the expressions (4.3) and (4.4). Selecting suitable u, d, p (we are not limited in any way here) and based on our finding from the previous section that a random variable of the form $K \cdot 1 + (N-K) \cdot (-1)$ can be described approximately by a standard normally distributed random variable w, we want to try to further reinforce this similarity:

For this purpose, we set $\log u = a + b$ and $\log d = a - b$ with a and b being any arbitrary value (4.4) then becomes

$$W\left(\log \frac{A}{S(0)} < N \cdot a + b\left(K \cdot 1 + (N-K) \cdot (-1)\right) < \log \frac{B}{S(0)}\right) \tag{4.5}$$

If we compare (4.5) with (4.3), that comparison would suggest that we set $\mu T = N \cdot a$, i.e. select $a = \mu \cdot \frac{T}{N} = \mu \cdot dt$.

What is more: Since w is a symmetrical random variable ($w \sim \mathcal{N}(0, 1)$), it is reasonable to assume that the random variable $K \cdot 1 + (N-K) \cdot (-1)$ is also symmetrical. So we choose $p = \frac{1}{2}$. If we choose $p = \frac{1}{2}$, however, we can use the central limit theorem from the beginning of Sect. 4.13 for $p = \frac{1}{2}$. That is,

$$K \cdot 1 + (N-K) \cdot (-1) = X^{(N)} = Y^{(N)} \cdot \sqrt{N}.$$

Equation (4.5) then becomes

$$W\left(\log \frac{A}{S(0)} < \mu T + b\sqrt{N} \cdot Y^{(N)} < \log \frac{B}{S(0)}\right).$$

Now (to reconcile this with (4.3)), we only need to set $b\sqrt{N} = \sigma\sqrt{T}$, i.e. $b = \sigma\sqrt{\frac{T}{N}} = \sigma\sqrt{dt}$, and let $N \to \infty$, which yields

$$\begin{aligned}
&\lim_{N\to\infty} W\left(A < BM(T) < B\right) = \\
&\lim_{N\to\infty} W\left(\log\frac{A}{S(0)} < \mu T + \sigma\sqrt{T}\cdot Y^{(N)} < \log\frac{B}{S(0)}\right) = \\
&W\left(\log\frac{A}{S(0)} < \mu T + \sigma\sqrt{T}w < \log\frac{B}{S(0)}\right) = \\
&= W\left(A < S(T) < B\right).
\end{aligned}$$

So, the specific choice of u, d, and p that gave us this convergence was

$$p = \frac{1}{2}, u = e^{a+b} = e^{\mu dt + \sigma\sqrt{dt}} \text{ and } d = e^{a-b} = e^{\mu dt - \sigma\sqrt{dt}}.$$

4.15 The Brownian Motion, Motivation, and Definition

In the following, we will discuss an alternative representation of the Wiener model, using the so-called Brownian motion. The Brownian motion is a stochastic process. A "stochastic process" is a concept that we will discuss in more detail and greater depth in Volume II Chapter 3. In the following, we will use the term only informally and use examples to illustrate it. Intuitively speaking, a stochastic process is a random process that evolves over time.

Let us start with a very simple example of a stochastic process:

Two players—A and B—play a game against each other. The two players master the game equally well, and the game is a fair game that offers both players the same odds of winning. Therefore, each player's probability to win the game is 50%. The odds of a player winning are always 50% regardless of previous outcomes. The two players play the game several times against each other. For each game won, the winner receives one euro from the one who lost.

Both players are interested in their "winning process". They are interested in how their winnings will evolve over time. There is no way to predict this, of course. Their winnings can evolve in many different ways. The winning process is therefore a stochastic process.

Each of the possible trajectories (each concrete realization) of this stochastic process is called a "path" of this process. In the following graphs, we want to illustrate possible paths of this winning process. To do this, we let players A and B play the game 10 times (100 times, 1000 times) against each other. Each game represents a time step. For example, the duration of a game can be 1 min. In the following Graphs 4.60, 4.61, 4.62, 4.63, 4.64, and 4.65, we plot the time steps on the x-axis and the current amount that player A has won on the y-axis.

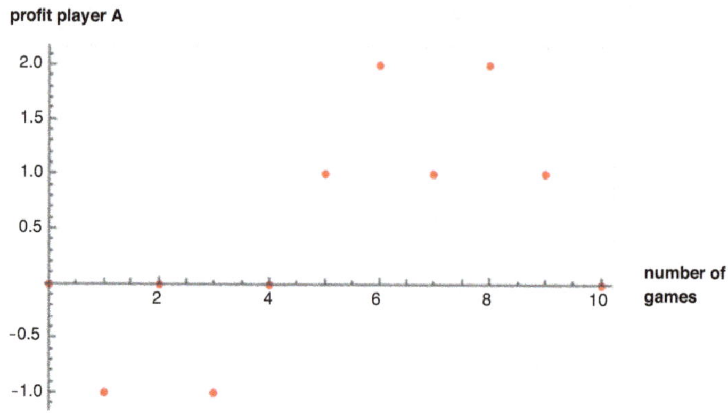

Fig. 4.60 One possible realization of the winning process for player A in ten games

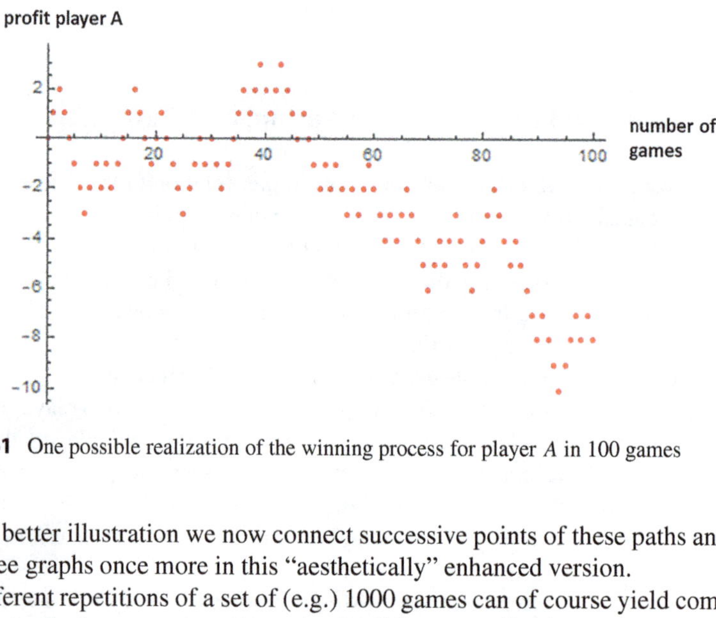

Fig. 4.61 One possible realization of the winning process for player A in 100 games

For better illustration we now connect successive points of these paths and show the three graphs once more in this "aesthetically" enhanced version.

Different repetitions of a set of (e.g.) 1000 games can of course yield completely different winning processes (completely different realizations of the underlying stochastic process).

In Fig. 4.66, we superimposed five such possible realizations.

This stochastic process is a **simple one-dimensional discrete random walk.**

If X_n denotes the value of the stochastic process (i.e. of the winning process, the amount won) at time n (after n games have been played), then

$X_0 = 0$ and X_{n+1} equals either $X_n + 1$ or $X_n - 1$, each with probability $\frac{1}{2}$.

The stochastic process is then given by $(X_n)_{n \in \{0,1,2,\ldots,N\}}$.

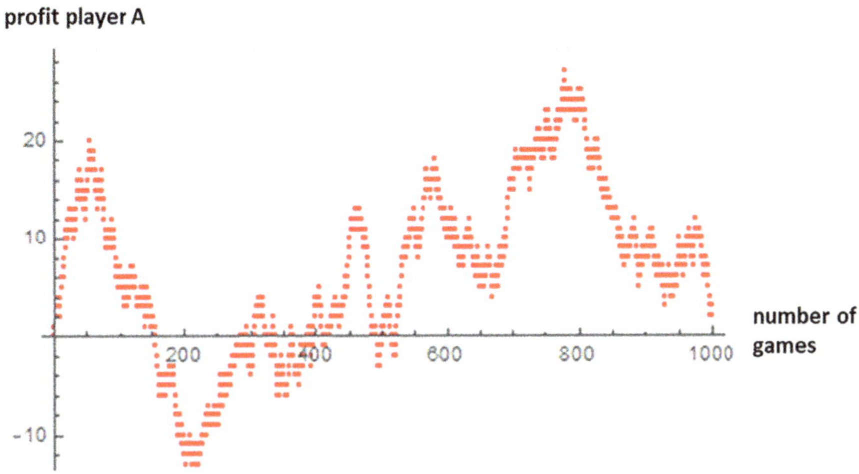

Fig. 4.62 One possible realization of the winning process for player A in 1000 games

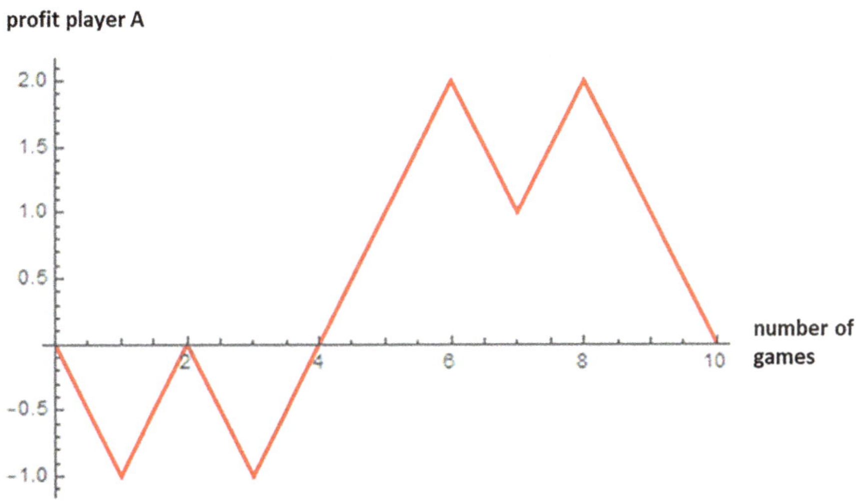

Fig. 4.63 One possible realization of the winning process for player A in 10 games, where values have been connected

By connecting all values X_n and X_{n+1}, respectively, we have mapped a value X_t to any real "point in time" t between 0 and N. We thus have a stochastic process of the form $(X_t)_{t\in[0,N]}$.

Another interesting trajectory is obtained when we independently generate two different paths, $(X_n)_{n\in\{0,1,2,\ldots,N\}}$ and $(Y_n)_{n\in\{0,1,2,\ldots,N\}}$, according to the above model and combine these points into two-dimensional pairs $(X_n, Y_n)_{n\in\{0,1,2,\ldots,N\}}$. These two-dimensional points can then be plotted in a two-dimensional coordinate

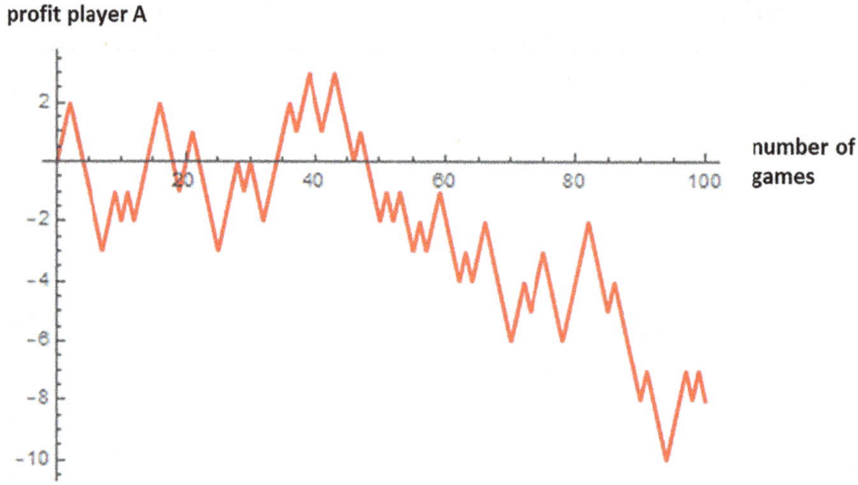

Fig. 4.64 One possible realization of the winning process for player A in 100 games, where values have been connected

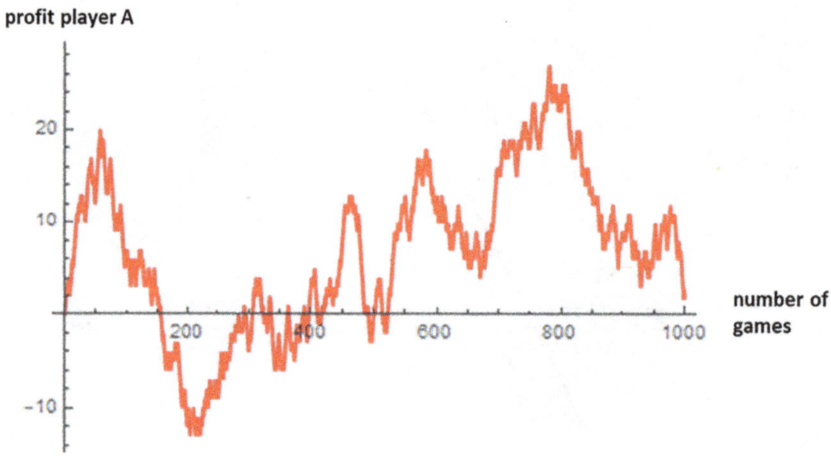

Fig. 4.65 One possible realization of the winning process for player A in 1000 games, where values have been connected

system and connected chronologically. If we do this for a sample path with 10 points, we typically get a graph like the one shown in Fig. 4.67, a **simple two-dimensional discrete random walk**.

Here, we deliberately plotted the sixth and the eighth point (both with coordinates $\{2, 4\}$) slightly offset from each other, for better visibility.

If we generate 1000 or 10,000 points of a path, we typically get images like the following two in Figs. 4.68 and 4.69.

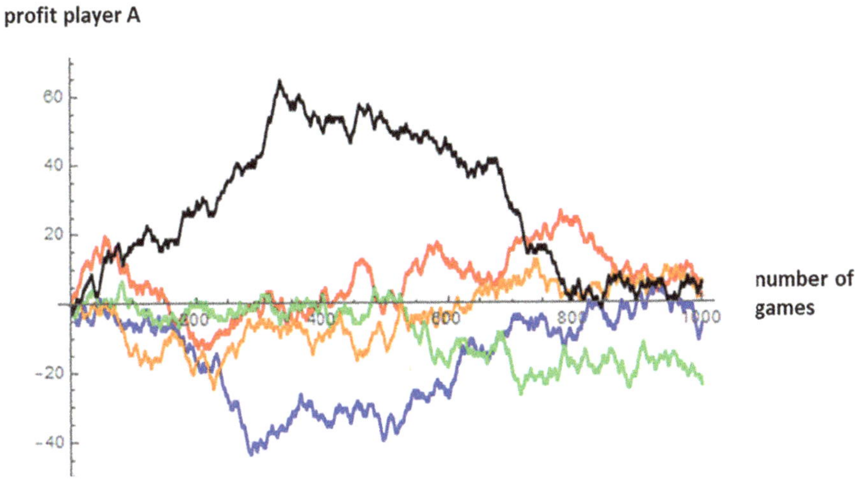

Fig. 4.66 Five possible realizations of the winning process for player A in 1000 games, where values have been connected

In the following we are going to look into the winning process of a somewhat more complex but still a fair game: In this game, both players have an equal probability of winning. However, the payoff per game (the value of the game) is not one euro. Instead, the payoff in the n-th game is now given by a normally distributed random variable w_n. A positive w_n means a payoff in the amount of w_n for player A, and a negative w_n means a payoff in the amount of $-w_n$ for player B.

The amount of the payoff w_n is independent of the amount of any previous payoffs.

So, for the winning process $(Z_n)_{n\in\{0,1,2,\ldots,N\}}$, we now have $Z_{n+1} = Z_n + w_{n+1}$.

For purposes of the winning process $(Z_n)_{n\in\{0,1,2,\ldots,N\}}$ or $(Z_n)_{t\in[0,T]}$ in this somewhat more complex game, we will display the counterparts to Figs. 4.60, 4.61, 4.62, 4.63, 4.64, 4.65, 4.66, 4.67, 4.68, and 4.69 in the subsequent Figs. 4.70, 4.71, 4.72, 4.73, 4.74, 4.75, 4.76, 4.77, 4.78, and 4.79.

We are now looking at a **one-dimensional discrete random walk with normally-distributed increments** and then at a **two-dimensional discrete random walk with normally-distributed increments**.

So, for the winning process $(Z_n)_{n\in\{0,1,2,\ldots,N\}}$, the relationship $Z_n = Z_{n-1} + w_n = Z_{n-2} + w_{n-1} + w_n = \ldots = w_0 + w_1 + \ldots + w_{n-1} + w_n$ holds for every $n \in \{0, 1, 2, \ldots, N\}$, with the w_k being independent standard normally distributed random variables.

Instead of the points in time $1, 2, 3, \ldots, N$, we now want to consider N different points in time but at a shorter distance Δt from one another. Thus, following our motivation example, a game is to take place at every point in time $n \cdot \Delta t$, where $n = 1, 2, 3, \ldots, N$ and where Δt is typically assumed to be a very small time interval (significantly smaller than 1). The last point in time $N \cdot \Delta t$ is denoted by T.

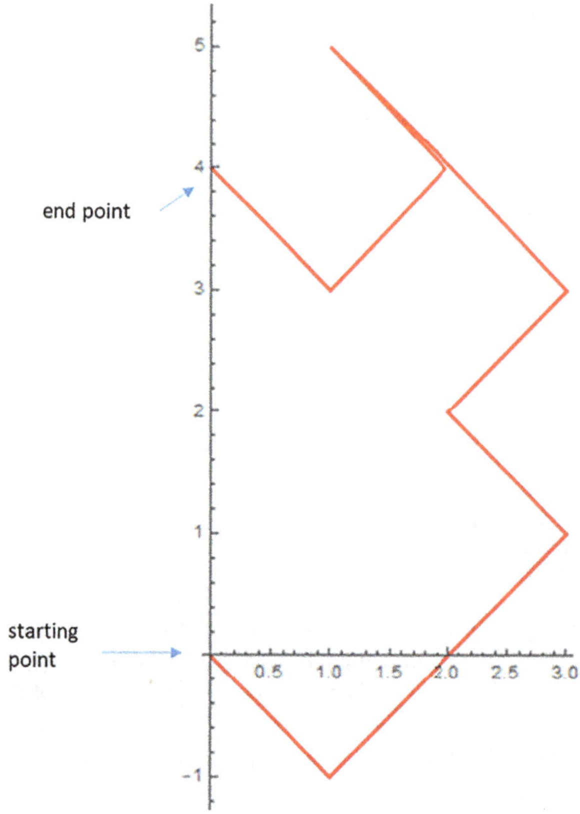

Fig. 4.67 Typical path of a two-dimensional discrete random walk with ten time steps

However, the payoff that can be achieved per game will now be adjusted. Instead of a payoff equal to a standard normally distributed (i.e. $\mathcal{N}(0, 1)$ distributed) random variable w_n, the attainable payoff per game is equal to an $\mathcal{N}(0, \Delta t)$ distributed random variable w_n.

Why do we choose an $\mathcal{N}(0, \Delta t)$ distributed random variable here, that is, why do we choose a variance of Δt for the payoff?

The reason for this is:

Let us suppose that the variance of the independent random variable is equal to a Δt dependent value $f(\Delta t)$. Later we are going to let Δt go to 0. So, in a time range of length 1, approximately $\frac{1}{\Delta t}$ games will be played. (We are going to assume that $M = \frac{1}{\Delta t}$ in this case is an integer.) Consequently, the payoff in the time range $[0, 1]$ is $w_0 + w_1 + \ldots + w_{M-1} + w_M$, and, as

(continued)

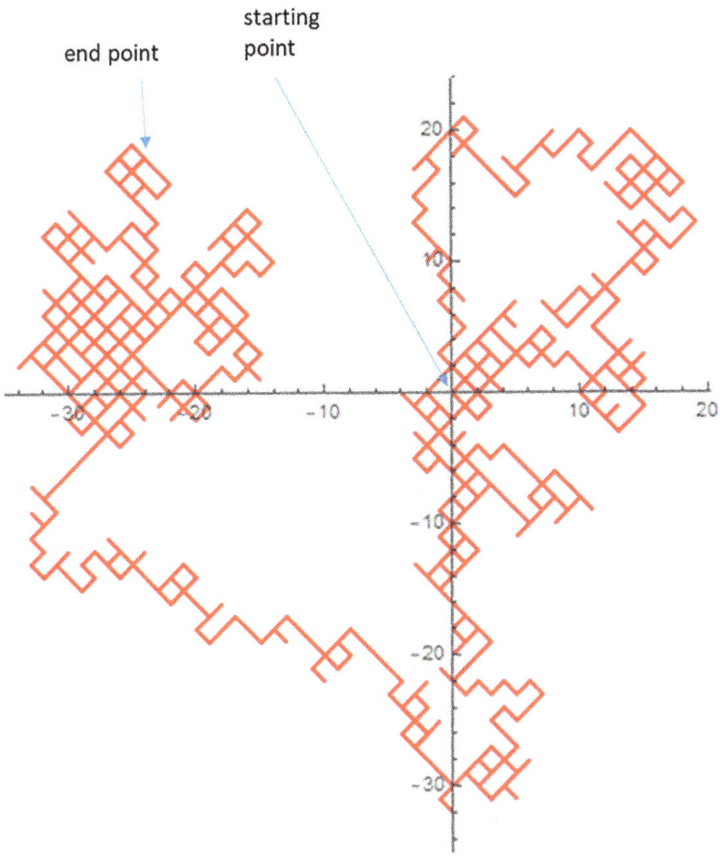

Fig. 4.68 Typical path of a two-dimensional discrete random walk with 1000 time steps

a sum of M mutually independent $\mathcal{N}(0, f(\Delta t))$ distributed random variables, that is an $\mathcal{N}(0, M \cdot f(\Delta t))$ distributed, i.e. an $\mathcal{N}\left(0, \frac{f(\Delta t)}{\Delta t}\right)$ random variable. So that, with Δt tending to zero, the variance of the winning process does not converge to zero nor to infinity, $f(\Delta t)$ **must** be (essentially) equal to Δt. If we choose a variance equal to Δt, then, especially in the time range $[0, 1]$, the winning process will converge to a standard normally distributed random variable and will thus correspond to the distribution of the winning process for the original game with standard normally distributed increments.

In the following, we are going to look at a few typical paths of this winning process over the time range $[0, 10]$ with successively decreasing Δt (compare Figs. 4.80, 4.81, 4.82, and 4.83).

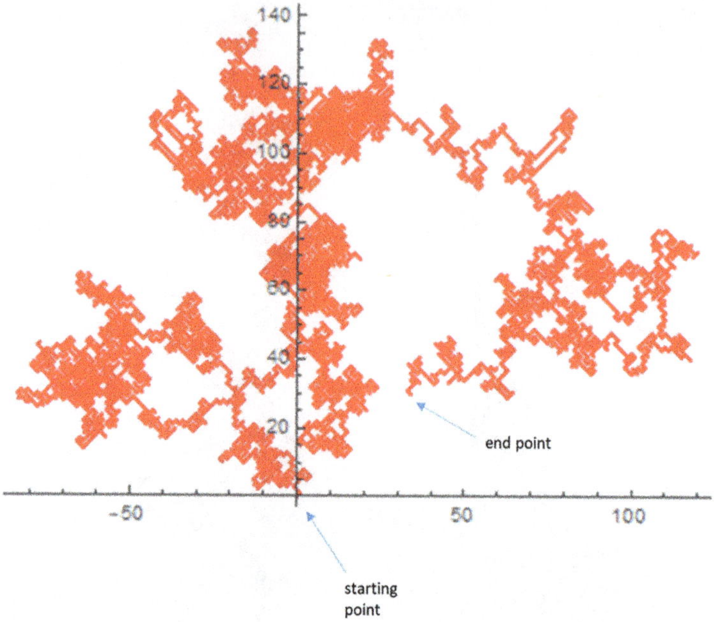

Fig. 4.69 Typical path of a two-dimensional discrete random walk with 10,000 time steps

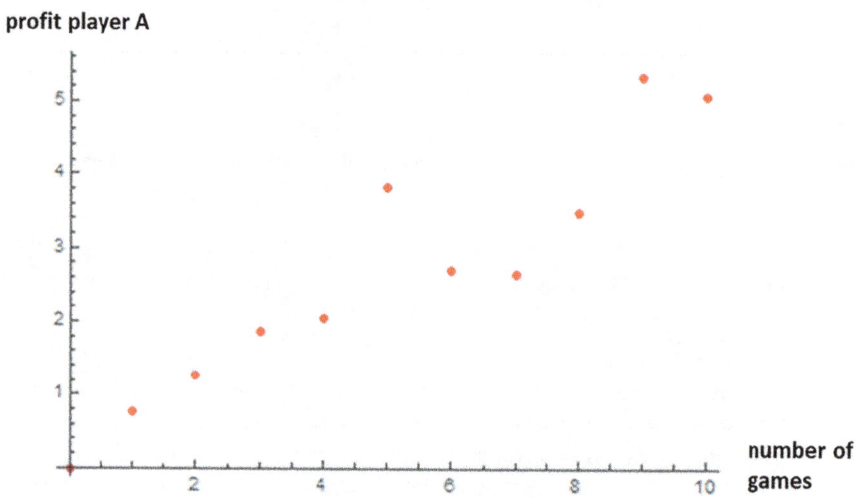

Fig. 4.70 Normally distributed increments: one possible realization of the winning process for player A in ten games

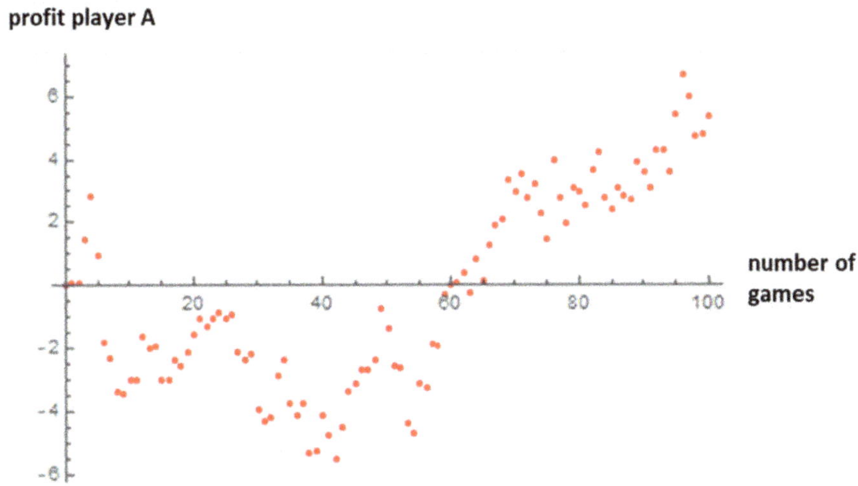

Fig. 4.71 Normally distributed increments: 1 possible realization of the winning process for player A in 100 games

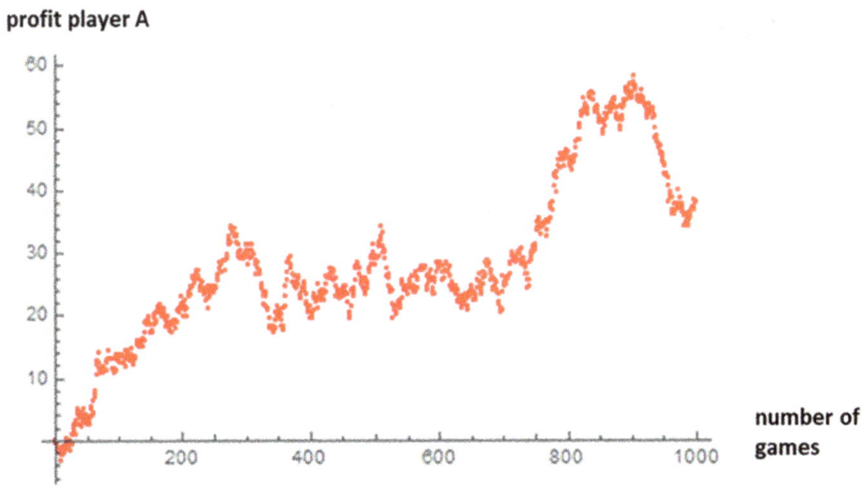

Fig. 4.72 Normally distributed increments: 1 possible realization of the winning process for player A in 1000 games

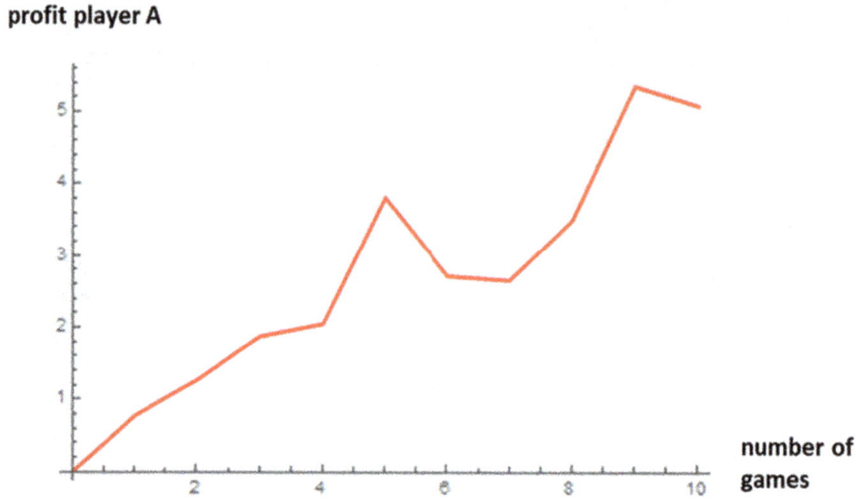

Fig. 4.73 Normally distributed increments: 1 possible realization of the winning process for player A in ten games, where values have been connected

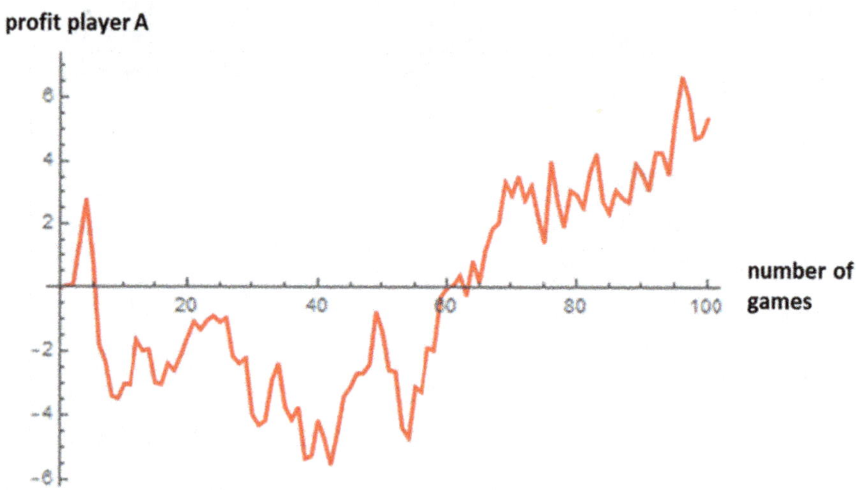

Fig. 4.74 Normally distributed increments: 1 possible realization of the winning process for player A in 100 games, where values have been connected

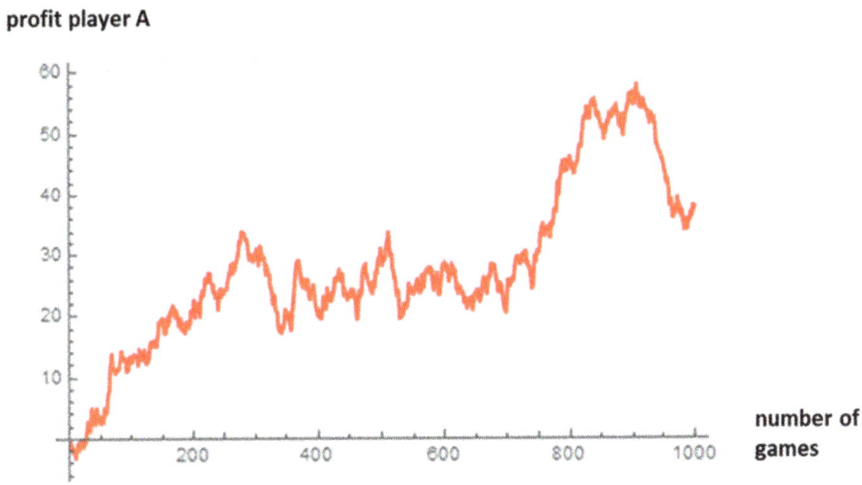

Fig. 4.75 Normally distributed increments: 1 possible realization of the winning process for player A in 1000 games, where values have been connected

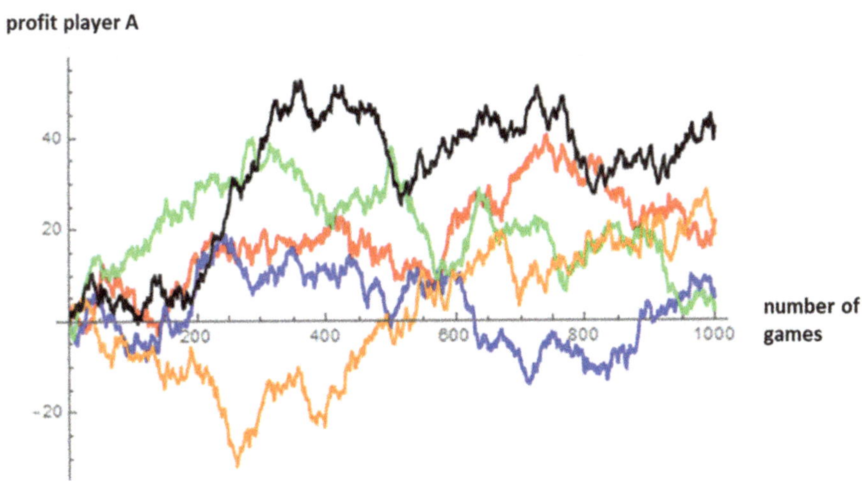

Fig. 4.76 Normally distributed increments: 5 possible realizations of the winning process for player A in 1000 games, where values have been connected

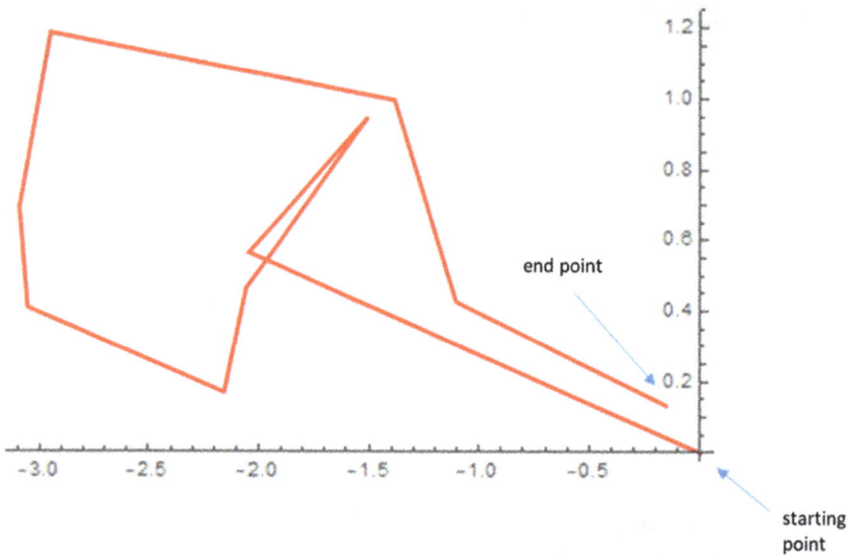

Fig. 4.77 Normally distributed increments: typical path of a two-dimensional discrete random walk with ten time steps

We see that the paths are very similar to those in Figs. 4.73, 4.74, and 4.75, except that the time range is now fixed at [0, 10] (or generally at $[0, T]$).

In Fig. 4.83 we also show a path for the case $\Delta t = \frac{1}{10{,}000}$ and note that there is not much of a visible difference compared to Fig. 4.82 with the time increment $\Delta t = \frac{1}{100}$.

We get a continuous, immensely chiseled and ramified curve. Even if we let Δt decrease further, the visible structure barely changes any more. It is only upon zooming deeply into the structure of the curve that the ever-finer ramifications become apparent. With Δt tending to zero (which we won't formalize strictly here), the zoomed-in section remains just as heavily chiseled, no matter how strong a zoom we apply.

We can again combine two independent samples of these time-normalized paths with an extremely small Δt in a two-dimensional graph and get a counterpart to Fig. 4.79 (see Fig. 4.84) but this time within a limited and invariable time range only.

Such typical motion graphs (as shown in Fig. 4.84) were observed by the Scottish botanist Robert Brown in 1827 as he tracked the thermal movement of minute particles in liquids and gases under the microscope.

The formation of these motion patterns was described and explained by Albert Einstein, among others, in 1905. In the 1920s, the motion patterns were then mathematically modeled (among others by the American mathematician Norbert Wiener, who is also and probably best known as the originator of cybernetics),

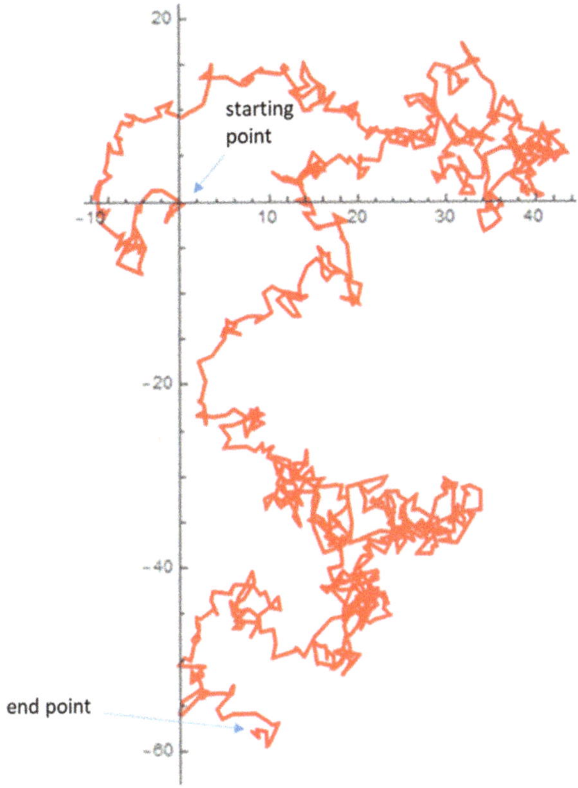

Fig. 4.78 Normally distributed increments: typical path of a two-dimensional discrete random walk with 1000 time steps

precisely with the aid of the above stochastic process with time increment Δt converging to zero.

In honor of Robert Brown, who was the first person to describe these motion patterns, the stochastic process depicted in Fig. 4.83 (with time increment Δt converging to zero) is called **one-dimensional standard Brownian motion**, and we denote it by $\boldsymbol{(B(t))_{t\in[0,T]}}$ (or later occasionally by $(W(t))_{t\in[0,T]}$).

The stochastic process depicted in Fig. 4.84 (with time increment Δt converging to zero) is referred to as **two-dimensional standard Brownian motion**.

4.16 The Brownian Motion: Basic Properties

The Brownian motion is a fascinating object with many "mysterious" properties. In the following, we want to make some of these characteristics plausible.

In plausibilizing properties of the Brownian motion, we are going to **conceptualize (!)** that motion in the following form:

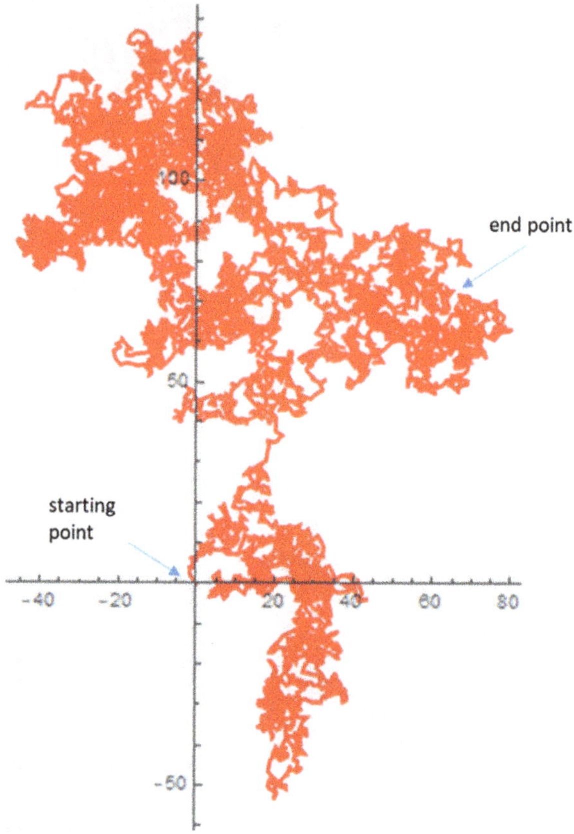

Fig. 4.79 Normally distributed increments: typical path of a two-dimensional discrete random walk with 10,000 time steps

We deem every point in time t that we are going to consider on our time range $[0, T]$ to be representable in the form $t = n \cdot \Delta t$ with "infinitesimal" Δt. The value of the Brownian motion is then given by $w_1 + w_2 + \ldots + w_n$ with mutually independent $\mathcal{N}(0, \Delta t)$ distributed random variables w_k.

(We derived the Brownian motion only intuitively here. Precise mathematical treatment, in particular for deriving the following properties, would require an exact formal definition; however, that would go beyond the scope of the mathematical knowledge required for purposes of this book.)

With this idea in mind, we infer the following basic properties of the Brownian motion:

Property (i)
It follows from the definition itself that $\boldsymbol{B(0) = 0}$.
This means that the standard Brownian motion always begins at zero.

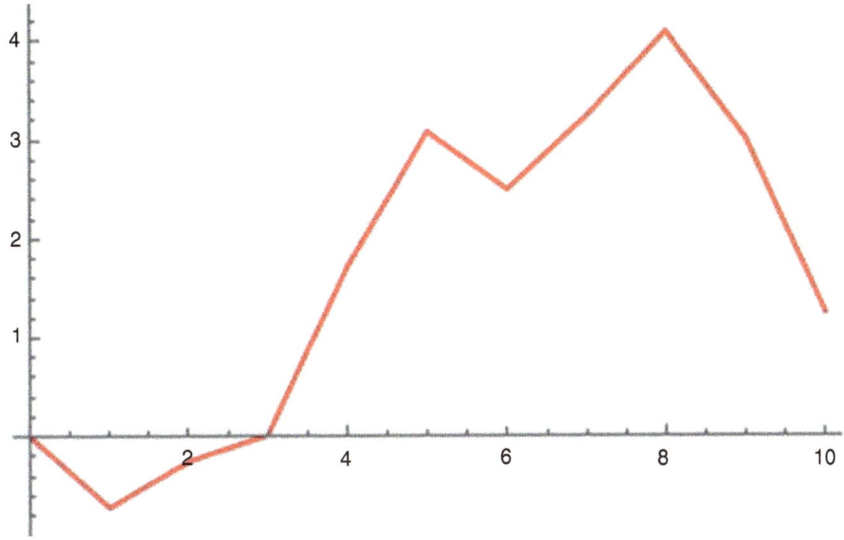

Fig. 4.80 Normally distributed increments: one possible realization of the winning process for player A in ten games, where values have been connected over the time range $[0, 10]$, $\Delta t = 1$

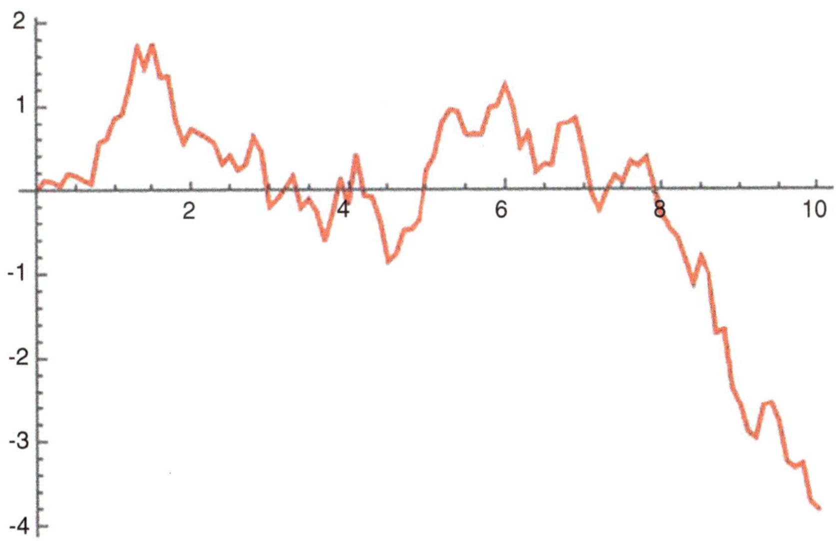

Fig. 4.81 Normally distributed increments: 1 possible realization of the winning process for player A in 100 games, where values have been connected over the time range $[0, 10]$, $\Delta t = \frac{1}{10}$

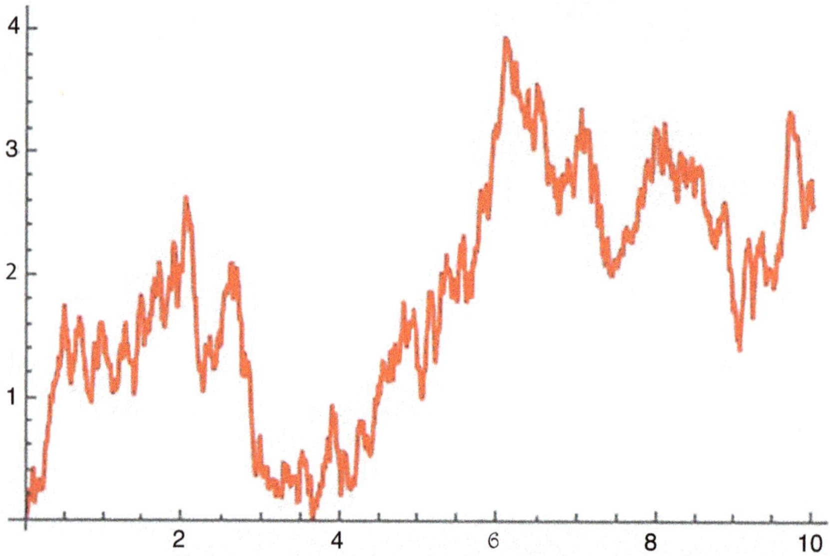

Fig. 4.82 Normally distributed increments: 1 possible realization of the winning process for player A in 1000 games, where values have been connected over the time range $[0, 10]$, $\Delta t = \frac{1}{100}$

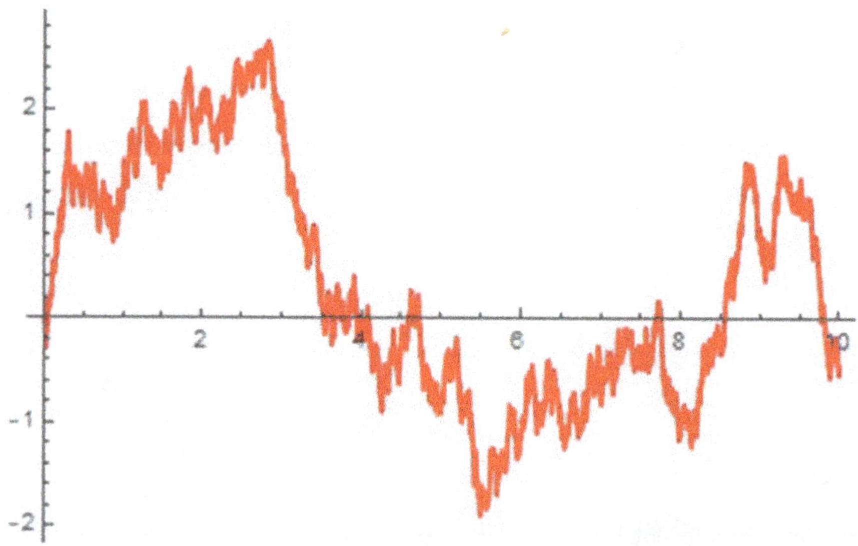

Fig. 4.83 Normally distributed increments: 1 possible realization of the winning process for player A in 100,000 games, where values have been connected over the time range $[0, 10]$, $\Delta t = \frac{1}{10{,}000}$

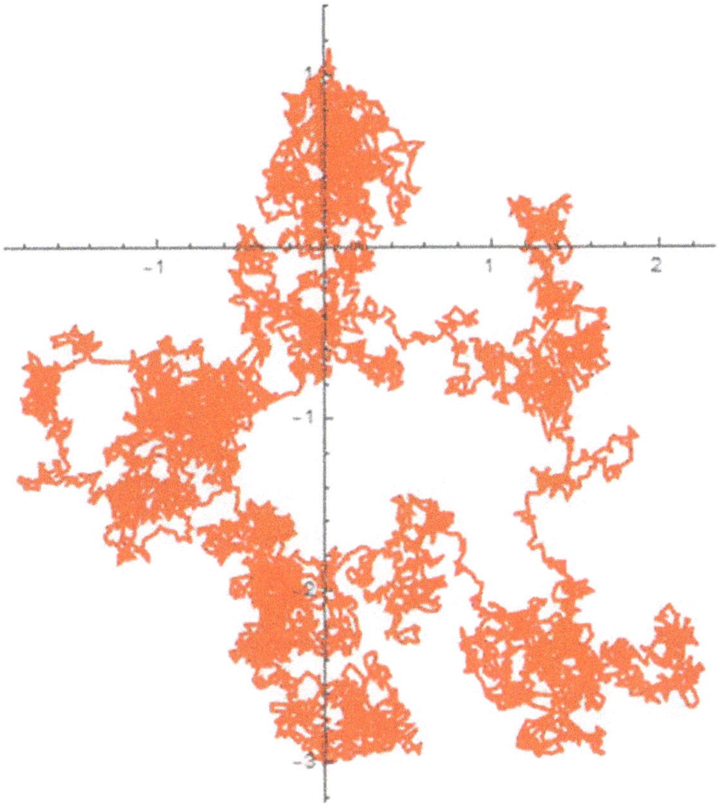

Fig. 4.84 Normally distributed increments: typical path of a two-dimensional discrete random walk with 10,000 time steps, time-normalized

Property (ii)
For each t, $B(t) = w_1 + w_2 + \ldots + w_n$ with mutually independent $\mathcal{N}(0, \Delta t)$ distributed random variables w_k, where $t = n \cdot \Delta t$. Thus, **for each** $\boldsymbol{t}$, $\boldsymbol{B(t)}$ is always a $\mathcal{N}(0, n \cdot \Delta t)$ distributed, i.e. **a** $\boldsymbol{\mathcal{N}(0, t)}$ **distributed random variable**.

In particular, therefore, the expectation of $B(t)$ is always equal to 0 and the variance of $B(t)$ is always equal to t!

Property (iii)
Let us look at two different points in time s and t where $s < t$. Here again we assume $t = n \cdot \Delta t$ and $s = m \cdot \Delta t$ (see Fig. 4.85).

We then have

$$B(t) = w_1 + w_2 + \ldots + w_n \quad \text{and} \quad B(s) = w_1 + w_2 + \ldots + w_m$$

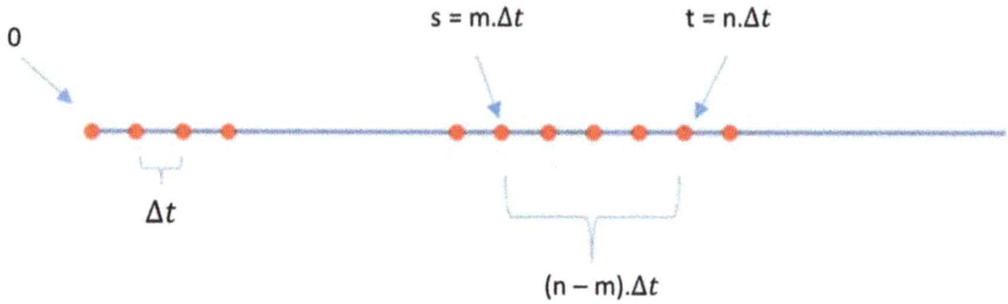

Fig. 4.85 Time range for Brownian motion

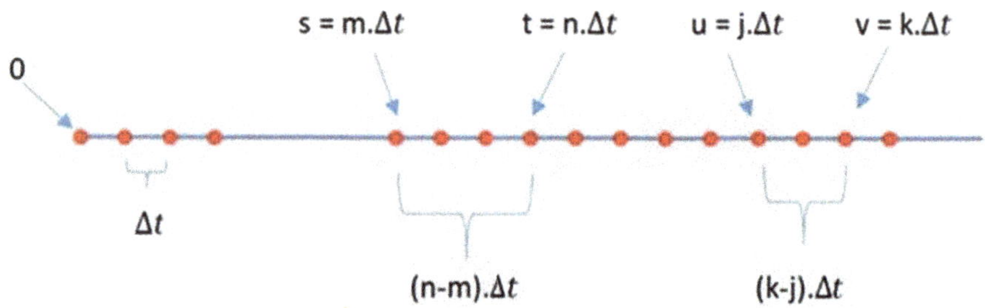

Fig. 4.86 Time range for Brownian motion

and therefore

$$B(t) - B(s) = w_{m+1} + w_{m+2} + \ldots + w_n,$$

i.e. the sum of $n - m$ mutually independent $\mathcal{N}(0, \Delta t)$ distributed random variables w_k.

From that follows, however, that $\boldsymbol{B(t) - B(s)}$ is $\mathcal{N}(0, (n - m) \cdot \Delta t)$ distributed, hence **always $\boldsymbol{\mathcal{N}(0, t - s)}$ distributed.**

Property (iv)

Now let us look at successive points in time $s < t \leq u < v$. Again we define $t = n \cdot \Delta t$ and $s = m \cdot \Delta t$ and $u = j \cdot \Delta t$ and $v = k \cdot \Delta t$ (see Fig. 4.86). We therefore have $m < n \leq j < k$.

Thus, $B(t) - B(s) = w_{m+1} + w_{m+2} + \ldots + w_n$ and $B(v) - B(u) = w_{j+1} + w_{m+2} + \ldots + w_k$.

All random variables w_i in the two preceding lines are different from one another and are therefore independent of one another. This means, however, that **the values $\boldsymbol{B(t) - B(s)}$ and $\boldsymbol{B(v) - B(u)}$ are also independent of one another.**

It is important to observe that:

It is not the **values of the Brownian motion**, for example, $B(t)$ and $B(v)$, that are independent of one another **but the increments of the Brownian motion** that are independent of one another.

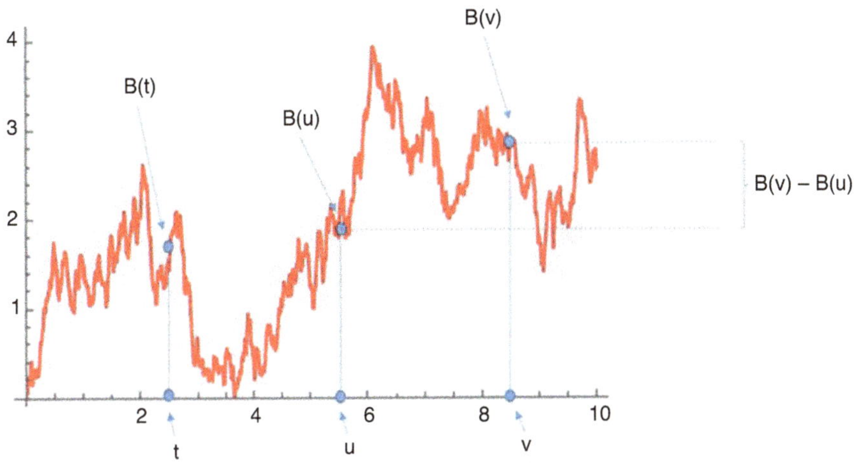

Fig. 4.87 Independence between $B(t)$ and $B(v) - B(u)$

How much the Brownian motion grows between time u and time v is independent of how much it grew between time s and time t! It is important of course that the time intervals $[s, t]$ and $[u, v]$ do not overlap.

On the other hand, value $B(v)$ is very much dependent on how large the value of $B(t)$ was.

A very large value $B(t)$ makes a higher value for $B(v)$ more likely than if $B(t)$ were a very low value. To find out how strong this dependence is, we are going to compute the correlation between $B(t)$ and $B(v)$ further below.

Note that $\boldsymbol{B(v) - B(u)}$ **in particular is independent from** $B(t) - B(0) = \boldsymbol{B(t)}$ (see Fig. 4.87).

Property (v)
Due to the original definition (linear connection of the values for $B(n \cdot \Delta t)$ and $B((n+1) \cdot \Delta t)$, the paths (realizations) of a Brownian motion are always continuous. The **Brownian motion** is therefore a **continuous stochastic process**.

(More precisely, using a formally exact definition of the Brownian motion, the correct wording would be *The paths of the Brownian motion are continuous with probability 1.*)

On the other hand, the paths of the Brownian motion are **nowhere differentiable**, so never smooth at any point. This too is plausible if you look at the illustrations: At any time t of the form $t = n \cdot \Delta t$, there is a spike in the graph because of the linear connection.

Much more surprising at first sight, however, is the following property of the paths of the Brownian motion: **The paths of the Brownian motion are of infinite length on any time interval** $[t, t + \varepsilon]$**, no matter how small that interval is.**

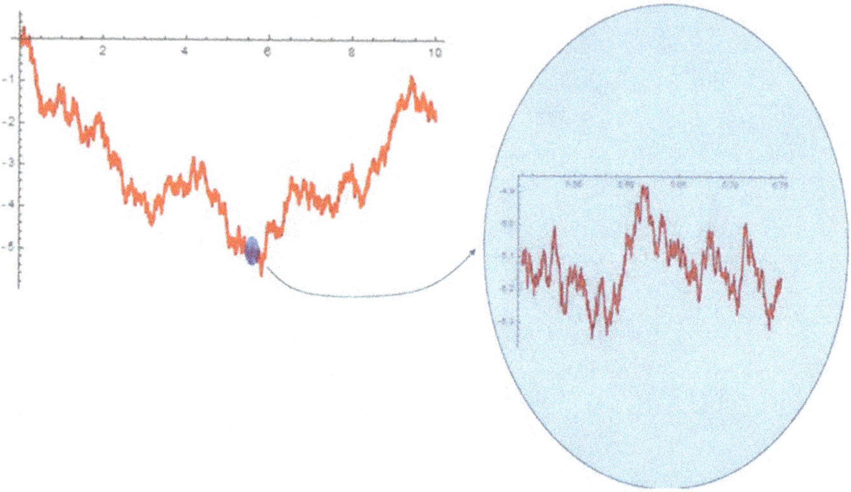

Fig. 4.88 Infinite length of the paths of the Brownian motion

So the tiny oval blue part of the path of the Brownian motion in Fig. 4.88 is supposed to be of infinite length!?

How is that supposed to be possible or even conceivable?

Here is just one small example to illustrate that in cases like these, we should not simply rely on our imagination for guidance. Let us consider a square with a side length of 1.

Question: How long is the red line in the square shown in Fig. 4.89?

The answer of course is: the length of the red line is 2.

Next question: How long are the red lines in the two graphs in Fig. 4.90?

Again, the answer is: The length of each of the red lines is 2.

Why is that? Well, the total length of the vertical red line segments added together is obviously 1, and the total length of the horizontal red line segments added together is obviously 1.

And this remains the case even if we subdivide the red line further and further: The total length of the vertical red curve segments added together will obviously always be 1, and the total length of the horizontal red curve segments added together will obviously always be 1. The total length of the red lines is therefore always exactly 2. This also applies to the following graphs: In each of the images in Fig. 4.91, the length of the red line is 2.

But if showed only the last picture to a person who is unaware of our above explanations and then asked that person:

(continued)

How long is the red line in this Picture 4.92? Then that person would probably argue that the length of the red line must surely be less than the sum of the two sides of the square, i.e. less than 2. Or they might cite the Pythagorean theorem and argue that the length of the red hypotenuse is equal to the square root of the sum of the squared lengths of the two legs, i.e. equal to $\sqrt{1^2+1^2}=\sqrt{2}=1.41\ldots$.

Claiming that the length of the red line is equal to 2 would massively contradict their view.

It is indeed hard to imagine that any part, however small, of a Brownian motion path is in fact of infinite length.

The following heuristic may be helpful in understanding the rationale behind this:

The path of a Brownian motion from time s to time t (with $s = m \cdot \Delta t$ and $t = n \cdot \Delta t$) is given by the connection of the points $\{s, B(s)\}$, $\{s+\Delta t, B(s)+w_{m+1}\}$, $\{s+2\cdot\Delta t, B(s)+w_{m+1}+w_{m+2}\}, \ldots, \{s+(n-m)\cdot\Delta t, B(s)+w_{m+1}+w_{m+2}+\ldots+w_n\}$.

So the path is longer than the sum of the length of the changes in the y-coordinates of these points. However, the length of the changes from the first to the

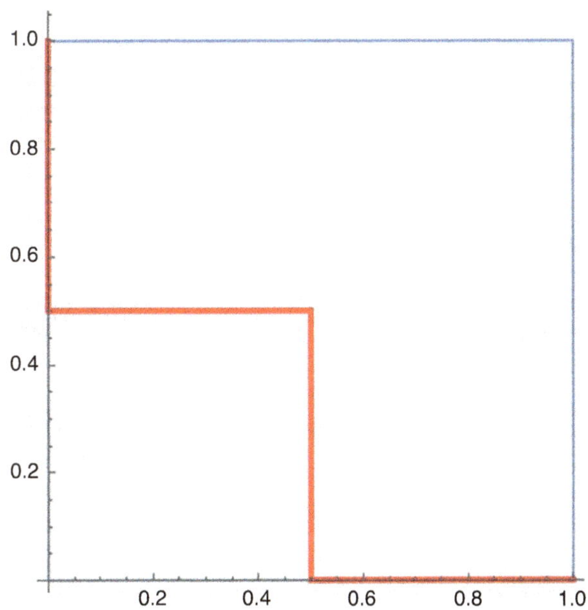

Fig. 4.89 How long is the red line?

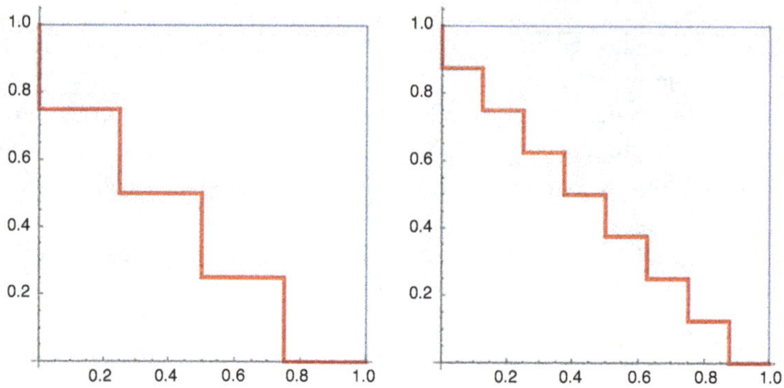

Fig. 4.90 How long are the red lines?

second point is precisely $|w_{m+1}|$, from the second to the third point $|w_{m+2}|, \ldots,$ from the last-but-one to the last point $|w_n|$.

So the length of the path from time s to time t is at least $|w_{m+1}| + |w_{m+2}| + \ldots + |w_n|$. It is the sum of $(n - m) = \frac{t-s}{\Delta t}$ lengths.

The average value of an $\mathcal{N}(0, \Delta t)$ distributed random variable w_i is roughly equal to the standard deviation of that random variable, i.e. approximately equal to $\sqrt{\Delta t}$.

On average, we can therefore at least expect a path length of approximately $\frac{t-s}{\Delta t} \cdot \sqrt{\Delta t}$.

However, this last expression is equal to $\frac{t-s}{\sqrt{\Delta t}}$, and that goes to infinity when Δt goes to zero.

Finally, we are going to determine the **correlation between B(s) and B(t)**, that is to say between the values of the Brownian motion at two different points in time s and t (with $s < t$). It is obvious that these values depend on each other, in a positive way: The higher $B(s)$, the higher the value that we can expect for $B(t)$ and vice versa.

We start by calculating the covariance $cov(B(s), B(t))$. We have:

$$cov(B(s), B(t)) = \\ = E((B(t) - E(B(t))) \cdot (B(s) - E(B(s)))) = E(B(t) \cdot B(s)) = E((B(t) - B(s)) \cdot B(s)) + E((B(s))^2).$$

We carried out this transformation for the following reason: The expected value $E(B(t) \cdot B(s))$ cannot be split into two factors $E(B(t)) \cdot E(B(s))$, since $B(t)$ and $B(s)$ are not independent of one another. Recall however, that due to property (iv) of the Brownian motion, $(B(t) - B(s))$ and $B(s)$ are independent of one another. We can thus factorize the expected value. Moreover, observe that $E((B(s))^2)$ is nothing other than the variance of $B(s)$, and as a result of property (ii) of the Brownian motion, this variance is exactly s. So we can continue our calculations (bearing in mind that $E(B(s)) = 0$):

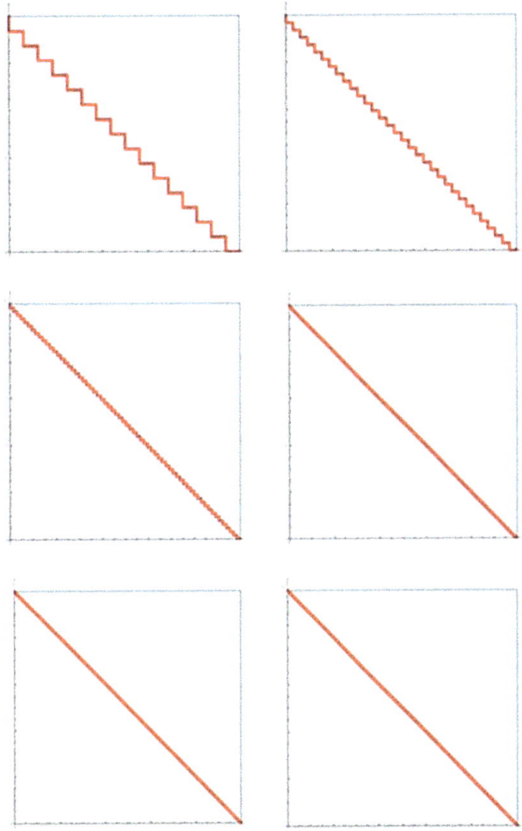

Fig. 4.91 Length of the red lines = 2

$$cov(B(s), B(t)) =$$
$$= E((B(t) - B(s)) \cdot B(s)) + E((B(s))^2) = E(B(t) - B(s)) \cdot E(B(s)) + s = s$$

The **correlation $\rho(B(s), B(t))$ between $B(s)$ and $B(t)$** is obtained by dividing the covariance by the standard deviations of $B(s)$ and of $B(t)$, i.e. by

$$\rho(B(s), B(t)) = \frac{cov(B(s), B(t))}{\sqrt{s}\sqrt{t}} = \frac{s}{\sqrt{s}\sqrt{t}} = \frac{\sqrt{s}}{\sqrt{t}}.$$

So, as we expected, the correlation is always positive and of course (since $s < t$) always less than 1. However, the closer the time point s is to time point t, the more the correlation approaches the value 1, and the higher (obviously) the dependence between $B(s)$ and $B(t)$.

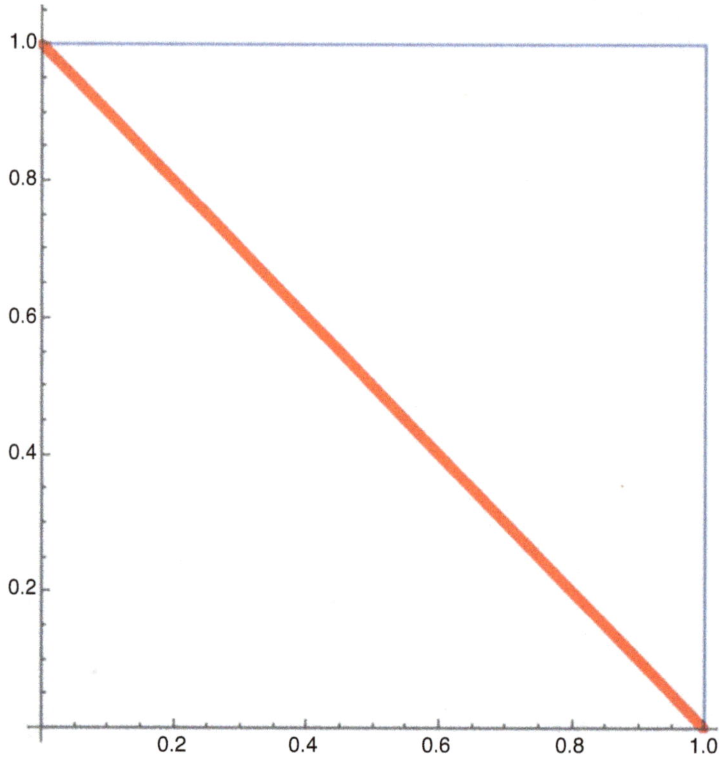

Fig. 4.92 How long is the red line?

A **final observation** for mathematically more experienced readers:

The mathematically sound definition of the standard Brownian motion is obtained "through the back door", so to speak:

It can be shown that there is exactly one stochastic process (on exactly one associated probability space) that exhibits the properties (i)–(v) formulated above. This follows essentially from a general theorem developed by Russian mathematician Andrey Kolmogorov, who is known, among other achievements, for his axiomatic foundation of the theory of probability.

This uniquely defined stochastic process is then referred to as a standard Brownian motion. So in fact, it is not even necessary to know this "explicit form" of the process, since all of its properties can be derived from its five basic properties.

4.17 The Wiener Model as a Geometric Brownian Motion and the Brownian Motion with Drift

We undertook this excursion into the world of Brownian motion with the aim of presenting an analogous representation of the Wiener model. Let us now revisit the definition (or heuristic derivation) of the Wiener model:

We recall the following statement made at the end of Sect. 4.4:

We consider the Wiener model for a stock price path $S(t)$ on a time interval $[0, T]$. We subdivide this time interval into N subintervals of (infinitesimally) small length dt. So, we have $T = N \cdot dt$.

Then we have N mutually independent $\mathcal{N}(0, dt)$ distributed random variables w_k for $k = 0, 1, \ldots, N-1$, so that for each $t \in [0, T]$ with $t = n \cdot dt$ we get

$$S(t) = S(n \cdot dt) = S(0) \cdot e^{n \cdot dt \cdot \mu + \sigma \cdot (w_0 + w_1 + \ldots + w_{n-1})} =$$
$$= S(0) \cdot e^{t \cdot \mu + \sigma \cdot (w_0 + w_1 + \ldots + w_{n-1})}$$

with mutually independent $\mathcal{N}(0, dt)$ distributed random variables w_i.

We note that the sums $w_0 + w_1 + \ldots + w_{n-1}$ correspond precisely to the standard Brownian motion (B_t). That is to say $w_0 + w_1 + \ldots + w_{n-1} = B(n \cdot dt) = B(t)$.

Thus, an **alternative representation of the Wiener model on $[0, T]$** is given by

$$\mathbf{S(t) = S(0) \cdot e^{t \cdot \mu + \sigma \cdot B_t}}$$

with a standard Brownian motion $(B_t)_{t \in [0,T]}$

This process is also called **Geometric Brownian Motion**, since here the Brownian motion occurs in the exponent.

The entire expression $t \cdot \mu + \sigma \cdot B_t$ in the exponent is also called a Brownian motion with drift and will come into play later (see Volume III Section 1.3).

4.18 The Black-Scholes Formula in the Wiener Model

The aim of this section is to derive the classic Black-Scholes formula for the valuation of derivatives on an underlying asset that moves according to the Wiener model.

So we have the following situation:

We have an underlying asset (a stock, a stock index, an exchange rate, ...) whose price $S(t)$ moves in the time interval $[0, T]$ according to a Wiener model with parameters μ and σ. Hence $S(T) = S(0) \cdot e^{\mu T + \sigma \sqrt{T} w}$. The underlying asset does not generate any (relevant) payments or (relevant) costs over the time interval $[0, T]$. (The case of underlying assets with payments will be dealt with later.) In addition, we have a (European) derivative D on S with expiration date T, which is given by its payoff function $\Phi : \mathbb{R} \to \mathbb{R}$. The payoff from the derivative D at time T is thus given by $\Phi(S(T))$.

Our interest now is to find the fair (arbitrage-free) price $F(0)$ of this derivative.

To obtain this fair price, we are going to proceed as follows:

We approximate the Wiener model by means of the N-step binomial model and by taking the limit for $N \to \infty$, just like we did in the previous section. In the N-step binomial model, we already know the formula $F^{(N)}$ for the fair price of derivative D from Chap. 3. If in this formula $F^{(N)}$, we let $N \to \infty$ (or $dt \to 0$), we get the fair price F of derivative D in the Wiener model.

The formula $F^{(N)}$ for the fair price in the N-step binomial model was stated in Theorem 4.10. For our purposes, however, we do not need the formula but only the last sentence given there:

f_0 (i.e. the fair price $F^{(N)}$) is the discounted expected payoff under the risk-neutral measure.

We now need to recall our understanding of "the risk-neutral measure" in an N-step binomial model. We understood it to mean a modified, artificial probability p' (specifically $p' = \frac{e^{r \cdot dt} - d}{u - d}$) with the following property:

Under the probability p' we have $E\left(e^{-rT} \cdot BM(T)\right) = BM(0)$.

(Here r denotes the risk-free interest rate for the period $[0, T]$.)

This property (risk-neutral measure) is retained when taking the limit from the binomial model to the Wiener model, i.e. from $BM(T)$ to $S(T)$. In taking the limit to the Wiener model, we get the following result for the fair price $F(0)$ of derivative D:

$F(0)$ is the discounted expected payoff under the risk-neutral measure in the Wiener model.

Now, what is the risk-neutral measure in the Wiener model? Just like in the binomial model, the risk-neutral measure is obtained by modifying the model's parameters (i.e. μ and/or σ), so that in the model with the new parameters, we get $E\left(e^{-rT} \cdot S(T)\right) = S(0)$.

What form do μ and/or σ need to have to satisfy the relationship $E\left(e^{-rT} \cdot S(T)\right) = S(0)$? To answer this question, we calculate the expected value $E\left(e^{-rT} \cdot S(T)\right)$:

$$\begin{aligned} E\left(e^{-rT} \cdot S(T)\right) &= \frac{1}{\sqrt{2\pi}} \cdot e^{-rT} \cdot S(0) \cdot \int_{-\infty}^{\infty} e^{\mu T + \sigma\sqrt{T}w} \cdot e^{-\frac{w^2}{2}} dw = \\ &= \frac{1}{\sqrt{2\pi}} \cdot e^{-rT} \cdot S(0) \cdot \int_{-\infty}^{\infty} e^{-\frac{\left(w - \sigma\sqrt{T}\right)^2}{2} + \frac{\sigma^2 T}{2} + \mu T} dw = \\ &= e^{-rT} \cdot S(0) \cdot e^{T\left(\frac{\sigma^2}{2} + \mu\right)} \cdot \frac{1}{\sqrt{2\pi}} \int_{-\infty}^{\infty} e^{-\frac{w^2}{2}} dw = \\ &= S(0) \cdot e^{T\left(\frac{\sigma^2}{2} + \mu - r\right)}. \end{aligned}$$

So, for this expected value to be $S(0)$, we only require that $\mu = r - \frac{\sigma^2}{2}$.

This gives us **one of the central tenets of financial mathematics, the classic Black-Scholes formula** in the Wiener model.

Theorem 4.10 (Black-Scholes Formula) *Let D be a European derivative with expiration date T and payoff function Φ on an underlying asset with price $S(t)$, evolving according to a Wiener model with parameters μ and σ over the time range $[0, T]$. (It is assumed that no payments or costs are generated by the underlying asset.) The fair price $F(0)$ of D at time 0 is then given by*

$$F(0) = e^{-rT} \cdot E\left(\Phi\left(\widetilde{S}(T)\right)\right)$$

where the price path of $\widetilde{S}$ is

$$\widetilde{S}(T) = S(0) \cdot e^{\left(r-\frac{\sigma^2}{2}\right)T+\sigma\sqrt{T}w}$$

with a standard normally distributed random variable w. "E" in this equation denotes the expected value and r is the risk-free interest rate $f_{0,T}$.

Comment 1 The formula can obviously be used to obtain the derivate's price $F(t)$ at any other point in time t within the time range $[0, T]$. Simply replace the 0 by t in the formula for $F(0)$, the time T until expiration by $T - t$, and r denotes the risk-free interest rate for the time range $[t, T]$. Thus $F(t) = e^{-r(T-t)} \cdot E\left(\Phi(\widetilde{S}(T))\right)$ with
$\widetilde{S}(T) = S(t) \cdot e^{\left(r-\frac{\sigma^2}{2}\right)(T-t)+\sigma\sqrt{T-t}\cdot w}$.

Comment 2 Since w is a standard normally distributed random variable, the expected value $E\left(\Phi(\widetilde{S}(T))\right)$ can always be calculated as follows:

$$E\left(\Phi\left(\widetilde{S}(T)\right)\right) = E\left(\Phi\left(S(t) \cdot e^{\left(r-\frac{\sigma^2}{2}\right)(T-t)+\sigma\sqrt{T-t}\cdot w}\right)\right) =$$

$$= \frac{1}{\sqrt{2\pi}} \int_{-\infty}^{\infty} \Phi\left(S(t) \cdot e^{\left(r-\frac{\sigma^2}{2}\right)(T-t)+\sigma\sqrt{T-t}\cdot w}\right) \cdot e^{-\frac{w^2}{2}} dw$$

It is rarely possible to calculate this integral explicitly. One therefore in most cases has to find approximate values for the integral. Such approximate values can be obtained by numerical integration methods but also by simple Monte Carlo simulation. We will deal with a number of examples of this later.

Comment 3 Above we denoted the fair price of a derivative at time t by $F(t)$. Yet F is of course a function $F(t, s, r, \sigma)$ of the variables t, s, r, and σ, where s denotes the current value $s = S(t)$ of the underlying asset. (The parameters K and T are fixed values and it is therefore not necessary to treat them as further variables.) In most cases however, we use the shortened representation $F(t)$, or $F(t, s)$.

From the integral representation of $E\left(\Phi(\widetilde{S}(T))\right)$ in Comment 2, it follows, based on fundamental analytical considerations, that for most (customary) payoff functions Φ, $F(t, s, r, \sigma)$ is, with respect to each of the variables, a function which can be continuously differentiated at least twice in the domain of definition. At this point, we will not go into the exact requirements that Φ needs to have for this property to be satisfied. Suffice it to observe that the differentiability of F, which is often assumed in the following, does not present a really relevant restriction of the payoff functions.

Comment 4 And finally, a fourth and very central comment, one that we will discuss in great detail later in this book: **The formula for the fair price of a derivative is independent of the underlying asset's trend parameter μ !**

Again, the formula has, in principle (!), exactly the form one would expect, namely, the value of the derivative is the discounted expected payoff of that derivative.

But: The expected value is not calculated on the basis of the model $S(T) = S(0) \cdot e^{\mu T + \sigma\sqrt{T}w}$, with an (individually) estimated trend μ but on the basis of the modified model $\widetilde{S}(T) = S(0) \cdot e^{\left(r - \frac{\sigma^2}{2}\right)T + \sigma\sqrt{T}w}$.

This means that pricing a derivative in the Black-Scholes model does not require us to estimate a future trend μ! The only parameter that needs to be estimated is the volatility σ of the underlying asset S. At first glance, this might seem counterintuitive: The value of a call option, for example, is supposed to be independent of whether we expect the stock's potential to be very positive (large trend μ) or very negative (low trend μ)? That doesn't seem logical.

At second glance, however, the formula turns out to be completely logical: We already agreed that the price of the derivative is the discounted expected payoff of that derivative. The only question is in relation to what trend μ. (We already agreed beforehand that we are using the Wiener model, so that is not up for discussion now.) It is clear that the "right" trend μ cannot be an individual estimate, as the derivate's fair price (which is what we are looking for) has to be a uniquely defined price that is valid for all market participants. So, the "right" trend μ must therefore be a trend that the market participants have implicitly (and tacitly) "agreed upon".

We now recall that the modified model $\widetilde{S}(T) = S(0) \cdot e^{\left(r - \frac{\sigma^2}{2}\right)T + \sigma\sqrt{T}w}$ was chosen precisely such that the expected discounted value $\widetilde{S}(T)$ is the current value $S(0)$. The trend of the model $\widetilde{S}(T)$ is given by $r - \frac{\sigma^2}{2}$.

Now, can the "right" mutually accepted trend μ be greater than $r - \frac{\sigma^2}{2}$? If this were the case, then the general average opinion of market participants would be

$$\begin{aligned} E\left(e^{-rT} S(T)\right) &= E\left(e^{-rT} \cdot S(0) \cdot e^{\mu T + \sigma\sqrt{T}w}\right) \\ &> E\left(e^{-rT} \cdot S(0) \cdot e^{\left(r - \frac{\sigma^2}{2}\right)T + \sigma\sqrt{T}w}\right) \\ &= S(0). \end{aligned}$$

So, the general average opinion of market participants is that the discounted future value of the underlying asset is higher than the current value $S(0)$! However, this would contradict the free dynamics of the markets: If market participants were indeed to believe, on average, that the discounted future value of an underlying asset is higher than its current value $S(0)$, they would increasingly seek to buy that asset at a price that is—per general consensus—"too cheap", and the price would then quickly level out at $E\left(e^{-rT} S(T)\right) = E\left(e^{-rT} \cdot S(0) \cdot e^{\mu T + \sigma\sqrt{T}w}\right)$, that is, a value greater than $S(0)$, or it would already have taken that value and would no longer be $S(0)$.

Likewise, by analogy, the "right" mutually accepted trend μ cannot be less than $r - \frac{\sigma^2}{2}$, so, **the "right" average mutually accepted trend μ must therefore be $r - \frac{\sigma^2}{2}$.**

With that, the Black-Scholes formula is now plausible:

The price of a derivative is the discounted expected value of the payoff, calculated with respect to the generally accepted average trend $\mu = r - \frac{\sigma^2}{2}$.

Investors are, of course, free to individually estimate the trend μ and be convinced of it, even if it deviates from the generally accepted average estimate $r - \frac{\sigma^2}{2}$. If their individual estimate μ is greater than $r - \frac{\sigma^2}{2}$, then their expected discounted payoff of, for example, a call option will also be higher than its fair price (as a call option's payoff function in $S(T)$ is monotonically increasing). Buying this call option—which would appear "too cheap" to such investors—would then seem to make sense and appear to have a higher likelihood of being profitable. By the same token, however, it would also make just as much sense for those investors to buy the underlying asset, as based on their estimated trend, that asset's discounted expected future value will be higher than its current value $S(0)$.

Comment 5 So, the only parameter to be estimated when applying the Black-Scholes formula is the volatility σ. We will look at methods of appropriately estimating volatility and the exact impact of volatility on the price of derivatives in detail in later chapters. At this point, however, we need to emphasize an assumption that we made in all of the foregoing without explicitly pointing it out:

In proving the Black-Scholes formula, we assumed constant volatility throughout the entire time range $[0, T]$. So, in valuing the option, the estimated current (!) volatility σ at any time $t = 0$ is assumed to be the volatility of the underlying asset at any time $t \in [0, T]$. Where in the proof of the Black-Scholes formula is this assumption taken into account? Answer: We took it into account when, in the N-step binomial model (with which we approximated the Wiener model), we always chose the same constant value σ for the values $u = e^{\mu dt + \sigma\sqrt{dt}}$ and $d = e^{\mu dt - \sigma\sqrt{dt}}$ at every time $i \cdot dt$. (We also always chose the same constant μ, which however, turned out to be irrelevant in the end.)

Comment 6 We recall that we already derived the fair price of a derivative on an underlying asset S with an average continuous return δ per anno over the life of the option in the binomial model. The result was

$$f_0 = e^{-rT} \cdot \sum_{k=0}^{N} \left(f_{u^k d^{N-k}} \cdot \binom{N}{k} \cdot (p')^k \cdot (1 - p')^{N-k} \right)$$

where we set

$$p' = \frac{e^{(r-\delta)dt} - d}{u - d} \text{ instead of } p' = \frac{e^{rdt} - d}{u - d}$$

(that is, r was to be replaced by $r - \delta$).

So, if we want to derive the Black-Scholes formula for an underlying asset S with an average continuous return δ over the option's life, we simply repeat the proof step by step, yet using the new p' instead of the previous p'.

(NB: In the discount factor e^{-rT}, the parameter r remains unchanged and is not replaced by $r - \delta$!) And as a result, we obviously get

Theorem 4.11 (Black-Scholes Formula for Underlying Assets with Continuous Return/Cost) *Let D be a European derivative with expiration date T and payoff function Φ on an underlying asset with price $S(t)$, evolving according to a Wiener model with parameters μ and σ over the time range $[0, T]$. Suppose the underlying asset S has an average continuous return δ over the option's life.*

The fair price $F(0)$ of D at time 0 is then defined as follows:

$$F(0) = e^{-rT} \cdot E\left(\Phi\left(\widetilde{S}(T)\right)\right)$$

where the price path of $\widetilde{S}$ is

$$\widetilde{S}(T) = S(0) \cdot e^{\left(r-\delta-\frac{\sigma^2}{2}\right)T + \sigma\sqrt{T}w}$$

with a standard normally distributed random variable w. "E" in this equation denotes the expected value and r is the risk-free interest rate $f_{0,T}$.

Comment 7 Finally, we would like to remind you once again that in the Black-Scholes formula we used the Wiener model to simulate the price path of the underlying asset. That is, a model that requires successive continuous returns to be independent of one another. (And which—as explained above—assumes constant volatility over the time interval $[0, T]$.) We will of course present and discuss other alternative models in detail later.

Since the rationale underlying the formulation of the theorem contains several heuristics, it is not completely acceptable as proof of the theorem. We therefore give the exact proof of the theorem in the following (only then will we discuss the result, put it into historical context, and after that, apply it extensively).

Proof of Theorem 4.10 We will now let $\widetilde{BM}$ denote the N-step binomial model with parameters $u = e^{\mu dt + \sigma\sqrt{dt}}$ and $d = e^{\mu dt - \sigma\sqrt{dt}}$ as above, and with the risk-neutral measure $p' = \frac{e^{r dt} - d}{u - d}$ instead of $p = \frac{1}{2}$.

$\widetilde{S}$ will again denote the risk-neutral Wiener model, hence $\widetilde{S}(T) = S(0) \cdot e^{\left(r - \frac{\sigma^2}{2}\right)T + \sigma\sqrt{T}w}$.

We are going to show below that $\widetilde{BM}$ for $N \to \infty$ converges to $\widetilde{S}$.

Since the fair price of the derivative in the binomial model BM is equal to the discounted expected value of the payoff function in the model $\widetilde{BM}$, the fair price of the derivative D in the Wiener model S must then also be equal to the discounted expected value of the payoff function in the model $\widetilde{S}$, and thus the theorem has been proven. To demonstrate the convergence of $\widetilde{BM}$ for $N \to \infty$ to $\widetilde{S}$, we proceed in the same way as we did in proving the convergence of BM to S further above. We will just have a little more computation to do. More specifically, we are going to need the following simple auxiliary result:

Lemma 1

(a) For $dt \to 0$, p' converges to $\frac{1}{2}$.
(b) For $dt \to 0$, $\sigma N \sqrt{dt}\left(2p' - 1\right)$ converges to $T\left(r - \mu - \frac{\sigma^2}{2}\right)$.

Proof of Theorem 4.10 For (a) we make use of the fact that $e^{x dt + y\sqrt{dt}} = 1 + y\sqrt{dt} + \mathcal{O}(dt)$ holds for all fixed $x, y \in \mathbb{R}$ and all dt with $|dt| \leq 1$.

(Here and below, $\mathcal{O}(Z)$ denotes a variable with $|\mathcal{O}(Z)| \leq c \cdot |Z|$ for a constant $c > 0$ and for all given Z.)

And for (b) we use the fact that $e^{x dt + y\sqrt{dt}} = 1 + x dt + y\sqrt{dt} + \frac{y^2}{2} dt + \mathcal{O}\left(dt^{\frac{3}{2}}\right)$ holds for all fixed $x, y \in \mathbb{R}$ and all dt with $|dt| \leq 1$.

Both representations follow directly from the series representation $e^x = 1 + x + \frac{x^2}{2!} + \frac{x^3}{3!} + \ldots$ of the exponential function.

(continued)

Ad (a):

$$p' = \frac{e^{rdt} - d}{u - d} = \frac{e^{rdt} - e^{\mu dt - \sigma\sqrt{dt}}}{e^{\mu dt + \sigma\sqrt{dt}} - e^{\mu dt - \sigma\sqrt{dt}}} =$$

$$= \frac{\sigma\sqrt{dt} + \mathcal{O}(dt)}{2\sigma\sqrt{dt} + \mathcal{O}(dt)} = \frac{1 + \mathcal{O}\left(\sqrt{dt}\right)}{2 + \mathcal{O}\left(\sqrt{dt}\right)} \overset{dt \to 0}{\longrightarrow} \frac{1}{2}.$$

Ad (b):

$$\sigma N\sqrt{dt} \cdot (2p' - 1) =$$

$$= \sigma N\sqrt{dt}$$

$$\times \left(2 \frac{(1 + rdt) - \left(1 + \mu dt - \sigma\sqrt{dt} + \frac{\sigma^2}{2}dt\right) + \mathcal{O}\left(dt^{\frac{3}{2}}\right)}{\left(1 + \mu dt + \sigma\sqrt{dt} + \frac{\sigma^2}{2}dt\right) - \left(1 + \mu dt - \sigma\sqrt{dt} + \frac{\sigma^2}{2}dt\right) + \mathcal{O}\left(dt^{\frac{3}{2}}\right)} - 1 \right) =$$

$$= \sigma N\sqrt{dt} \left(2 \frac{rdt - \mu dt + \sigma\sqrt{dt} - \frac{\sigma^2}{2}dt + \mathcal{O}\left(dt^{\frac{3}{2}}\right)}{2\sigma\sqrt{dt} + \mathcal{O}\left(dt^{\frac{3}{2}}\right)} - 1 \right) =$$

$$= \sigma N\sqrt{dt} \, \frac{rdt - \mu dt - \frac{\sigma^2}{2}dt + \mathcal{O}\left(dt^{\frac{3}{2}}\right)}{\sigma\sqrt{dt}} =$$

$$= Ndt\left(r - \mu - \frac{\sigma^2}{2} + \mathcal{O}\left(\sqrt{dt}\right)\right)$$

$$\overset{dt \to 0}{\longrightarrow} T\left(r - \mu - \frac{\sigma^2}{2}\right).$$

□

The remainder of the theorem's proof is now easily done. For calculating the probability $W\left(A < \widetilde{BM}(T) < B\right)$ in the following, we will proceed in the exact same way as we did in calculating $W(A < BM(T) < B)$ of the result in Sect. 4.14. By using $\widetilde{K}$ instead of K, we indicate that we are now going to work with p' instead of $p = \frac{1}{2}$, and for that reason we need to use the more

(continued)

general form of the central limit theorem. Thus

$$W\left(A < \widetilde{BM}(T) < B\right) =$$

$$= W\left(A < T\mu + \sigma\sqrt{dt}\,(\widetilde{K}\cdot 1 + (N-\widetilde{K})\cdot(-1)) < B\right) =$$

$$= W\left(A < T\mu + \sigma\sqrt{dt}\left(N\,(2p'-1) + 2\sqrt{Np'\,(1-p')}\cdot y^{(N)}\right) < B\right).$$

For $dt \to 0$ we can now apply Lemma 1 in the last expression and thus obtain

$$W\left(A < \widetilde{BM}(T) < B\right) \overset{dt\to 0}{\longrightarrow} W\left(A < T\left(r - \frac{\sigma^2}{2}\right) + \sigma\sqrt{T}w < B\right)$$

with an $\mathcal{N}(0, 1)$ distributed random variable w.

Thus we have demonstrated that $\widetilde{BM}$ for $N \to \infty$ does indeed converge to $\widetilde{S}$ and this concludes the proof of the general Black-Scholes formula in the Wiener model.

□

Before we discuss this formula any further, and present historical comments on it, we are going to demonstrate an application to illustrate how this formula, which may appear rather abstract at first, is used in practice.

4.19 The Fair Price of a European Call Option and a European Put Option in the Wiener Model

The first example we are going to look at for the application of the Black-Scholes formula is the fair price of a European call option.

So we calculate $C(t) = e^{-r(T-t)} \cdot E\left(\Phi\left(\widetilde{S}(T)\right)\right)$, where $\widetilde{S}$ evolves as $\widetilde{S}(T) = S(t)\cdot e^{\left(r-\frac{\sigma^2}{2}\right)(T-t)+\sigma\sqrt{T-t}\cdot w}$, for the specific case of the call option's payoff function, that is, for $\Phi(x) = \max(x-K, 0)$. Here K denotes the strike price of the call option. We get

$$E\left(\Phi\left(\widetilde{S}(T)\right)\right) =$$

$$= E\left(\Phi\left(S(t)\cdot e^{\left(r-\frac{\sigma^2}{2}\right)(T-t)+\sigma\sqrt{T-t}w}\right)\right) =$$

(continued)

$$= E\left(\max\left(S(t)\cdot e^{\left(r-\frac{\sigma^2}{2}\right)(T-t)+\sigma\sqrt{T-t}w}-K,0\right)\right)=$$

$$=\frac{1}{\sqrt{2\pi}}\int_{-\infty}^{\infty}\max\left(S(t)\cdot e^{\left(r-\frac{\sigma^2}{2}\right)(T-t)+\sigma\sqrt{T-t}w}-K,0\right)\cdot e^{-\frac{w^2}{2}}dw=$$

$$=\frac{1}{\sqrt{2\pi}}\int_{L}^{\infty}\left(S(t)\cdot e^{\left(r-\frac{\sigma^2}{2}\right)(T-t)+\sigma\sqrt{T-t}w}-K\right)\cdot e^{-\frac{w^2}{2}}dw=$$

$$\left(\text{where } L := \frac{\log\frac{K}{S(t)}-\left(r-\frac{\sigma^2}{2}\right)(T-t)}{\sigma\sqrt{T-t}}\right)$$

$$=\frac{1}{\sqrt{2\pi}}\int_{L}^{\infty}S(t)\cdot e^{\left(r-\frac{\sigma^2}{2}\right)(T-t)+\sigma\sqrt{T-t}w-\frac{w^2}{2}}dw-\frac{1}{\sqrt{2\pi}}\int_{L}^{\infty}K\cdot e^{-\frac{w^2}{2}}dw=$$

$$=S(t)\cdot\frac{1}{\sqrt{2\pi}}\int_{L}^{\infty}e^{-\frac{\left(w-\sigma\sqrt{T-t}\right)^2}{2}}\cdot e^{r(T-t)}dw-K\cdot\frac{1}{\sqrt{2\pi}}\int_{-\infty}^{-L}e^{-\frac{w^2}{2}}dw=$$

$$=S(t)\cdot e^{r(T-t)}\cdot\mathcal{N}\left(\sigma\sqrt{T-t}-L\right)-K\cdot\mathcal{N}(-L),$$

where $\mathcal{N}(x) := \frac{1}{\sqrt{2\pi}}\int_{-\infty}^{x}e^{-\frac{w^2}{2}}dw$ denotes the distribution function of the standard normal distribution.

If we also insert in that $-L = \frac{\log\frac{S(t)}{K}+\left(r-\frac{\sigma^2}{2}\right)(T-t)}{\sigma\sqrt{T-t}}$, we get

Theorem 4.12 *For a call option's fair price $C(t)$ at time $t \in [0, T]$ on an underlying asset S which evolves according to a Wiener model with parameters μ and σ (and which does not generate any payments or cost in $[0, T]$), we find that*

$$C(t) = S(t)\cdot\mathcal{N}(d_1) - e^{-r(T-t)}\cdot K\cdot\mathcal{N}(d_2)$$

where $d_1 = \frac{\log\frac{S(t)}{K}+\left(r+\frac{\sigma^2}{2}\right)(T-t)}{\sigma\sqrt{T-t}}$ and $d_2 = \frac{\log\frac{S(t)}{K}+\left(r-\frac{\sigma^2}{2}\right)(T-t)}{\sigma\sqrt{T-t}}$ and $\mathcal{N}$ denotes the distribution function of the standard normal distribution.

The fair price of a put option is now simply derived from the put-call parity equation in Sect. 3.2. We had shown there that for the fair prices $C(t)$ and $P(t)$ of a call and a put option on the same underlying asset S with the same expiration date T and the same strike price K, we have

$C(t) + K \cdot e^{-r(T-t)} = S(t) + P(t)$. Hence $P(t) = C(t) + K \cdot e^{-r(T-t)} - S(t)$. If we substitute the formula we derived above for $C(t)$, we get

$$\begin{aligned} P(t) &= S(t) \cdot \mathcal{N}(d_1) - e^{-r(T-t)} \cdot K \cdot \mathcal{N}(d_2) + K \cdot e^{-r(T-t)} - S(t) = \\ &= S(t) \cdot (\mathcal{N}(d_1) - 1) + e^{-r(T-t)} \cdot K\,(1 - \mathcal{N}(d_2)) = \\ &= e^{-r(T-t)} \cdot K \cdot \mathcal{N}(-d_2) - S(t) \cdot \mathcal{N}(-d_1) \end{aligned}$$

And this gives us:

Theorem 4.13 *For the fair price $P(t)$ of a put option with expiration T and strike K at time $t \in [0, T]$ on an underlying asset with price $S(t)$ (that does not generate payments or costs in $[0, T]$), which evolves according to a Wiener model with parameters μ and σ, we have*

$$P(t) = e^{-r(T-t)} \cdot K \cdot \mathcal{N}(-d_2) - S(t) \cdot \mathcal{N}(-d_1)$$

with d_1 und d_2 as given in Theorem 4.12 and $\mathcal{N}$ being the distribution function of the standard distribution function.

Note As is easily verified, the fair prices $C = C(t, s, r, \sigma)$ and $P = P(t, s, r, \sigma)$ of call and put options can be differentiated arbitrarily often with respect to each variable.

The exact (analytical) behavior of the two pricing formulas will be studied in great detail in Sects. 4.24–4.26.

If we derive the explicit formulas for the fair price of call and put options from the general Black-Scholes formula for derivatives on underlying assets **with a continuous payout or cost rate** q, then the explicit call and put price formulas can be obtained in exactly the same way for this case as well. The theorems then read as follows:

Theorem 4.14 *For the fair price $C(t)$ of a call option with expiration T and strike K at time $t \in [0, T]$ on an underlying asset with price $S(t)$, which generates payments or costs in $[0, T]$ in the amount of a continuous return q p.a. and evolves according to a Wiener model with parameters μ and σ, we have*

$$C(t) = e^{-q(T-t)} \cdot S(t) \cdot \mathcal{N}\left(\tilde{d}_1\right) - e^{-r(T-t)} \cdot K \cdot \mathcal{N}\left(-\tilde{d}_2\right)$$

with

$$\tilde{d}_1 = \frac{\log\left(\frac{S(t)}{K}\right) + \left(r - q + \frac{\sigma^2}{2}\right)(T-t)}{\sigma\sqrt{T-t}}$$

and

$$\tilde{d}_2 = \frac{\log\left(\frac{S(t)}{K}\right) + \left(r - q - \frac{\sigma^2}{2}\right)(T - t)}{\sigma\sqrt{T - t}}$$

and $\mathcal{N}$ being the distribution function of the standard normal distribution.

Theorem 4.15 *For the fair price $P(t)$ of a put option with expiration T and strike K at time $t \in [0, T]$, on an underlying asset with price $S(t)$ which generates payments or costs in $[0, T]$ in the amount of a continuous return q p.a. and evolves according to a Wiener model with parameters μ and σ, we have*

$$P(t) = e^{-r(T-t)} \cdot K \cdot \mathcal{N}\left(-\tilde{d}_2\right) - e^{-q\cdot(T-t)} S(t) \cdot \mathcal{N}\left(-\tilde{d}_1\right),$$

with $\tilde{d}_1$ and $\tilde{d}_2$ as in the theorem above and N being the distribution function of the standard normal distribution.

4.20 A (Very) Short History of the Black-Scholes Formula

As regards the history of the Black-Scholes formula, we are only going to include some remarks here and refer, for example, to the articles by Fisher Black "How we came up with the option formula" [2] and by Darrell Duffie "Black, Merton, and Scholes—Their Central Contributions to Economics" [3] for longer expositions. In addition, Walter Schachermayer's Lecture Notes "Introduction to the Mathematics of Financial Markets", which can be found at https://www.mat.univie.ac.at/~schachermayer/preprnts/prpr0104.pdf, are also highly recommended. In Chap. 1 of these Lecture Notes, the author states that a version—accurate in a certain sense—of an option pricing formula (using an equivalent to "a risk-neutral measure") had already been given by Louis Bachelier in his dissertation in 1900. The essential difference between Bachelier's results and modern option pricing theories, however, is that Bachelier did not model the underlying asset's pricing process $S(t)$ as a geometric Brownian motion (that is, he did not use a Wiener model) but used the representation $S(t) = c + \sigma \cdot B(t)$ with a standard Brownian motion B instead. Yet we already know that the Brownian motion can take arbitrarily strong negative values, meaning that also an $S(t)$ modeled in this way has positive probability of taking negative values.

What is most astonishing in Bachelier's argumentation is, however, his observation that in pricing an option, it is in fact irrelevant whether the underlying asset is modeled in the above-mentioned form $S(t) = c + \sigma \cdot B(t)$ or in the more general form of a Brownian motion with drift, i.e. in the form $S(t) = c + \mu \cdot t + \sigma \cdot B(t)$. More precisely, Bachelier argues, based on certain "equilibrium arguments", that using models of the form $S(t) = c + \mu \cdot t + \sigma \cdot B(t)$ unconditionally implies that $\mu = 0$. Bachelier's discovery essentially corresponds to the finding presented 70 years later in the formulas developed by Black, Scholes, and Merton, namely, that μ has no relevance. Fischer Black probably did not know about Bachelier's work and the findings it contained when he set out to develop pricing formulas for convertible bonds in the late 1960s. At that time, the market for convertible bonds (which come with an embedded option) was significantly more liquid and more attractive for investors than the options market. Black was considering underlying securities that evolve according to a geometric Brownian motion. He actually succeeded in deriving a differential equation for pricing convertible bonds that already bore all the distinguishing marks of the later Black-Scholes differential equation. However, he was unable to provide an explicit solution to this differential equation, and so, for some time, his work on convertible bonds fell into oblivion. It was not until around 1970 that Black, together with Myron Scholes and in a lively discourse with Robert Merton, resumed his work on options pricing. The two researchers again started by first deriving the corresponding differential equation for the fair price of a call option, and again they did not succeed in solving the equation explicitly, at least not at first. However, due to the fact that the differential equation did not contain the trend parameter μ, it was clear that this parameter could be chosen arbitrarily in the model. Based on this realization, Black and Scholes were now able to come up with an alternative way for developing an explicit solution formula. It turned out that this formula did indeed provide a solution to Black's earlier differential equation and thus began the advent of the Black-Scholes theory that was to revolutionize the world of derivatives trading. Although at first there was yet another hurdle to overcome: Black and Scholes' work was initially rejected twice by journals to which it had been submitted for publication, and it was only accepted on the third submission, appearing under the title "The Pricing of Options and Corporate Liabilities" in the Journal of Political Economy [4].

In footnotes in this work, the two authors acknowledged Robert Merton's essential contribution, in the form of in-depth discussions and helpful observations that Merton had provided to the authors; and Merton also contributed significantly to further refining the Black-Scholes theory in the following years. It was only a logical consequence then that Robert Merton was also awarded the Nobel Prize in economics, alongside Myron Scholes, in 1997. Fischer Black had already died in 1995 at the age of only 58.

4.21 Perfect Hedging in the Black-Scholes Model

We recall the method of hedging a derivative D, which we developed in the N-step binomial model. The principle of a perfect hedge of a (European) derivative D on an underlying asset S is the following:

Find a self-financing trading strategy that is based solely on trading in the underlying asset S and on investing cash at the risk-free interest rate, and that is designed to deliver a payoff that is exactly equal to the payoff of the derivative D upon expiry T in each case. "Self-financing" in this context means that we are making an investment only at the start time 0. All other trading transactions are to be financed by rebalancing the hedging portfolio. So, no new funds will be injected at any other time for the duration of the strategy. This means that the amount of money required at the beginning of the strategy—"the replicating trading strategy"—is exactly equal to the fair price of derivative D. An investment firm that sells a derivative D and wishes to fully hedge against any losses uses the proceeds from selling the derivative to implement the replicating trading strategy. The money received through this trading strategy can then be used upon maturity of the derivative for payment of its payoff. We know from Sects. 3.10, 3.14, and 3.18 what the replicating trading strategy for derivatives in the N-step binomial model looks like.

We now ask ourselves: Can this principle of hedging a derivative also be applied to underlying assets in the Wiener model? And if so, how do we proceed?

The answer is: "**Perfect** hedging in the Wiener model is possible **in theory**, but **not in practice**."

In the one-step binomial model (see Fig. 4.93), static hedging was all we needed for a perfect hedge. We only had to do the following:

We calculated the derivative's fair price f_0 at time 0 and then used f_0 euros for trading the replicating portfolio.

We used that to buy $x = \frac{f_u - f_d}{S_0(u-d)} = \frac{f_u - f_d}{S_0 u - S_0 d}$ units of the underlying asset (see Formula (3.3) in Sect. 3.9) and invested the rest of the money at the risk-free interest rate. This approach gave us exactly the replicating portfolio.

In the N-step binomial model, it was then necessary to trade dynamically and suitably at each time $i \cdot dt$. For this purpose, in Sect. 3.18 it was specified exactly how many units $x = x_{i \cdot dt}$ of the underlying asset the hedger had to hold in each of the time intervals $i \cdot dt$ to $(i+1) \cdot dt$ We will take a closer look at this again below.

When passing to the limit, this means that, to obtain a perfect dynamic hedge in the Wiener model, we would need to **trade continuously**, that is, at every point in time.

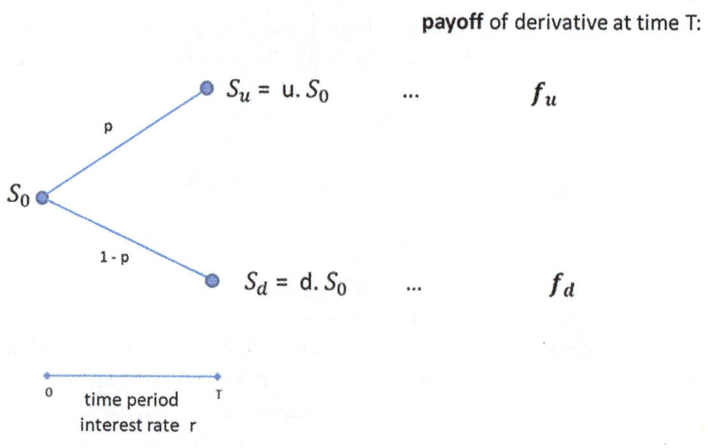

Fig. 4.93 The binomial one-step model, revisited

So we have to find a function $x(t)$ that tells us how many units of the underlying asset need to be held at *any* point in time t to ensure perfect replication. Realistically speaking, *perfect* hedging is therefore not possible in the Wiener model, as in reality we can only execute trades at a finite number of discrete points in time. So we will have to settle for near-perfect (approximate) hedging. Nevertheless, it is essential (also for the approximative hedging) to know what $x(t)$ should theoretically look like. Once again we will surely be able to find this information through information about the $x_{i \cdot dt}$ in the N-step binomial model and by taking the limit for N to infinity.

Let us have another look at how hedging was done in the N-step binomial model (see Sects. 3.15 and 3.18):

- If the current time is $i \cdot dt$ and the underlying asset has a certain current value $S(i \cdot dt)$, the underlying asset can take the value $S((i+1) \cdot dt) = u \cdot S(i \cdot dt)$ or $S((i+1) \cdot dt) = d \cdot S(i \cdot dt)$ by time $(i+1) \cdot dt$.
- We are now going to consider precisely this one-step binomial model created from the points in time $i \cdot dt$ and $(i+1) \cdot dt$ and the values $S(i \cdot dt)$, $u \cdot S(i \cdot dt)$ and $d \cdot S(i \cdot dt)$.
- Let $F((i+1) \cdot dt, u \cdot S(i \cdot dt))$ and $F((i+1) \cdot dt, d \cdot S(i \cdot dt))$ denote the derivative's fair price at time $(i+1) \cdot dt$ if S has the value $u \cdot S(i \cdot dt)$ or $d \cdot S(i \cdot dt)$ at time $(i+1) \cdot dt$.
- Observe that in this notation, we consider the fair price F of a derivative as a function of the underlying asset's time and current value!
- We know from Formula (3.3) in Sect. 3.9 how to choose $x_{i \cdot dt}$, namely,

$$x_{i \cdot dt} = \frac{F((i+1) \cdot dt, u \cdot S(i \cdot dt)) - F((i+1) \cdot dt, d \cdot S(i \cdot dt))}{S(i \cdot dt) \cdot u - S(i \cdot dt) \cdot d}$$

- As we know from Comment 3 in Sect. 4.18, we can assume continuity of F in time t and continuous differentiability of F at least with respect to the underlying asset's value s.
- For heuristic purposes, let us set $t := (i+1) \cdot dt$ and $s = S(t) := S(i \cdot dt) \cdot d$, as well as $S(i \cdot dt) \cdot u = S(i \cdot dt) \cdot d + h$. Then $x_{i \cdot dt} = \frac{F(t,s+h)-F((t,s))}{h}$. For $dt \to 0$, h goes to 0, and $i \cdot dt$ goes to t, and so $x(t) = \lim_{h \to 0} \frac{F(t,s+h)-F((t,s))}{h}$, which is nothing other than the derivative $\frac{\partial F}{ds}(t,s)$ of the fair price F with respect to the underlying asset's price s at time t. We will later refer to this term $\frac{\partial F}{ds}(t,s)$ as the *delta of F at time t*.
- What we deduced heuristically here can also be demonstrated exactly by way of precise limiting process considerations.

To summarize:

Theorem 4.16 *Let $F(t,s)$ be the fair price (continuous in t and continuously differentiable with respect to (s) of the derivative D on an underlying asset that* *follows a Wiener process. The following continuous self-financing trading strategy provides a dynamic replicating portfolio for D:*

- *Start with an initial capital of $F(0, S(0))$.*
- *At any time $t \in [0, T]$, hold $x(t) := \frac{\partial F}{ds}(t, s)$ units of the underlying asset and invest the remaining cash at the risk-free interest rate.*

Example 4.17 We calculate the perfect hedging portfolio at time $t \in [0, T]$ for a call option and for a put option when the underlying asset has the value s at time t. To do this, as per Theorem 4.16, we simply determine the delta of the call price C (or the put price P, as the case may be), i.e. we differentiate C (or P) with respect to s.

So, we differentiate $C(t, s) = s \cdot \mathcal{N}(d_1) - e^{-r(T-t)} \cdot K \cdot \mathcal{N}(d_2)$ with respect to s, while keeping in mind that d_1 and d_2 also depend on s. In the following, h' will always denote the derivative of a function h with respect to s. We find that

- using the product rule and the chain rule for differentiation, and
- since $\frac{\partial \mathcal{N}(d)}{\partial d} = \frac{1}{\sqrt{2\pi}} \cdot e^{-\frac{d^2}{2}}$, and
- since $d_2 = d_1 - \sigma\sqrt{T-t}$ and hence $d_2{}' = d_1{}'$, and
- by substituting in $C'(t, s) = \mathcal{N}(d_1) + s\frac{\partial \mathcal{N}(d_1)}{\partial d_1} \cdot d_1{}' - e^{-r(T-t)} \cdot K \cdot \frac{\partial \mathcal{N}(d_2)}{\partial d_2} \cdot d_2{}'$,

the last two summands cancel each other out, giving us $C'(t, s) = \mathcal{N}(d_1)$.

Hence we got: **The perfect hedging portfolio for a call option in a Wiener model consists of holding $x(t) = \mathcal{N}(d_1)$ units of the underlying asset at time t.**

To determine the "delta" of a put option, we will once again resort to the put-call parity equation. Recall that—in our current terminology—the equation went as follows:

$$
\begin{aligned}
P(t, s) + s &= C(t, s) + K \cdot e^{-r(T-t)}, \text{ i.e. } P(t, s) \\
&= C(t, s) + K \cdot e^{-r(T-t)} - s \text{ and hence} \\
P'(t, s) &= C'(t, s) - 1 = \mathcal{N}(d_1) - 1
\end{aligned}
$$

Hence we got: **The perfect hedging portfolio for a put option in a Wiener model consists of holding $x(t) = \mathcal{N}(d_1) - 1$ units of the underlying asset at time t.**

Note $x(t)$ is of course also a function of the underlying asset's current price s at time t!

Comment $\mathcal{N}$ is a distribution function that always assumes values between 0 and 1. As a result, the delta $\mathcal{N}(d_1)$ of a call option is always between 0 and 1, and the delta $\mathcal{N}(d_1) - 1$ of a put option is always between -1 and 0.

In a perfect hedge of a call option, we are always long on the underlying asset, and in a perfect hedge of a put option, we are always short on the underlying asset.

These relatively narrow limits for the volume of the underlying asset to be held in each case are also relevant in that, if we were required to hold excessively large quantities of the underlying asset, the actual execution would become unrealistic.

The function d_1 is dependent on s and grows as s grows. Therefore, the delta $\mathcal{N}(d_1)$ of a call option, just like the delta $\mathcal{N}(d_1) - 1$ of a put option, grows as s grows.

4.22 Another Example of the Application of the Black-Scholes Formula and of Perfect Hedging as Well as Its Implementation in Discrete Hedging

We now want to illustrate the perfect hedge that we described above, using a real-world example, and then carry out and test its approximate implementation through discrete trading. In particular, we want to get a sense of how real-market executions can cause deviations from a perfect hedge.

For this purpose, we are going to choose a rather simple example of a derivative D with the payoff function $\Phi(x) = x^2$. In the next section, we will then create a discrete hedging strategy for the more realistic case of a call option.

First, let us calculate this derivative's fair value $F(0)$. To do so, we can just as well consider the more general case of a payoff $\Phi(x) = x^a$ for an arbitrarily chosen exponent a.

So, derivative D makes a payment in the amount of $(S(T))^a$ at time T. To answer the question as to the derivative's fair price $F(0)$ at time 0, our first impulse would likely be "$F(0) = (S(0))^a$". This answer is certainly correct where $a = 1$, since in that case, one unit of the derivative will definitely have the same payoff at time T as one share of the stock, namely, $S(T)$. Because the stock share perfectly replicates the derivative, they must both have the same value $S(t)$ at any moment in time t.

But we see already for the case where $a = 0$ that this "first-impulse answer" cannot be right. Because in this case, the derivative's payoff will definitely be $(S(T))^0 = 1$. The derivative's value at an earlier time t is therefore $e^{-r(T-t)}$, which is different from $(S(t))^0$.

The Black-Scholes formula now yields

$$
\begin{aligned}
\boldsymbol{F(t,s)} &= e^{-r(T-t)} \cdot E\left(\Phi\left(\widetilde{S}(T)\right)\right) = \\
&= e^{-r(T-t)} \cdot \frac{1}{\sqrt{2\pi}} \int_{-\infty}^{\infty} \Phi\left(s \cdot e^{\left(r-\frac{\sigma^2}{2}\right)(T-t)+\sigma\sqrt{T-t}w}\right) \cdot e^{-\frac{w^2}{2}} dw = \\
&= e^{-r(T-t)} \cdot \frac{1}{\sqrt{2\pi}} \int_{-\infty}^{\infty} s^a \cdot e^{a\left(r-\frac{\sigma^2}{2}\right)(T-t)+a\sigma\sqrt{T-t}w} \cdot e^{-\frac{w^2}{2}} dw =
\end{aligned}
$$

$$= e^{-r(T-t)} \cdot s^a \cdot \frac{1}{\sqrt{2\pi}} \int_{-\infty}^{\infty} e^{-\frac{\left(w-a\sigma\sqrt{T-t}\right)^2}{2}+\frac{a^2\sigma^2(T-t)}{2}+a\left(r-\frac{\sigma^2}{2}\right)(T-t)} dw =$$

$$= e^{-r(T-t)} \cdot s^a \cdot e^{\frac{a^2\sigma^2(T-t)}{2}+a\left(r-\frac{\sigma^2}{2}\right)(T-t)} =$$

$$\mathbf{= s^a \cdot e^{(a-1)r(T-t)+\frac{a(a-1)\sigma^2(T-t)}{2}}}$$

For $a = 0$ and for $a = 1$, the result is consistent with the results we obtained in our considerations above.

For $a = 2$ we get $F(t, s) = s^2 \cdot e^{(r+\sigma^2)(T-t)}$, and for the delta of F, we therefore get $\frac{\partial F}{ds}(t, s) = 2s \cdot e^{(r+\sigma^2)(T-t)}$.

In times of negative interest rates—in order to simplify the expression and increase clarity (after all, this example is meant to serve as illustration)—we take the liberty of assuming that $r = -\sigma^2$. In that case $F(t, s) = s^2$ and $\frac{\partial F}{ds}(t, s) = 2s$.

Perfect hedging in our example therefore means holding $2 \cdot S(t)$ units of the underlying asset at any point in time t. More precisely: We start with a cash amount that is equal to the derivative's fair price, i.e. $(S(0))^2$. Then, through continuous trading, we create a stock portfolio that contains exactly $2 \cdot S(t)$ units of the underlying asset at any time t. Money that remains at our disposal in the course of these trading activities or that has to be borrowed to finance those trades bears continuously compounded interest at the risk-free rate r. As per our above considerations, our total assets from this procedure would then be exactly $(S(T))^2$ euros at time T.

We will now execute this hedging process in a realistic manner, by subdividing the time range $[0, T]$ into N parts of equal length dt and rebalance the portfolio only at times $i \cdot dt$ for $i = 0, 1, 2, \ldots, N$. More precisely:

0 (a) We start, at time zero, with a cash amount ("current assets at time **0**") of $F(0, S(0)) = \mathbf{(S(0))^2}$.

0 (b) At time 0 we are supposed to be holding $2 \cdot S(0)$ units of the underlying asset. However, that would cost $2 \cdot (S(0))^2$. We therefore need to take out a (virtual) loan and borrow $2 \cdot S(0) \cdot S(0) - F(0, S(0)) = (S(0))^2$ euros.

0 (c) At time dt the price of the underlying asset has changed to $S(dt)$. The value of our stock portfolio is now $2 \cdot S(0) \cdot S(dt)$. Due to the interest incurred, our loan debt is now $e^{rdt} \cdot (S(0))^2$.

1 (a) Our "current assets at time dt" is thus $2 \cdot S(0) \cdot S(dt) - e^{rdt} \cdot (S(0))^2$.

1 (b) At time dt we are supposed to be holding $2 \cdot S(dt)$ units of the underlying asset. We therefore need to rebalance our stock portfolio. The cost of rebalancing is $2(S(dt) - S(0)) \cdot S(dt)$ euros. Our **current assets at time** $\boldsymbol{dt}$ after rebalancing is therefore

$$\mathbf{2 \cdot S(dt) \cdot S(dt) - \left(e^{rdt} \cdot (S(0))^2 + 2(S(dt) - S(0)) \cdot S(dt)\right)}$$

1 (c) At time $2 \cdot dt$ the price of the underlying asset has changed to $S(2 \cdot dt)$. The value of our stock portfolio is now $2 \cdot S(dt) \cdot S(2 \cdot dt)$. Due to the interest incurred, our loan debt is now $e^{r2dt} \cdot (S(0))^2 + e^{rdt} \cdot 2(S(dt) - S(0)) \cdot S(dt)$.

2 (a) Our "current assets at time $2 \cdot dt$" is thus $2 \cdot S(dt) \cdot S(2 \cdot dt) - (e^{r2dt} \cdot (S(0))^2 + e^{rdt} \cdot 2(S(dt) - S(0)) \cdot S(dt))$.

2 (b) At time $2 \cdot dt$ we are supposed to be holding $2 \cdot S(2 \cdot dt)$ units of the underlying asset. So we need to rebalance our stock portfolio once again. The cost of this rebalancing is $2(S(2 \cdot dt) - S(dt)) \cdot S(2 \cdot dt)$ euros.
Our **current assets at time** $\boldsymbol{2 \cdot dt}$ after rebalancing is therefore $\boldsymbol{2 \cdot S(2 \cdot dt) \cdot S(2 \cdot dt) - \left(e^{r2dt} \cdot (S(0))^2 + e^{rdt} \cdot 2(S(dt) - S(0)) \cdot \right.}$ $\boldsymbol{\left. \cdot S(dt) + 2(S(2 \cdot dt) - S(dt)) \cdot S(2 \cdot dt)\right)}$.

If we continue our hedging procedure in this way until time $N \cdot dt = T$, we will obtain **current assets in the discrete hedging strategy at time** $\boldsymbol{T = N \cdot dt}$ in the amount of

$$2 \cdot S(N \cdot dt) \cdot S(N \cdot dt) - \Big(e^{rNdt} \cdot (S(0))^2 + e^{r(N-1)dt} \cdot 2(S(dt) - S(0)) \cdot S(dt) +$$

$$+ e^{r(N-2)dt} \cdot 2(S(2 \cdot dt) - S(dt)) \cdot S(2 \cdot dt)\Big) + \ldots + e^{rdt} \cdot 2(S((N-1)dt) -$$

$$- S((N-2)dt)) \cdot S((N-1)dt) + 2(S(N \cdot dt) - S((N-1)dt)) \cdot S(N \cdot dt)) =$$

$$\boldsymbol{= 2 \cdot S(T) \cdot S(T) - e^{rT} \cdot (S(0))^2 - e^{rT} \cdot \sum_{i=1}^{N} ((2 \cdot S(i \cdot dt) -}$$

$$\boldsymbol{-2 \cdot S((i-1)dt)) \cdot S(i \cdot dt)) \cdot e^{-r \cdot i \cdot dt}.}$$

(For later purposes (next section), we note two things regarding the formula set out above: The second factor of the second summand in the formula given in item 0 b) of the above algorithm was obtained by simplifying the term $\Delta_0 \cdot S(0) - F(0, S(0)) = 2 \cdot S(0) \cdot S(0) - F(0, S(0)) = (S(0))^2$, and the expressions $2 \cdot S(T)$ and $2 \cdot S(i \cdot dt)$ and $2 \cdot S((i-1) \cdot dt)$ in that formula are precisely the delta values of the derivative at the respective points in time, i.e. Δ_T and $\Delta_{i \cdot dt}$ and $\Delta_{(i-1)dt}$, respectively. Our formula can therefore also be written in the form

current assets in the discrete hedging strategy at time $\boldsymbol{T =}$
$\boldsymbol{= \Delta_T \cdot S(T) - e^{rT} \cdot (\Delta_0 \cdot S(0) - F(0, S(0))) -}$
$\boldsymbol{-e^{rT} \cdot \sum_{i=1}^{N} \left(\Delta_{i.dt} - \Delta_{(i-1)dt}\right) \cdot S(i \cdot dt) \cdot e^{-r \cdot i \cdot dt}}$

This should be as close as possible to the payoff of the derivative, i.e. to $(S(T))^2$. The difference in the results between this realistic hedging process and that of $(S(T))^2$ in a perfect hedging case naturally depends on various factors. The most obvious factors are length of the time intervals (fineness of time discretization) and volatility of the underlying asset.

Another major factor in real trade settings is, of course, any transaction costs that may occur.

Yet another aspect that can be relevant (though generally to a lesser extent, especially when trading larger volumes) is the fact that we can only trade whole-number quantities of the underlying asset, while the values of Δ_t are not generally integers.

An advantage in this example is that the derivative's delta is always positive ($\Delta_t = 2 \cdot S(t)$), which means that we don't have to worry about how to best go short on the underlying stock.

In the following, we are going to carry out a couple of simulations for our hedging example. The test programs for this can be found on our website, where you can execute an unlimited number of additional tests. See: https://app.lsqf.org/book/call-put-hedging

Caveat: Note that the results of our simulations will be of somewhat limited value: That is because in the following simulations, we are going to assume **simulated prices for the underlying asset**, more precisely, simulations with respect to the Wiener model **with constant volatility**. This does not, of course, reflect the general reality of price movements in financial products. But it does give us a sense of how the hedging process works and how to develop hedging strategies.

Later (see Volume III Section 3.13), we are going to investigate how we can proceed in the case of real price data. To do this, however, we first have to study the concept of volatility in more detail.

Test examples We are now going to simulate three typical sample paths of geometric Brownian motions, using the following parameters (which remain the same for all three examples):

- Life to expiry $T = 1$ year
- Trend 10%
- Volatility 20%
- Initial stock value 100
- $r = -\sigma^2 = -4\%$
- Assumed transaction costs 0.25% of purchase price (no minimum fees)

For each sample path, we are going to apply 1 static hedging process, 1 with a one-time adjustment, 1 with 12 (approximately monthly) adjustments, and 1 with 50 (approximately weekly) adjustments. The four images for simulation path 1 (Fig. 4.94) show the path of the derivative's current payoff (blue) and the respective value of the (approximate) replicating trading strategy at the beginning of trading, at the end of trading, and at the adjustment times (red).

At the end of the trading period, the value of the hedging strategy should be as close as possible to the payoff of the derivative. Discrepancies between the hedging strategy and the derivative can of course have both positive and negative values.

Remember: At the beginning, the derivative and the hedging strategy have exactly the same value $F(0)$.

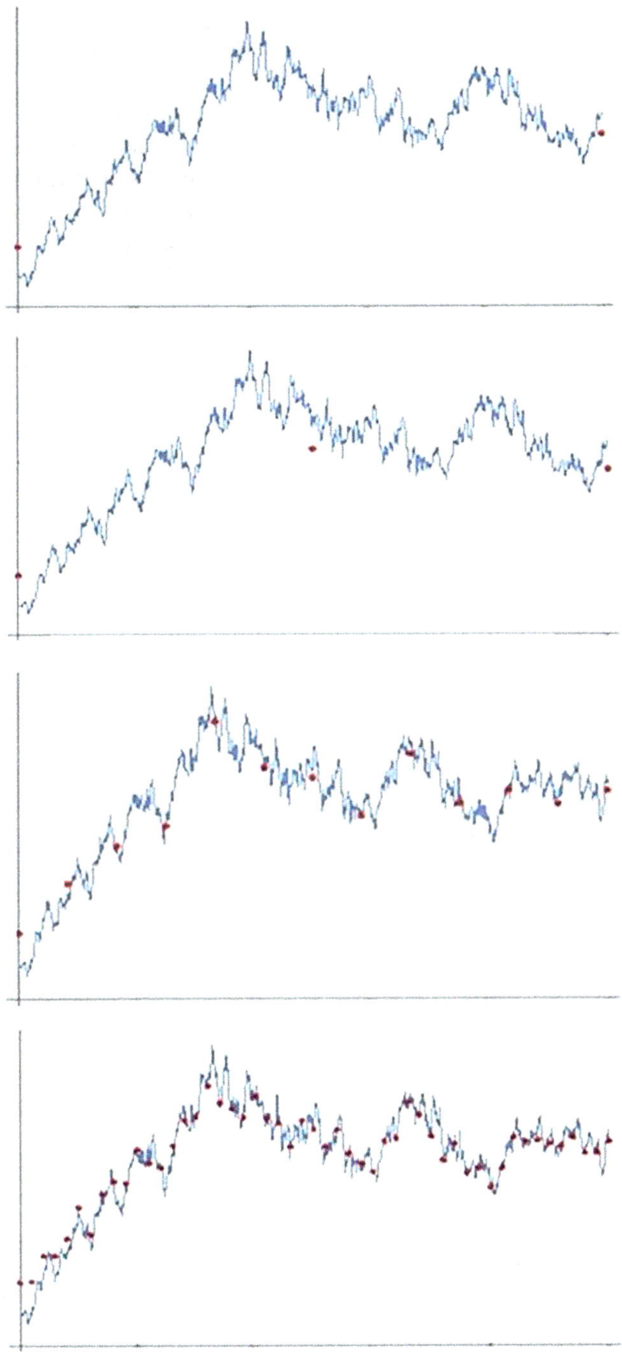

Fig. 4.94 Hedging results, simulation example 1

If we assume that we sell the derivative at $F(0)$ and start our hedging strategy at price $F(0)$, then the positive difference means that we, the hedger, make a profit. A negative discrepancy means a loss to the hedger.

In the tables following the graphs, we state the discrepancies first without transaction costs, then with all transaction costs (i.e. including the cost for the first trade upon initiating the hedging strategy as well as the cost for the last trade, when the entire hedging portfolio is liquidated), and finally, we also specify the value of the discrepancy inclusive of transaction cost only for the adjustments of the hedging strategy. (This can be relevant if the hedger—a large bank, for instance—already holds a large volume in the underlying asset and wants to keep holding it beyond the derivatives expiration. To hedge the derivative in that case, the bank only needs to make adjustments to the stock portfolio it already holds and doesn't have to make an initial trade or liquidate any positions).

In addition, we provide a graph (Fig. 4.95) showing the quantities that need to be traded in the underlying asset for the adjustments in simulation example 1 in the case of weekly adjustments (positive value, purchase; negative value, sale).

Hedging programs are available on the website and can be used to carry out and test any number of further simulation examples with any chosen payoff! These programs automatically calculate the delta, i.e. the number of shares that need to be held of the underlying asset for a perfect hedge.

The results are given in the subsequent Figs. 4.94 and 4.95 and the corresponding Table 4.4.

Results of simulation example 1:

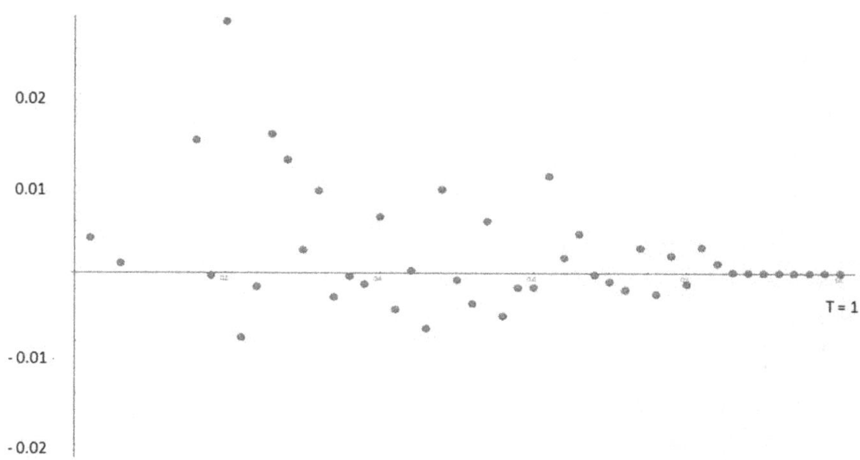

Fig. 4.95 Volumes to be traded in the case of weekly adjustments in simulation example 1

Table 4.4 Hedging results for the three simulation examples

Entries for Simulation 1 Simulation 2 Simulation 3	Final value strategy	Payoff derivative	Final value strategy minus derivative, without transaction costs, absolute and in percent of the derivative's price	Final value strategy minus derivative, with all transaction costs, absolute and in percent of the derivative's price	Final value strategy minus derivative, with only adjustment costs absolute and in percent of the derivative's price
Static	16,275	16,748	−473 // −4.73%	−606 // −6.06%	−473 // −4.73%
	10,837	10,450	+387 // +3.87%	+284 // +2.84%	+387 // +3.87%
	8681	8362	+319 // +3.19%	+227 // +2.27%	+319 // +3.19%
1 Adjustment	16,182	16,748	−565 // −5.65%	−700 // +7.00%	−566 // −5.66%
	10,417	10,450	−33 // −0.33%	−145 // −1.45%	−43 // −0.43%
	8440	8362	+78 // +0.78%	−18 // −0.18%	+74 // +0.74%
12 Adjustments	16,556	16,748	−191 // −1.91%	−361 // −3.61%	−227 // −2.27%
	10,744	10,450	+295 // +2.95%	+173 // +1.73%	+275 // +2.75%
	8499	8362	+137 // +1.37%	+22 // +0.22%	+114 // +1.14%
50 Adjustments	16,594	16,748	−154 // −1.54%	−362 // −3.62%	−228 // −2.28%
	10,514	10,450	+64 // +0.64%	−98 // −0.98%	+4 // +0.04%
	8334	8362	−28 // −0.28%	−181 // −1.81%	−89 // −0.89%

4.23 Discretely Approximated Perfect Hedging for European Derivatives, Especially for European Call and Put Options

In the previous section, we illustrated a near-perfect hedging strategy using an artificial yet instructive example. This approach can of course be used in general, for **any European derivatives**, and especially for European calls and puts. This is what we are going to do and discuss in this section (initially however only for the case of artificially simulated stock prices and constant volatility). The procedure can be copied from the previous section, substituting the relevant variables accordingly:

We are now looking at an arbitrarily chosen European derivative D whose fair price at time 0 and given the current underlying stock price of S_0 has the value $F(0, S_0)$. At any time t within the time interval $[0, T]$ and the then current stock price S_t, the fair price is $F(t, S_t)$.

The delta Δ_t of the derivative D at time t is then given through $\Delta_t = \Delta(t, S_t) = \frac{\partial F}{ds}(t, s)\big|_{s=S_t}$.

Let us recall: For a perfect hedge, we need to hold exactly Δ_t units of the underlying asset at any point in time t. Now we are again going to rebalance the portfolio (like we did in the previous section) at each point in time $i \cdot dt$ (for $i = 0, 1, 2, \ldots, N$) so that we hold exactly $\Delta_{i \cdot dt}$ shares of the underlying stock in the time interval $(i \cdot dt, (i+1) \cdot dt)$. In our example in the previous section, we got exactly $\Delta_t = 2 \cdot S(t)$ for Δ_t. For $F(0, S_0)$ in the previous section we got the value $F(0, S_0) = (S(0))^2$. If we substitute the values $2 \cdot S(0), 2 \cdot S(dt), 2 \cdot S(2.dt), \ldots, 2 \cdot S(N \cdot dt)$ in the algorithm of the previous section by $\Delta_0, \Delta_{dt}, \Delta_{2 \cdot dt}, \ldots, \Delta_{N \cdot dt}$, respectively, and if we substitute (in Step 0 a) of the previous section) the value $(S(0))^2$ by $F(0, S_0)$, we get precisely the formula given in our comment in the previous section in general form:

If we execute the above hedging strategy at the fixed points in time $i \cdot dt$ (for $i = 0, 1, 2, \ldots, N$) for an initial investment of $F(0, S_0)$, the value of the total investment at time T will be exactly

$$\Delta_T \cdot S(T) - e^{rT} \cdot (\Delta_0 \cdot S(0) - F(0, S(0))) \\ - e^{rT} \cdot \sum_{i=1}^{N} \left(\Delta_{i \cdot dt} - \Delta_{(i-1) \cdot dt}\right) \cdot S(i \cdot dt) \cdot e^{-i \cdot r \cdot dt}$$

To obtain hedging efficiency, this value should be as close as possible to the derivative's payoff. We are now going to test this hedging process (proceeding in the same way as we did in our tests for the derivative with payoff $(S(T))^2$ in the previous section) for a plain vanilla call option with prices generated artificially using the Wiener model and assuming constant volatility. Put options can be treated in the same (analogous) way, of course. Again, you can use the simulation program available on this book's website for this and any further tests. See: https://app.lsqf.org/book/call-put-hedging.

We are going to use the same parameters for our illustrative examples as in the previous section, yet with two differences: We will now assume the more realistic value $r = 0$ for the risk-free interest rate and choose a higher volatility of 40% for the underlying asset. We will run three simulations for a call option with strike price $K = 90$. Again, we are only going to graphically illustrate one of the simulations, i.e. compare the Plots 4.96 and 4.97, as well as the obtained Table 4.5.

Sample illustration: simulation of a call option hedge

Comment

1. The fair price of this call option is 20.57. The volumes to be traded for purposes of rebalancing the portfolio are very low (in the range of mostly 0 to 0.02 units per option, see Fig. 4.97). The results suggest that in this case, the hedging strategy should be adjusted at least on a weekly basis.
2. In the third simulation example, the payoff is 0. Figure 4.98 shows the evolution of the associated hedging portfolio.

The process and analysis of a put option hedge can be derived directly from the above analyses of the call option hedge and from the put-call parity equation. The put-call parity equation is $P(t) = C(t) + K \cdot e^{-r(T-t)} - S(t)$. This means that for

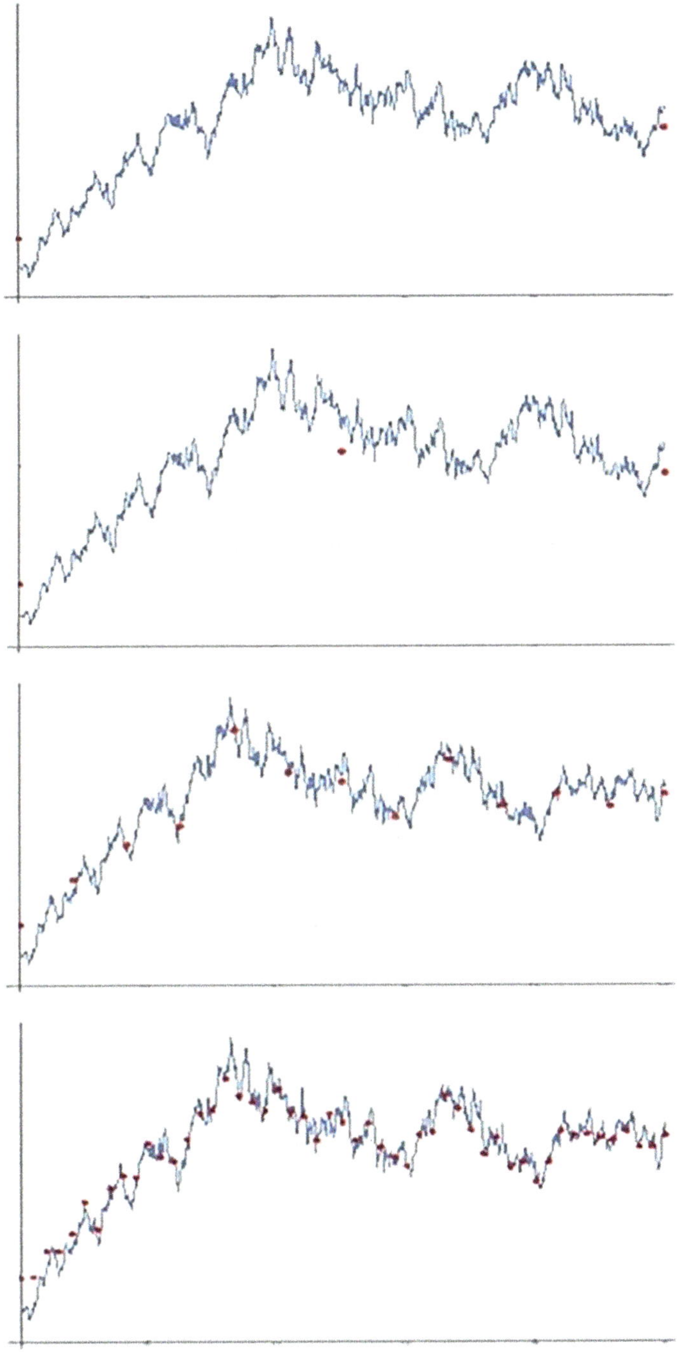

Fig. 4.96 Hedging results for call option, simulation example 1

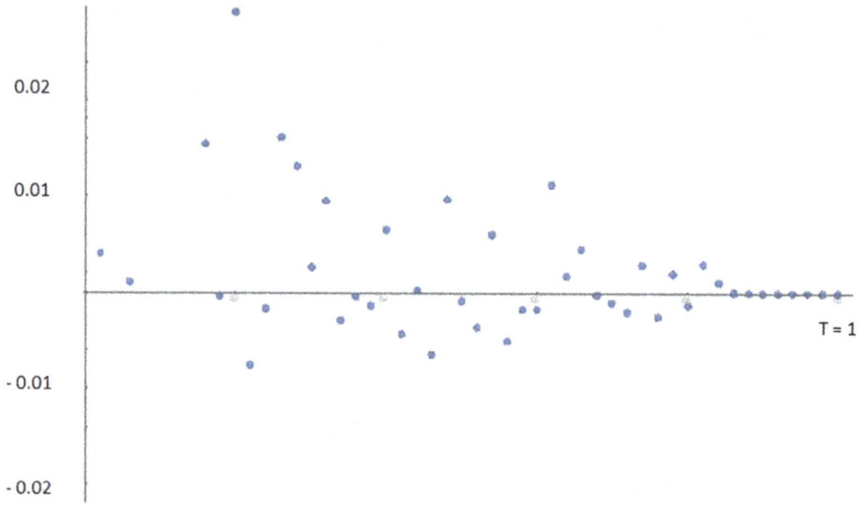

Fig. 4.97 Volumes to be traded in the case of adjustments in simulation example 1, *call option*

Table 4.5 Hedging results for the three simulation examples, *call option*

Call option Entries for: Simulation 1 Simulation 2 Simulation 3	Final value strategy	Payoff derivative	Final value strategy minus derivative, w/o transaction costs, absolute and in percent of the derivative's price	Final value strategy minus derivative, with all transaction costs, absolute and in percent of the derivative's price	Final value strategy minus derivative, with adjustment costs only, absolute and in percent of the derivative's price
Static	59	67	−7.72 // −38%	−8.28 // −40%	−7.72 // −38%
	128	169	−40.59 // −197%	−41.4 // −201%	−40.59 // −197%
	2.10	0	+2.10 // +10.2%	+1.93 // +9.38%	+2.10 // +10.2%
1 Adjustment	58	67	−8.99 // −44%	−9.68 // −47%	−9.12 // −44%
	139	169	−30.29 // −147%	−31.29 // −152%	−30.47 // −148%
	3.67	0	+3.67 // +17.8%	+3.43 // +16.7%	+3.60 // +17.5%
12 Adjustments	65	67	−1.82 // −9%	−2.50 // −12%	−1.94 // −9%
	166.28	169.11	−2.83 // −13.75%	−3.76 // −18.27%	−2.94 // −14.30%
	1.89	0	+1.89 // +9.18%	+1.51 // +7.36%	+1.68 // +8.19%
50 Adjustments	66.36	66.89	−0.53 // −3%	−1.26 // −6%	−0.70 // −3%
	168.49	169.11	−0.62 // −2.99%	−1.58 // −7.67%	−0.76 // −3.69%
	0.39	0	+0.39 // +1.92%	−0.24 // −1.17%	−0.07 // −0.24%

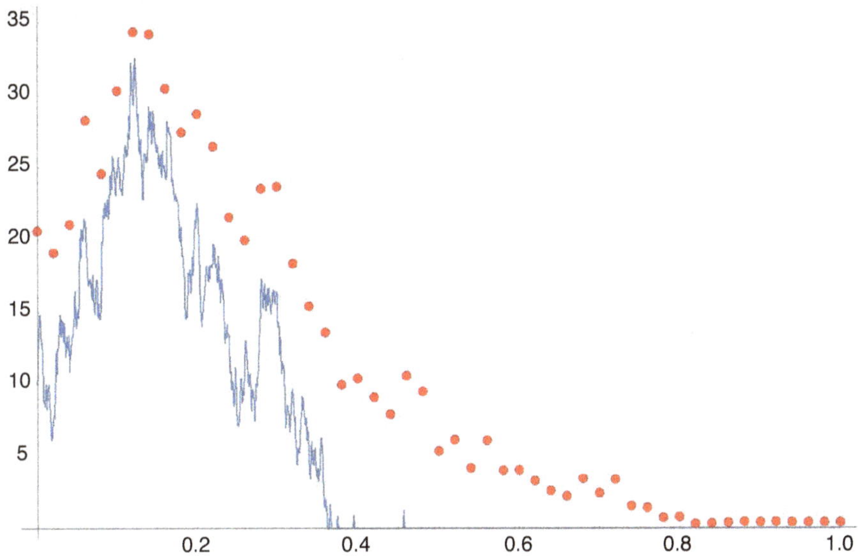

Fig. 4.98 Simulation example: Evolution of a hedging portfolio for a call option with payoff 0, weekly adjustment

the replication of a put option, the number of underlying stock shares that need to be held at any given moment is equal to the number of underlying stock shares required for the replication of the call option, minus 1.

4.24 Detailed Discussion of the Black-Scholes Formula for European Call Options I (Dependence on S and t, Intrinsic Value, Fair Value)

For anyone venturing into professional options trading, it is crucial to get a very precise "feel" for the dynamics of options prices and of prices in option strategies. This feel is difficult to impart theoretically (despite the fact that we already have all the theoretical tools—the Black-Scholes formulas—at our disposal); it is best and primarily gained by trading actively and by constantly analysing processes in the option markets.

The analyses of option pricing formulas and their dynamics that we are going to run on a few select examples in this and in the following two sections (and in later chapters) are therefore only an initial—yet necessary (!)—step toward providing an understanding of option price movements.

It is also important to reiterate the following:

In these three sections, we are going to carry out price analyses on the basis of the (no-arbitrage) formulas for option prices and option strategies. When working with real options data, it is to be expected that real price movements will often deviate substantially from theoretical price movements.

In Sect. 4.19, we derived the formula for the fair price of a call option.

For a given

- strike price K
- risk-free interest rate r
- remaining life to expiration T
- current point in time t
- volatility of the underlying asset σ over the option's life
- and price S of the underlying asset at time t

the fair price $C(t)$ of a call option with the above parameters is

$$C(t) = S \cdot \mathcal{N}(d_1) - K \cdot e^{-r(T-t)} \cdot \mathcal{N}(d_2)$$

where

$$d_1 = \frac{\log\left(\frac{S}{K}\right) + \left(r + \frac{\sigma^2}{2}\right)(T-t)}{\sigma\sqrt{T-t}} \text{ and } d_2 = \frac{\log\left(\frac{S}{K}\right) + \left(r - \frac{\sigma^2}{2}\right)(T-t)}{\sigma\sqrt{T-t}},$$

(i.e. $d_2 = d_1 - \sigma\sqrt{T-t}$) and where $\mathcal{N}(x) := \frac{1}{\sqrt{2\pi}}\int_{-\infty}^{x} e^{-y^2} dy$ (i.e. $\mathcal{N}$ is the distribution function of the standard normal distribution).

If $t = T$, then $C(t)$ is of course identical with the option's payoff. This is not immediately clear from the above formula, since the values d_1 and d_2 are not defined for $t = T$ (denominator = 0).

The following graph (Fig. 4.99), however, clearly shows how the pricing formula for t converges to T. In this graph, we plotted the movements of $C(t)$ for the four time points $t_1 = 0, t_2 = 0.3, t_3 = 0.6$, and $t_4 = 0.9$ for a fixed choice of $K = 100, r = 0.01, T = 1$, and $\sigma = 0.5$ as a function of different values $S = S(t)$ of the underlying asset at time t (graphs in red colours). For comparison, the payoff function is also shown as a function of $S = S(T)$ (in blue). The uppermost red line reflects the price movements for t_1, the second line from the top the price movements for t_2, the third line the price movements for t_3, and the bottom red line the price movements for t_4.

We see how the pricing formula approximates the payoff function as time t progresses (i.e. as time to expiration $T - t$ decreases). Again we see, as observed earlier, that the call option prices stay above the payoff function throughout. As pointed out earlier, this is due to the fact that we assumed a positive risk-free interest rate in this example. If we run the same example with a negative interest rate $r = -0.1$ (which is an unrealistically strong negative rate but gives us a graphically clearer picture), we obtain the following Plot 4.100:

Here we see that, where the underlying asset's price is well above the strike price, the call option's price falls below the payoff before expiration.

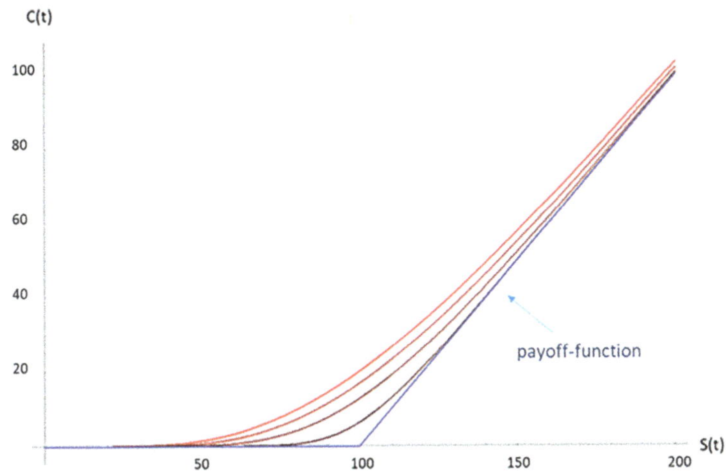

Fig. 4.99 Call option price as a function of the underlying asset's price for four different points in time (assuming a positive interest rate)

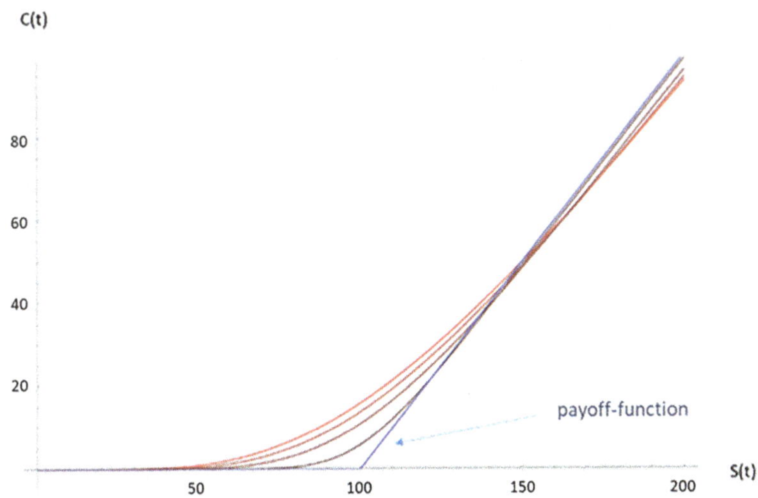

Fig. 4.100 Call option price as a function of the underlying asset's price for four different points in time (assuming a negative interest rate)

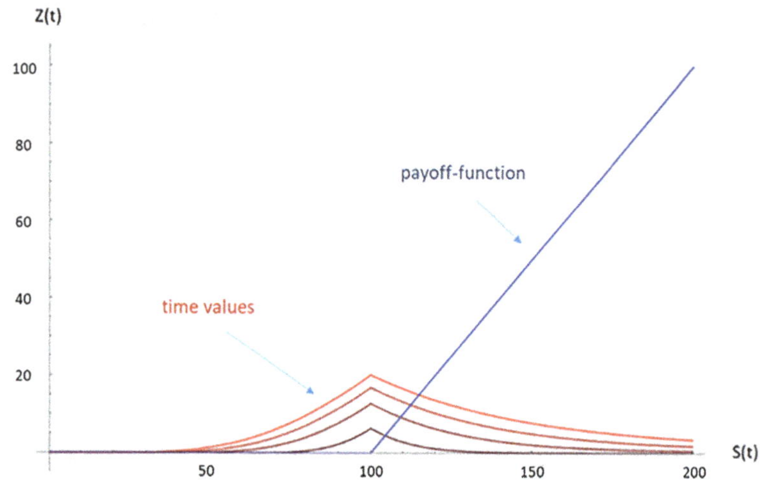

Fig. 4.101 Time value of a call option as a function of the underlying asset's price for four different time points (assuming a positive interest rate)

The **difference $Z(t)$ between the current price of an option at time t before expiration and the payoff function** at this moment (i.e. at the current price of the underlying asset) is called the **time value of the option**.

The value of the payoff function at this moment (i.e. at the current price of the underlying asset) is called the **intrinsic value of the option**. The intrinsic value of a call option at time t is therefore $\max(S(t) - K, 0)$ and $C(t) = \max(S(t) - K, 0) + Z(t)$.

The time value of an option for t going to T converges to 0.

In an interest rate environment with non-negative interest rates, the time value $Z(t)$ of a call option is always greater than zero. Figure 4.101 shows the time value for the example of Fig. 4.99.

We see, of course, that the lower t is (i.e. the longer the remaining time until expiration $T - t$), the higher the (positive) time value is. For a given t, the time value is at its highest when $S(t)$ is near K.

The key properties of an option when traded dynamically (i.e. as opposed to expiration-focused static trading) reside in the interplay between its intrinsic value and its time value.

So, we have seen that the price of a call option and the time value of a call option decrease (assuming non-negative interest rates) over time t (i.e. the shorter the remaining time to expiration $T - t$). We support this observation by further graphs (compare Figs. 4.102 and 4.103) where we are now going to plot $C(t)$ as a function of time t (for different but fixed values $S = S(t)$ and otherwise the same parameters as above).

However, in situations with a negative interest rate r, we may be looking at a totally different picture. For Fig. 4.103 we have selected $r = -0.05$ and depicted

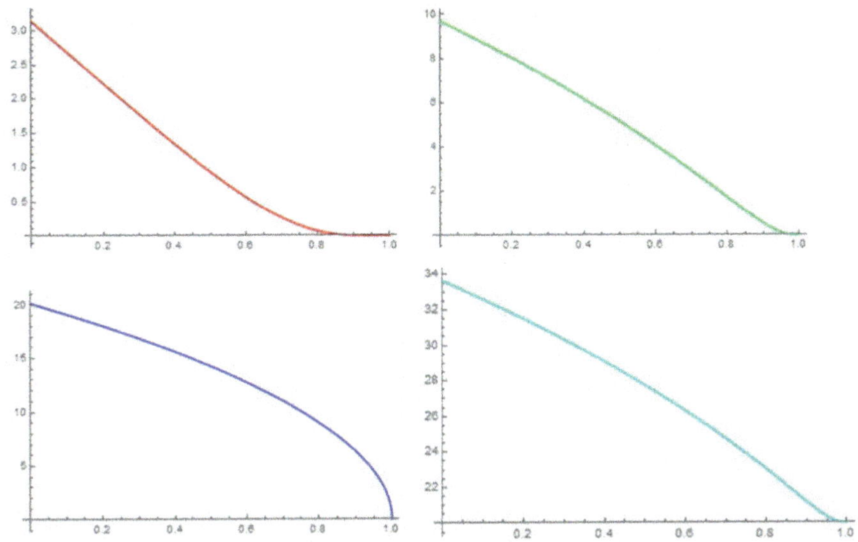

Fig. 4.102 Call option price over time for $K = 100$ and for $S(t)$ of 60 (red), 80 (green), 100 (blue), and 120 (turquoise), $r = 0.01$

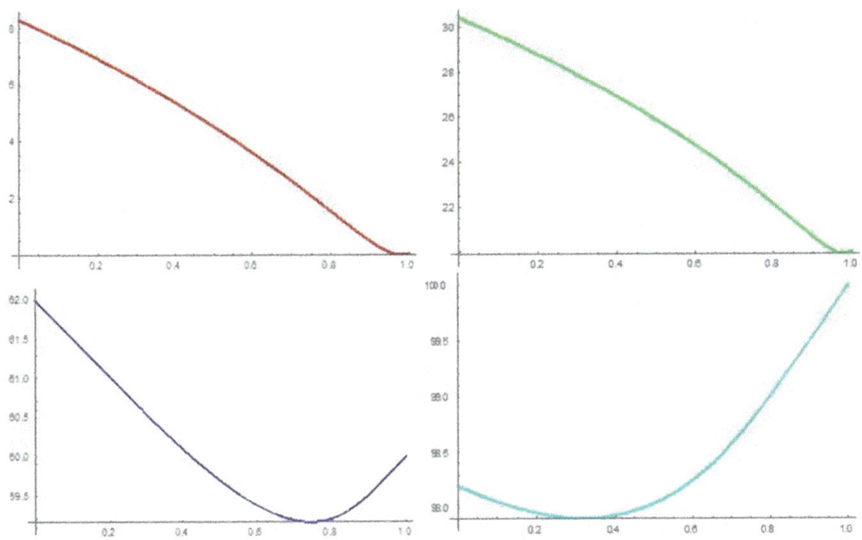

Fig. 4.103 Call option price over time for $K = 100$ and for $S(t)$ of 80 (red), 120 (green), 100 (blue), and 200 (turquoise), $\boldsymbol{r = -0.05}$

the call price curve over time with strike price $K = 100$ and the values $S(t) = 80, 120, 160$, and 200.

Here we see, at least for higher $S(t)$ values, that the call option price increases over time as it nears expiration.

The fact that $C(t)$ (for non-negative r (!)) is monotonically decreasing in t is easily verified by computation. To this end, we will differentiate the Black-Scholes formula for the call price $C(t)$ with respect to time t.

In the following, ϕ denotes the density function of the standard normal distribution, i.e. $\phi(x) := \frac{1}{\sqrt{2\pi}} e^{-\frac{x^2}{2}}$, while C' and $d_i{}'$ denote the derivative of C and d_i with respect to t. We also recall that $d_2 = d_1 - \sigma\sqrt{T-t}$ and thus $d_2{}' = d_1{}' + \sigma \frac{1}{2\sqrt{T-t}}$.

Thus we also have

$$\phi(d_2) := \frac{1}{\sqrt{2\pi}} e^{-\frac{d_2^2}{2}} = \frac{1}{\sqrt{2\pi}} e^{-\frac{\left(d_1 - \sigma\sqrt{T-t}\right)^2}{2}} = \phi(d_1) \cdot e^{d_1 \sigma \sqrt{T-t} - \frac{\sigma^2(T-t)}{2}} =$$

$$= \phi(d_1)\, e^{\log\left(\frac{S}{K}\right) + \left(r + \frac{\sigma^2}{2}\right)(T-t) - \frac{\sigma^2(T-t)}{2}} = \phi(d_1) \frac{S}{K} e^{r(T-t)}.$$

This gives us

$$\begin{aligned}
C'(t) &= \frac{d}{dt}\left(S \cdot \mathcal{N}(d_1) - K \cdot e^{-r(T-t)} \cdot \mathcal{N}(d_2)\right) = \\
&= S\phi(d_1) \cdot d_1{}' - K \cdot e^{-r(T-t)} \cdot \phi(d_2) \cdot d_2{}' - K \cdot r \cdot e^{-r(T-t)} \cdot \mathcal{N}(d_2) = \\
&= S\phi(d_1) \cdot d_1{}' - K \cdot e^{-r(T-t)} \cdot \phi(d_1) \cdot \frac{S}{K} \cdot e^{r(T-t)} \cdot d_2{}' - \\
&\quad - K \cdot r \cdot e^{-r(T-t)} \cdot \mathcal{N}(d_2) = \\
&= S\phi(d_1) \cdot d_1{}' - S\phi(d_1) \cdot d_2{}' - K \cdot r \cdot e^{-r(T-t)} \cdot \mathcal{N}(d_2) = \\
&= -S\phi(d_1) \cdot \frac{\sigma}{2\sqrt{T-t}} - K \cdot r \cdot e^{-r(T-t)} \cdot \mathcal{N}(d_2)
\end{aligned}$$

Hence $C'(t) = -\left(S \cdot \phi(d_1) \cdot \frac{\sigma}{2\sqrt{T-t}} + K \cdot r \cdot e^{-r(T-t)} \cdot \mathcal{N}(d_2)\right)$.

All values within the bracket are greater than 0 (assuming r is positive), which means that $C'(t)$ is negative and the function $C(t)$ is strictly monotonically decreasing. So as time progresses, with otherwise constant parameters, the price of a call option falls.

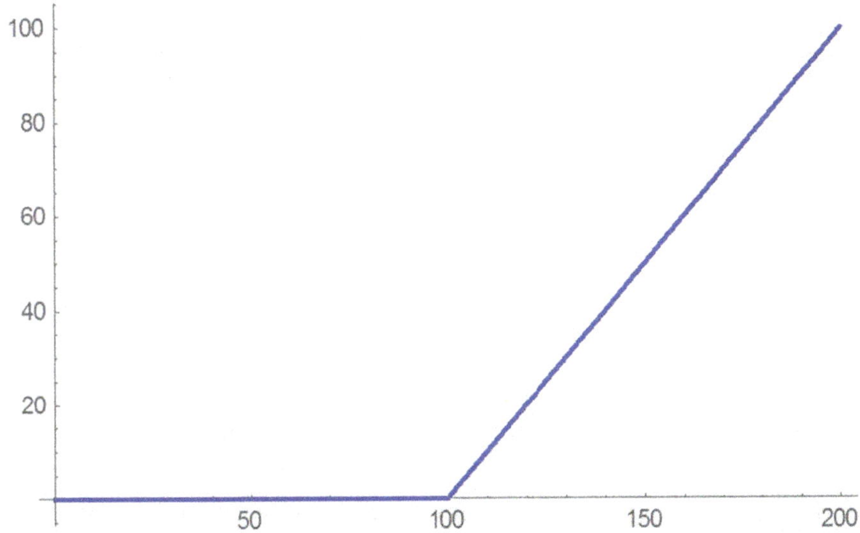

Fig. 4.104 Call option payoff function

A frequently used argument to underpin this result ("the price of a call option is monotonically decreasing in t") as being intuitively obvious goes as follows:

- The fair price of a call option is, as we know, the discounted expected payoff under a certain probability measure.
- The payoff of a call option at expiration (typically) has the form given in the subsequent Fig. 4.104:
- If the price $S(t)$ of the underlying asset is at a certain point at time t, e.g. at around 70 ("case 1") or around 130 ("case 2") as shown in Fig. 4.105, and if the price of the underlying asset moves down or up x points by time T, the following will happen:

  In case 1, the payoff remains at 0, but there is a chance (if there is enough time left until expiration) that $S(T)$ might rise to over $K = 100$ (in our example), resulting in a positive payoff. The expected payoff is thus greater than the payoff (= 0) that would result if the option were exercised immediately. The longer the remaining time to expiration, the greater the chance for that to happen.

  In case 2, the payoff can both increase and decrease linearly by value x. However, while the linear upside potential is unlimited, the linear decrease is bounded below by the minimum payoff 0. On average, therefore, we can

(continued)

expect a higher payoff than if we exercised the option immediately. With a longer time to expiration, we can usually expect to see more swings, so that the above effect is more likely to come into play the more time is left until expiration. Figure 4.105 displays the options payoffs and price movements.

Conclusion:

With a longer time remaining to expiration, we can expect to see a higher average payoff, so a longer remaining time to expiration therefore implies a higher call price.

While this argumentation supports our intuitive hunch, it has two major **logical flaws**:

Firstly:

The payoff has not been discounted. A longer time to expiration reduces the value of the future payoff due to discounting (assuming a positive interest rate r). This effect needs to be taken into account in the above argumentation.

Secondly:

In the argument for case 2, it was implicitly presumed that a price increase of the underlying by at least x points (and thus an increase in the payoff by at least x points) was as likely to occur as a decrease in the price by at least x points (and thus a decrease in the payoff by at least x points, or in the worst case even to 0). Only when an increase by at least x points has a constantly equal or even higher probability to occur than a decrease by at least x points, only then can we conclude so intuitively and simply that an average higher payoff is to be expected.

Excursus:

Out of curiosity, given the above considerations about the plausibility of the call price falling over time, as well as to deepen our understanding of the Wiener model, we ask ourselves the following question:

As we know, the Wiener model satisfies $S(T) = S(t) \cdot e^{\mu(T-t)+\sigma\sqrt{T-t}w}$. Under what conditions is the probability that $S(T) > S(t) + x$ greater than the probability that $S(T) < S(t) - x$? Put differently: When does $W(S(T) > S(t) + x) > W(S(T) < S(t) - x)$ hold and what requirements do μ and σ have to satisfy so that this holds for all x, t, and $S(t)$?

(continued)

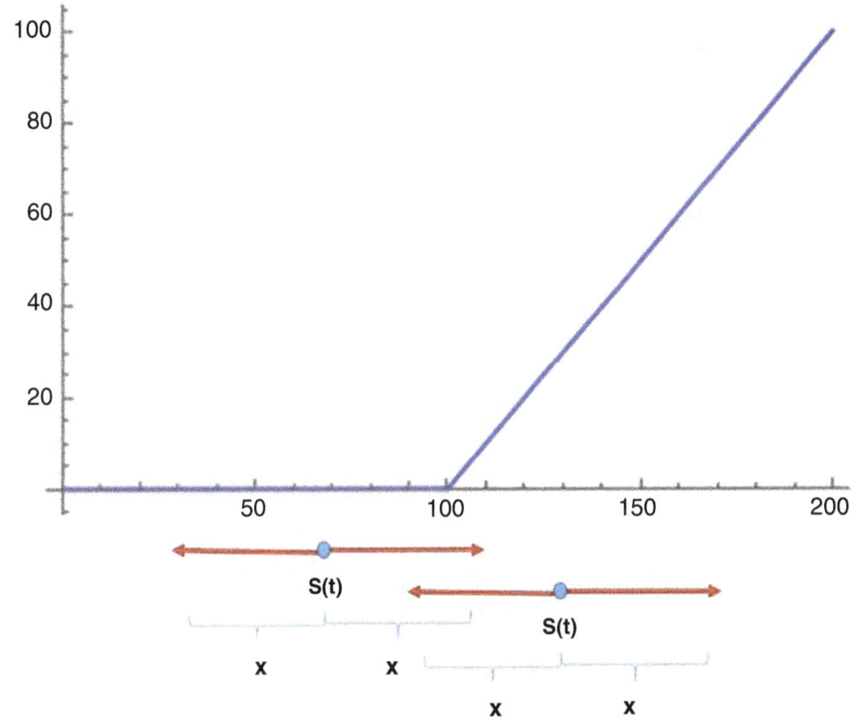

Fig. 4.105 Call option payoff and price movements

We get the answer to this through the following simple calculation:

In the following, we let $\tau = T - t$ denote the option's remaining time to expiration and set $y = \frac{x}{S(t)}$:

$$W\left(S(T) > S(t) + x\right) \geq W\left(S(T) < S(t) - x\right)$$

$$\Leftrightarrow W\left(S(t)\left(e^{\mu\tau+\sigma\sqrt{\tau}w} - 1\right) > x\right) \geq W\left(S(t)\left(e^{\mu\tau+\sigma\sqrt{\tau}w} - 1\right) < -x\right)$$

$$\Leftrightarrow W\left(\left(e^{\mu\tau+\sigma\sqrt{\tau}w} - 1\right) > y\right) \geq W\left(\left(e^{\mu\tau+\sigma\sqrt{\tau}w} - 1\right) < -y\right)$$

(continued)

$$\Leftrightarrow W\left(e^{\sigma\sqrt{\tau}w} > \frac{1+y}{e^{\mu\tau}}\right) \geq W\left(e^{\sigma\sqrt{\tau}w} < \frac{1-y}{e^{\mu\tau}}\right)$$

$$\Leftrightarrow W\left(w > \frac{1}{\sigma\sqrt{\tau}} \cdot \log\frac{1+y}{e^{\mu\tau}}\right) \geq W\left(w < \frac{1}{\sigma\sqrt{\tau}} \log\frac{1-y}{e^{\mu\tau}}\right).$$

Due to the symmetry of the standard normally distributed random variables w, the last inequality is satisfied precisely when

$$\frac{1}{\sigma\sqrt{\tau}} \cdot \log\frac{1+y}{e^{\mu\tau}} + \frac{1}{\sigma\sqrt{\tau}} \log\frac{1-y}{e^{\mu\tau}} \leq 0,$$

that is, precisely when

$$\frac{1-y^2}{e^{2\mu\tau}} \leq 1$$

$$\Leftrightarrow y^2 \geq 1 - e^{2\mu\tau}$$

Now, if $\mu\tau \geq 0$, that is, if $\mu \geq 0$, then the last inequality is satisfied for all y.

If $\mu < 0$, then the inequality is satisfied precisely for those y for which $y \geq \sqrt{1-e^{2\mu\tau}}$, that is, for those x for which $x \geq S(t)\sqrt{1-e^{2\mu\tau}}$.

For the risk-neutral model, i.e. for the case $\mu = r - \frac{\sigma^2}{2}$, this means:

It is only when $r \geq \frac{\sigma^2}{2}$ that the above reasoning regarding the plausibility of the call price monotony (leaving discounting aside) is in fact correct for all x.

4.25 Detailed Discussion of the Black-Scholes Formula for European Call Options II (Dependence on Volatility)

We are now going to examine the dependence of a call option's fair price C on volatility. If we are still pondering the intuitive but not entirely consistent argumentation in the previous section that the call price is monotonically decreasing over time, then the following heuristic consideration would also seem obvious:

The more the price of the underlying asset fluctuates, the higher the likelihood of obtaining a higher average payoff. The obvious conclusion would therefore be that a call price is strictly monotonically increasing with volatility σ.

In the Extreme Case that $\sigma = 0$

The option price in this case ($\sigma = 0$) is given by

$$\begin{aligned} C(t) &= e^{-r(T-t)} \cdot E\left(\max\left(S(t) \cdot e^{\left(r-\frac{\sigma^2}{2}\right)(T-t)+\sigma\sqrt{T-t}w} - K, 0\right)\right) = \\ &= e^{-r(T-t)} \cdot E\left(\max\left(S(t) \cdot e^{r(T-t)} - K, 0\right)\right) = \\ &= \max\left(S(t) - K \cdot e^{-r(T-t)}, 0\right) \end{aligned}$$

So in this case, with volatility 0, $C(t)$ (as a function of $S(t)$) has the form of the payoff curve, just slightly translated to the left (for positive r) or slightly translated to the right (for negative r) as one can see in Fig. 4.106. In the case that $r = 0$, the curve for C as a function of $S(t)$ coincides with the payoff curve.

In the following representation (Fig. 4.107), we plotted the call price C for the four different volatilities $\sigma_1 = 0, \sigma_2 = 0.3, \sigma_3 = 0.6$, and $\sigma_4 = 0.9$ as a function of different values $S = S(0)$ of the underlying asset at time 0 for a fixed choice of $K = 100, r = 0.05, T = 1$, and $t = 0$ (graphs in red). For comparison, the payoff function is also shown as a function of $S = S(T)$ (in blue). The lowermost price line reflects the price movements for σ_1, the second line from below the price movements for σ_2, the third line the price movements for σ_3, and the uppermost red line the price movements for σ_4.

For a negative r $(r = -0.05)$ the corresponding graph looks as follows (Fig. 4.108):

So apparently, the price of a call option increases as volatility σ increases. We support this observation by further graphs (see Fig. 4.109) where we are now going to plot C as a function of volatility σ (for different but fixed values $S = S(0)$ and otherwise the same parameters as above).

We now want to discuss the pricing formula for call options as a function of volatility σ and prove the monotony analytically. In the following, $C'(\sigma)$ is the derivative of $C(\sigma)$ with respect to σ.

We recall the relations we used in the previous section: $d_2 = d_1 - \sigma\sqrt{\tau}$ and $\phi(d_2) = \phi(d_1) \cdot \frac{S}{K} \cdot e^{r\tau}$. (And again: $\tau = T - t$.)

For the derivatives $d_1'(\sigma)$ and $d_2'(\sigma)$, we therefore have the relationship $d_2'(\sigma) = d_1'(\sigma) - \sqrt{\tau}$, and we get

$$\begin{aligned} C'(\sigma) &= \frac{d}{d\sigma}\left(S \cdot \mathcal{N}(d_1) - K \cdot e^{-r\tau} \cdot \mathcal{N}(d_2)\right) = \\ &= S \cdot \phi(d_1) \cdot d_1'(\sigma) - K \cdot e^{-r\tau} \cdot \phi(d_2) \cdot d_2'(\sigma) = \end{aligned}$$

(continued)

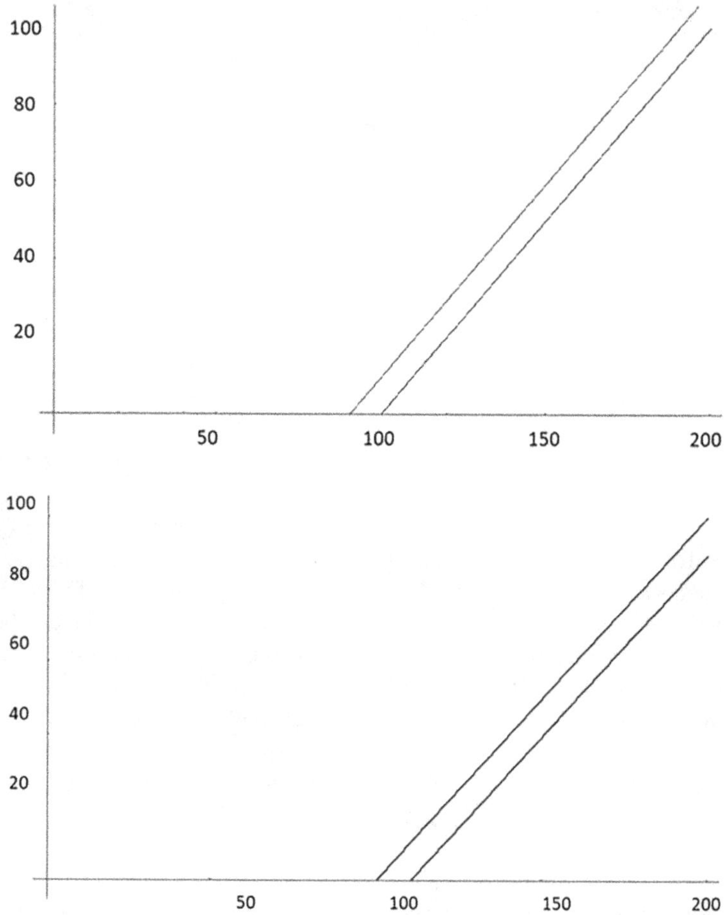

Fig. 4.106 Price of call option as a function of $S(t)$ for volatility 0 (red) compared to payoff function (blue), higher for $r > 0$, lower for $r < 0$

$$= S \cdot \phi(d_1) \cdot d_1{}'(\sigma) - K \cdot e^{-r\tau} \cdot \phi(d_1) \cdot \frac{S}{K} \cdot e^{r\tau} \cdot \left(d_1{}'(\sigma) - \sqrt{\tau}\right) =$$

$$= \sqrt{\tau} \cdot S \cdot \phi(d_1).$$

By substituting, using the relationship between $\phi(d_2)$ and $\phi(d_1)$, we get the alternative representation $C'(\sigma) = K \cdot e^{-r\tau} \sqrt{\tau} \phi(d_2)$.

We thus have $C'(\sigma) = \sqrt{T-t} \cdot S \cdot \phi(d_1) = K \cdot e^{-r(T-t)} \cdot \sqrt{T-t} \cdot \phi(d_2)$.

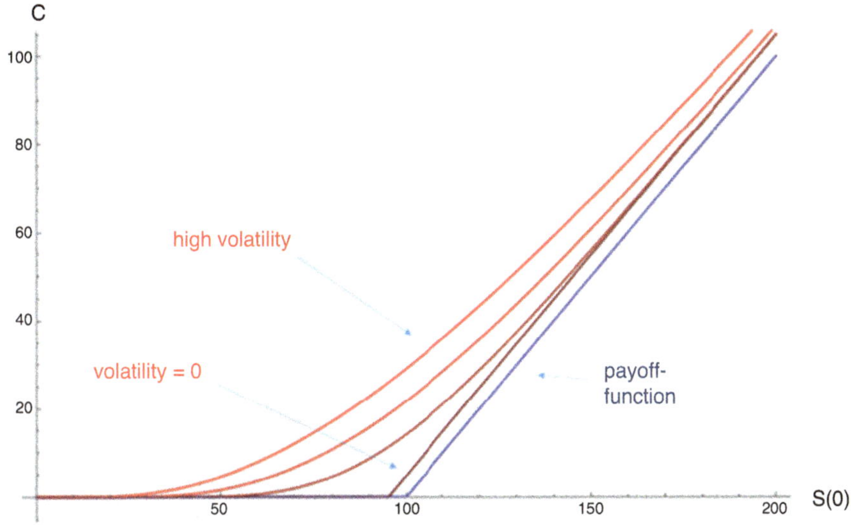

Fig. 4.107 Call option price as a function of the underlying asset's price for four different volatilities (assuming a positive interest rate)

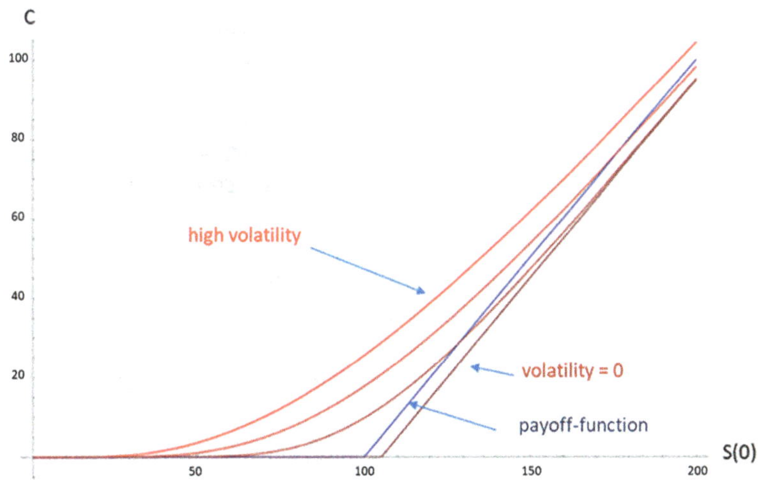

Fig. 4.108 Call option price as a function of the underlying asset's price for four different volatilities (assuming a negative interest rate)

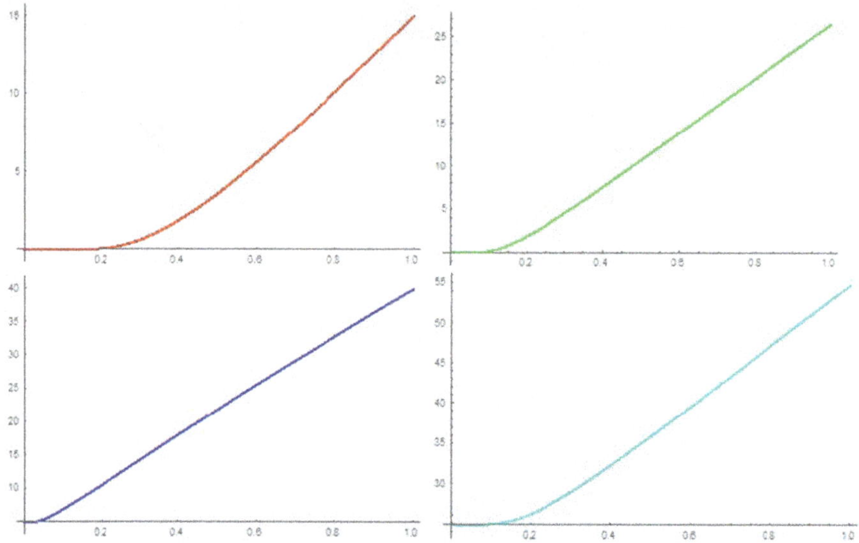

Fig. 4.109 Call option price depending on volatility, for $K = 100$ and for $S(t)$ of 60 (red), 80 (green), 100 (blue), and 120 (turquoise), $r = 0.05$

All factors (in both representations) are positive. The price of a call option is therefore monotonically increasing with volatility in all circumstances.

4.26 Detailed Discussion of the Black-Scholes Formula for European Call Options III (Dependence on Risk-Free Interest Rate)

We start our discussion of the call price formula with respect to the risk-free interest rate r by going straight to the analytical part, differentiating $C(r)$ with respect to r. This is easily done if we again observe that $\phi(d_2) = \phi(d_1) \cdot \frac{S}{K} \cdot e^{r(T-t)}$ and $d_2 = d_1 - \sigma\sqrt{T-t}$ and thus $d_2{}'(r) = d_1{}'(r)$. Because then we get

$$\begin{aligned} C'(r) &= \frac{d}{dr}\left(S \cdot \mathcal{N}(d_1) - e^{-r\tau} \cdot K \cdot \mathcal{N}(d_2)\right) = \\ &= S \cdot \phi(d_1) \cdot d_1{}' - e^{-r\tau} \cdot K \cdot \phi(d_2) \cdot d_2{}' + \tau \cdot e^{-r\tau} \cdot K \cdot \mathcal{N}(d_2) = \\ &= \tau \cdot e^{-r\tau} \cdot K \cdot \mathcal{N}(d_2)\,. \end{aligned}$$

So the derivative of C with respect to r is given by $C'(r) = \tau \cdot e^{-r\tau} \cdot K \cdot \mathcal{N}(d_2)$ and is therefore obviously always positive. ***C* is thus strictly monotonically increasing in *r*.**

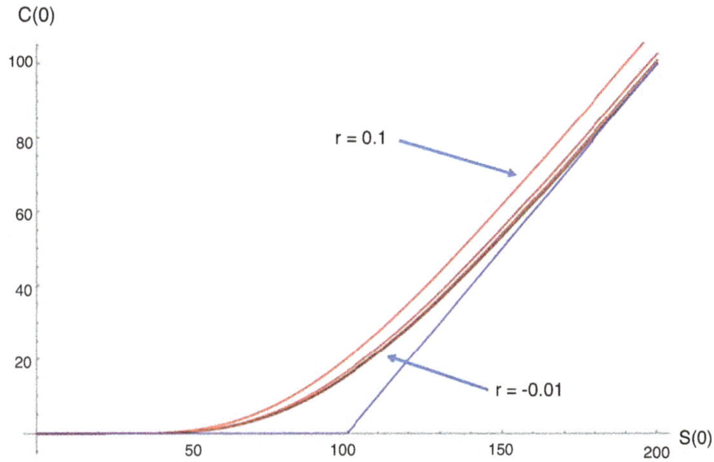

Fig. 4.110 Call option price as a function of the underlying asset's price for four different risk-free interest rates $r = -0.01$, $r = 0$, $r = 0.02$ and $r = 0.1$

To visually illustrate this result, we provide the corresponding graphs in analogy to the previous two sections.

Figure 4.110 shows the call price curves for four different interest rates (and the parameters $K = 100$, $T = 1$, $t = 0$, $\sigma = 0.4$). The bottom price curve corresponds to the lowest interest rate, the highest price curve to the uppermost interest rate.

We also see that call prices are relatively stable in relation to changes in the risk-free interest rate r. Changes in the interest rate by 1% have only a very minor effect on call prices. In addition, over normal—not overly long—time spans until expiration, it is generally not to be expected that interest rates experience major changes. The impact of interest rate changes on call option prices (i.e. the interest rate risk) is therefore generally relatively small.

We are now adding the other Graph 4.111, where we are going to plot C as a function of the risk-free interest rate r (for different but fixed values $S = S(0)$ and otherwise the same parameters as above).

Again, it would be nice to get an intuitive insight, to find a heuristic explanation for why call option prices are monotonically increasing in r. Using the Black-Scholes formula for the call price, this explanation can be found immediately (even without differentiating C with respect to r):

(continued)

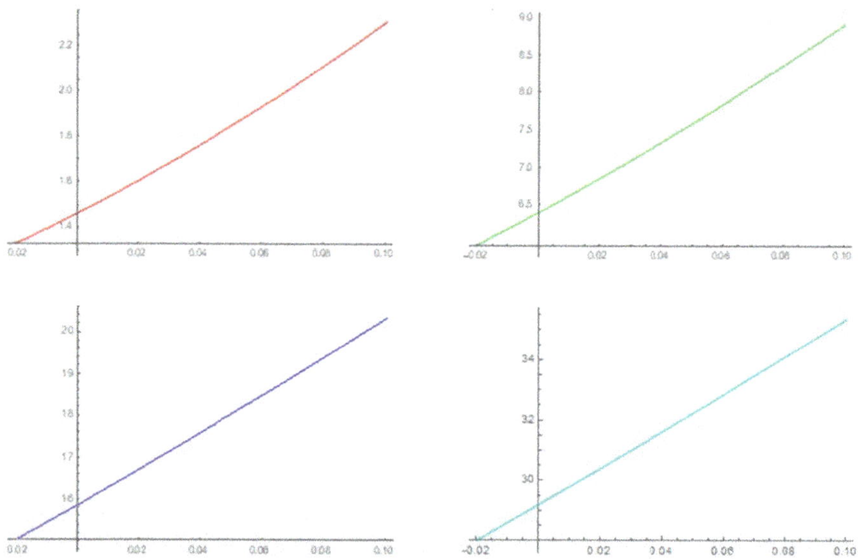

Fig. 4.111 Call price with respect to risk-free interest rate r for $K = 100$ and for $S(t)$ of 60 (red), 80 (green), 100 (blue), and 120 (turquoise), $\sigma = 0.4$, r from - 2% to + 10%

After all, according to Black-Scholes

$$C(t) = e^{-r(T-t)} \cdot E\left(\max\left(S(t) \cdot e^{\left(r-\frac{\sigma^2}{2}\right)(T-t)+\sigma\sqrt{T-t}w} - K, 0\right)\right) =$$

$$= E\left(\max\left(S(t) \cdot e^{-\frac{\sigma^2}{2}(T-t)+\sigma\sqrt{T-t}w} - K \cdot e^{-r(T-t)}, 0\right)\right)$$

In this notation of the formula, the parameter r occurs only in the term $-K \cdot e^{-r(T-t)}$, which is monotonically increasing in r, and so C is monotonically growing in r. However, this is not an immediately intuitive explanation, as we draw on knowledge of the Black-Scholes formula for it.

Another—truly heuristic—approach would be as follows:

A call option, as we know, offers a **relatively inexpensive** opportunity to bet **on increasing prices** of an underlying asset. In a replication, this can surely be achieved only by **buying a (not too small) quantity of the underlying asset** and constantly adjusting the quantity after that). Since the call option is relatively inexpensive, the calculated option price alone will not be sufficient to buy the required quantity of the underlying asset. **For the replication, it will therefore be necessary to take out a loan.** A loan

(continued)

is all the more expensive the higher the risk-free interest rate is. And so the replication and thus the call option's price increase as the risk-free interest rate r increases. Consequently and logically, the price of a call option is thus monotonically increasing in r.

We get further support for this argumentation if we think back to how an option is priced in the one-step binomial model:

In Sect. 3.9 we derived the replicating portfolio for a derivative with two possible payoffs f_u and f_d in a one-step binomial model. The cash amount in the replicating portfolio was $y = \frac{u \cdot f_d - d \cdot f_u}{u-d}$. In the case of a call option, $f_d = 0$ and $f_u > 0$, i.e. $y = -\frac{d \cdot f_u}{u-d} < 0$. This means: To create a replicating portfolio, we do indeed need to take out a loan, and this supports our heuristic reasoning above.

4.27 Some Brief Remarks on the Use of the Black-Scholes Formula and Its Parameters r and σ

At this point, a few parenthetical remarks are in order:

1. It is important that we be aware that, in all of the call option pricing formula analyses that we ran above, we used the **formula** for the fair price of these options (and assumed a Wiener model for the underlying asset). **All the conclusions that we drew (e.g. monotony) apply to the options pricing formula but need not necessarily always hold for real options prices.** For instance, it cannot be ruled out that, despite an (assumed) positive interest rate r, the price of a particular call option, at otherwise unchanged parameters (unchanged r, unchanged σ), moves up over a short period of time, even though it should theoretically fall over time. (As an aside, we are going to discuss in depth at a later point how "unchanged σ" is to be understood.) But what does it mean if real option prices do not behave in the way the theoretical conclusions from the Black-Scholes formula would suggest? For example, what are the implications if the price of a call option rises over a certain period of time $[t_1, t_2]$, despite a positive interest rate r with all other parameters remaining the same?
   It means that the price of the option has to deviate from the fair price either at time t_1 or at time t_2 and that (provided the deviation is sufficiently significant) arbitrage would therefore be possible either as from time t_1 or as from time t_2 (again assuming that the underlying asset actually evolves according to a Wiener model with the assumed parameters).
2. Two of the parameters occurring in the Black-Scholes formula are variable, namely, the interest rate r and the volatility σ. It is quite possible, and even probable in the case of volatility, that these parameters will change during the remaining life of the option. **In our approach so far**, in the form in which we

have derived and used the Black-Scholes formula so far, **we assumed that the interest rate r and volatility σ do not change over the life of the derivative!** What are the consequences of these assumptions, and how do they affect the relevance of the Black-Scholes formula? What alternative approaches could we use to get around these restrictions? We will address these questions provisionally (!) in the next two paragraphs.

3. Regarding the interest rate r:
   With regard to the interest rate r in the Black-Scholes formula, another question arises—in addition to the question on the admissibility of the assumption that the interest rate remains unchanged. Namely: **What interest rate should be used in the Black-Scholes formula?** Is it the interest rate for the time range $[0, T]$ or the shortest-term (overnight) interest rate? To clarify this question, let us recall once again the derivation of the Black-Scholes formula (via the N-step binomial model). We did so by designing a perfect hedging strategy for the derivative. The initial investment required for this hedging strategy then became the fair price of the derivative. In this hedging strategy, it was necessary to continuously rebalance the portfolio and, among other things, to **continuously** invest or borrow cash (in amounts and for periods that were not known in advance). The only relevant interest rate in this setting is the **shortest-term (overnight) interest rate**. At time 0 (i.e. on the valuation date), this interest rate is known only for time 0, but not for later points in time. If it were known, i.e. if the shortest-term interest rate were already known at time 0 as a deterministic function $r(t)$ for the entire time range $[0, T]$, we would have to use, in deriving the Black-Scholes formula or using the approximation formula (obtained through a very finely meshed N-step binomial model approximating the Wiener process), the interest rate $r(t)$ applicable at every single one of those steps. Valuation in a two-step binomial model would then look like the following Fig. 4.112:

   For the possible fair values f_u and f_d of the derivative at time dt, the following is obtained when using interest rate $r(dt)$:

$$f_u = e^{-r(dt)dt} \cdot \left(p'(dt) \cdot f_{u^2} + \left(1 - p'(dt)\right) \cdot f_{ud}\right) \text{ resp.}$$

$$f_d = e^{-r(dt)dt} \cdot \left(p'(dt) \cdot f_{ud} + \left(1 - p'(dt)\right) \cdot f_{d^2}\right)$$

where $p'(dt) = \frac{e^{r(dt)dt} - d}{u-d}$
and subsequently for the derivative's fair value f_0 at time 0:

$$\begin{aligned} f_0 &= e^{-r(0)dt} \cdot \left(p'(0) \cdot f_u + \left(1 - p'(0)\right) \cdot f_d\right) = \\ &= e^{-(r(0)+r(dt))dt} \cdot \left(p'(0) \cdot p'(dt) \cdot f_{u^2} + \left(p'(0) \cdot \left(1 - p'(dt)\right) + \right.\right. \\ &\quad \left.\left. + \left(1 - p'(0)\right) \cdot p'(dt)\right) \cdot f_{ud}\right) + \left(1 - p'(0)\right) \cdot \left(1 - p'(dt)\right) \cdot f_{d^2} \end{aligned}$$

where $p'(0) = \frac{e^{r(0)dt} - d}{u-d}$.

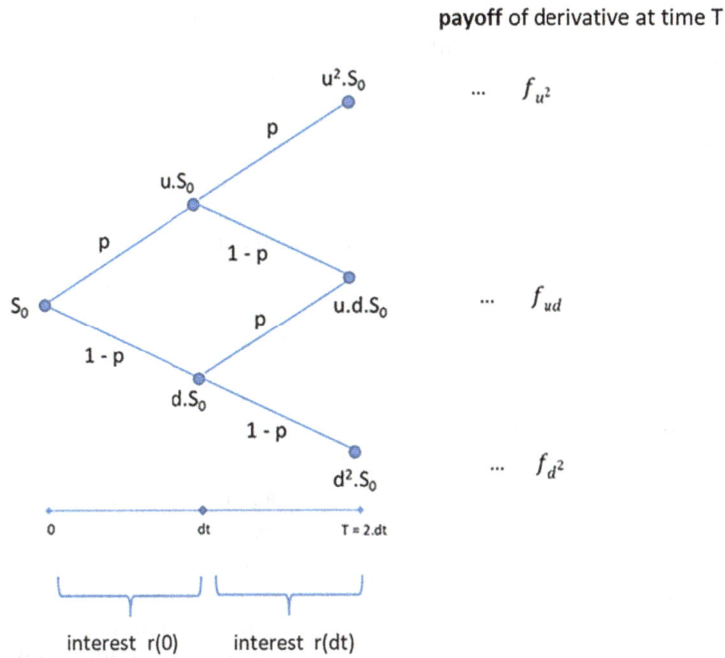

Fig. 4.112 Two-step binomial model with variable interest rate

This procedure can then, of course, be expanded algorithmically to any N-step models.

In reality, the risk-free shortest-term interest rate r will not remain constant over the time range $[0, T]$, nor will its movements be known deterministically in advance. A priori, the evolution of r is again a stochastic random process. An attempt could now be made to simulate the risk-free interest rate by means of a suitable model and then work with a Wiener model with a stochastic interest rate. To do this, however, we would first need (in addition to a suitable model) the relevant valuation theory, which we have so far only derived for the case of fixed interest rates.

As shown above, option prices are relatively insensitive to changes in the risk-free interest rate. In addition, interest rates usually do not change sharply over the life of an option (provided that life is not overly long). Thus, in general, the assumption of a fixed risk-free interest rate for the life of the option to be valued is not an excessive restriction. Another note in this context: If the risk-free shortest-term interest rate r is assumed to be constant on the time interval $[0, T]$, then, given the argument of no-arbitrage, all longer-term risk-free interest rates $f_{0,t}$ for $0 < t \leq T$ will of course also have the same value r.

4. Regarding the assumption of constant volatility of the underlying asset:
The underlying asset's volatility is the essential variable parameter in derivatives valuations using the Wiener model. A derivative's fair price is substantially

dependent on this volatility. In our model so far, we have assumed constant volatility σ over a derivative's life. This is generally an unrealistic assumption. It is equally unrealistic to assume that a certain deterministic evolution of volatility in $[0, T]$ is known in advance.

And, more importantly: What does "current volatility of the underlying asset" even mean? How can we measure it? We are going to address this question in detail in Volume II Chapter 1. For now, let us assume that we have found a reasonably satisfactory answer to this fundamental question and agreed on a methodology for estimating the underlying asset's current volatility σ. This methodology will yield a currently valid value σ, which, however, will generally not remain constant over the life of the derivative, nor will its movements be known deterministically in advance. A priori, the evolution of σ is again a stochastic random process. An attempt could now be made to simulate the underlying asset's volatility by means of a suitable model and then to work with a Wiener model with stochastic volatility. To do this, however, we would first need (in addition to a suitable model) the relevant valuation theory, which we have so far only derived for the case of constant volatility. Not to worry: We will discuss such a valuation theory that assumes stochastic volatility at a later point.

One possible approach to adapting the assumption of constant volatility toward a somewhat more realistic model is based on the **observation of rising volatilities with falling prices of the underlying asset and falling volatility with rising prices of the underlying asset** (this fact will also be discussed in more detail later when we delve into a more in-depth discussion of volatility). Here again, think back to the valuation of a derivative using a Wiener model through valuation using an approximating N-step binomial model. In light of the above-mentioned dependence between volatility and price of the underlying asset, it would make sense therefore to assign different volatilities to the nodes in the binomial model (the lower the price of the underlying asset, the higher the volatility at this node). This would then affect the current parameters u and d for this node and thus the artificial probability p' to be applied at this node.

In the following section, we want to carry out such an approach, explain the pricing program available on our website, and run some tests. See: https://app.lsqf.org/book/pricing-binomial

4.28 Program and Test: Valuation of Derivatives by Approximation Using an N-Step Binomial Model Where Volatility Is Correlated with the Price of the Underlying Asset

The problem with this approach, however, lies in the fact that due to changing volatility and thus changing u and d values, the values of the underlying asset at the individual nodes depend not only on the **number** of upward or downward movements in the binomial model (i.e. on the respective node) but also on the particular path that led to this node. The approach can still be carried out, but it

Table 4.6 Volatility of the SPX for different choices of a

Change in SPX from 2900 to	a = 4 Volatility change from 15% to	a = 5 Volatility change from 15% to	a = 6 Volatility change from 15% to
2800	17.26	17.88	18.52
2700	19.96	21.44	23.03
2600	23.22	25.89	28.88
2500	27.16	31.51	36.55
3000	13.10	12.66	12.24
3100	11.49	10.75	10.05
3200	10.12	9.17	8.31

is numerically considerably more complex. To better illustrate the principle, we are going to explain it using a two-step binomial model only. It should then be clear how the principle can subsequently be applied to any N-step models.

Regarding a potential correlation between price and volatility of an underlying asset, the following approach would be a plausible one:

A change in a stock's price from S_0 to $S_t = S_0 \cdot (1+x)$, i.e. by x percent, implies a change in volatility from σ to $\sigma_t = \frac{\sigma}{(1+x)^a}$ for an appropriately selected parameter a. This is equivalent to the relationship $\sigma_t = \sigma \cdot \left(\frac{S_0}{S_t}\right)^a$.

So, for different choices of parameter a, (short-term) changes (over the life of a derivative) in, for example, the S&P500 Index from initially 2900 points and an initial volatility of 15% would yield the volatility changes shown in the following Table 4.6:

Historical data show that such volatility reactions to (short-term) price changes are well within a realistic range. Further analyses can be found in Volume II Chapter 1. The value $a = 0$, by the way, leads to constant volatility.

For the sake of simplicity, we choose a risk-free interest rate $r = 0$ below. As said, we are now going to deal in detail with a two-step binomial model that we want to (roughly) approximate a given Wiener model.

The stock price S_0 may change to $u \cdot S_0$ or to $d \cdot S_0$ in the first step. Volatility at time 0 is σ. Due to the change in the stock price, volatility changes to $\sigma_u = \frac{\sigma}{u^a}$ or $\sigma_d = \frac{\sigma}{d^a}$.

To approximate the Wiener model using the binomial model, we have to choose (as we know from Sect. 4.14) $u = e^{\mu dt + \sigma\sqrt{dt}}$ and $d = e^{\mu dt - \sigma\sqrt{dt}}$.

Since we want to approximate the Wiener model (for valuation of derivatives) for the risk-neutral measure, we have to choose $\mu = r - \frac{\sigma^2}{2} = -\frac{\sigma^2}{2}$, and thus

$$\boldsymbol{u} = e^{-\frac{\sigma^2}{2}dt + \sigma\sqrt{dt}} \text{ and } \boldsymbol{d} = e^{-\frac{\sigma^2}{2}dt - \sigma\sqrt{dt}}.$$

Hence $\boldsymbol{\sigma_u} = \frac{\sigma}{u^a} = \sigma \cdot e^{a\frac{\sigma^2}{2}dt - a\sigma\sqrt{dt}}$ and $\boldsymbol{\sigma_d} = \frac{\sigma}{d^a} = \sigma \cdot e^{a\frac{\sigma^2}{2}dt + a\sigma\sqrt{dt}}$.

Due to the changed volatilities at time dt, the u or d values (as the case may be) will also change at time dt. We denote the new u or d values by u_u and d_u (if they originate at $u \cdot S_0$) and by u_d and d_d (if they originate at $d \cdot S_0$).

If at time dt we are at node $u \cdot S_0$:

Starting from here, we can assume for the price of the underlying asset that the values at time $T = 2 \cdot dt$ are $u_u \cdot u \cdot S_0$ or $d_u \cdot u \cdot S_0$.

If at time dt we are at node $d \cdot S_0$:

Starting from here, we can assume for the price of the underlying asset that the values at time $T = 2 \cdot dt$ are $u_d \cdot d \cdot S_0$ or $d_d \cdot d \cdot S_0$.

A total of four different payoffs are therefore possible.

The values u_u, d_u, u_d, and d_d are calculated in the same way as the values u and d, yet using the volatilities σ_u and σ_d instead of σ, so

$$u_u = e^{-\frac{\sigma_u^2}{2}dt+\sigma_u\sqrt{dt}}, d_u = e^{-\frac{\sigma_u^2}{2}dt-\sigma_u\sqrt{dt}}, u_d = e^{-\frac{\sigma_d^2}{2}dt+\sigma_d\sqrt{dt}},$$

$$d_d = e^{-\frac{\sigma_d^2}{2}dt-\sigma_d\sqrt{dt}}.$$

To price options in this two-step model, the only other input we need are the artificial probabilities p', or p'_u and p'_d (at the nodes $u \cdot S_0$ and $d \cdot S_0$), which can of course be determined as follows (recalling that we have set $r = 0$):

$p' = \frac{1-d}{u-d}$, $p'_u = \frac{1-d_u}{u_u-d_u}$ and $p'_d = \frac{1-d_d}{u_d-d_d}$.

If we let f be the payoff function of the derivative that we want to price, we get the following result for the fair value f_0 of that derivative:

$$f_0 = p' \cdot \left(p'_u \cdot f\left(u_u \cdot u \cdot S_0\right) + \left(1 - p'_u\right) \cdot f\left(d_u \cdot u \cdot S_0\right)\right) + \left(1 - p'\right) \cdot$$

$$\cdot \left(p'_d \cdot f\left(u_d \cdot d \cdot S_0\right) + \left(1 - p'_d\right) \cdot f\left(d_d \cdot d \cdot S_0\right)\right)$$

This procedure can now be extended to any N-step models, of course, although it becomes a lot more intricate.

That this f_0 is indeed the derivative's fair price in this model follows directly from the fact that at the price f_0 at time 0, we can start a perfect hedging strategy in the obvious way for the derivative in this model.

We are now going to apply this model to a realistic example in order to get an approximate sense of how a volatility that correlates (negatively) with the stock price affects the price of derivatives. This example is based on the S&P500 index data assumed above, which roughly corresponded to reality in August 2018. For simplicity, we do not take dividends into account here and below:

$S_0 = 2,900$
$r = 0$
$\sigma = 15\% \ldots 0.15$

Table 4.7 Parameters for derivative pricing under the given assumptions

	a = 0	a = 4	a = 5	a = 6
σ	0.15	0.15	0.15	0.15
u	1.03	1.03	1.03	1.03
d	0.97	0.97	0.97	0.97
σ_u	0.15	0.133	0.129	0.125
σ_d	0.15	0.17	0.175	0.181
u_u	1.03	1.027	1.026	1.026
d_u	0.97	0.973	0.974	0.974
u_d	1.03	1.035	1.036	1.037
d_d	0.97	0.965	0.964	0.963
p'	0.5	0.5	0.5	0.5
p'_u	0.5	0.5	0.5	0.5
p'_d	0.5	0.5	0.5	0.5
$u_u \cdot u \cdot S_0$	3080.25	3069.86	3067.45	3065.12
$d_u \cdot u \cdot S_0$	2897.28	2907.67	2910.08	2912.41
$u_d \cdot d \cdot S_0$	2897.28	2908.67	2911.74	2914.91
$d_d \cdot d \cdot S_0$	2725.18	2713.80	2710.72	2707.55

Table 4.8 Resulting prices under the consideration of the parameters from the previous Table 4.7

	a = 0	a = 4	a = 5	a = 6	Black Scholes
Call option	45.06	46.55	47.32	48.11	50.09
Put option	37.45	40.30	41.07	41.86	38.37

For the option's life to expiration T, we choose a month, so $T = \frac{1}{12}$ and thus $dt = \frac{1}{24}$. For the three values of $a = 4, 5, 6$ and, for comparison, for the value $a = 0$ (i.e. constant volatility), we will now calculate the parameters required for pricing the derivative and present them in the following Table 4.7:

We will now use these values to price a call option with strike 2900 and a put option with strike 2875 in this two-step model with correlated volatility, and we will also compare the results with the prices determined using the Black-Scholes formula in the Wiener model (with constant volatility). The resulting prices for the a values of the two options are shown in the following Table 4.8.

It seems that a higher negative correlation between the price of the underlying asset and volatility tends to lead to slightly higher prices both in the call option and in the put option case. The prices obtained in the two-step binomial model do not differ much from the Black-Scholes prices.

4.29 Break-Even for Call-Only Strategies

For option traders it is an indispensable prerequisite that they gain a strong sense of the price dynamics of options combinations. Time and again in the following, we will discuss thoughts and observations in order to give our readers such a sense of basic strategies. In this section, we are going to look at a call-only strategy.

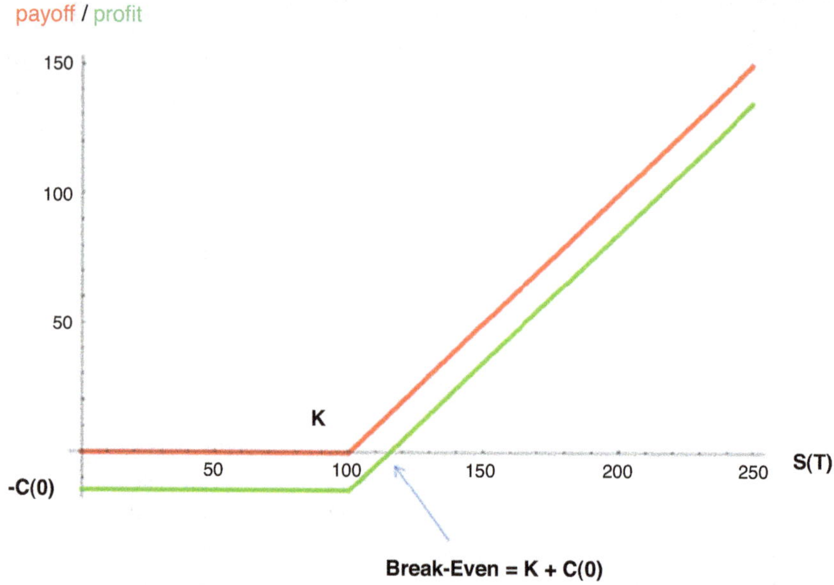

Fig. 4.113 Payoff, profit, and break-even point of a call option

Purchasing a long call with expiration T, strike K, and price $C(0)$ on an underlying asset with the initial value S_0 generates a payoff function $\max(S(T) - K, 0)$ and a profit function $\max(S(T) - K, 0) - C(0)$ for date T.

The break-even point for the expiration date T is therefore $K + C(0)$ (compare Fig. 4.113). So we see that the long position in the call option generates a profit only if the price of the underlying asset at time T is above the value of $K + C(0)$.

However, as holders of a long position, we also have the option to sell the call before expiration and—if the underlying asset is adequately priced—pocket a profit from this early sale. The question we therefore have to ask ourselves is: Where is the call option's break-even point at any one time $t \in [0, T]$?

In other words: What is the minimum price that the underlying asset needs to have at time $t \in [0, T]$ in order for the seller of the call option to make a profit at time t?

We denote the break-even point at time t by $BE(t)$. To answer the question, we assume that the call option can always be traded at the fair Black-Scholes price $C(t)$.

At time 0, the break-even point is at S_0, of course: If we were to resell the call option immediately after buying it, we would (given no change in the underlying asset's price (S_0)) immediately get back the call option's price $C(0)$. We wouldn't have made any profit nor any loss. If the price of the underlying asset had risen to a value greater than S_0 (e.g. to S_{0+}) immediately after buying the call option, then the price of the call option (at unchanged volatility) would also have risen (e.g. to $C(0+)$), so that an immediate sale of the option would have generated a profit.

When volatility remains unchanged over the life of the option, the break-even point thus moves from S_0 (at time 0) to $S_0 + C(0)$ (at time T).

For any time $t \in [0, T]$, the break-even point (assuming unchanged volatility over the life of the option) can be calculated from the following equation (where $BSCall(t, s)$ denotes the fair price of the call option at time t for the value s of the underlying asset, i.e. $BSCall(t, s)$ is the Black-Scholes formula for call options with the fixed parameters K, r and σ and the variable parameters t and s:

$$C(0) = BSCall(t, s)$$

So, in a nutshell: We need to determine $s = s(t)$ for every $t \in [0, T]$ so that $C(0) = BSCall(t, s)$. This $s(t)$ then is the break-even point $BE(t) = s(t)$. It is clear that this equation determines the break-even point, since the sale of the call option at time t needs to yield a payoff that is at least equal to what it costs to buy the call option at time 0. (For $s > s(t)$ we know that $C(0) < BSCall(t, s)$, given the monotony of the Black-Scholes formula in s)

Written explicitly, the conditional equation reads

$$C(0) = s \cdot \mathcal{N}\left(\frac{\log\left(\frac{s}{K}\right) + \left(r + \frac{\sigma^2}{2}\right)(T - t)}{\sigma\sqrt{T - t}}\right) - \\ -K \cdot e^{-r(T-t)} \cdot \mathcal{N}\left(\frac{\log\left(\frac{s}{K}\right) + \left(r - \frac{\sigma^2}{2}\right)(T - t)}{\sigma\sqrt{T - t}}\right) \tag{4.6}$$

From this equation, we need to determine for every $t \in [0, T]$ the value s that solves the equation. This $s = s(t)$ then is the break-even point $BE(t)$ at time t. It is quite obvious that s cannot be expressed explicitly based on this equation. The equation therefore has to be solved for each t by means of an approximation method.

This is what we are going to do in the following example, and we will then graphically represent the result, i.e. the trajectory of the break-even point over time.

Example

underlying asset S&P500
$S_0 = 2,900$
$r = 0$
$\sigma = 0.15$
$K = 2,900$
$T = \frac{1}{12}$

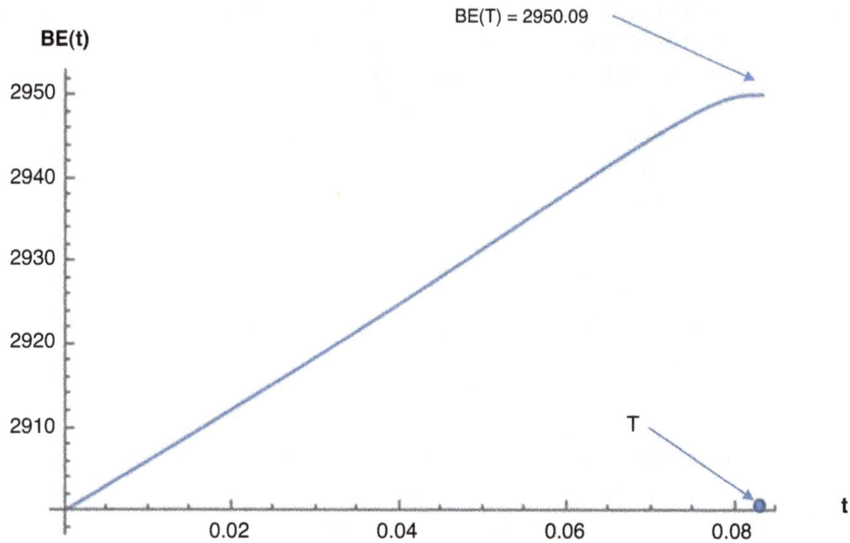

Fig. 4.114 Break-even points over the life of a call option

To write the conditional Eq. (4.6) for the break-even point explicitly, we also need the price $C(0)$ of the option at time 0. As a result of the Black-Scholes formula, $C(0) = 50.09$. The break-even point at time T is thus 2950.09. Solving the Eq. (4.6) by approximation for each $t \in [0, T]$ using Mathematica yields the break-even point curve shown in Fig. 4.114 (in the case of unchanged volatility over the call option's life).

We see an almost linear path over time that only flattens noticeably toward the end of the option's life. One week in, for example, the break-even point is at around 2912 points. So, if the S&P500 moves up to more than 2912 points in the first week of the call option's life, selling the call option (in this setting) can already generate a profit.

In most real-world cases, it is to be expected, however, that as prices of underlying assets rise, their volatility generally declines. Yet decreasing volatility implies lower call option prices. Selling the call option early is therefore less lucrative than when volatility remains unchanged. Consequently, the break-even point will be higher than in the case of unchanged volatility.

If we model volatility as a (deterministic) function of the underlying asset's price, like we did in the previous Sect. 4.28, that is, using $\sigma_t = \sigma \cdot \left(\frac{S_0}{S_t}\right)^a$ for a suitable

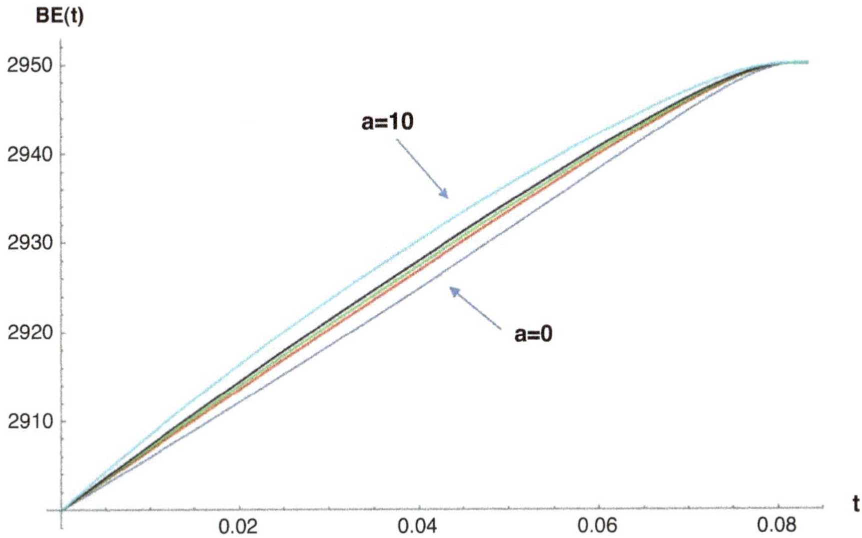

Fig. 4.115 Break-even points at variable volatility

value of a (e.g. $a = 4, 5, 6$), then Eq. (4.6) adapts to

$$C(t) = s \cdot \mathcal{N}\left(\frac{\log\left(\frac{s}{K}\right) + \left(r + \frac{\sigma(s)^2}{2}\right)(T-t)}{\sigma(s)\sqrt{T-t}}\right) - \tag{4.7}$$

$$-K \cdot e^{-r(T-t)} \cdot \mathcal{N}\left(\frac{\log\left(\frac{s}{K}\right) + \left(r - \frac{\sigma(s)^2}{2}\right)(T-t)}{\sigma(s)\sqrt{T-t}}\right)$$

where $\sigma(s) = \sigma \cdot \left(\frac{S_0}{s}\right)^a$.

If we again solve this equation for every t with respect to s, we get the break-even points for the case where volatility changes with the price of the underlying asset. Figure 4.115 depicts the break-even points for $a = 0$ (constant volatility, blue, lowest line), $a = 4$ (red line), $a = 5$ (green line), $a = 6$ (black line), and $a = 10$ (turquoise line).

In the case of $a = 10$, the break-even point is at approximately 2917 points after 1 week, at a volatility of 0.14 (as opposed to 2912 in the case of constant volatility of 0.15).

On the website associated with this book, you will find the software to run any amount of further tests like these. See: https://app.lsqf.org/book/pricing-with-black-scholes

The following comparison is also very interesting (although the result is quite obvious): Suppose we buy call options with the same basic parameters as in the

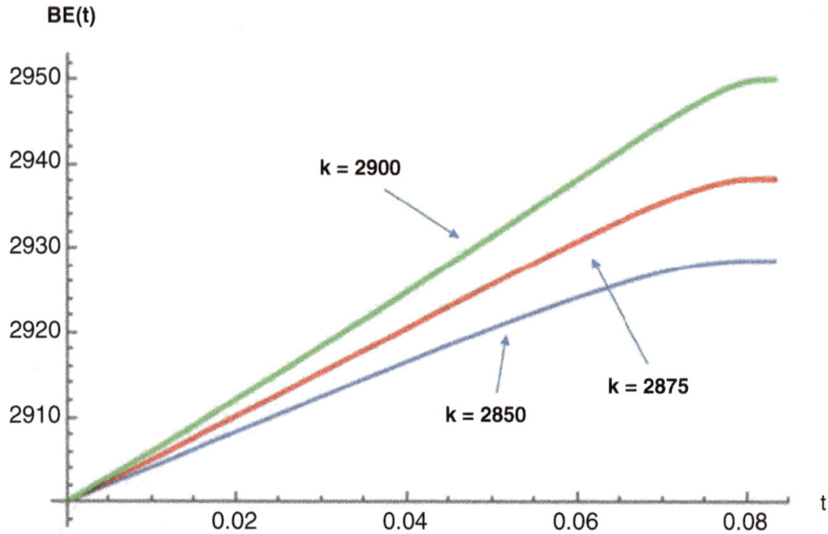

Fig. 4.116 Break-even points, constant volatilities, various strike prices

above example, also with a life of 1 month but with different strikes K = 2850, 2875, and 2900. How do the break-even point trajectories (assuming constant volatility) for these three options compare? The result is shown in Fig. 4.116. The green line represents the case for K = 2900, the red line the case for K = 2875, and the blue line the case for K = 2850.

The higher the strike price of the call option is, the higher the break-even point is, at any time. (For example, when buying a call option with strike price K = 2850, the break-even point is at approximately 2908 points after 1 week.) This is clear in that the break-even point at time T is precisely $K + C(0) = K+$ (intrinsic value of the option) $+$ (time value of the option at time 0) $= S_0+$ (time value of the option at time 0).

Remember however, as noted earlier, that the time value of the call option gets smaller the higher the difference between the strike price and S_0 is. This means, as is easily understood, that even at earlier times, the break-even points for options with a smaller strike price are lower than the break-even points for options with a higher strike price.

It may be helpful to compare the profit functions of the three call options again at this point (Fig. 4.117).

To summarize: The lower the strike price, the lower the break-even point (in T, yet also at earlier times) and the higher the potential profits but also the potential losses from this option (and also: the more money is needed to buy the option).

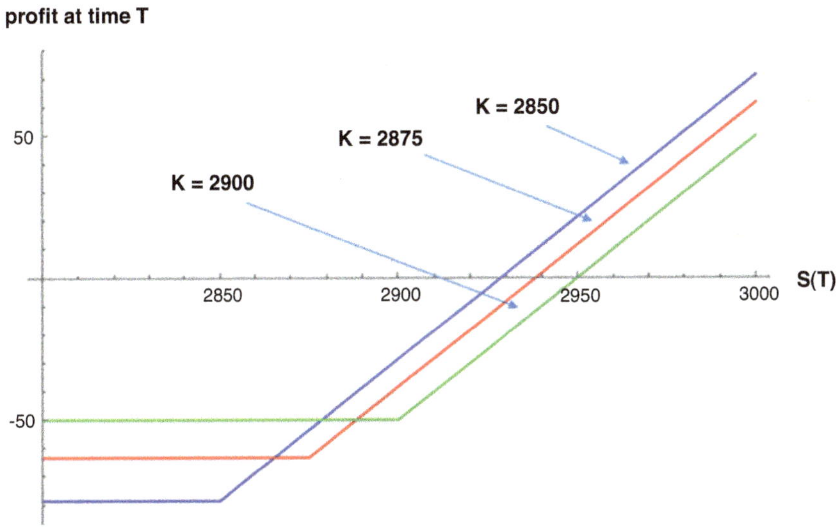

Fig. 4.117 Profit functions of the three call options

To conclude this simple analysis, we ask ourselves the following question:

Is it possible to make a profit with a long position in a call option even if the price falls over the life of the option?
The answer is: "Yes, that is indeed possible".

We want to illustrate this statement with just one example. Again, let's start from the situation in our previous example:

Suppose that the S&P500 at time 0 is again $S_0 = 2900$
Strike price of the call option $K = 2900$
Life of option 1 month, $T = \frac{1}{12}$
$r = 0$
Volatility at time $t = 0$ is $\sigma = 0.15$.

For the price of the option at time 0, we again have $C(0) = 50.09$.

We now assume that the volatility of the S&P500 has risen after 1 week (e.g. due to falling prices) (e.g. from 0.15 to 0.175, or 0.2, or 0.225, or 0.25). Figure 4.118 shows the difference between the call option price $C(t)$ after 1 week and the initial option price $C(0)$ as a function of the price of the S&P500 at time t (1 week after we started out). If this difference is positive, then selling the call option after 1 week yields a profit, even though the price of the underlying asset has fallen.

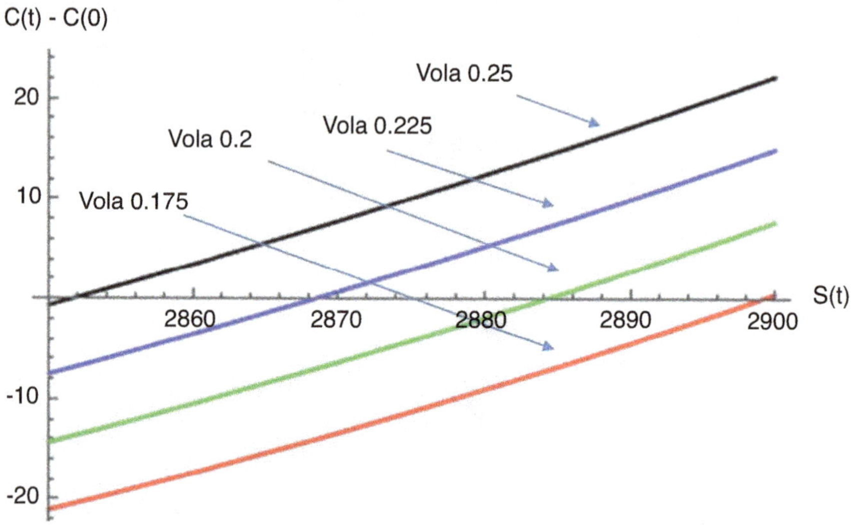

Fig. 4.118 Difference between call price after 1 week and initial call price at increased volatility as a function of the underlying asset's price

This means, for example, if volatility rises to 0.2 in the course of a week (green line) and the price of the S&P500 is above 2884 points, selling the call option will yield a profit (even though the price of the underlying asset has fallen).

4.30 Analysis of the Black-Scholes Price of Put Options

In Sects. 4.26–4.29, we dealt extensively with the analysis of the Black-Scholes price of call options. In this section and in the next, we are, by analogy with that, going to derive the **most essential properties of the fair price of put options**. We will purposely keep it brief, as we can mostly apply very similar arguments as in the case of call options or proceed in the same way as we did for our analysis of call option prices. Some of the following results can also easily be derived directly from the results for the call option prices and the put-call parity equation $P(t) = C(t) + K \cdot e^{-r(T-t)} - S(t)$.

Especially the remarks in Sect. 4.27 apply analogously to put options. The specific example of price movements of derivatives in a two-step binomial model where volatility is correlated with the price of the underlying asset was already carried out in Sect. 4.27 for both call and put options.

For a given

- strike price K
- risk-free interest rate r
- remaining life to expiration T

- current point in time t
- volatility of the underlying asset σ over the option's life
- and price S of the underlying asset at time t

the fair price $P(t)$ of a put option with the above parameters is

$$P(t) = e^{-r(T-t)} \cdot K \cdot \mathcal{N}(-d_2) - S(t) \cdot \mathcal{N}(-d_1)$$

where

$$d_1 = \frac{\log\left(\frac{S}{K}\right) + \left(r + \frac{\sigma^2}{2}\right)(T-t)}{\sigma\sqrt{T-t}} \text{ and } d_2 = \frac{\log\left(\frac{S}{K}\right) + \left(r - \frac{\sigma^2}{2}\right)(T-t)}{\sigma\sqrt{T-t}}$$

(i.e. $d_2 = d_1 - \sigma \cdot \sqrt{T-t}$) and $\mathcal{N}(x) := \frac{1}{\sqrt{2\pi}} \int_{-\infty}^{x} e^{-y^2} dy$ (i.e. $\mathcal{N}$ is the distribution function of the standard normal distribution).

In Fig. 4.119 we plotted the evolution of $P(t)$ for the four time points $t_1 = 0$, $t_2 = 0.3$, $t_3 = 0.6$, and $t_4 = 0.9$ for a fixed choice of $K = 100$, $r = 0.01$, $T = 1$, and $\sigma = 0.5$ as a function of different values $S = S(t)$ of the underlying asset at time t (graphs in red colours). For comparison, the payoff function is also shown as a function of $S = S(T)$ (in blue). The uppermost red line reflects the price movements

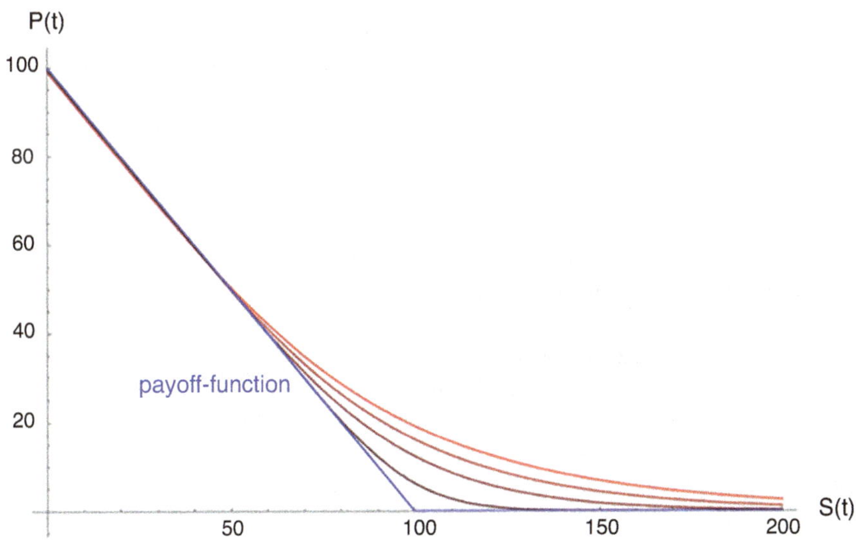

Fig. 4.119 Put option price as a function of the underlying asset's price for four different points in time (assuming a positive interest rate)

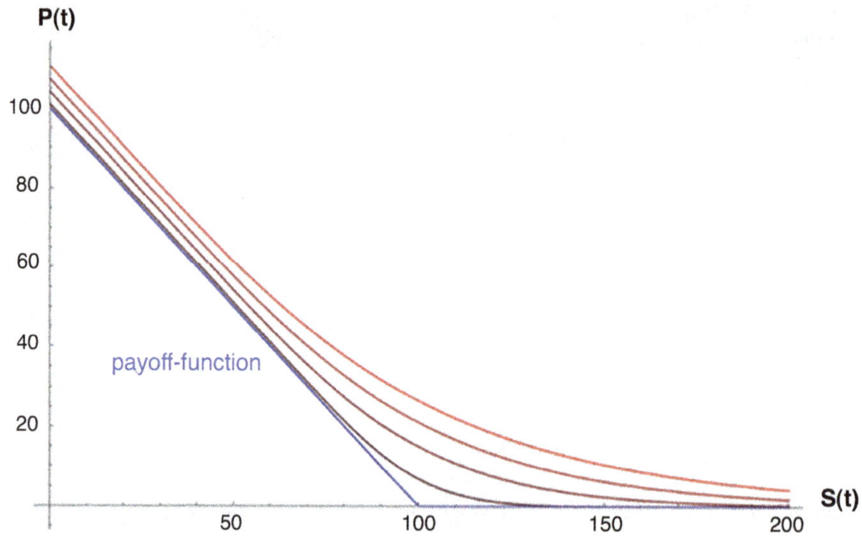

Fig. 4.120 Put option price as a function of the underlying asset's price for four different points in time (assuming a negative interest rate)

for t_1, the second line from the top the price movements for t_2, the third line the price movements for t_3, and the bottom red line the price movements for t_4.

If we run the same example with a negative interest rate $r = -0.1$ (which is an unrealistically strong negative rate but gives us a graphically clearer picture), we see the Plot 4.120:

As noted earlier, it is only in the case of a negative interest rate that the price of the put option will always be above the payoff function.

The intrinsic value of a put option at time t is $\max(K - S(t), 0)$, and the time value $Z(t)$ is given by the equation $P(t) = \max(K - S(t), 0) + Z(t)$. The time value of an option for t going to T converges to 0.

Figure 4.121 shows the time value for the example of Fig. 4.119 and for the example of Fig. 4.120:

Again, we see that the time value is always positive in the case of negative interest rates and that the time value is slightly negative for small $S(t)$ at a positive interest rate. For a given t the time value is at its highest when $S(t)$ is near K.

So, we have seen that the price of a put option, as well as the time value of a put option, decreases (assuming negative interest rates) over time t (i.e. the shorter the remaining time to expiration $T - t$). However, this is not necessarily the case with positive interest rates.

We support this observation by further graphs where we are now going to plot $P(t)$ as a function of time t (for different but fixed values $S = S(t)$ and otherwise the same parameters as above). In the first Fig. 4.122 we consider the case of a positive

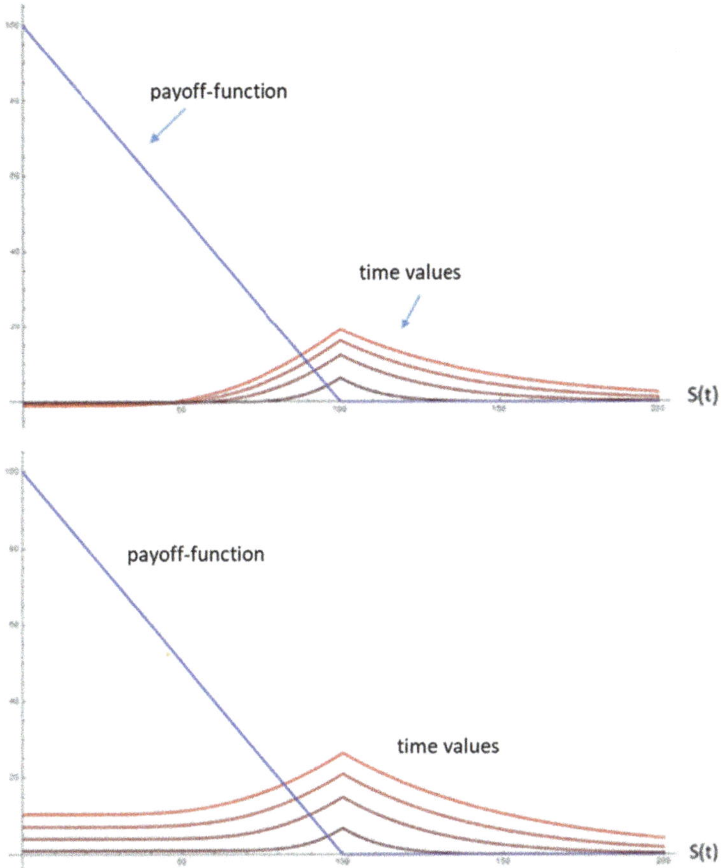

Fig. 4.121 Time value of a put option as a function of the underlying asset's price for four different time points (upper figure, at a positive interest rate; lower figure, at a negative interest rate)

interest rate $r = 0.01$ and in the second Fig. 4.123 the case of a negative interest rate $r = -0.05$.

We now calculate the derivative of $P(t)$ with respect to time t, i.e. $P'(t)$.

We already know that $C'(t) = -\left(S \cdot \phi\left(d_1\right) \cdot \frac{\sigma}{2\sqrt{T-t}} + K \cdot r \cdot e^{-r(T-t)} \cdot \mathcal{N}\left(d_2\right)\right)$.

From the put-call parity equation, we obtain $P(t) = C(t) + K \cdot e^{-r(T-t)} - S$ (the value of S is fixed in this case, since we want to calculate the derivative of P with respect to t with constant S) and thus

$$\begin{aligned}\boldsymbol{P'(t)} &= C'(t) + K \cdot r \cdot e^{-r(T-t)} = \\ &= -\left(S \cdot \phi\left(d_1\right) \cdot \frac{\sigma}{2\sqrt{T-t}} + K \cdot r \cdot e^{-r(T-t)} \cdot \mathcal{N}\left(d_2\right)\right) +\end{aligned}$$

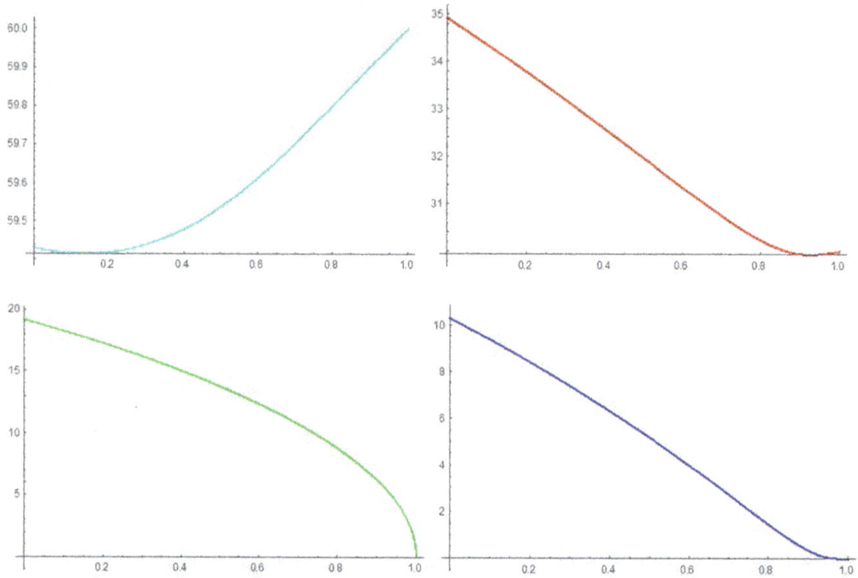

Fig. 4.122 Put option price over time for $K = 100$ and for $S(t)$ of 40 (turquoise), 70 (red), 100 (green), and 130 (blue), $r = 0.01$

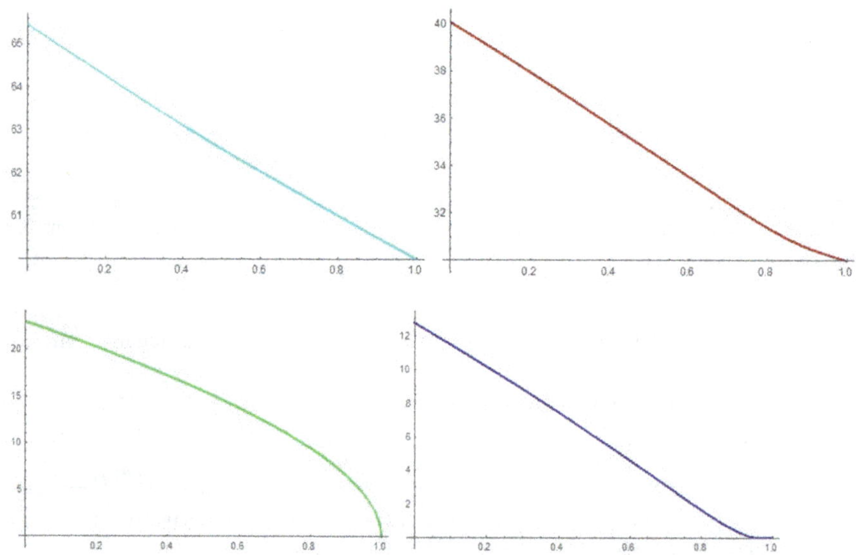

Fig. 4.123 Put option price over time for $K = 100$ and for $S(t)$ of 40 (turquoise), 70 (red), 100 (green), and 130 (blue), $\boldsymbol{r = -0.05}$

$$+K \cdot r \cdot e^{-r(T-t)} =$$

$$= -S \cdot \phi(d_1) \cdot \frac{\sigma}{2\sqrt{T-t}} + K \cdot r \cdot e^{-r(T-t)} \cdot (1 - \mathcal{N}(d_2)) =$$

$$= \boldsymbol{-S \cdot \phi(d_1) \cdot \frac{\sigma}{2\sqrt{T-t}} + K \cdot r \cdot e^{-r(T-t)} \cdot \mathcal{N}(-d_2)}$$

In the case of a negative interest rate r, both summands are negative, so $P'(t)$ is always negative in this case and **$P(t)$ therefore strictly monotonically decreasing in t.**

So as time progresses, with otherwise unchanged parameters and negative r, the price of a put option falls.

We are now going to analyse the dependence of the put option's price on volatility. For this purpose, we will first look at two graphs.

In the following representation, we plotted the evolution of P for the four different volatilities $\sigma_1 = 0, \sigma_2 = 0.3, \sigma_3 = 0.6$, and $\sigma_4 = 0.9$ as a function of different values $S = S(0)$ of the underlying asset at time 0 for a fixed choice of $K = 100, r = 0.05$ (we chose this higher interest rate to better illustrate the evolution) as well as $T = 1$ and $t = 0$ (graphs in red). For comparison, the payoff function is also shown as a function of $S = S(T)$ (in blue). The lowermost price line in Fig. 4.124 reflects the price movements for σ_1, the second line from below the price movements for σ_2, the third line the price movements for σ_3, and the uppermost red line the price movements for σ_4.

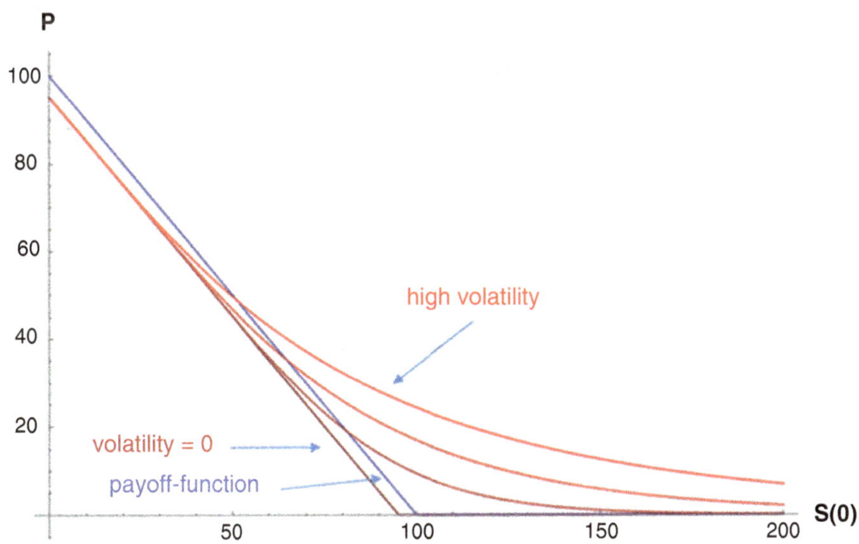

Fig. 4.124 Put option price as a function of the underlying asset's price for four different volatilities (assuming a positive interest rate)

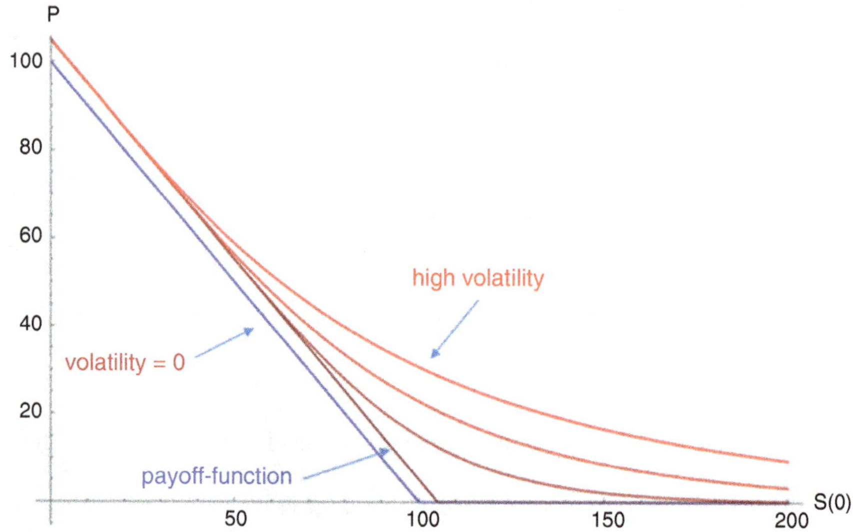

Fig. 4.125 Put option price as a function of the underlying asset's price for four different volatilities (assuming a negative interest rate)

For a negative r ($r = -0.05$), the corresponding graph looks as follows (Fig. 4.125):

So, the price of a put option evidently increases as volatility σ increases. We support this observation by further Graph 4.126 where we are now going to plot P as a function of volatility σ (for different but fixed values $S = S(0)$ and otherwise the same parameters as above). (Here, we limit ourselves to the case of a positive interest rate.)

We now want to discuss the pricing formula for put options as a function of volatility σ and prove the monotony analytically. In the following, $P'(\sigma)$ denotes the derivative of $P(\sigma)$ with respect to σ. Determining $P'(\sigma)$ is very simple now, as it immediately follows from the put-call parity equation $P(\sigma) = C(\sigma) + K \cdot e^{-r(T-t)} - S$ (we write P and C as a function of σ here) by differentiating with respect to σ, that $P'(\sigma) = C'(\sigma)$.

And from Sect. 4.26, we already know that $C'(\sigma) = K \cdot e^{-r(T-t)} \cdot \sqrt{T-t} \cdot \phi(d_2)$. From that follows that $P'(\sigma) = K \cdot e^{-r(T-t)} \cdot \sqrt{T-t} \cdot \phi(d_2)$. This expression is always positive and the monotonic growth of P in σ follows.

Finally, we look at the dependence of the put price P on the risk-free interest rate r.

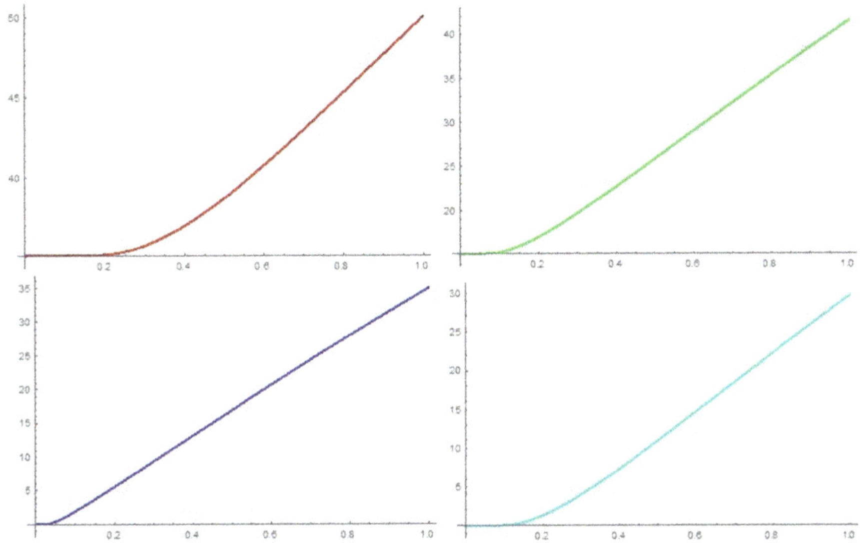

Fig. 4.126 Evolution of the put option price with volatility for $K = 100$ and for $S(t)$ of 60 (red), 80 (green), 100 (blue), and 120 (turquoise), $r = 0.05$

From 4.26 we already know that $C'(r) = \tau \cdot K \cdot e^{-r\tau} \cdot \mathcal{N}(d_2)$. Here we have set $T - t = \tau$. The put-call parity equation is $P(r) = C(r) + K \cdot e^{-r \cdot \tau} - S$ and thus we get

$$\begin{aligned} \boldsymbol{P'(r)} &= C'(r) - \tau \cdot K \cdot e^{-r\tau} = \tau \cdot K \cdot e^{-r\tau} \cdot \mathcal{N}(d_2) - \tau \cdot K \cdot e^{-r\tau} = \\ &= \tau \cdot K \cdot e^{-r\tau} \cdot (\mathcal{N}(d_2) - 1) = \boldsymbol{-\tau \cdot K \cdot e^{-r\tau} \cdot (\mathcal{N}(-d_2))} \end{aligned}$$

So, $P'(r)$ is always negative and **the price of a put option** is therefore always **strictly monotonically decreasing in r**.

We conclude this brief analysis of the fair price of put options with the corresponding graph for the dependence of the put price on r.

Again we see only a moderate dependence on the interest rate r. It is only when the interest rate gets much higher ($r = 0.1$ in Fig. 4.127) that the put prices change noticeably.

4.31 Break-Even for Put-Only Strategies

We are going to analyse the break-even points over time for a put option in the same way as we did for call options in Sect. 4.29. We can keep this brief as the details of the procedure were described in principle in 4.29.

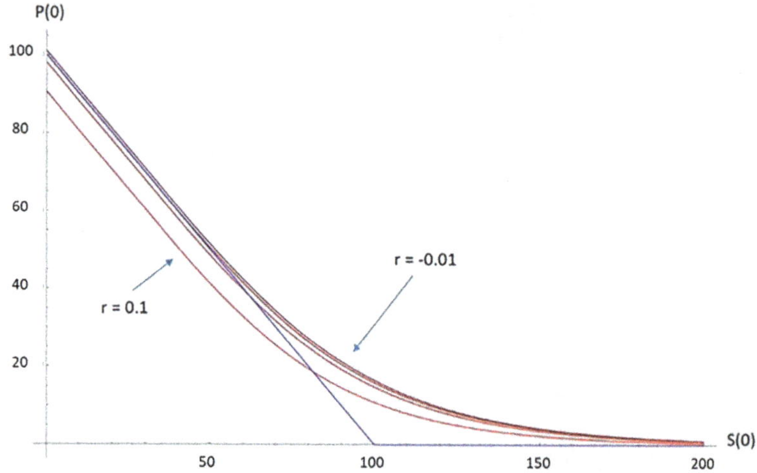

Fig. 4.127 Put option price as a function of the underlying asset's price for four different risk-free interest rates $r = -0.01$, $r = 0$, $r = 0.02$ und $r = 0.1$

The purchase of a long put option with expiration T, strike K, and price $P(0)$ on an underlying asset with the initial value S_0 generates a payoff function for time T of $\max(K - S(T), 0)$ and a profit function for time T of $\max(K - S(T), 0) - P(0)$.

The break-even point for the expiration date T is therefore $K - P(0)$ (see Fig. 4.128). So we see that the long position in the put option generates a profit only if the price of the underlying asset at time T is below the value of $K - P(0)$.

For any time $t \in [0, T]$, the break-even point (assuming constant volatility over the option's life) can be calculated from the following equation (where $BSPut(t, s)$ denotes the put option's fair price at time t for the value s of the underlying asset, i.e. $BSPut(t, s)$ is the Black-Scholes formula for put options with the fixed parameters K, r and σ and the variable parameters t and s):

$$P(0) = BSPut(t, s)$$

So, in a nutshell: We need to determine $s = s(t)$ for every $t \in [0, T]$ so that $P(0) = BSPut(t, s)$. This $s(t)$ then is the break-even point $BE(t) = s(t)$.

It is clear that this equation determines the break-even point, since the sale of the put option at time t needs to yield a payoff that is at least equal to what it costs to buy the put option at time 0. (For $s < s(t)$, we know that $P(0) < BSPut(t, s)$, given the monotony of the put price in s.)

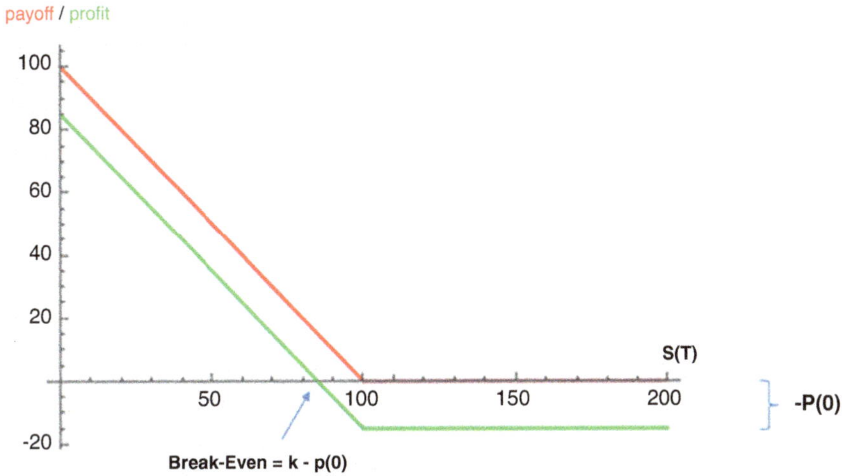

Fig. 4.128 Payoff, profit, and break-even point of a put option

Written explicitly, the conditional equation reads

$$P(0) = s \cdot \mathcal{N}\left(\frac{\log\left(\frac{s}{K}\right) + \left(r + \frac{\sigma^2}{2}\right)(T-t)}{\sigma\sqrt{T-t}}\right) - K \cdot e^{-r(T-t)} \cdot \tag{4.8}$$

$$\cdot \mathcal{N}\left(\frac{\log\left(\frac{s}{K}\right) + \left(r - \frac{\sigma^2}{2}\right)(T-t)}{\sigma\sqrt{T-t}}\right) + K \cdot e^{-r(T-t)} - s$$

From this equation, we need to determine for every $t \in [0, T]$ the value s that solves the equation. This $s = s(t)$ then is the break-even point $BE(t)$ at time t. It is quite obvious that s cannot be expressed explicitly based on this equation. The equation therefore has to be solved for each t by means of an approximation method. This is what we are going to do, using Mathematica in the following example, and we will then graphically represent the result, i.e. plot the break-even points over time.

Example

underlying asset S&P500
$S_0 = 2900$
$r = 0$

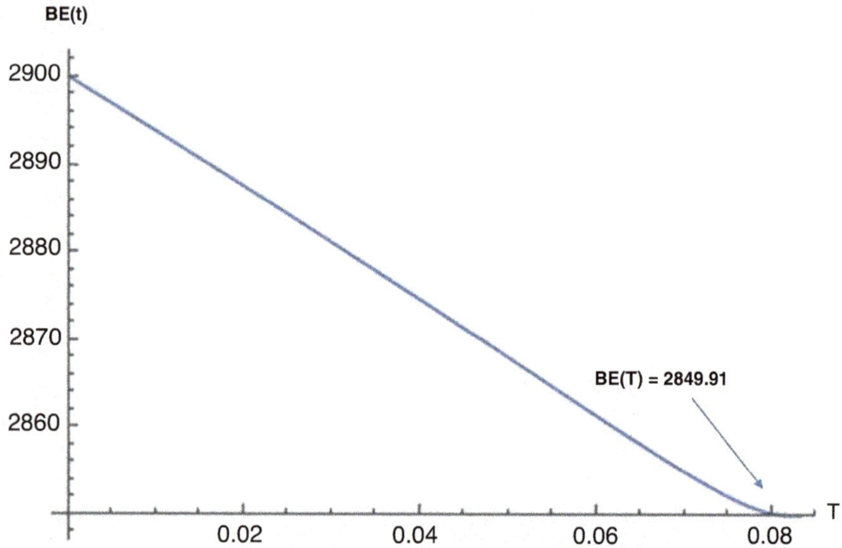

Fig. 4.129 Break-even points over the life of a put option

$\sigma = 0.15$
$K = 2900$
$T = \frac{1}{12}$

To write the conditional Eq. (4.8) for the break-even point explicitly, we also need the price $P(0)$ of the option at time 0. As a result of the Black-Scholes formula, $P(0) = 50.09$. (The price of the put option in this case is equal to the price of the call option. This also follows from the put-call parity, since we chose $r = 0$.) The break-even point at time T is thus 2849.91.

Solving Eq. (4.8) by approximation for each $t \in [0, T]$ using Mathematica yields the break-even point curve shown in Fig. 4.129 (in the case of constant volatility over the option's life).

We see an almost linear path over time that only flattens noticeably toward the end of the option's life. One week in, for example, the break-even point is at around 2885 points. So, if the S&P500 drops to under 2885 points in the first week of the option's life, selling the put option (in this setting) can already generate a profit.

If we model volatility as a (deterministic) function of the underlying asset's price, again like we did in Sect. 4.28, that is, using $\sigma_t = \sigma \cdot \left(\frac{S_0}{S_t}\right)^a$ for a suitable value of

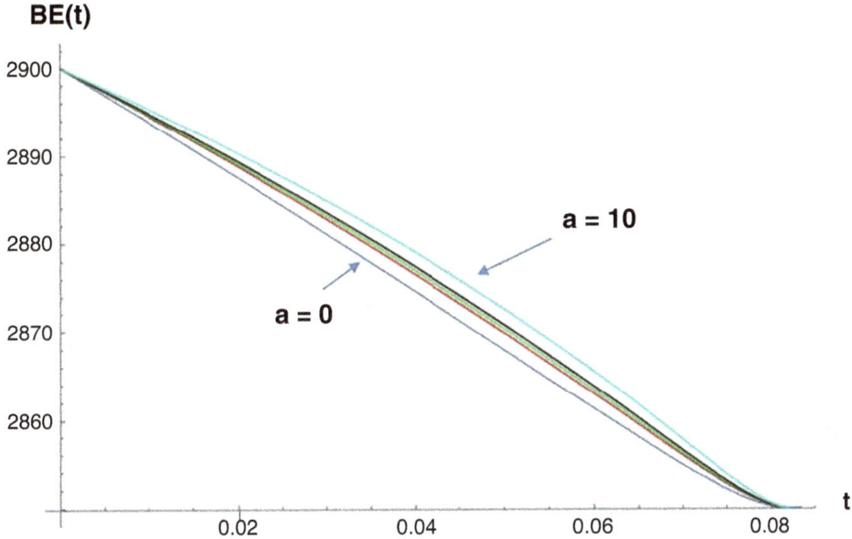

Fig. 4.130 Break-even points at variable volatility

a (e.g. $a = 4, 5, 6$), then the Eq. (4.8) adapts to

$$P(0) = s \cdot \mathcal{N}\left(\frac{\log\left(\frac{s}{K}\right) + \left(r + \frac{\sigma(s)^2}{2}\right)(T-t)}{\sigma(s)\sqrt{T-t}}\right) - K \cdot e^{-r(T-t)} \cdot \tag{4.9}$$

$$\cdot \mathcal{N}\left(\frac{\log\left(\frac{s}{K}\right) + \left(r - \frac{\sigma(s)^2}{2}\right)(T-t)}{\sigma(s)\sqrt{T-t}}\right) + K \cdot e^{-r(T-t)} - s$$

where $\sigma(s) = \sigma \cdot \left(\frac{S_0}{s}\right)^a$.

If we again solve this equation for every t with respect to s, we get the break-even points for the case where volatility changes with the price of the underlying asset. Figure 4.130 plots the path of the break-even points for $a = 0$ (constant volatility, blue, lowest line), $a = 4$ (red line), $a = 5$ (green line), $a = 6$ (black line), and $a = 10$ (turquoise line).

In the case of $a = 10$, the break-even point is at approximately 2890 points after 1 week, at a volatility of 0.156 (as opposed to 2885 at constant volatility of 0.15).

On the website of this book, you will find the software to run any number of further tests like these.

Suppose we buy put options with the same basic parameters as in the above example, also with a life of 1 month but with different strikes $K = 2900, 2925$, and

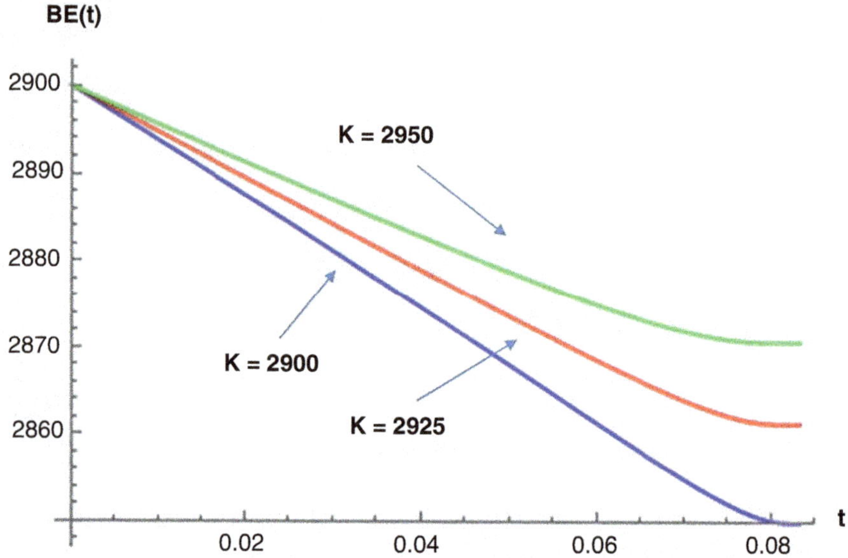

Fig. 4.131 Break-even points, constant volatilities, various strike prices

2950. How do the break-even point curves (assuming constant volatility) for these three options compare?

The result is shown in Fig. 4.131. The blue line represents the case for K = 2900, the red line for K = 2925, and the green line for K = 2950.

The higher the put option's strike price is, the higher the break-even point is at any time. (For example, when buying a put option with strike price $K = 2950$, the break-even point after 1 week is at approximately 2892 points.) This is clear in that the break-even point at time T is precisely $K - P(0) = K$—(intrinsic value of the option)—(time value of the option at time 0) $= S_0-$ (time value of the option at time 0). Remember however, as noted earlier, that the time value of the put option decreases as the distance between the strike price and S_0 increases.

This means, as is easily understood, that even at earlier times, the break-even points for options with a smaller strike price are smaller than the break-even points for options with a higher strike price.

It may be helpful to compare the profit functions of the three put options again at this point (Fig. 4.132).

To summarize: The higher the strike price, the higher the break-even point (in T, yet also at earlier times) and the higher the potential profits but also the potential losses from this option (and also, the more money is needed to buy the option).

Finally, just like we did for the case of call options, we ask ourselves the following question: Is it possible to make a profit with a long position in a put option even if the price rises over the option's life?

The answer is: "Under reasonably normal circumstances, that's not possible".

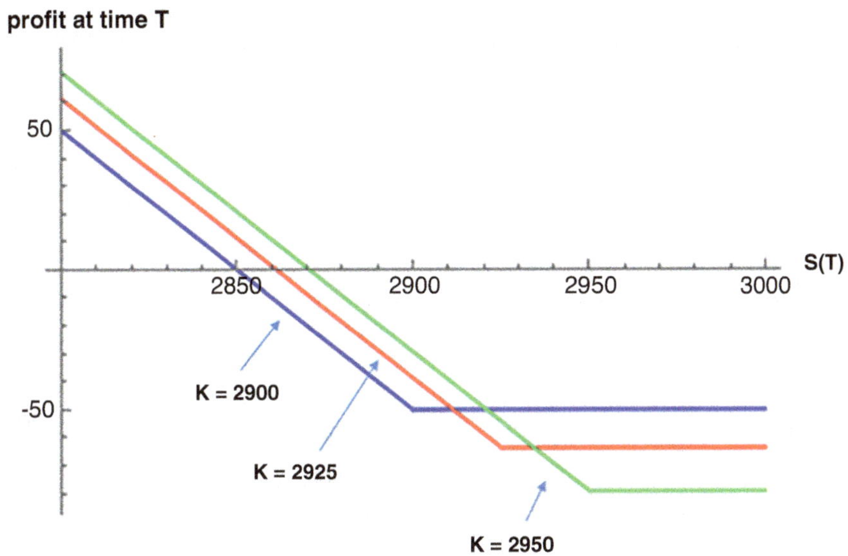

Fig. 4.132 Profit functions of the three put options

The price of a put option moves down as the price of the underlying asset moves up. When an underlying asset increases in price, its volatility generally falls and with that the price of the put option. To make a profit, the price of the put option would have to rise, however.

Using the analysis software on our website, readers can analyse further options combinations of any kind with regard to their price movements and their break-even points. In the following sections, we are going to briefly analyse just a few more basic strategies. Later we are going to analyse and discuss more complex trading strategies in detail, on the basis of realistic assumptions and partly on the basis of historical options data.

4.32 Analysis of the Price Paths of a Few Other Basic Option Strategies: Short Iron Butterfly

A Short Iron Butterfly is created from a short call and a short put with the same near at-the-money strike price K_S as well as a long call with strike $K_{LC} > K_S$ and a long put with strike $K_{LP} < K$. All options have the same expiration date. The distance between the two long position strikes and the short position strike can be equal, but doesn't have to be equal.

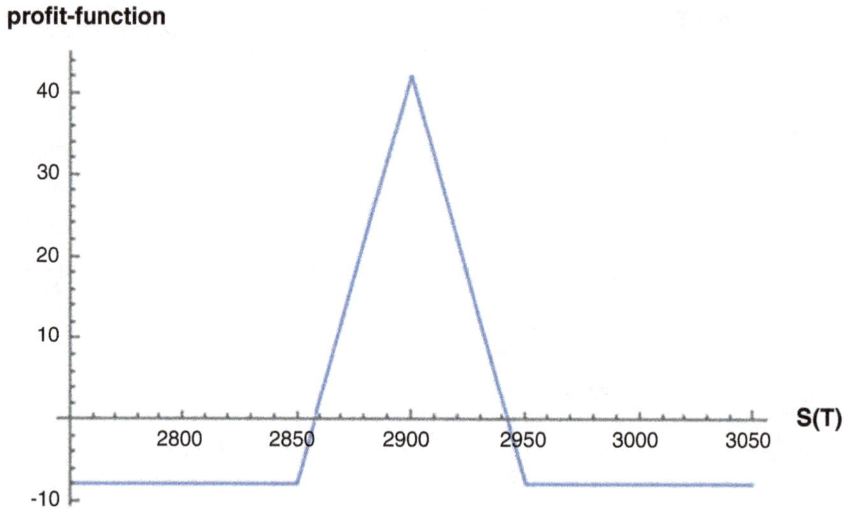

Fig. 4.133 Profit function for Short Iron Butterfly

The profit function typically looks like the following Fig. 4.133:

In this graph and below, we use the same parameters as we used for the numeric examples above:

underlying asset S&P500
$S_0 = 2900$
$r = 0$
$\sigma = 0.15$
$T = \frac{1}{12}$,

We are interested specifically in the strike prices $K_S = 2900$, $K_{LC} = 2950$, and $K_{LP} = 2850$ and assume options with price paths following the Black-Scholes formula.

The main motivation for entering into such a strategy is to be able to lock in some of the initially collected option premiums in the hope that the underlying asset's price will move only marginally over time (i.e. price of the underlying asset at expiration near the short strike K_S). The total premium received on initiating the position is

$$\text{Premium}(0) = P_S(0) + C_S(0) - P_L(0) - C_L(0).$$

P_S, C_S, P_L, and C_L denote the prices of the short put, short call, long put, and long call positions.

The maximum profit that can potentially be made with this strategy is precisely this premium (in the event that the price of the underlying asset is exactly K_S at expiration). In this specific example, that would be $50.09 + 50.09 - 29.41 - 26.61 = 42.16$ dollars.

The potential loss from the strategy is clearly limited. If the underlying asset is below K_{LP} at time T, the loss would be "Premium(0) $-$ $(K_S - K_{LP})$". If the underlying asset is above K_{LP} at time T, the loss would be "Premium(0) $-$ $(K_{LC} - K_S)$".

In our specific example, this would mean a maximum potential loss of $42.16 - 50 = -7.84$ dollars in both cases.

The break-even points of the strategy are K_S+ Premium(0) (= 2942.16 in our example) or K_S- Premium(0) (= 2857.84 in our example), respectively.

Above we said that the main motivation for running this strategy is the investor's hope that the underlying asset's price will be close to the short strike K_S at expiration T. However, the investor can also expect to possibly make a profit if *at some point* during the life of the options (later than $t = 0$ but not necessarily at $t = T$), the price of the underlying asset happens to be K_S or near K_S.

In our example, this means:

If during the life of the options, the S&P500 happens to take a value near 2900 at a (noticeably) later time than $t = 0$, the investor can expect to make a profit by closing the positions right then.

This is illustrated in the following Graph 4.134:

The blue curve represents the profit function of the strategy (options combination) at expiration.

The purple curve—slightly below—represents the price function of the strategy 3 days before expiration.

The turquoise curve represents the price function of the strategy 1 week before expiration.

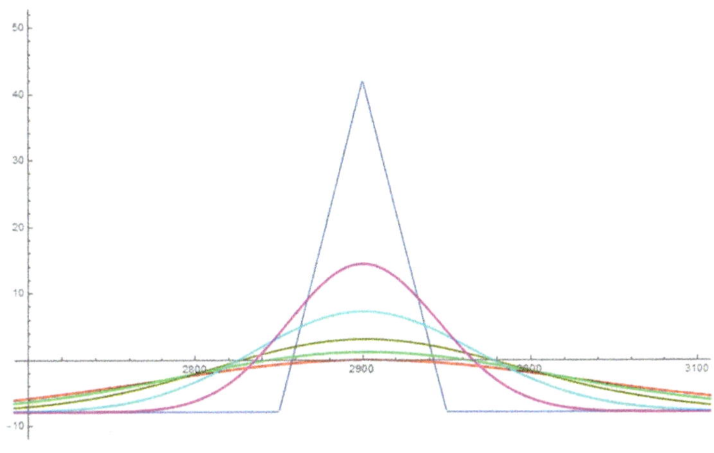

Fig. 4.134 Price curves of Short Iron Butterfly strategy, constant volatility

The olive green curve represents the price function of the strategy 2 weeks before expiration.
The light green curve represents the price function of the strategy 3 weeks before expiration.
The red curve represents the price function of the strategy at time $t = 0$.

Volatility was assumed to be constant throughout.

Let us look at the turquoise curve first. The intersections of this curve with the x-axis (break-even points) are at approximately 2826 and 2977, respectively. Thus, if the S&P500 takes a value between 2826 and 2977 1 week before expiration, a profit can be made by closing the positions. However, this profit is barely more than 7 dollars at best.

We see on all curves (except the red one, of course) that the profit range of these price curves (as from approximately 1 week after opening the positions) includes the profit range of the profit function at expiration.

This means: If the S&P500 happens to take a value somewhere between approximately 2858 and 2942 even just once after the first week of entering the positions, then these positions can be closed for a profit.

Incidentally, the curves of the break-even points over time show a very interesting form. This is shown in Fig. 4.135.

On opening the strategy, the break-even points are of course exactly at the strike price. The profit zone then widens over approximately two-thirds of the option's life and then narrows again to the break-even points of the payoff function at 2857.84 and 2942.16, respectively.

Figure 4.136 also illustrates the width of the profit range during the 1-month life of the option. This area has its largest width, namely, 154.46, extending from 2824.45 to 2979, at approximately 10–11 days before expiration.

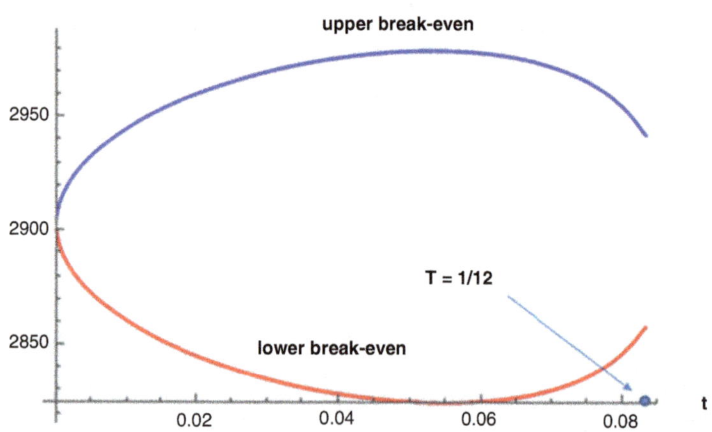

Fig. 4.135 Break-even points, Short Iron Butterfly

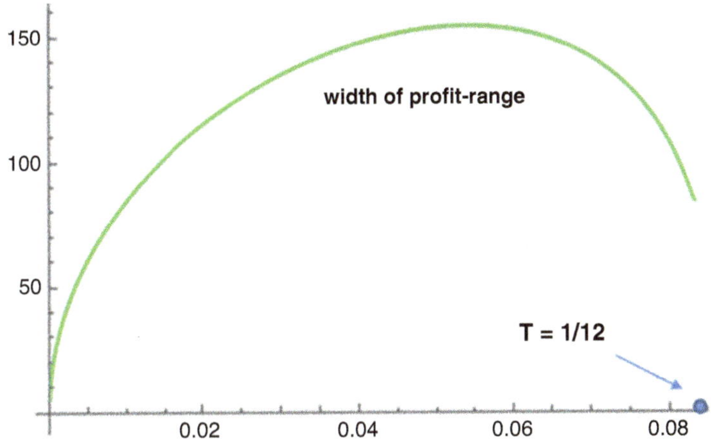

Fig. 4.136 Profit range width as a function of time, Short Iron Butterfly

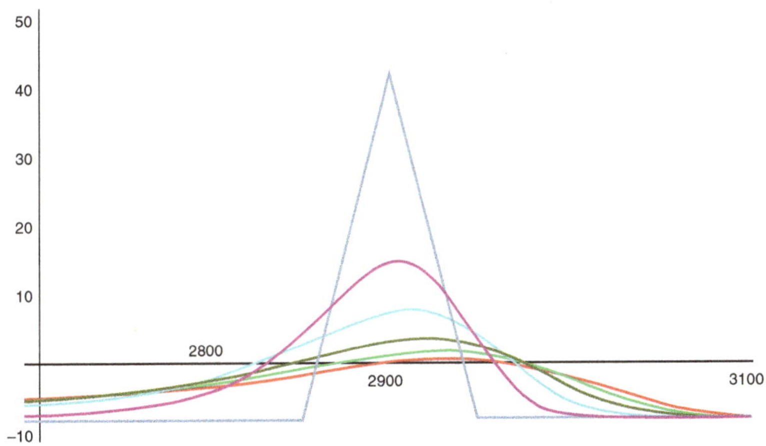

Fig. 4.137 Price curve Short Iron Butterfly strategy, with negatively correlated volatility

If we proceed from the more realistic assumption of a negative correlation between the underlying asset's volatility and price, we get a slightly different but essentially similar picture. Figure 4.137 again shows the situation where we assume a correlation between volatility and the price of the underlying asset in the form $\sigma(s) = \sigma \cdot \left(\frac{S_0}{s}\right)^a$ (with $a = 10$ in this case).

It is interesting to see here that there exists a profit range of positive width already at time $t = 0$. For illustration, we show the red price curve for $t = 0$ from Fig. 4.137 as a standalone graph in Fig. 4.138.

The graphic is to be understood as follows: If for example, the S&P500 were to jump to 2940 points immediately after opening the strategy, with a simultaneous

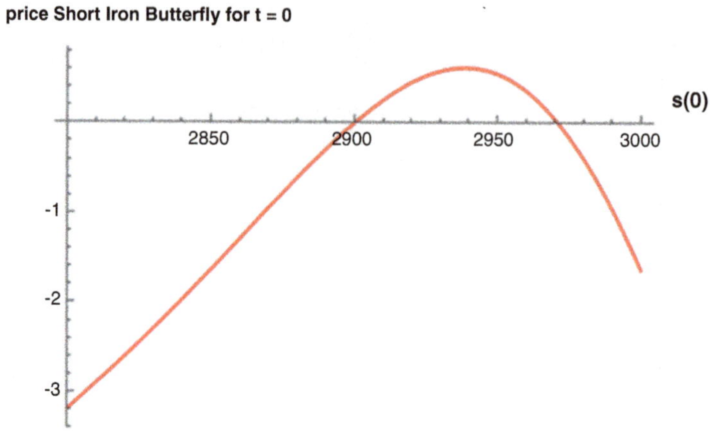

Fig. 4.138 Price curve Short Iron Butterfly for $t = 0$ with negatively correlated volatility

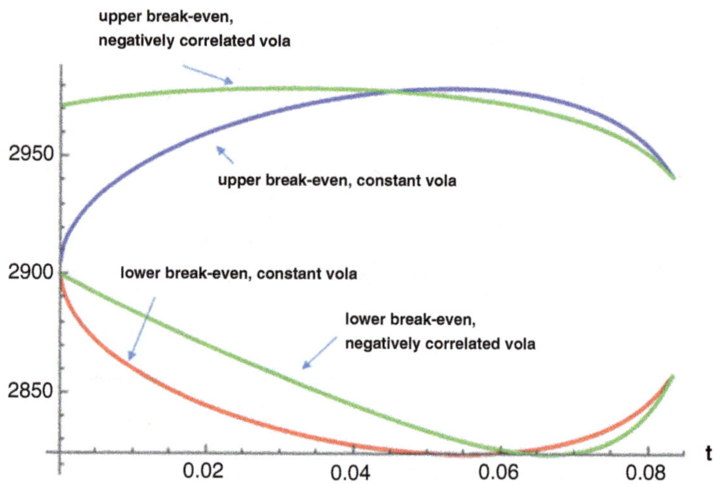

Fig. 4.139 Break-even points, Short Iron Butterfly (constant volatility, blue/red; negatively correlated volatility, green)

decrease in volatility, instantly closing the options contracts would yield a profit of around 0.80 dollars. The lower break-even point is 2900, of course. Figure 4.137 also illustrates that the upper break-even point remains in the range between 2970 and 2980 until close to the expiration date and only falls back to 2942 at expiration. The curves of the upper and lower break-even points are reproduced once more in Fig. 4.139 (green lines) and compared with the evolution of the break-even points in the case of constant volatility (blue and red lines).

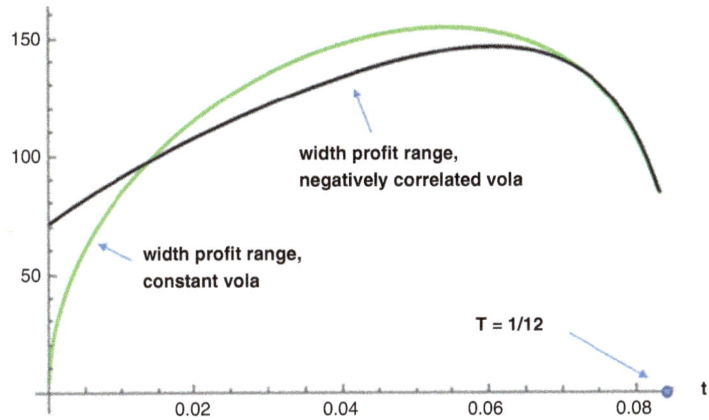

Fig. 4.140 Comparison of profit range width, Short Iron Butterfly, constant volatility vs. negatively correlated volatility

For the analysis of this example, we have chosen fixed parameters for the strike prices of the two long positions, their volatility and expiration T (compare Fig. 4.140). To get a deeper sense of the behavior and possible applications of a Short Iron Butterfly and the effects of longer expiration dates, other volatilities, and other (not necessarily symmetrical) long levels, we recommend that these analyses be carried out and thought through for many other parameter choices too. This can be done using the programs on our website. See: https://app.lsqf.org/optionskombinationen

4.33 Analysis of the Price Curves of a Few Other Basic Option Strategies: Naked Short Butterfly

A Naked Short Butterfly is created from a short call and a short put with the same near at-the-money strike price K_S and the same expiration date. The profit function typically looks like the subsequent Fig. 4.141:

In this graph and below, we use the same parameters as we used for the numeric examples in our Short Iron Butterfly analysis:

underlying asset S&P500
$S_0 = 2900$
$r = 0$
$\sigma = 0.15$
$T = \frac{1}{12}$,
we are interested specifically in the strike price $K_S = 2900$
and assume options with price paths following the Black-Scholes formula.

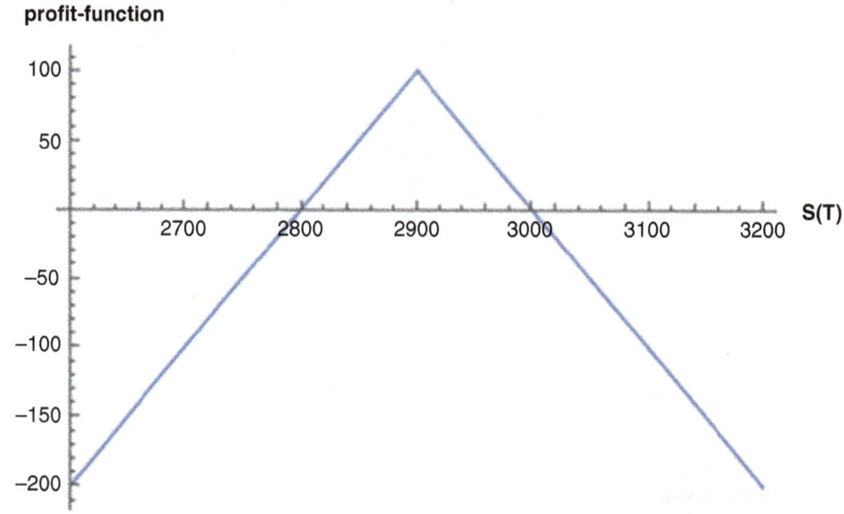

Fig. 4.141 Profit function Naked Short Butterfly

Again the main motivation for entering into such a strategy is to be able to retain part of the initially received option premiums in the hope that the underlying asset's price will move only slightly during the life of the options (i.e. price of the underlying asset at expiration near the short strike K_S). The total premium received on initiating the position in this case is

$$\text{Premium}(0) = P_S(0) + C_S(0)$$

and is therefore significantly higher than in the case of the Short Iron Butterfly.

The maximum profit that can potentially be made with this strategy is precisely this premium (in the event that the price of the underlying asset is exactly K_S at expiration). In this specific example, it would be $50.09 + 50.09 = 100.18$ dollars (compared to 42.16 dollars for the Short Iron Butterfly).

However: The potential loss from the strategy is now unlimited!

The break-even points of the strategy are $K_S +$ Premium(0) ($= 3000.18$ in our example) and $K_S -$ Premium(0) ($= 2799.82$ in our example), respectively.

Again, the investor can potentially expect to make a profit if *at some point* during the life of the options (later than $t = 0$ but not necessarily at $t = T$), the price of the underlying asset happens to be K_S or near K_S.

In our example, this means:

If during the life of the options, the S&P500 happens to take a value near 2900 at a (noticeably) later time than $t = 0$, the investor can expect to make a profit by closing the positions right then.

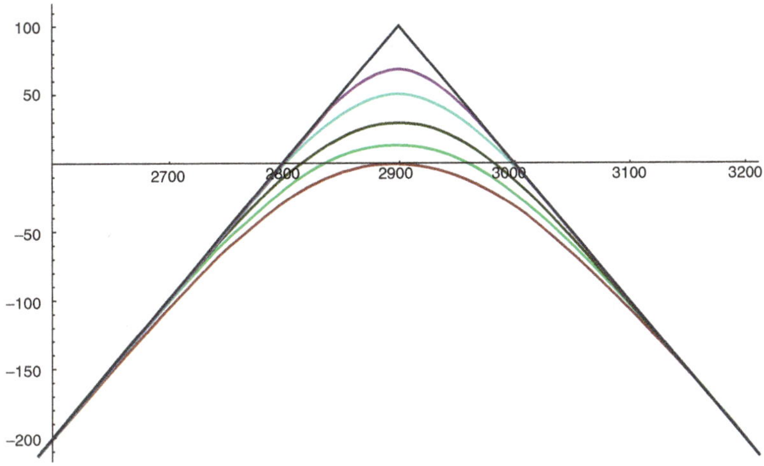

Fig. 4.142 Price curves of a Naked Short Butterfly strategy, constant volatility

This is illustrated in the following graph:

The blue curve represents the profit function of the strategy (options combination) at expiration.
The purple curve—slightly below—represents the price function of the strategy 3 days before expiration.
The turquoise curve represents the price function of the strategy 1 week before expiration.
The olive green curve represents the price function of the strategy 2 weeks before expiration.
The light green curve represents the price function of the strategy 3 weeks before expiration.
The red curve represents the price function of the strategy at time $t = 0$.

For Fig. 4.142, volatility over the option's life was assumed to remain constant throughout, and for Fig. 4.143, volatility over the option's life was assumed to be negatively correlated with the underlying asset's price, with parameter $a = 10$.

Neither graphic holds any major surprises. The way that the break-even points change here is only logical. The width and potential amount of the profit that can be taken (due to the significantly higher initial premium received) are generally significantly larger in the Naked version than in the Iron version. Yet the potential losses can also be considerably higher in the Naked version than in the Iron version.

It stands to reason therefore that in the case of Naked positions, i.e. combinations of options that allow a basically (practically) unlimited (or at least potentially very high) loss, other loss limitation strategies must be used (e.g. stop-loss orders, use of futures contracts for hedging purposes).

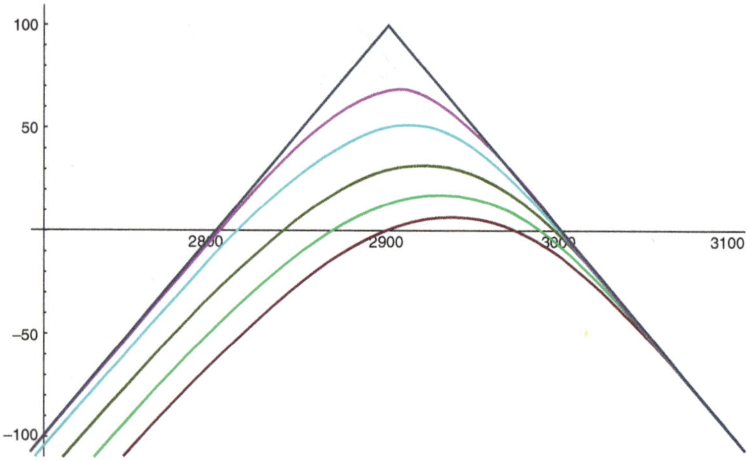

Fig. 4.143 Price curves of a Naked Short Butterfly strategy, negative correlation of volatility with price of underlying asset

A detailed analysis of such a trading strategy and its variants is presented in Volume III Section 3.14.

In the case of a Naked Short Butterfly, one approach to avoid or at least delay losses, if at all possible, is to continuously roll the positions until a profit occurs. This procedure, which is not possible for a Short Iron Butterfly, will be discussed below. For that, we also take into account the required margin. Section 2.15 describes the relevant margin requirements. To make the process clearer and easier to follow, we use a version below that has been slightly simplified compared to the actual requirements:

Margin = *current price of the two options +10% of the strike price*

- We assume that an investment of 100,000$ is available to the investor.
- That investor now enters into a contract of the Naked Short Butterfly strategy—i.e. one short call contract with strike 2900 and one short put contract with strike 2900—and receives the current time value of the two option contracts for a total of approximately 10,000$. This puts the investor's cash assets at 110,000$.
- The required margin is 29,000$ (= 10% of strike) + 10,000$ (= current price of the two options) = 39,000$.
- As long as the S&P500 stays in the range of approximately 2100 to 3700 (= strike +/- 27.5%), the investor's investment is sufficient margin. (For example, if the index were to drop to 2100, the price of the call option would be slightly more than 0$, and the price of the put option would be slightly more than 800$ (referring to the each option's intrinsic value, as the time value is practically equal to 0 in both cases). The required margin then is 29,000$+80,000$ = 109,000$.)
- If the S&P500 is somewhere in the range of approximately 2800–3000 at expiration, then a profit is made through the butterfly strategy.

- If the S&P500 is somewhere in the range between 2100–2800 and 3000–3700 at expiration (time T), then in principle a loss would occur. However, realizing this loss can be delayed by rolling the same option contracts from time T to time $2 \cdot T$.
- For example: If the S&P500 stands at 2700 at time T, the call contract expires worthless. A payoff of 20,000$ would have to be paid for the put. At time T, however, the investor enters into another call and another put, each with strike 2900 and run-time T (all expire at $2 \cdot T$)). For this, the investor receives the time value of the two option contracts plus the intrinsic value of 20,000$ of the put option. With that, the investor's cash at hand increases to slightly more than 110,000$. Margin requirements remain essentially unchanged, meaning that the margin will be sufficient for at least as long as the S&P500 stays in a range of approximately 2100–3700.
- This procedure can be continued for as long as the investor chooses. With every rolling operation, the investor's cash increases slightly by the time values of the two option contracts.
- If at some point, the S&P500 returns to around 2900 points, the strategy can be ended by closing the option contracts. The strategy will have produced a positive profit. This profit was achieved in that, *at some point in the future* (not necessarily in the time range $[0, T]$), the S&P500 moved back into the range it had when the strategy was initiated.

The reason why this procedure is **not possible in the case of a Short Iron Butterfly** is that, to buy the long positions required for each rolling operation, the trader needs to pay the time value of these long positions, and (since the S&P500 is far from the short strike upon rolling) that time value can be significantly higher than the time value received for the short positions.

4.34 Excursus: Brief Remark on the "Asymmetry of Call and Put Prices"

It is often erroneously thought that call options on stocks or stock indices with a strike of $S_0 + Y$ have substantially the same prices as put options with a strike of $S_0 - Y$. (Here, S_0 denotes the current price of the underlying asset and Y denotes any kind of positive or negative deviation from S_0.) This misconception may then lead to misjudgments in that people assume symmetry of certain trading strategies.

This incorrect assumption is straightened out by simply looking at the prices of S&P500 options with approximately 1 month to expiration (see Fig. 4.144) or approximately 1 year to expiration (see Fig. 4.145) on 24 September 2018. The S&P500 stood at 2925 points at the time of these quotations.

Calls		Puts
76.30 x 78.80	2860	12.70 x 13.60
72.10 x 74.50	2865	13.40 x 14.30
67.70 x 70.10	2870	14.20 x 15.10
63.80 x 66.10	2875	15.00 x 15.90
59.70 x 62.00	2880	15.90 x 16.80
55.60 x 57.80	2885	16.90 x 17.80
51.80 x 54.00	2890	17.90 x 18.90
48.00 x 50.10	2895	19.10 x 20.00
44.30 x 46.30	2900	20.30 x 21.30
40.70 x 42.60	2905	21.60 x 22.60
37.10 x 38.90	2910	23.00 x 24.10
33.80 x 35.60	2915	24.50 x 25.70
30.60 x 32.20	2920	26.30 x 27.50
27.60 x 28.80	2925	27.90 x 29.7 0
24.70 x 25.90	2930	29.90 x 31.80
22.00 x 23.10	2935	32.10 x 34.10
19.40 x 20.50	2940	34.60 x 36.70
17.10 x 18.10	2945	37.00 x 39.30
15.00 x 15.90	2950	39.80 x 42.10
13.10 x 13.90	2955	42.70 x 45.20
11.30 x 12.10	2960	45.90 x 48.50
9.80 x 10.50	2965	49.30 x 52.00
8.40 x 9.10	2970	53.00 x 55.60
7.30 x 7.90	2975	56.50 x 59.50
6.20 x 6.80	2980	60.40 x 63.50
5.40 x 5.90	2985	64.50 x 67.60
4.60 x 5.10	2990	68.90 x 71.90

Fig. 4.144 Quotation S&P500 options on 24 September 2018 with expiration date 22 October 2018

It can be clearly seen (from the 1-month options) that for positive y, put options with a strike price of $S_0 - Y$ are consistently significantly more expensive than call options with strike $S_0 + Y$. In the following, we will describe such puts and calls as symmetrical to each other.

For example, the

put, October 2018, strike 2875 (i.e. $2,925 - 50$) is quoted at 15.00 // 15.90, and the call, October 2018, strike 2,975 (i.e. $2,925 + 50$) is quoted at 7.30 // 7.90.

Calls				Puts
	414.70 x 419.40	2600	69.70 x 71.50	
	394.60 x 399.10	2625	73.70 x 75.60	
	374.60 x 379.10	2650	78.00 x 80.00	
	354.90 x 359.10	2675	82.50 x 84.40	
	335.50 x 339.60	2700	87.40 x 89.40	
	316.30 x 320.30	2725	92.40 x 94.50	
	297.40 x 301.30	2750	97.50 x 99.70	
	278.80 x 282.60	2775	103.30 x 105.50	
	260.60 x 264.60	2800	109.20 x 111.50	
	242.60 x 246.20	2825	115.30 x 117.60	
	225.10 x 228.70	2850	121.90 x 124.50	
	207.90 x 211.50	2875	129.00 x 131.60	
	191.10 x 194.60	2900	136.50 x 139.10	
	174.70 x 178.10	2925	144.40 x 147.20	
	159.00 x 162.10	2950	152.70 x 155.50	
	143.90 x 147.00	2975	161.80 x 164.80	
	129.40 x 132.40	3000	171.10 x 174.40	
	115.40 x 118.20	3025	181.40 x 184.80	
	102.30 x 105.00	3050	192.80 x 196.20	
	90.00 x 92.50	3075	204.40 x 208.20	
	78.80 x 81.10	3100	217.10 x 221.10	
	68.10 x 70.30	3125	230.90 x 234.90	
	58.60 x 60.70	3150	245.50 x 249.60	
	49.90 x 51.90	3175	260.80 x 265.30	
	42.30 x 44.20	3200	277.30 x 282.00	
	35.60 x 37.40	3225	294.60 x 299.50	
	29.70 x 31.30	3250	312.90 x 318.00	
	24.70 x 26.20	3275	332.10 x 337.10	

Fig. 4.145 Quotation S&P500 options on 24 September 2018 with expiration date 19 September 2019

For Y values up to around 100, the 1-year call options are more expensive than the symmetrical puts. For Y values greater than 100, however, the puts are more expensive than the symmetrical calls.

For example, the

put, September 2019, strike 2,725 (i.e. $2,925 - 200$) is quoted at 92.40 // 94.50, and the

call, September 2019, strike 3125 (i.e. $2925 + 200$) is quoted at 68.10 // 70.30.

A strategy that regularly sells puts with a strike at a certain percentage below S_0 will therefore show a significantly different dynamic than a strategy that regularly sells calls with a strike at a certain percentage above S_0. This is because the receipts from selling the puts are significantly higher than from selling the calls.

In this context, one occasionally hears statements like this:

The price of a put with a strike K below S_0 is significantly higher than the symmetrical call because, by shorting such a put, one exposes oneself to the risk of high losses if the price of the underlying asset crashes, and one obviously wants to collect a higher premium as compensation for taking this risk. There is generally no risk of the price "crashing upward", so a symmetrical short call position is not nearly as risky.

Or, in other words:

There is a great demand for put options with a strike price in a certain out-of-the-money range (strike below S_0) for the purpose of hedging stock holdings. At a minimum we can say that this demand is significantly greater than the demand for symmetrical call options. This increased demand results in a higher price for the put options on the options exchanges.

We are not going to rule out that this argumentation may be partially correct and plausible, nor will we analyse to what extent it is conclusive and accurate. Our only interest in the following is to examine to what extent the asymmetries of these option prices have already been taken into account in the Black-Scholes formula. For that purpose, we are going to compare the Black-Scholes formula for the fair price of symmetrical put and call options.

We start with strike prices right at the money, hence $K = S_0$. How do the prices of a put option and a call option with an at-the-money strike compare to one another? To find the answer to this first question, there is no need for the Black-Scholes formula just yet; all we need to do is inspect the put-call parity equation.

For this purpose we are going to use the more general form of the put-call parity equation, which can also be used for underlying instruments with costs and returns at an interest rate d. The put-call parity for $K = S_0$ then is (see Sect. 3.6):

$$C(0) + S_0 \cdot e^{-rT} = P(0) + e^{-dT} \cdot S_0$$

This version is of course applicable even where there are no payments, in which case we simply set $d = 0$. Thus,

$$C(0) + S_0 \left(e^{-rT} - e^{-dT}\right) = P(0)$$

and therefore:

$$\begin{aligned} C(0) &= P(0) \text{ if } r = d \\ C(0) &> P(0) \text{ if } r > d \\ C(0) &< P(0) \text{ if } r < d \end{aligned}$$

Consider, for example, the at-the-money options (i.e. strike 2925) on the SPX with approximately 1 year to expiration as shown in Fig. 4.145.

Call, 19 September 2019, strike 2925: quotes 174.40 // 177.30, midpoint = 176 dollars
Put, 19 September 2019, strike 2925, quotes 144.50 // 147.30, midpoint = 145.90 dollars

Based on the above considerations, we should therefore get (approximately)

$$2925 \cdot (e^{-r} - e^{-d}) = P(0) - C(0) = -30.10.$$

The risk-free interest rate $r = f_{0.1}$ in dollars was approximately $r = 0.029$ at the time of quotation on 24 September 2018. Substituting this value in the above equation and solving for d, we get

$$d = -\log\left(e^{-0.029} + \frac{30.10}{2925}\right) = 0.0185,$$

which suggests an anticipated SPX dividend yield of 1.85%. This is consistent with the values in Fig. 3.5.

For $Y > 0$, we perform some tests using the Black-Scholes formulas and get the results shown in the graphs below.

In all three examples below, we assumed the S&P500 as being the underlying asset, standing at 2925 points, options with 1 month to expiration $\left(T = \frac{1}{12}\right)$ and volatility of 0.12. These are essentially exactly the parameters that existed on 24 September 2018, when the above option price tables were created. In the first example, we chose the risk-free interest rate $r = 2.2\%$ (= approximately the 1-month dollar Libor on 24 September 2018).

In the first Picture 4.146, we chose a fixed $Y = 100$ and illustrated the price curve for the call option with strike 3025 (blue) and for the put option with strike 2825 (red) over the options' remaining time to expiration. In the second Picture 4.147 (left), we set the time $t = 0$ and illustrated the prices of the call options (with strikes $2,925 + Y$, in blue) and of the put options (with strikes $2,925 - Y$, in red). The picture on the right shows the price difference (green) between call and put.

Contrary to what one would expect based on real prices, we see a slightly higher price for the call options compared to the symmetrical put options. Yet we would of course have had to work with the Black-Scholes formula for underlying assets with payments. We leave this correction to the readers as an exercise.

In the second example, we chose interest rate $r = 0$, and in the third example, a strong negative interest rate $r = -0.1$. We then see that for ($r = 0$), the call price is only slightly higher than the put price, while for $r = -0.1$, the put price is significantly higher than the call price. The corresponding Figs. 4.148, 4.149, 4.150, and 4.151 are given.

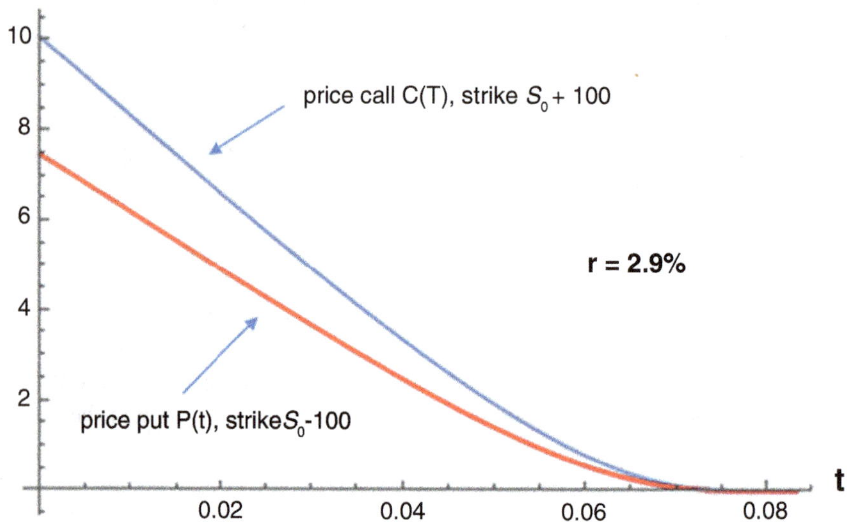

Fig. 4.146 Price curves of symmetrical call (blue) and put (red) options over time, $r = 2.9\%$

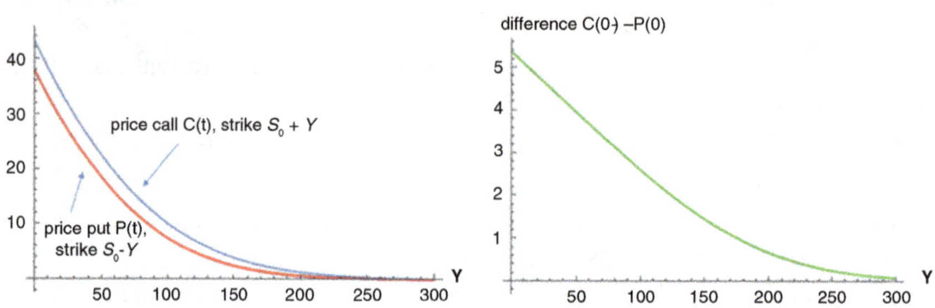

Fig. 4.147 Prices of call (blue) and symmetrical put (red) options for varying strikes $S_0 + Y$ and difference between the two prices (green, right), $r = 2.9\%$

The results become plausible when the Black-Scholes prices of the call and put options are again considered in the general form as discounted expected payoffs with respect to the artificial (risk-neutral) Wiener model, i.e. for the call as

$$C(0) = e^{-rT} \cdot E\left(\max\left(S(0) \cdot e^{\left(r-\frac{\sigma^2}{2}\right)T+\sigma\sqrt{T}w} - K, 0\right)\right)$$

(continued)

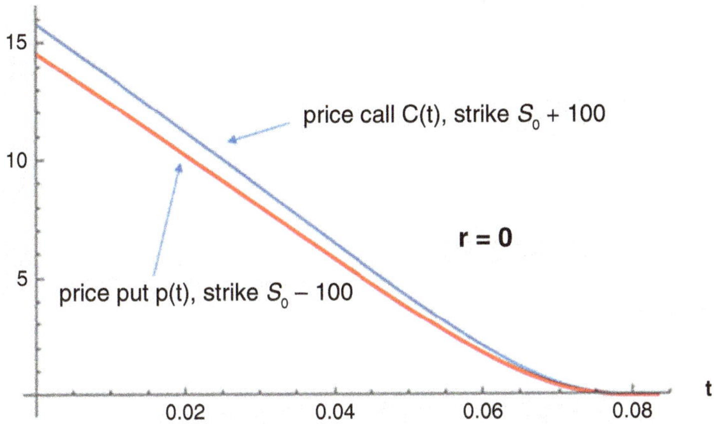

Fig. 4.148 Price curves of symmetrical call (blue) and put (red) options over time, $r = 0$

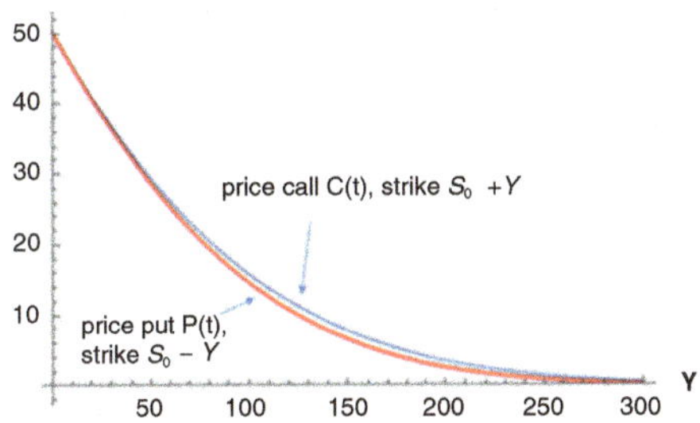

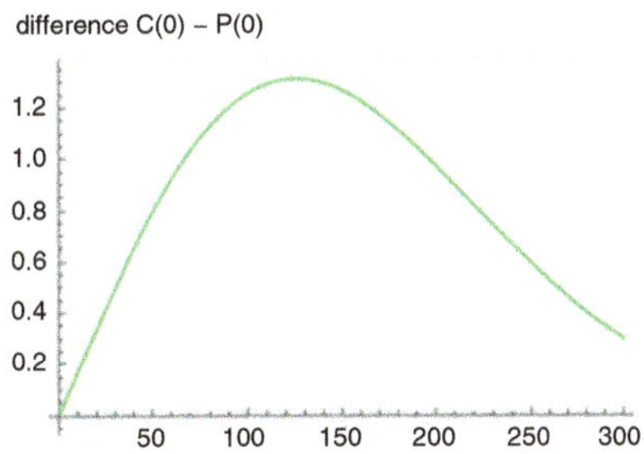

Fig. 4.149 Prices of call (blue) and symmetrical put (red) options for varying strikes $S_0 + Y$ and difference between the two prices (green, right), $r = 0$

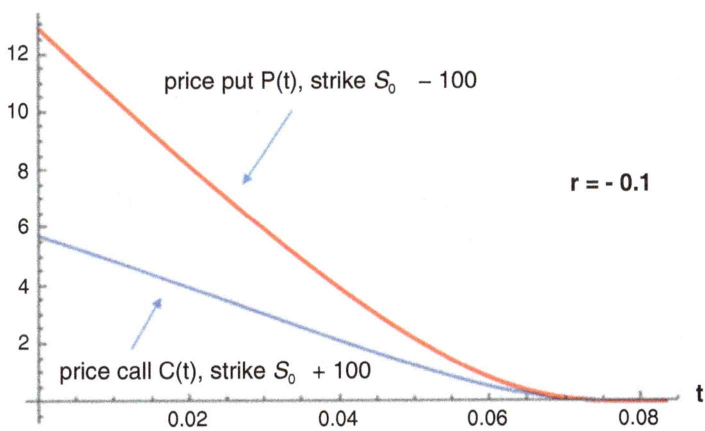

Fig. 4.150 Price curves of symmetrical call (blue) and put (red) options over time, $r = -0.1$

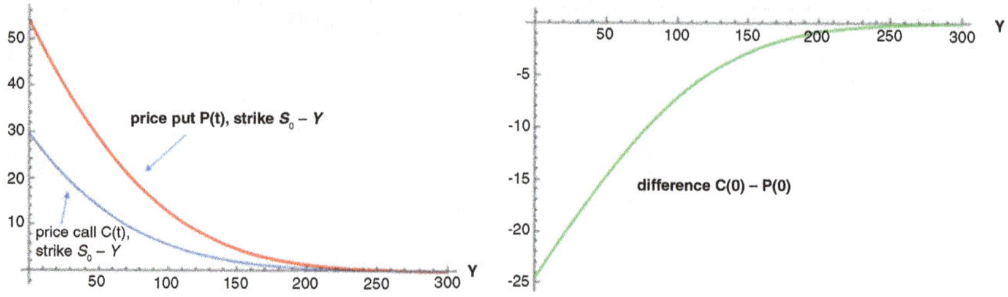

Fig. 4.151 Prices of call (blue) and symmetrical put (red) options for varying strikes $S_0 + Y$ and difference between the two prices (green, right), $r = -0.1$

and for the put as

$$P(0) = e^{-rT} \cdot E\left(\max\left(K - S(0) \cdot e^{\left(r-\frac{\sigma^2}{2}\right)T+\sigma\sqrt{T}w}, 0\right)\right).$$

Setting $K = S(0) - Y$ for the put and $K = S(0) + Y$ for the call, we get

$$e^{rT} \cdot C(0) = E\left(\max\left(S(0) \cdot \left(e^{\left(r-\frac{\sigma^2}{2}\right)T+\sigma\sqrt{T}w} - 1\right) - Y, 0\right)\right)$$

$$e^{rT} \cdot P(0) = E\left(\max\left(S(0) \cdot \left(1 - e^{\left(r-\frac{\sigma^2}{2}\right)T+\sigma\sqrt{T}w}\right) - Y, 0\right)\right).$$

Now it is obvious that for a large r, the expression $\left(e^{\left(r-\frac{\sigma^2}{2}\right)T+\sigma\sqrt{T}w} - 1\right)$ will, on the average, greatly outweigh the expression $\left(1 - e^{\left(r-\frac{\sigma^2}{2}\right)T+\sigma\sqrt{T}w}\right)$ and that for a small r the opposite will be the case.

4.35 Analysis of the Price Curves of a Few Other Basic Option Strategies: Simple Time Spreads

Before we proceed with further expanding our knowledge of quantitative finance fundamentals, we are going to perform a brief analysis of another option strategy and its price dynamics, this time looking at a calendar spread (also referred to as a time spread). More specifically, and by way of example, we are going to analyse a bear time spread. (Readers are again encouraged to test any other types of time spreads by themselves, using the software on our website.)

Rapidly recognizing price trends and with that the effects of time spreads requires some experience in options trading (especially when it comes to more complex types). To start with, there is the difficulty that in the case of time spreads, there is no final payoff function (or profit function), since the components (legs) of a time spread do not have a common expiration date. However, any one time spread obviously comes to an end when the option in the spread with the shortest expiration T_1 expires, because then the time spread loses its original character. If we assume that all other components of the spread are also closed (or at least priced) at this time, then an approximate payoff/profit function can be calculated for time T_1 using the Black-Scholes formula.

Let us look at the most basic example of a **bear time spread of call options**. In such a bear time spread, one long call option C_1 with a shorter time to expiration T_1 is combined with a short call option C_2 with a longer time to expiration T_2 (but both having the same strike price K). Since the time value of C_2 is higher than the time value C_1, a positive premium is received at time 0.

Figure 4.152 shows the typical price curves of a bear time spread for the following parameters:

$S_0 = 2900$
$K = 2900$
$T_1 = \frac{1}{12}$
$T_2 = \frac{2}{12}$
$r = 0$
$\sigma = 0.15$

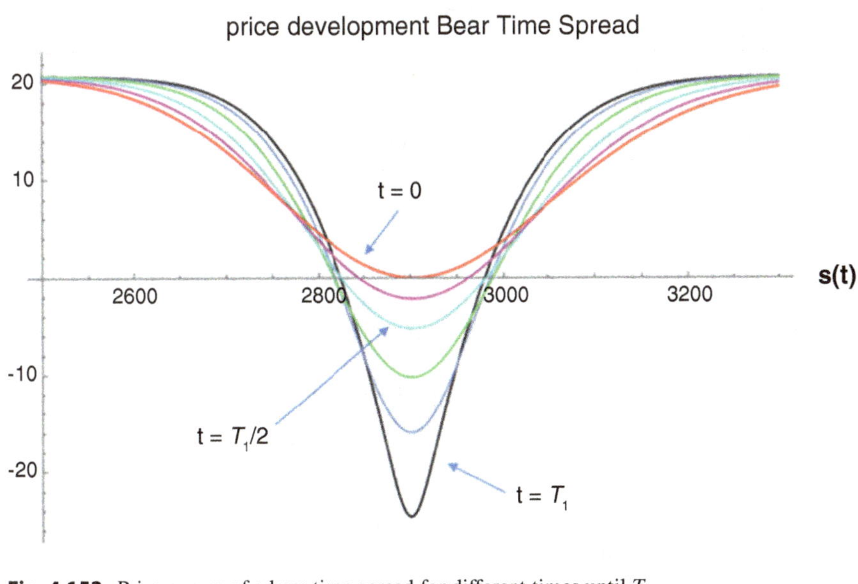

Fig. 4.152 Price curves of a bear time spread for different times until T_1

The black curve represents the price of the combination at time T_1, i.e. the profit function of the strategy in the above sense. The other curves represent the price movements of the strategy at earlier times t (red; $t = 0$; magenta, $t = 1$ week; turquoise, $t = 2$ weeks; green, $t = 3$ weeks; blue, 3 days before expiration) as a function of the underlying asset price $S(t)$.

The premium received on entering into the strategy is 20.74 dollars. The premium is calculated as the difference between the price of approximately 71$ for C_2 and 50$ for C_1. This premium also represents the highest possible profit in this strategy. How does this shape of the profit function come about?

- If the underlying asset remains essentially unchanged until time T_1, then C_1 essentially expires worthless. We are still short C_2. If we closed C_2 right now, we would have to pay the time value for C_2. All else being equal, the price of C_2 at time T_1 will correspond approximately to the price of C_1 at time 0, i.e. approximately 50$. This means that the strategy will produce a loss.
- If the underlying price drops sharply by time T_1, then C_1 will essentially expire worthless. We are still short C_2. C_2 currently has no intrinsic value and only a low time value. This is therefore a good time to close C_2, as we will get to keep most of the initially received premium as a profit.
- If the underlying price rises sharply by time T_1, we receive payment of the option's intrinsic value through C_1. At this moment, option C_2 has the same intrinsic value as C_1 plus a small time value. To close C_2 we need the intrinsic value of C_1 that we just collected and the small time value. We therefore get to keep most of the initially received premium as a profit.

The price curves are strongly reminiscent of the smoothed price curves of a conventional (i.e. non-calendar) short butterfly spread, or more specifically, of a short butterfly with expiration in T_1, cusp in 2900 and corner points in about 2700 and 3100. Such a short butterfly is created from one long put and one long call with strike 2900, one short call with strike 3100, and one short put with strike 2700. However, it turns out that while such a combination has approximately the same shape as a bear time spread, its extension along the y-axis is approximately five times greater. So if we want a short butterfly to generate a profit profile similar to that of a bear time spread, we have to compare the bear time spread with about one-fifth of a short butterfly (or five bear time spreads with one short butterfly). Figure 4.153 shows the comparison of the two profit functions. Figure 4.154 depicts the price curves of the bear time spread (left) and $\frac{1}{5}$ of the short butterfly next to one another.

As can be seen from Fig. 4.153—comparison of the two profit curves of $\frac{1}{5}$ of a short butterfly and a bear time spread—the difference between the two strategies, i.e. $\frac{1}{5}$ *of a short butterfly and shorting a bear time spread* yields quite an interesting profit curve for time T_1 (see Fig. 4.155).

We will now briefly discuss another time spread, the so-called **diagonal bull spread**, without commenting it any further.

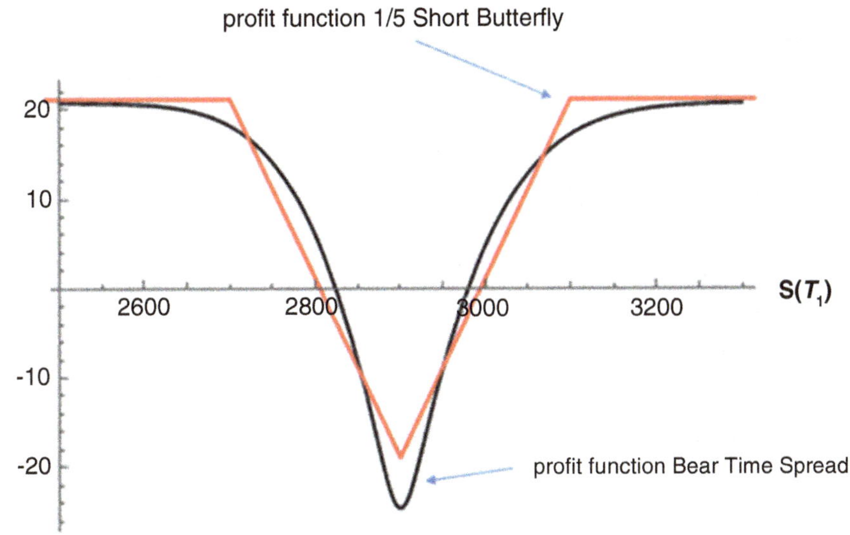

Fig. 4.153 Comparison of profit functions in T_1 of bear time spread (black) and $\frac{1}{5}$ short butterfly (red)

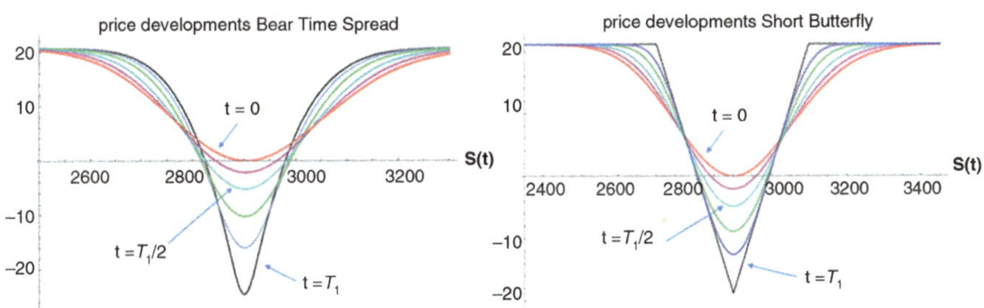

Fig. 4.154 Comparison of price curves of bear time spread (left) and $\frac{1}{5}$ short butterfly (right) at different times

A diagonal bull spread consists of a longer-dated long call (C_2, time to expirationT_2) with a strike price K_2 and a shorter-dated short call C_1 (time to expiration T_1) with a strike price K_1 that is higher than K_2 ($K_1 > K_2$).

We are going to illustrate this with an example, for which we again assume the parameters from our previous examples:

$$S_0 = 2900$$
$$T_1 = \frac{1}{12}$$
$$T_2 = \frac{2}{12}$$
$$r = 0$$
$$\sigma = 0.15$$

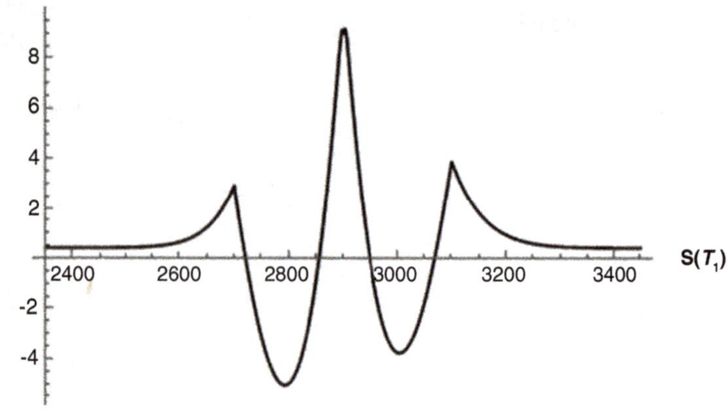

Fig. 4.155 Profit curve for $\frac{1}{5}$ short butterfly plus short bear time spread

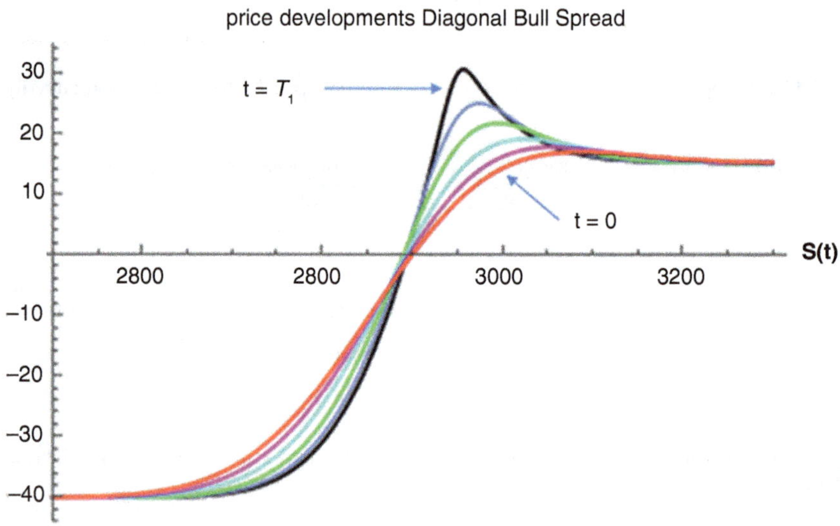

Fig. 4.156 Price curves of diagonal bull spread for different times until T_1

For the strike prices, we now choose

$$K_1 = 2950$$
$$K_2 = 2900$$

In the following Graph 4.156, we are again depicting the price curve of this diagonal bull spread for different times over the life of the combination, i.e. up to time T_1. The black curve represents the price of the combination at time T_1, i.e. the profit function of the strategy in the above sense. The other curves represent the price movements of the strategy at earlier times t (red, $t = 0$; magenta, $t = 1$ week; turquoise, $t = 2$ weeks; green, $t = 3$ weeks; blue, 3 days before expiration) as a function of the underlying asset price $S(t)$.

Again we encourage readers to use our software to graphically depict and get a good grasp of the price curves over time for a wide variety of options combinations (including time spreads). See: https://app.lsqf.org/optionskombinationen

4.36 The Greeks

In previous chapters, in particular in Sects. 4.24–4.26 and in 4.30, we already discussed how and in what ways the fair prices of call and put options depend on the parameters occurring in the formulas for these prices and how sensitively these prices respond to changes in these parameters.

Knowledge of the dependencies of fair prices of derivatives (and of derivative trading strategies) is essential to gaining a deeper understanding of options trading strategies, their risks and rewards, and consequently for their design and management. The basic tools for analysing these dependencies are of course the first (and in some cases also the higher) derivatives of the pricing formulas (the prices of the trading strategies to be analysed) with respect to the relevant parameters. These derivatives are essential not only for the fundamental analysis of price formulas and strategies—they also take on a central role in a wide variety of other contexts (for example in perfect hedges of derivatives, see Sect. 4.21).

Given the relevance of these derivatives, the financial trading world has adopted a specific set of notations. The derivatives are essentially represented by certain letters of the Greek alphabet and are therefore referred to as "the Greeks".

In this chapter, we will discuss only the most essential (!) Greeks and calculate, compile, and briefly discuss the properties of these Greeks for call and put options (except where we have already done so in previous chapters). There are plenty of other Greeks, but they often play only a minor role. Should any of them nevertheless occur in our future analyses, they will be introduced there and then.

Let D be any derivative with life to expiration $[0, T]$ on an underlying asset S, where S has the price $S(t)$ at time t. D can also be a portfolio of multiple derivatives on the same underlying asset. We are referring to this as a derivative trading strategy.

We assume that a formula for the fair price is known for D. We denote the fair price of the derivative at time t by $F(t)$. We assume of course that $F(t)$ depends on the parameter t (or on the remaining time to expiration $\tau = T - t$), and additionally on the underlying asset's price $s = S(t)$ at time t, on the underlying asset's volatility σ at time t, and on the risk-free interest rate r at time t. If we want to explicitly point to this dependency, we will write $F(\tau, s, \sigma, r)$ instead of $F(t)$ in the following.

So we have

1. $\mathbf{\Delta_D(t)} = \Delta_D(\tau, s, \sigma, r) := \frac{dF(\tau,s,\sigma,r)}{ds}$, i.e. the **derivative of F with respect to the price** of the underlying asset, **the delta of D**.

The current value of $\Delta_D(t)$ provides a preliminary and approximate answer to the following question:
How much does the value of the derivative change if the price of the underlying rises by one point (with all other parameters remaining the same)?
The answer is: *The value changes by approximately* $\boldsymbol{\Delta_D(t)}$ *currency units.*
The particular significance of $\Delta_D(t)$ therefore lies, on the one hand, in the area of risk management, where it is essential to be able to accurately estimate the impact of changes in the underlying asset price on currently contemplated derivative trading strategies and, on the other hand, in the area of hedging (we already know that the delta indicates the exact number of underlying asset units that need to be held at any one moment to obtain a perfect hedging strategy (see 4.21).

2. $\boldsymbol{\theta_D(t)} = \theta_D(\tau, s, \sigma, r) := \frac{dF(\tau,s,\sigma,r)}{d\tau}$, i.e. the **derivative of F with respect to the remaining time to expiration τ, the Theta of D.**
The current value of $\theta_D(t)$ provides a preliminary and approximate answer to the following question:
How much does the value of the derivative change if the remaining time to expiration decreases by a short time span of dt (years) (with all other parameters remaining the same)?
The answer is: *The value changes by approximately* $-dt \cdot \theta_D(t)$ *currency units.*
Among other things, the theta value of a trading strategy's derivative is a key measure when it comes to deciding at what point (during the life of a derivative/strategy) the respective derivative position(s) should be entered into or closed. The theta value often shows a significantly different behavior toward the end of a derivative's life than at earlier times during its life (with all other parameters remaining the same).
3. $\boldsymbol{\Upsilon_D(t)} = \Upsilon_D(\tau, s, \sigma, r) := \frac{dF(\tau,s,\sigma,r)}{d\sigma}$, i.e. the **derivative of F with respect to the volatility σ, the Vega of D.**
The current value of $\Upsilon_D(t)$ provides a preliminary and approximate answer to the following question:
How much does the value of the derivative change if volatility increases by 0.01, that is, by 1% (with all other parameters remaining the same)?
The answer is: *The value changes by approximately* $0.01 \cdot \Upsilon_D(t)$ *currency units.*
Vega is not a Greek letter, by the way. Changes in volatility often have a strong impact on the price of derivatives and are often underestimated when it comes to analysing trading strategies in real applications.
4. $\boldsymbol{\rho_D(t)} = \rho_D(\tau, s, \sigma, r) := \frac{dF(\tau,s,\sigma,r)}{dr}$, i.e. the **derivative of F with respect to the risk-free interest rate r, the Rho of D.**
The current value of $\rho_D(t)$ provides a preliminary and approximate answer to the following question:
How much does the value of the derivative change if the risk-free interest rate increases by 0.01, that is, by 1% (with all other parameters remaining the same)?
The answer is: *The value changes by approximately* $0.01 \cdot \rho_D(t)$ *currency units.*
5. $\boldsymbol{\Gamma_D(t)} = \Gamma_D(\tau, s, \sigma, r) := \frac{d^2F(\tau,s,\sigma,r)}{ds^2}$, i.e. the **second derivative of F with respect to the price** of the underlying asset, **the Gamma of D.**

So, gamma is a second derivative, in that it is the first derivative of what is probably the most important Greek, the delta of the derivative. The gamma tells us, for example, how much the proportion of the underlying asset changes in a perfect hedging portfolio for a derivative.
The current value of $\Gamma_D(t)$ provides a preliminary and approximate answer to the following question:
How much does the delta of the derivative change (= by how many units does the portion of the underlying asset in the perfect hedging portfolio for the derivative change) if the price of the underlying asset increases by one point (with all other parameters remaining the same)?
The answer is: *The value changes by approximately* $\Gamma_D(t)$ *units.*

4.37 The Greeks for Call Options and Put Options

The formulas for the Greeks $\Delta, \theta, \Upsilon, \rho$ of call options and put options were already derived in Sects. 4.21, 4.24–4.26, and in 4.30. (In Sect. 4.24 we did not calculate the theta of the options directly, i.e. the derivative of the option price with respect to τ, but instead the derivative with respect to t. Since $\tau = T - t$, however, it immediately follows from the chain rule for differentiation that the derivative with respect to τ is nothing else than the negative of the derivative with respect to t.) Now we just calculate the gamma of call and put options and summarize all the formulas for these Greeks in the Tables 4.9 and 4.10.

The delta of a call option is (see Sect. 4.21) given by $\mathcal{N}(d_1)$ with

$$d_1 = \frac{\log\left(\frac{s}{K}\right) + \left(r + \frac{\sigma^2}{2}\right)(T-t)}{\sigma\sqrt{T-t}}.$$

The gamma is the derivative of the delta with respect to s and therefore (applying the chain rule):

$$\boldsymbol{\Gamma_C(t)} = \frac{d\Delta_C}{ds} = \frac{d\mathcal{N}(d_1)}{ds} = \mathcal{N}'(d_1)\cdot d_1'(s) = \boldsymbol{\phi(d_1)\cdot\frac{1}{s\sigma\sqrt{\tau}}}.$$

Table 4.9 Greeks of a call option

	Call option
Fair price	$s\cdot\mathcal{N}(d_1) - K\cdot e^{-r(T-t)}\cdot\mathcal{N}(d_2)$
Delta	$\mathcal{N}(d_1)$
Theta	$S\cdot\phi(d_1)\cdot\frac{\sigma}{2\sqrt{T-t}}+$ $+K\cdot r\cdot e^{-r(T-t)}\cdot\mathcal{N}(d_2)$
Vega	$K\cdot e^{-r(T-t)}\cdot\sqrt{T-t}\cdot\phi(d_2)$
Rho	$(T-t)\cdot e^{-r(T-t)}\cdot K\cdot\mathcal{N}(d_2)$
Gamma	$\phi(d_1)\cdot\frac{1}{s\sigma\sqrt{T-t}}$

Table 4.10 Greeks of a put option

	Put option
Fair price	$e^{-r(T-t)} \cdot K \cdot \mathcal{N}(-d_2) - S(t) \cdot \mathcal{N}(-d_1)$
Delta	$\mathcal{N}(d_1) - 1$
Theta	$S \cdot \phi(d_1) \cdot \frac{\sigma}{2\sqrt{T-t}} -$ $-K \cdot r \cdot e^{-r(T-t)} \cdot \mathcal{N}(-d_2)$
Vega	$K \cdot e^{-r(T-t)} \cdot \sqrt{T-t} \cdot \phi(d_2)$
Rho	$-(T-t) \cdot e^{-r(T-t)} \cdot K \cdot \mathcal{N}(-d_2)$
Gamma	$\phi(d_1) \cdot \frac{1}{s\sigma\sqrt{T-t}}$

The delta of a put option is (see 4.21) given by $\mathcal{N}(d_1) - 1$. By differentiating this delta with respect to s, one obviously obtains the same gamma value as for the call option.

So, to summarize, we have

where

$$\mathcal{N}(x) := \frac{1}{\sqrt{2\pi}} \int_{-\infty}^{x} e^{-y^2} dy \text{ and } \phi(x) := \frac{1}{\sqrt{2\pi}} e^{-\frac{x^2}{2}}$$

and

$$d_1 = \frac{\log\left(\frac{s}{K}\right) + \left(r + \frac{\sigma^2}{2}\right)(T-t)}{\sigma\sqrt{T-t}} \text{ and } d_2 = \frac{\log\left(\frac{s}{K}\right) + \left(r - \frac{\sigma^2}{2}\right)(T-t)}{\sigma\sqrt{T-t}}.$$

4.38 Graphical Illustration of the Greeks of Call Options

In the following graphs, we illustrate the typical movements of these Greeks (for call options in this section, for put options in the next). For each of the graphs, we choose the following parameter standard values, all of which are fixed and not variable (!):

$K = 100$,
$S(t)$ in the range from 0 to 200
$T = 1$,
$r = 0.02$,
$\sigma = 0.5$
$t = 0, \frac{T}{4}, \frac{T}{2}, \frac{3T}{4}, 0.9T$ and $t = T$

Delta
Figure 4.157 depicts the evolution of the call price as a function of $S(t)$ for the six different values of t in the upper graph and the evolution of the associated delta value (in the same colours) in the graph below. If we bear in mind the definition of

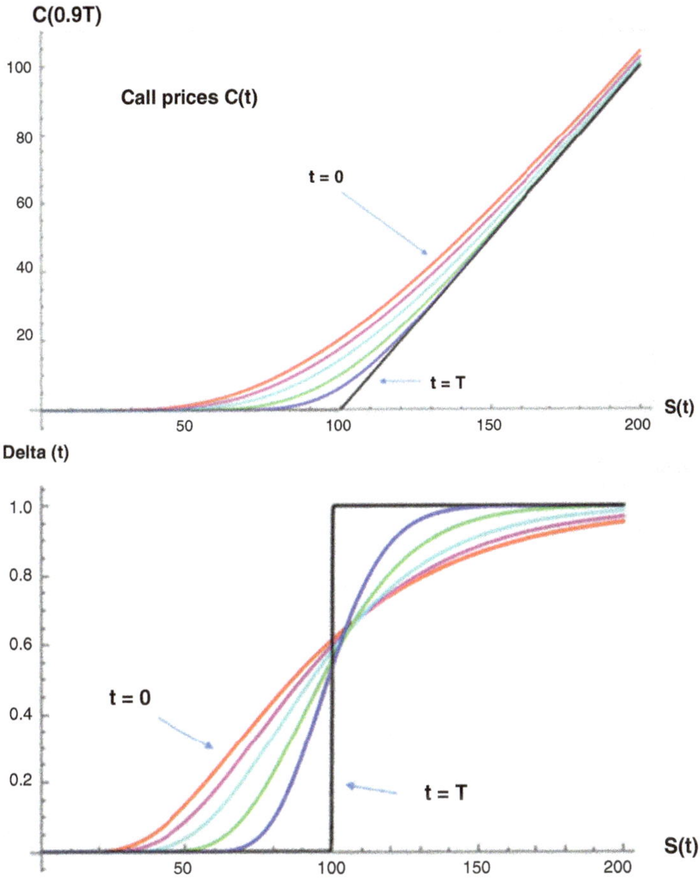

Fig. 4.157 Call price and associated delta as a function of the underlying asset price for different times t (red, $t = 0$; magenta, $t = \frac{T}{4}$; turquoise, $t = \frac{T}{2}$; green, $t = \frac{3T}{4}$; blue, $t = 0.9T$; black, $t = T$)

delta as a derivative of the call-price C with respect to parameter s, then the depicted form of the delta is obvious:

- Delta is always positive (or 0), since C is always monotonically increasing in S.
- Delta is monotonically increasing because C' keeps increasing at all times. (C is a convex function (curved to the left)).
- The delta always has values less than or equal to 1, since the increase in C is never more than 1 at most.
- If $t = T$, that is to say if C is exactly the payoff function, then the delta is equal to 0 below K and equal to 1 above K (C is not differentiable in K, so there is no delta there).

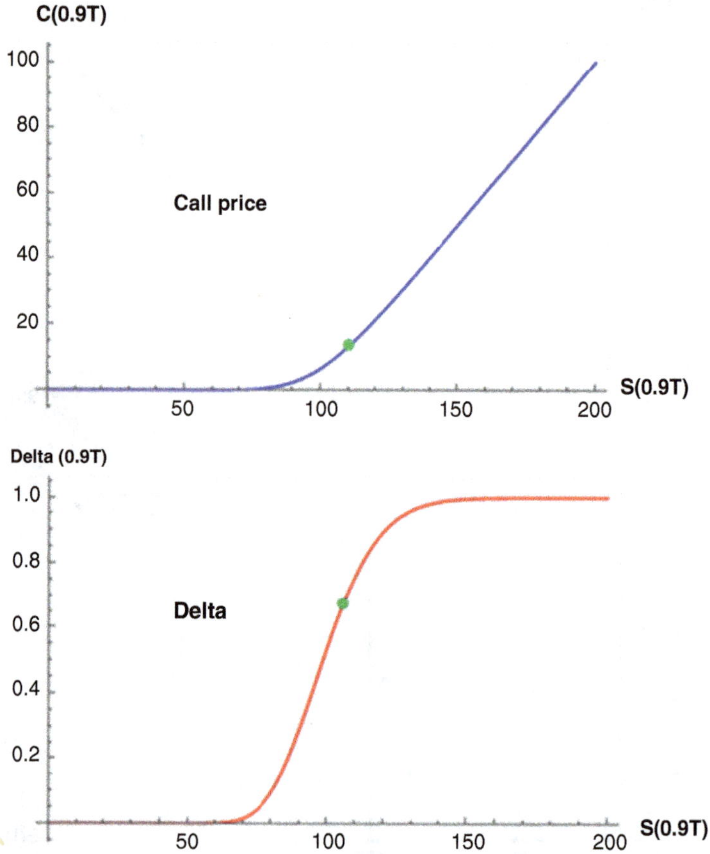

Fig. 4.158 Call price (above) and associated delta (below) for $t = 0.9T$

To illustrate this even more clearly, we show $C(t)$ and the associated delta for one explicit time $t = 0.9T$ in Fig. 4.158. Here we see once again the typical form of the delta curve.

The green dot indicates the values for $C(t) \approx 15$ and for $\Delta(t) \approx 0.7$ for $S(t) = 110$. Consequently, a change of about one point in the underlying price implies a change of about 0.7 in the call price at this moment.

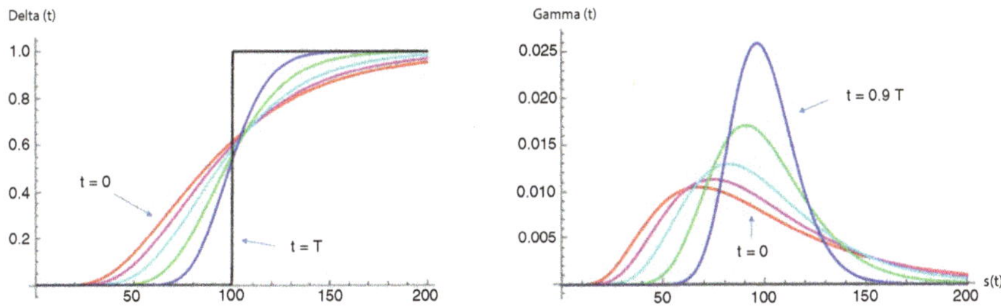

Fig. 4.159 Delta and associated gamma as a function of the underlying asset price for different times t (red, $t = 0$; magenta, $t = \frac{T}{4}$; turquoise, $t = \frac{T}{2}$; green, $t = \frac{3T}{4}$; blue, $t = 0.9T$; black, $t = T$)

Gamma

Next we are going to bring forward the gamma (as a derivative of the delta) and discuss the gamma value of a call option.

Figure 4.159 depicts the evolution of the delta as a function of $S(t)$ for the six different values of t on the left and the evolution of the associated gamma value (in the same colours) on the right.

At first glance, the gamma value for $t = T$ (black curve) appears to be missing in the picture on the right. But looking at the delta for $t = T$ in the left graph and bearing in mind that the gamma is precisely the derivative of the delta with respect to s, we realize that:

The increase in the delta is equal to 0 throughout, except directly at strike $K (= 100)$, where the increase is infinitely large. So, when $t = T$, the gamma is equal to 0 for $S(T) \neq K$ and equal to infinity for $S(T) = K$. If you look closely, the values for which the gamma is equal to 0 are actually plotted in the graph on the right (blackened x-axis). The depicted form of the gamma is also obvious for the other values of t:

- Gamma is always positive, because delta is always monotonically increasing.
- Gamma always has a maximum where delta has an inflection point (i.e. where it changes from a left curvature to a right curvature). These delta inflection points move further and further to the right toward $S(t) = K$, and so the gamma maximums also move further and further to the right toward $S(t) = K$.
- The gamma maximums get higher and higher, since the delta near $S(t) = K$ grows more and more rapidly as t grows.

We will again illustrate the meaning of a gamma value using a specific example in our setting:

If, for example, $t = 0.9T$ and if $S(t)$ is, for example, at $K = 100$, a one-point change in the price of the underlying asset implies a change in the delta by approximately 0.025. (In this situation, a perfect hedging strategy would therefore require reallocation of 0.025 units of the underlying asset in the hedging portfolio.)

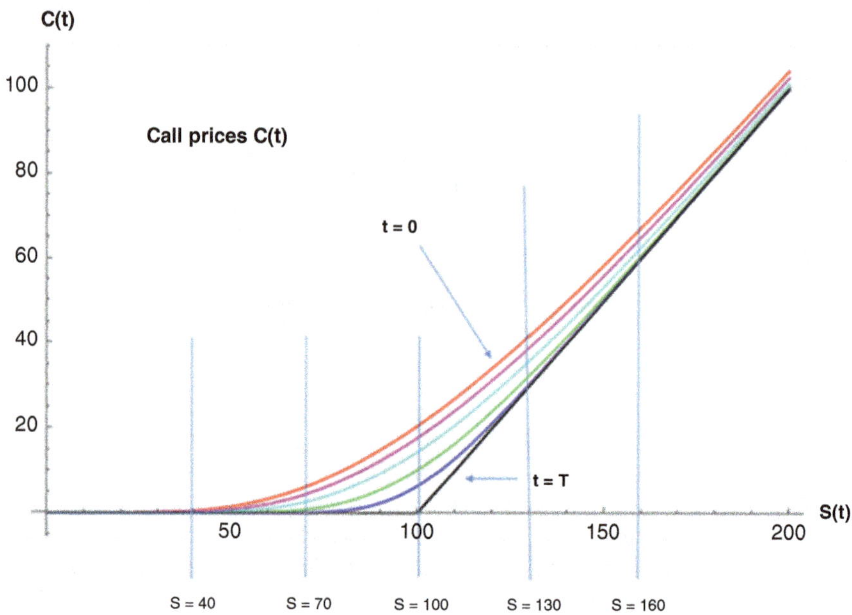

Fig. 4.160 Call price as a function of the underlying asset price for different times t (red, $t = 0$; magenta, $t = \frac{T}{4}$; turquoise, $t = \frac{T}{2}$; green, $t = \frac{3T}{4}$; blue, $t = 0.9T$; black, $t = T$)

Theta

The shape of the theta curve is much more subtle. In the following graphs, we prefer to depict the call price derivative with respect to t rather than with respect to the amount of time left to expiration τ. This means: In the following graphs, you will see the value $-\theta$ instead of θ.

In Fig. 4.160, we are again depicting the evolution of call prices as a function of $S(t)$ for various values of t. We will now pick various specific values of $S(t)$ (namely $S(t) = 40, 70, 100, 130, 160$) and look at the evolution of $C(t)$ over time for each of these values of $S(t)$, each one in its own graph. In each of these graphs, an additional curve in orange represents the derivative of this evolution for $C(t)$ with respect to t (i.e. $-\theta$).

- The (negative) theta values are negative throughout (as previously noted). So, for non-negative r (!), $C(t)$ is monotonically decreasing over time.
- The absolute values of the theta increase the closer $S(t)$ is to the strike price K.

$-\theta(t)$ takes a highly interesting shape when $S(t) = K$ or gets very close to K. We see a strong decrease in the call price as it nears expiry, i.e. a very high $\theta(t)$ (a very negative $-\theta(t)$). In other words, when $S(t)$ is very close to K, a call option has a clearly positive value practically right up until expiry, dropping to 0

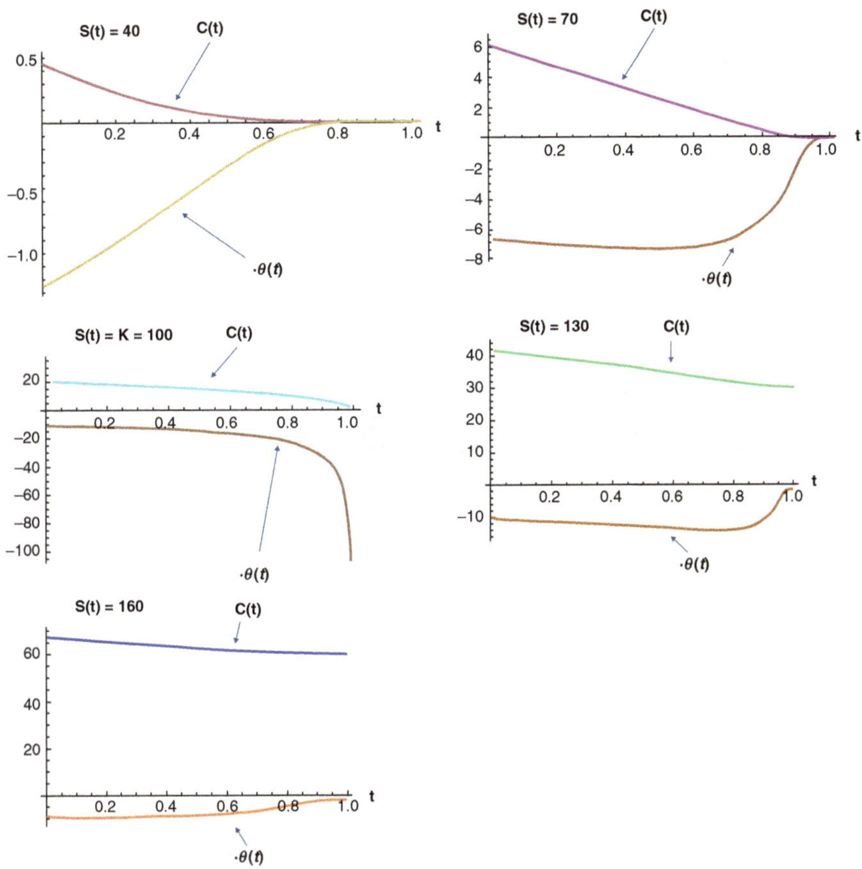

Fig. 4.161 $C(t)$ as a function of time t for different values of $S(t)$ and associated (negative) theta values (orange)

only immediately before it expires. We illustrate this in more detail by taking the section in Fig. 4.161 that relates to $S(t) = K = 100$ and creating a zoom for t from 0.95 to 1 (see Fig. 4.162; $C(t)$ and $-\theta(t)$ are shown here as two separate graphs).

Let us again look at a specific value: Where $S(t) = K = 100$, the theta value is approximately $-\theta(t) \approx -100$ at time $t = 0.99$. We recall the interpretation of theta that we gave above:

> How much does the value of the derivative change if the remaining time to expiration decreases by a short time span of dt (years) (with all other parameters remaining the same)?

The answer is: *The value changes by approximately* $-dt \cdot \theta$.

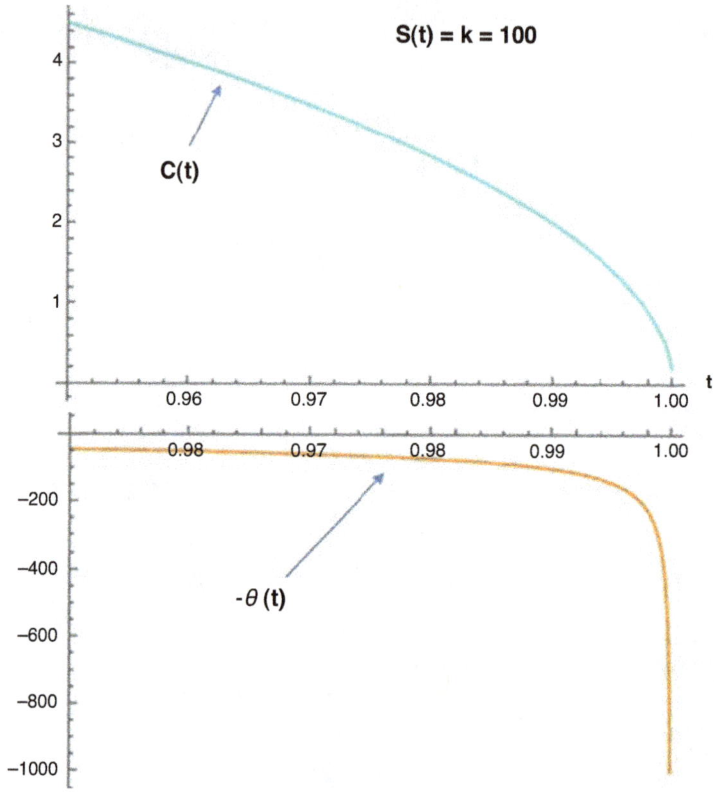

Fig. 4.162 $C(t)$ (turquoise, above) as a function of time t for $S(t) = K = 100$ and associated (negative) theta values (orange, below)

$t = 0.99$ means that there are about 3–4 days left before the option expires. Consider the small time span $dt = \frac{1}{360}$ ($\approx$ 1 day). We then have $-dt \cdot \theta(t) \approx -\frac{100}{360} \approx 0.28$. From that we conclude:

> 3 to 4 days before expiry, the value of a call option drops by approximately 0.28 currency units within 1 day (assuming the above volatility, interest rate and $S(t) \approx K = 100$).

The theta of a call option is the only one of the Greeks that we are discussing where the shape of the curve is substantially contingent on whether the interest rate r is positive or negative. In the following, we will therefore present (only briefly) the essential information and graphics (compare Fig. 4.163) for the theta value of call options when r is negative:

When r is negative, the call prices $C(t)$ for t near T and large $S(t)$ lie **below the payoff curve** $C(T)$. Because $C(t)$ approaches $C(T)$ for t converging to T, $C(t)$ is monotonically increasing in t in these areas (t near T and sufficiently large $S(t)$).

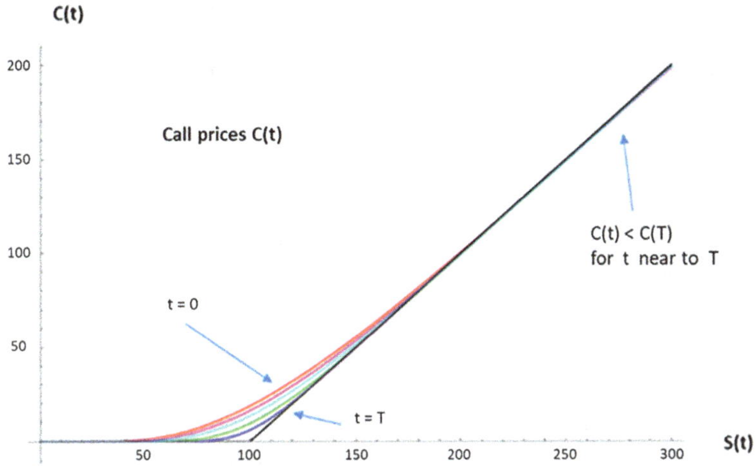

Fig. 4.163 Call price as a function of the underlying asset price for different times t (red, $t = 0$; magenta, $t = \frac{T}{4}$; turquoise, $t = \frac{T}{2}$; green, $t = \frac{3T}{4}$; blue, $t = 0.9T$; black, $t = T$) for **negative r**

This fact can also be clearly seen in the following Graph 4.164 (the counterparts to the graphs in Fig. 4.161): In the last two graphs (where $S(t)$ is sufficiently large) the (negative), theta is positive for t near T!

Vega

For a graphical presentation (see Fig. 4.165) of the vega, we proceed as follows:

We are going to pick out three different values of $S(t)$ (i.e. $S(t) = 50, 100, 150$) and three values of t (i.e. $t = 0, \frac{T}{2}, 0.9T$) and plot the evolution of $C(\sigma)$ as a function of volatility σ for all the combinations of these values of $S(t)$ and t, each one in a separate graph. In each of these graphs, an additional curve in orange represents the derivative of $C(\sigma)$ with respect to the volatility σ, i.e. the Vega. For these representations, we select a range for σ from 0 to 1 (0% to 100%).

The graphs visually confirm that:

- The vega of a call option is always positive, and $C(\sigma)$ is therefore monotonically increasing in σ.
- What is striking is an almost linear increase of $C(\sigma)$ (for σ in realistic value ranges) where $S(t) = K$.
- The $\Upsilon(\sigma)$ has the highest values for $S(t)$ near K and increases with the time left to expiration (falls with t). Call options with longer expiration dates and $S(t)$ near K exhibit the highest sensitivity to changes in volatility.

To provide a clearer idea of the orders of magnitude, we present yet another example:

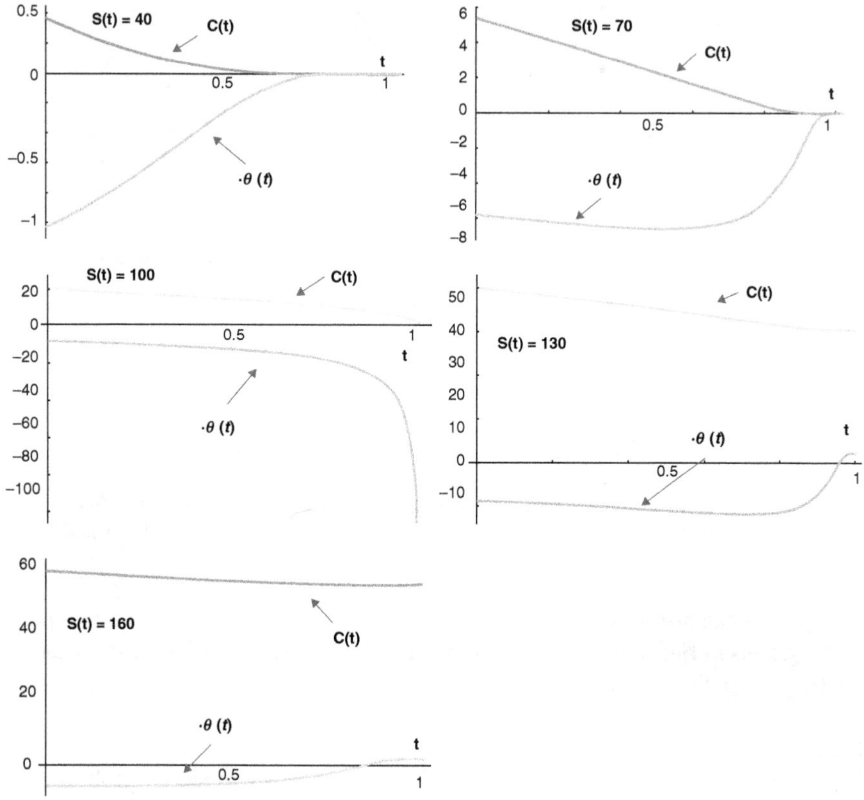

Fig. 4.164 $C(t)$ as a function of time t for various values of $S(t)$ and associated (negative) theta values (orange) for negative r

For example, for the parameters $t = 0.5$ and $S(t) = 100$, the value of the vega for $\sigma = 0.4$ is approximately $\Upsilon(\sigma) = 27$. Again, we recall the interpretation given further above:

How much does the value of the derivative change if volatility increases by 0.01, that is, by 1% (with all other parameters remaining the same)?

The answer is: *The value changes by approximately* $0.01 \cdot \Upsilon$ *currency units.*

Applied to our example, this means that, for an underlying asset price near $K = 100$, half a year's time left to expiration, and volatility of approximately 40%, the price of the call option changes by approximately $0.01 \cdot \Upsilon(t) \approx 0.27$ currency units when volatility changes by 1%.

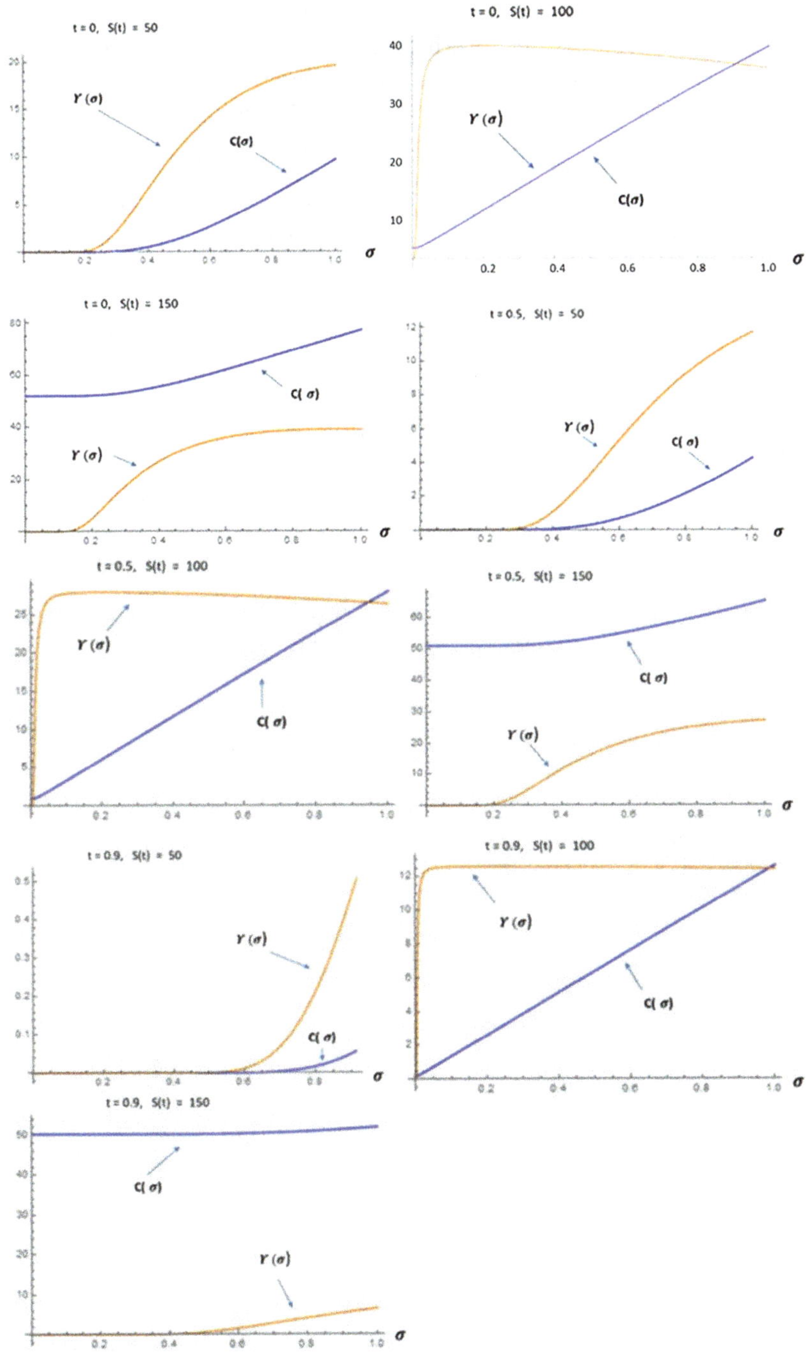

Fig. 4.165 Call price $C(\sigma)$ (blue) and vega $\Upsilon(\sigma)$ (orange) for various values of t and $S(t)$ as function of volatility σ

Rho

For the graphical representation of Rho, we proceed in the same way as we did for Vega: We are again going to pick out three different values of $S(t)$ (i.e. $S(t) = 50, 100, 150$) and three values of t (i.e. $t = 0, \frac{T}{2}, 0.9T$) and plot $C(r)$ as a function of the risk-free interest rate r for all the combinations of these $S(t)$ and t values, each one in a separate graph. In each of these graphs, an additional curve in orange represents the derivative of $C(r)$ with respect to the interest rate r, i.e. the Rho. For these representations, we select a range for r from -0.02 to 0.1 (-2% to 10%).

We see that:

- Rho $\rho(r)$ is always positive, and so $C(r)$ is continuously growing in r.
- Changes in $C(r)$ as a function of r largely occur approximately linearly.

To provide a clearer idea of the orders of magnitude, we present yet another example:

For example (see graph in the middle in Fig. 4.166), for the parameters $t = 0.5$ and $S(t) = 100$, the value of the rho for $r = 0$ is approximately $\rho(r) = 15$. Again, we recall the interpretation given further above:

> How much does the value of the derivative change if the risk-free interest rate increases by 0.01, that is, by 1% (with all other parameters remaining the same)?

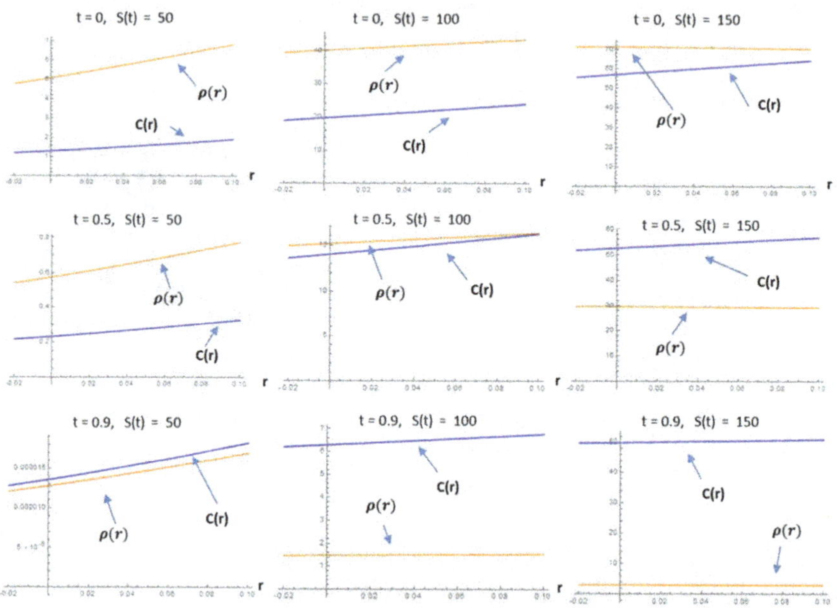

Fig. 4.166 Call price $C(r)$ (blue) and rho $\rho(r)$ (orange) for various values of t and $S(t)$ as a function of risk-free interest rate r

The answer is: *"The value changes by approximately* $0.01 \cdot \rho$ *currency units."*

Applied to our example, this means that, for an underlying asset price near $K = 100$, half a year's time left to expiration, and a risk-free interest rate of 0%, the price of the call option changes by approximately $0.01 \cdot \rho \approx 0.15$ currency units when the interest rate changes by 1%.

4.39 Graphical Illustration of the Greeks of Put Options

We are going to keep the discussion very brief here, essentially limiting ourselves to just presenting the graphics and generally proceeding in the same way as we did to illustrate the Greeks of call options.

For all of the graphs, we choose the following standard parameter values, all of which are fixed and not variable (!):

$K = 100$,
$S(t)$ in the range from 0 to 200
$T = 1$,
$r = 0.02$,
$\sigma = 0.5$
$t = 0, \frac{T}{4}, \frac{T}{2}, \frac{3T}{4}, 0.9T$ and $t = T$

Delta

- Delta is always negative (or 0), since $P(t)$ is always monotonically decreasing.
- Delta is monotonically increasing, as the speed of decline of $P(t)$ keeps decreasing ($P(t)$ is curved to the left).
- The delta always has values greater than or equal to -1, since the speed of decline of $P(t)$ is never more than 1 at most.
- If $t = T$, that is to say if $P(t)$ is exactly the payoff function, the delta equal to 0 is below K and equal to -1 above K ($P(t)$ is not differentiable in K, so the delta does not exist there).

Figure 4.167 illustrates the put price and associated Delta as a function of the underlying asset price for different times t.

Gamma

The put option gamma has the same formula as the call option gamma (see Fig. 4.168).

- Where $t = T$, the gamma is equal to 0 for $S(T) \neq K$ and equal to infinity for $S(T) = K$. The values for which gamma equals 0 are in fact plotted in the lower graph (blackened x-axis).
- Gamma is always positive, as delta is always monotonically growing.
- Gamma always has a maximum where delta has an inflection point (i.e. where it changes from a left curvature to a right curvature). These delta inflection points

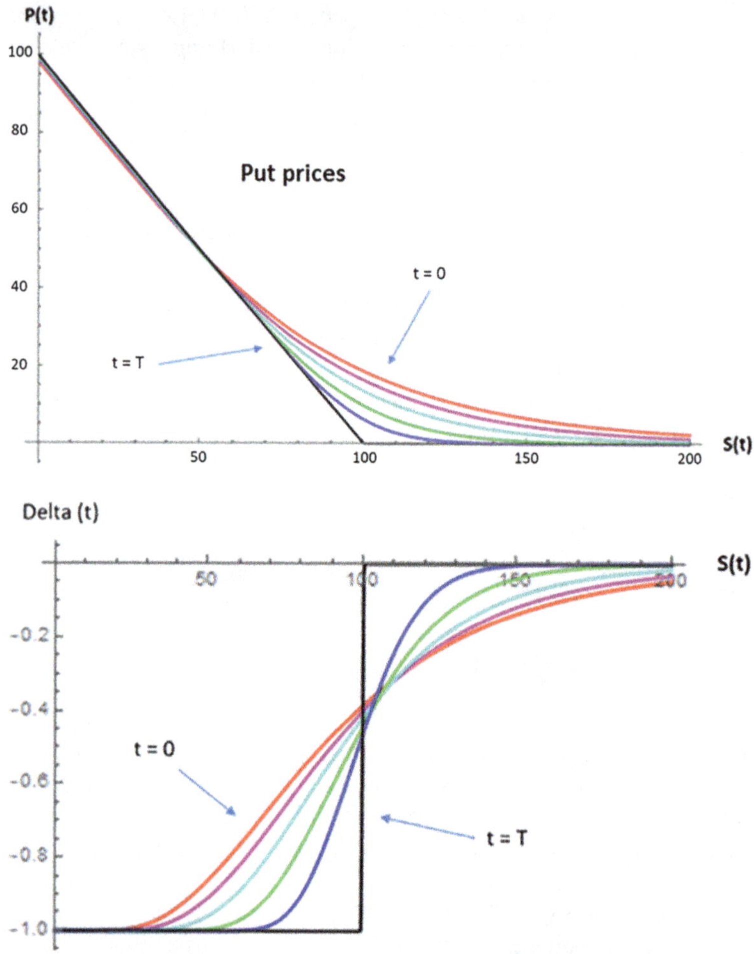

Fig. 4.167 Put price and associated delta as a function of the underlying asset price for different times t (red, $t = 0$; magenta, $t = \frac{T}{4}$; turquoise, $t = \frac{T}{2}$; green, $t = \frac{3T}{4}$; blue, $t = 0.9T$; black, $t = T$)

move further and further to the right toward $S(t) = K$, and so the gamma maximums also move further and further to the right toward $S(t) = K$.

- The gamma maximums get higher and higher, since the delta at $S(t) = K$ grows more and more rapidly as t grows.

Theta

- With put options, the (negative) theta values are not always negative throughout. For non-negative r (!), $P(t)$ is occasionally monotonically increasing over time. This is the case when $S(t)$ is substantially smaller than K.

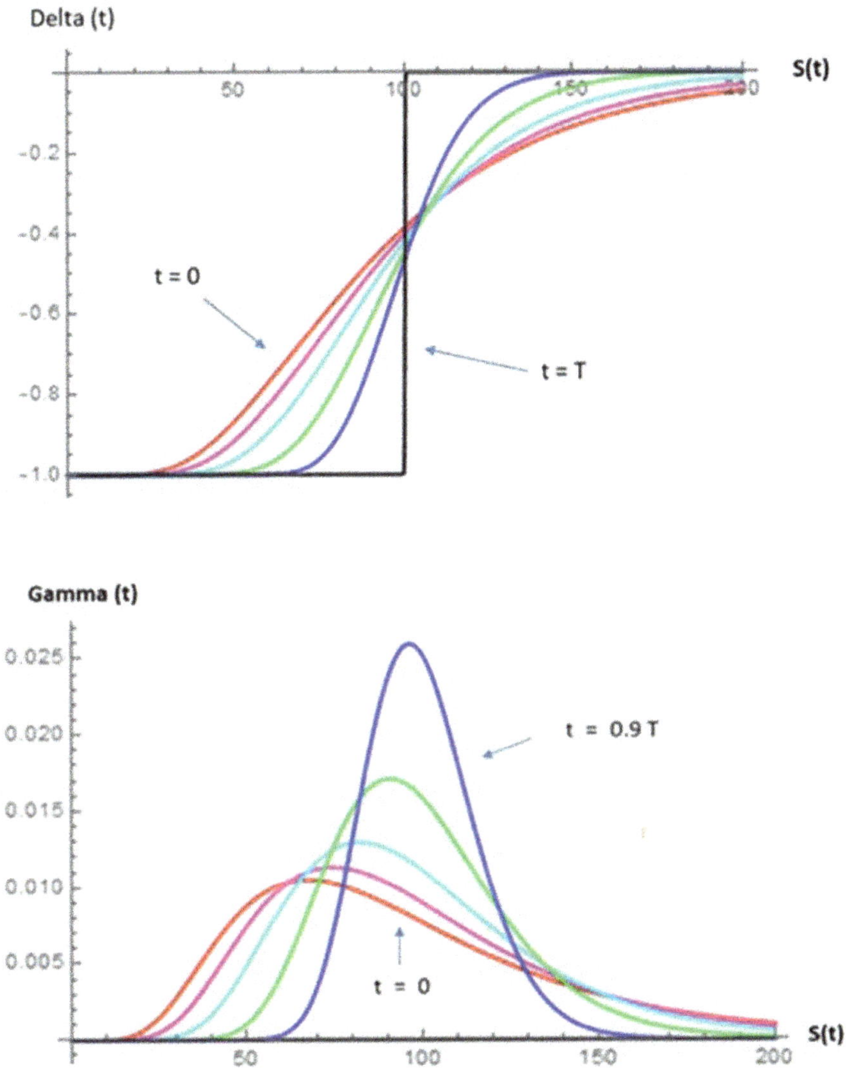

Fig. 4.168 Delta and associated gamma as a function of the underlying asset price for different times t (red, $t = 0$; magenta, $t = \frac{T}{4}$; turquoise, $t = \frac{T}{2}$; green, $t = \frac{3T}{4}$; blue, $t = 0.9T$; black, $t = T$)

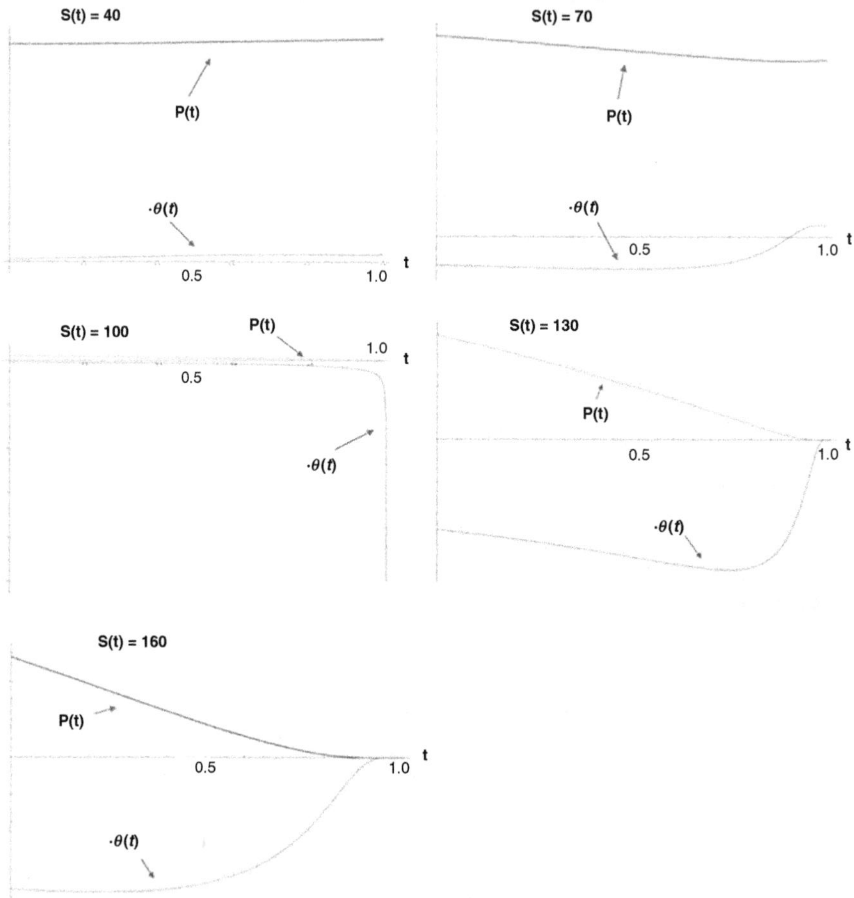

Fig. 4.169 $P(t)$ as a function of time t for various values of $S(t)$ and associated (negative) theta values (orange)

Of particular interest in the above Graph 4.169 is the case where $S(t) = 70$. Here, the (negative) theta value changes from negative to positive just before the option expires.

- The absolute values of the theta increase the closer $S(t)$ is to the strike price K.

Just like we saw for the call option, the basic shape of the theta changes for the put option too, as soon as the **risk-free interest rate *r*** is **negative**. In that case, the (negative) theta values are then actually always negative, and ***P*(*t*) is always monotonically decreasing over time** when ***r*** is negative **!**

Vega

For the graphical representation of a put option's vega and its properties, we can simply refer to Fig. 4.165. The vega of a put option has the same value as the vega

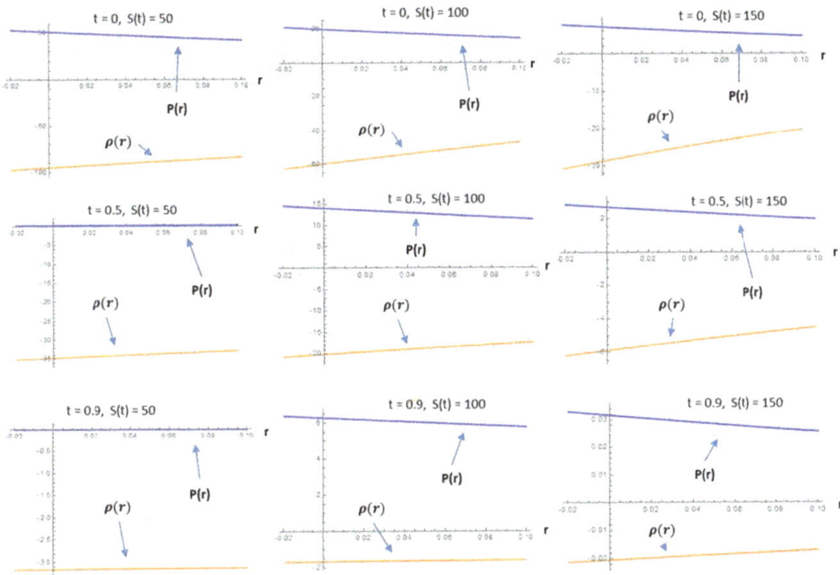

Fig. 4.170 Put price $P(r)$ (blue) and rho $\rho(r)$ (orange) for various values of t and $S(t)$ as a function of the risk-free interest rate r

of a call option. The obvious reason for this can be found, for example, in the put-call parity equation. After all, based on this equation, $P(t) = C(t)+K\cdot e^{-r(T-t)} - S(t)$.

At fixed values for $t, T, r, S(t) = s$, the prices P of put options and C of call options differ for any value of σ by the fixed constant $K \cdot e^{-r(T-t)} - S(t)$ (which is independent of σ). Therefore, the derivative of P with respect to σ is equal to the derivative of C with respect to σ.

Rho

To represent the rho of put options, we will proceed in the same way as we did for the call options and simply create Fig. 4.170 as a counterpart to Fig. 4.166.

- The rho $\rho(r)$ of a put option is always negative, i.e. $P(r)$ is always monotonically decreasing in r.
- Changes in $P(r)$ as a function of r largely occur approximately linearly.

4.40 Delta and Gamma: Analysis of a Put Bull Spread

In addition to its role as a general risk measure of a trading strategy and its importance for hedging derivative positions, the delta also plays an essential role in fund management regulations by official supervisory bodies and those of investment companies. Certain types of investment funds, for example, are not allowed to exceed a certain delta exposure at any point during an option's life. If it is exceeded, then (usually after a certain grace period) positions must be closed or new positions must be opened, so that the delta of the strategy drops back to a level below the stipulated limit.

Using the programs on our website, readers can determine and graphically illustrate the current delta and gamma for any combinations of options and futures on an underlying asset and anticipate potential further delta and gamma movements.

In the following, by way of example, we are going to analyse the delta of just one very basic options combination on the basis of the Black-Scholes formula. The combination we want to analyse is a **Bull Put Spread** (= short put P_1 with strike K_1 and long put P_2 with strike K_2, same expiration T and $K_1 > K_2$).

We perform the **delta/gamma analysis of a Bull Put Spread** again on a specific numerical example. For this purpose, we assume the following parameters as fixed at time $t = 0$:

$$S_0 = 2900$$
$$T = \frac{1}{12}$$
$$r = 0.02$$
$$\sigma = 0.15$$
$$K_1 = 2800$$
$$K_2 = 2750$$

Moreover, we assume an existing investment of 100,000 dollars that we want to utilize to the full.

For these parameters, the Black-Scholes formula gives us a fair price of the short position P_1 of 13.58$ per unit (i.e. 1358$ per 100-unit-contract) and a fair price of the long position P_2 of 5.98$ per unit (i.e. 598$ per contract). The premium received per combination contract is 760$. The required margin per combination contract is 5000$, since the maximum potential loss in the worst-case scenario for such a short/long combination (if the underlying asset falls below 2750 points) would be 50$ per unit ($= K_1 - K_2$) or 5000$ per contract. Since a premium of 760$ is received per combination contract, the required additional margin is $5000 - 760 = 4240$$.

Since we already have a 100,000$ margin, we can enter into $\left[\frac{100{,}000}{4{,}240}\right] =$ 23 combination contracts. This ensures that there is always sufficient margin throughout the entire life of the options. ($[x]$ denotes the floor function of any number x, i.e. $[x]$ denotes the largest integer that is less than or equal to x. For example, $\left[\frac{100{,}000}{4{,}240}\right] = [23.5859\ldots] = 23$).

However, caution should be exercised if the margin is deposited in a foreign currency amount equivalent to 100,000$ (e.g. in euros). If the euro exchange rate were to fall against the dollar during the life of the options, the deposited margin would fall below 100,000 dollars, triggering a margin call or forcing the investor to close contracts. So, the premium received is $23 \times 760 = 17{,}480$$. The payoff and profit functions of the 23 combination contracts have the following form, given in Fig. 4.171:

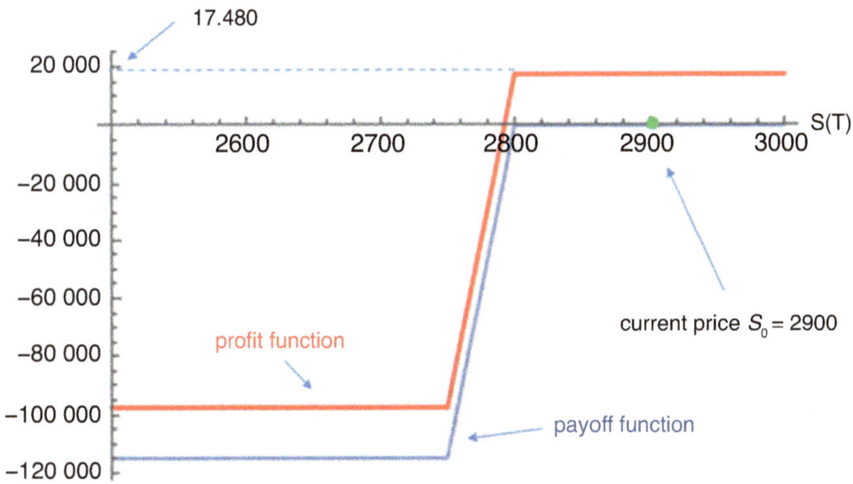

Fig. 4.171 Profit and payoff function Bull Put Spread

If, at expiration, the underlying asset remains above the strike K_1 of the short position, then both option types expire worthless and the investor pockets 17,480$ in profit (= 17.48% of the investment used). If the S&P500 falls below 2750 points by the expiration date and if contracts to limit losses have not been closed in time beforehand (!), then a payment of 5000$ ($= 100 \times (K_1 - K_2)$) will be due per combination contract, i.e. a total payment of 115,000$, which is almost the entire sum of the 100,000$ investment and the 17,480$ premium received. The investor practically loses the entire investment.

Let us first visualize an approximate price curve for this strategy over time based on the Black-Scholes formulas for the fair price of the two options that make up the combination. Again, we will do this (like we did in Sects. 4.32, 4.33, and 4.35) for the times $t = 0$ (red), $t = \frac{T}{4} \approx 1$ week (magenta), $t = \frac{T}{2} \approx 2$ weeks (turquoise), $t = 3\frac{T}{4} \approx 3$ weeks (green), $t = 0.9T \approx 3$ days before expiration (blue), and $t = T$ = expiration (payoff function) (black) and plot the curves in a graph. In Fig. 4.172, we do this assuming constant volatility $\sigma = 0.15$, and in Fig. 4.173, we assume volatility that is negatively correlated with the underlying asset's price. Again we assume a relationship of the form $\sigma_t = \sigma \cdot \left(\frac{S_0}{S_t}\right)^a$, where $a = 6$, between the volatility at time t and the price of the underlying asset at time t. (σ_t denotes the volatility of the underlying asset at time t. σ_t therefore increases as the underlying asset's price S_t falls at time t.)

Obviously, the price curves increasingly approach the strategy's profit function as t approaches T. However, the curves also reveal a dilemma faced by everyone implementing this strategy: It stands to reason and makes good sense that, in addition to the Bull Put Spread combination strategy, a profit-taking strategy and

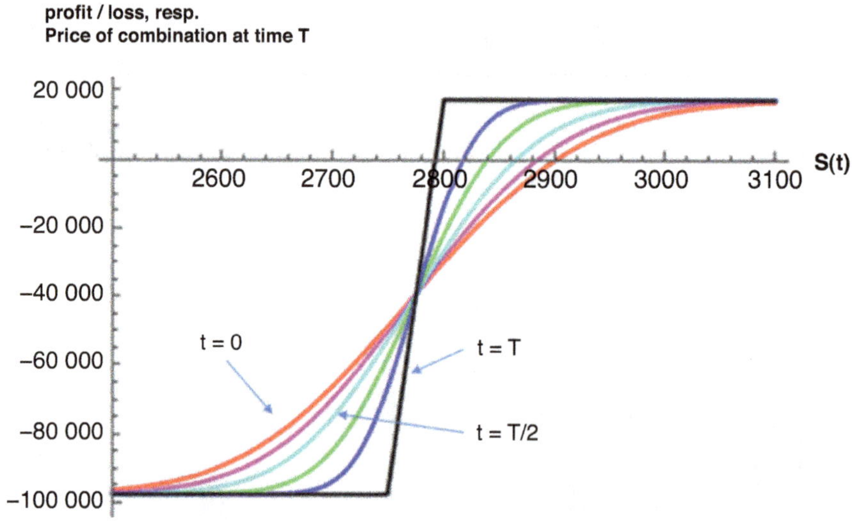

Fig. 4.172 Bull put spread price as a function of $S(t)$ for various times t, assuming constant volatility over time until expiration

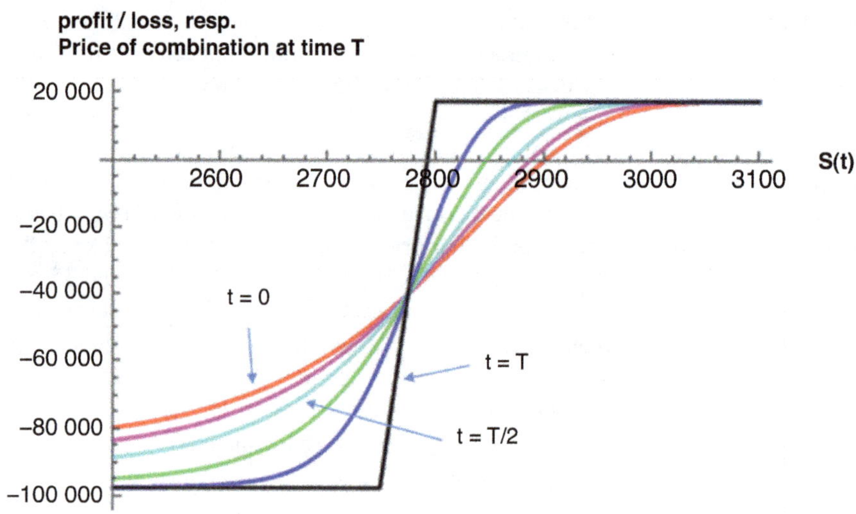

Fig. 4.173 Bull put spread price as a function of $S(t)$ for various times t, assuming volatility that is negatively correlated with the underlying asset's price

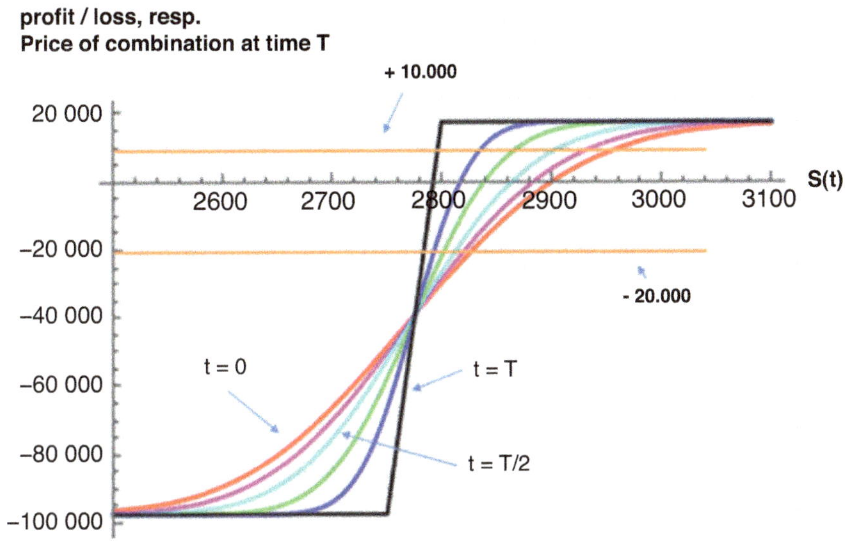

Fig. 4.174 Bull put spread price as a function of $S(t)$ for various times t, assuming constant volatility over time until expiration and profit-taking barrier at 10,000\$ and stop-loss barrier at $-20{,}000\$$

stop-loss strategy be considered, that is, to close the combination if a certain minimum profit or maximum loss level is exceeded, and to then stick to that strategy without fail.

In our example, one possible approach would be as follows: Take profit (i.e. by closing the strategy) as soon as a profit of 10,000\$ can be realized, and limit losses (by closing the strategy) as soon as a loss of 20,000\$ would occur. In Fig. 4.174, these two barriers are shown in orange and for better clarity, the critical area where closing the strategy is necessary is shown as a zoom in Fig. 4.175.

Figure 4.175 clearly shows that early profit-taking is only possible at a relatively high price of the S&P500, while on the other hand, it is necessary to pull the plug relatively early to limit losses in the event of price losses. For example, after approximately 1 week's time in (magenta-coloured curve), it is possible to take a profit of approximately 10,000 euros with the S&P500 at approximately 2940 points, while at approximately 2820 points, the strategy needs to be closed at $-20,000\$$ to limit losses. One week before expiration (green line), these limits are approximately 2870 points (profit-taking) and approximately 2805 points (loss limitation). Now of course this strategy comes at the risk that in the event of a temporary drop of the S&P500, a loss of 20,000\$ could potentially be incurred only one week in, while the same price of the S&P500 at expiration would result in a maximum profit, so there is the risk of closing too hastily. On the other hand, however, the profit/loss barriers are clearly defined.

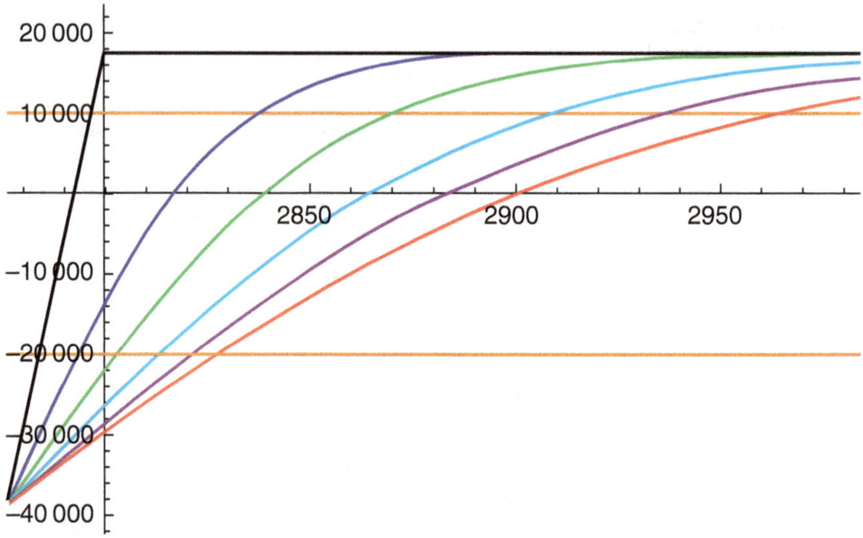

Fig. 4.175 Bull put spread price as a function of $S(t)$ for various times t, assuming constant volatility over time until expiration and profit-taking barrier at 10,000$ and stop-loss barrier at −20,000$, zoom of critical area

Figures 4.176 and 4.177 illustrate the same situation for the (more realistic) case of volatility that is negatively correlated with the underlying asset's price (correlation parameter $a = 6$).

Here, the barriers for profit-taking and loss limitation are at approximately 2930 and 2830 points after 1 week and at approximately 2875 and 2810 points after approximately 3 weeks.

Another loss limitation method would consist in closing the strategy at exactly the time when the S&P500 drops into the loss range of the payoff function, i.e. below 2800 points, for the first time. The drawback of this method obviously is that the actual loss in this case cannot be accurately calculated beforehand and can in fact be very high, for example, in case of greatly increased volatility. We see the situation in Fig. 4.178 (zoom in Fig. 4.179) for the case of constant volatility and in Fig. 4.180 (zoom in Fig. 4.181) for the case of volatility that is negatively correlated with the price.

In the event of constant volatility, this approach would result in the following losses in our setting:

After 1 week approximately −28,000$
After 2 weeks approximately −26,000$
After 3 weeks approximately −22,000$
Approximately 3 days before expiration −14,000$
Profit at expiration of approximately 17,400$

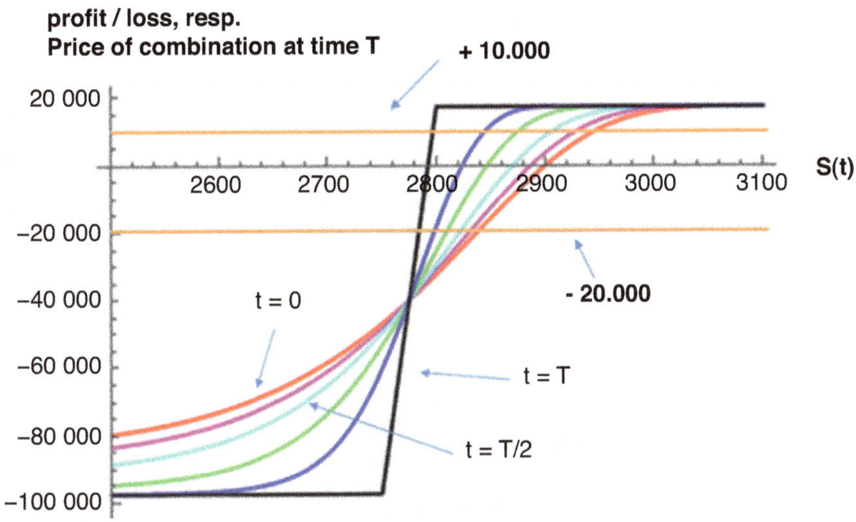

Fig. 4.176 Bull put spread price as a function of $S(t)$ for various times t, assuming volatility that is negatively correlated with the underlying asset's price over time until expiration, with profit-taking barrier at 10,000$ and stop-loss barrier at $-20{,}000$$

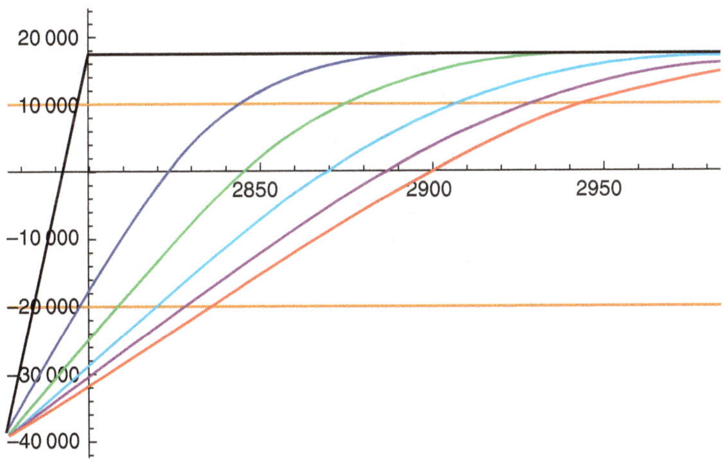

Fig. 4.177 Bull put spread price as a function of $S(t)$ for various times t, assuming volatility that is negatively correlated with the underlying asset's price over time until expiration, with profit-taking barrier at 10,000$ and stop-loss barrier at $-20{,}000$$, zoom of critical area

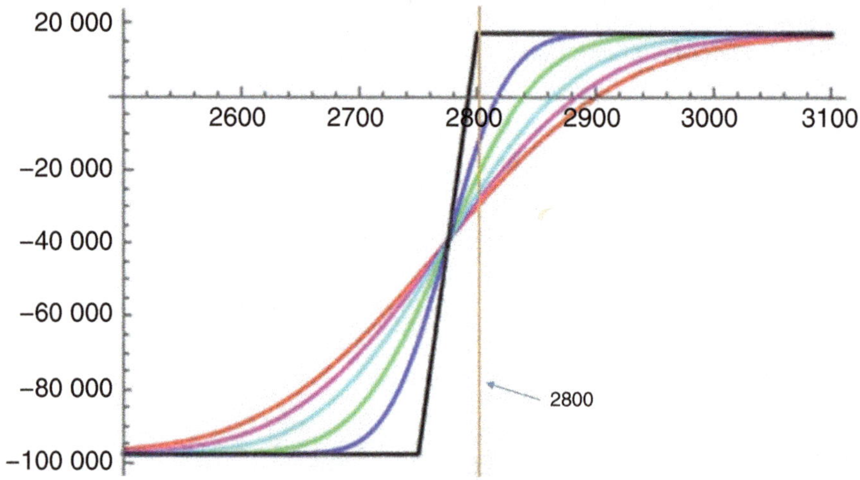

Fig. 4.178 Bull put spread price as a function of $S(t)$ for various times t, assuming constant volatility over time and loss amounts upon closing the positions at 2800 points

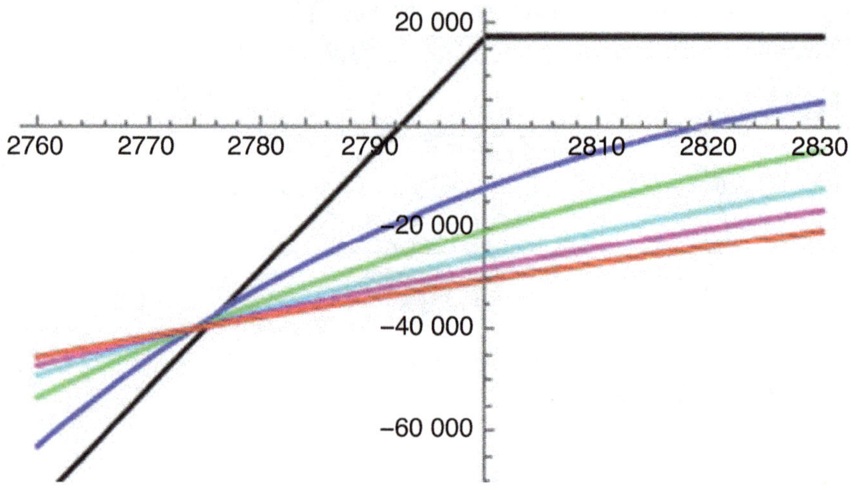

Fig. 4.179 Bull put spread price as a function of $S(t)$ for various times t, assuming constant volatility over time and loss amounts upon closing the positions at 2800 points, zoom of critical zone

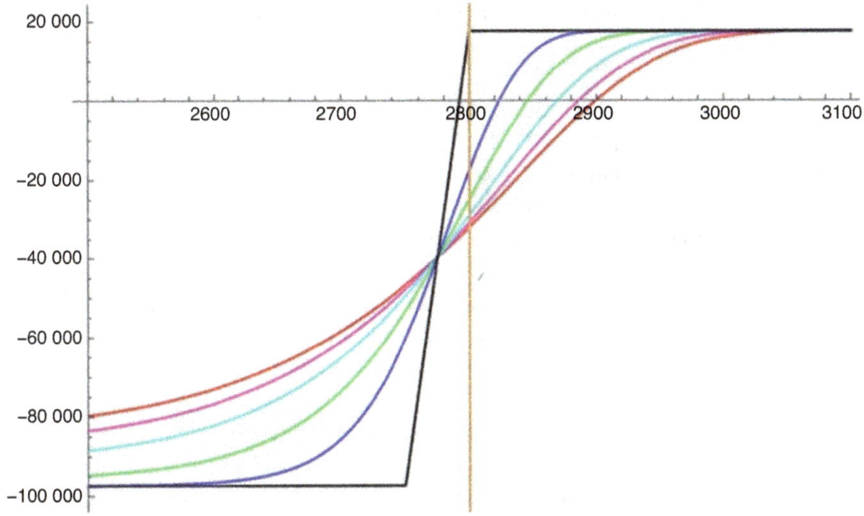

Fig. 4.180 Bull put spread price as a function of $S(t)$ for various times t, assuming volatility that is negatively correlated with the underlying asset's price over time and loss amounts upon closing the positions at 2800 points

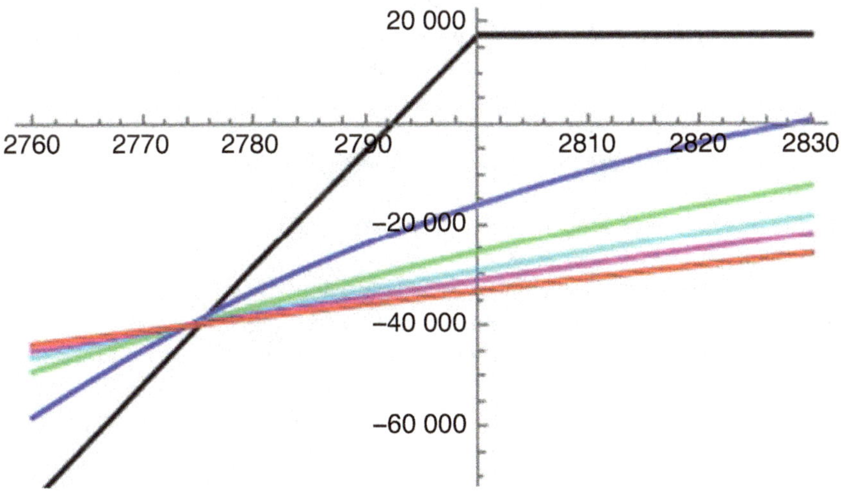

Fig. 4.181 Bull put spread price as a function of $S(t)$ for various times t, assuming volatility that is negatively correlated with the underlying asset's price over time and loss amounts upon closing the positions at 2800 points, zoom of critical zone

In the event of correlated volatility, this approach would result in the following losses in our setting:

After 1 week approximately −32,000$
After 2 weeks approximately −31,000$
After 3 weeks approximately −29,000$
Approximately 3 days before expiration −18,000$
Profit at expiration of approximately 17,400$

We are going to test such strategies again in more detail, using real data and real trading conditions, in our sample case in Volume III Section 3.5!

Clear guidelines are indispensable and essential when it comes to risk management of derivative trading strategies. However, every trader should also be aware that loss limitation rules based on stop-loss orders, i.e. rules on closing positions (or opening hedging positions) when certain price events occur, may not always be executable as planned. In his role as a strategy manager, the author of this book had a somewhat agonizing experience in this sense on 10 October 2008, at the height of the financial crisis. In early October, he had opened a large number (approximately 20,000 contracts) of bull put spreads on the S&P500 with expiry 17 October 2008, strike of the short put position at $K_1 = 900$, and strike of the long put position at $K_2 = 825$, both in strategies he managed and in his own trading strategies. The "exit strategy" was to close all contracts upon the S&P500 falling below the limit of 900 points. In Sect. 1.22, we already presented the actual movements of the S&P500 index in the days around 10 October 2008, and we show the corresponding chart with the S&P500 tick data again in Fig. 4.182, from the time shortly before close of trading on 9 October 2008 until shortly after start of trading on 13 October 2008.

On 9 October 2008, the S&P500 had closed at 909.92 points. Therefore, due to the exit strategy, none of the contracts had yet been closed on 9 October 2008. Within a matter of just a few minutes after the exchange opened on 10 October 2008, the index plunged below 900 points and straight down to 839.80 points. The order to close all contracts had been placed upon start of trading. By the time the closings could actually be carried out, however, the price of the S&P500 had already dropped below 850 points. Figure 4.183 shows the price movements of the short put position with strike 900. At the close of the exchange on 9 October 2008 (S&P500 at 909 points), the option price was still around 40$ (lower red dot in Fig. 4.183). Immediately after the exchange opened on 10 October 2008 (S&P500 at around 850 to 840 points), the option price jumped up to around 70–80 dollars. The put options were then closed at an average price of approximately 78 dollars. If closing the options

(continued)

Fig. 4.182 The S&P500 Index around 10 October 2008, tick data, (source: Bloomberg)

instantly when the S&P500 hit 900 points had been possible, the closing price would likely have been approximately 45 dollars. And a closing price in this region is what had been expected from the outset. Although this extremely unfortunate constellation had come about at a time of immense irritation in the financial markets, it serves as a warning against relying too unreservedly on the effectiveness of stop-loss orders.

Another example where stop-loss orders failed to work as expected (in the area of currencies) will be studied in detail in one of the case studies that we are going to discuss in one of the later real-world application chapters (see Volume III Section 3.7).

Deciding which exit strategy will work best (and we will now assume in the following that the exit strategies do indeed work precisely as expected, i.e. that it will be possible to close positions at the predefined point) is very difficult and depends of course on the strategy and on the actual movements of the underlying asset's price. As an example of how to approach the study of this question through simulations, as well as in order to gain an idea of the quality of the various exit strategies, we will discuss this question again in more detail in the next section,

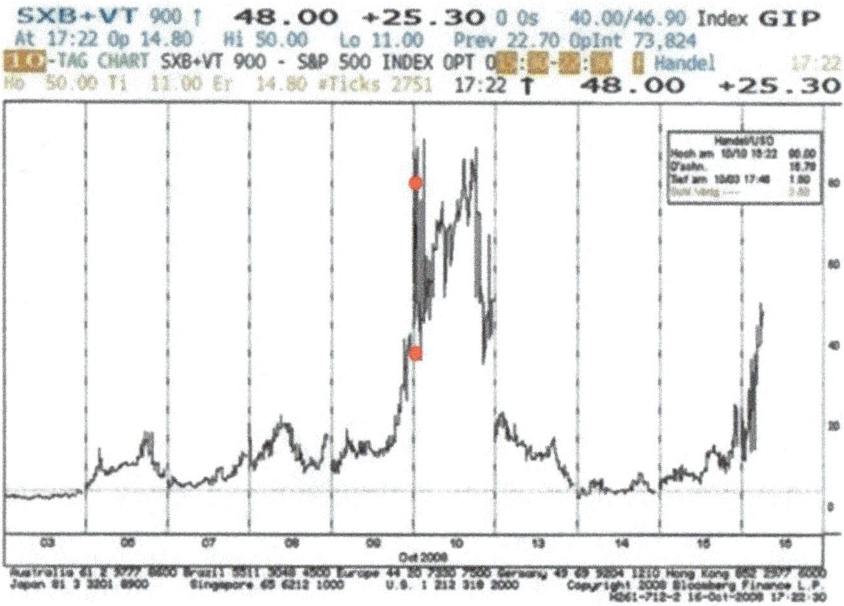

Fig. 4.183 Put price movements, strike 900, expiry 21 October 2008, from 3 October 2008 to 16 October 2008, (source: Bloomberg)

for the case of a bull put spread strategy (in the above setting and with the above parameter choices).

Before we do so, however, let us get back to our initial question in this section, namely: How do the delta and gamma of this strategy evolve over time as a function of the underlying asset's price?

The delta and gamma of the bull put spread strategy simply result as the sum of the deltas and gammas of the individual options positions. Recall that the strategy consisted of 23 contracts (i.e. 2300 units) of short puts (P_1) and of 23 contracts (i.e. 2300 units) of long puts (P_2). If we let Δ denote the delta of the overall strategy and Γ the gamma of the overall strategy and Δ_1, Δ_2 and Γ_1, Γ_2 the delta and gamma of the options P_1 and P_2 we get $\Delta = 2300 \times (\Delta_2 - \Delta_1)$ and $\Gamma = 2300 \times (\Gamma_2 - \Gamma_1)$. Substituting in the formulas to compute Δ_1, Δ_2 and Γ_1, Γ_2, again for the times $t = 0, \frac{T}{4}, \frac{T}{2}, 3\frac{T}{4}$ and $0.9T$ yields the following graphs for Δ and Γ (Fig. 4.184; each upper chart for constant volatility and each lower chart for volatility that is negatively correlated with the underlying asset's price).

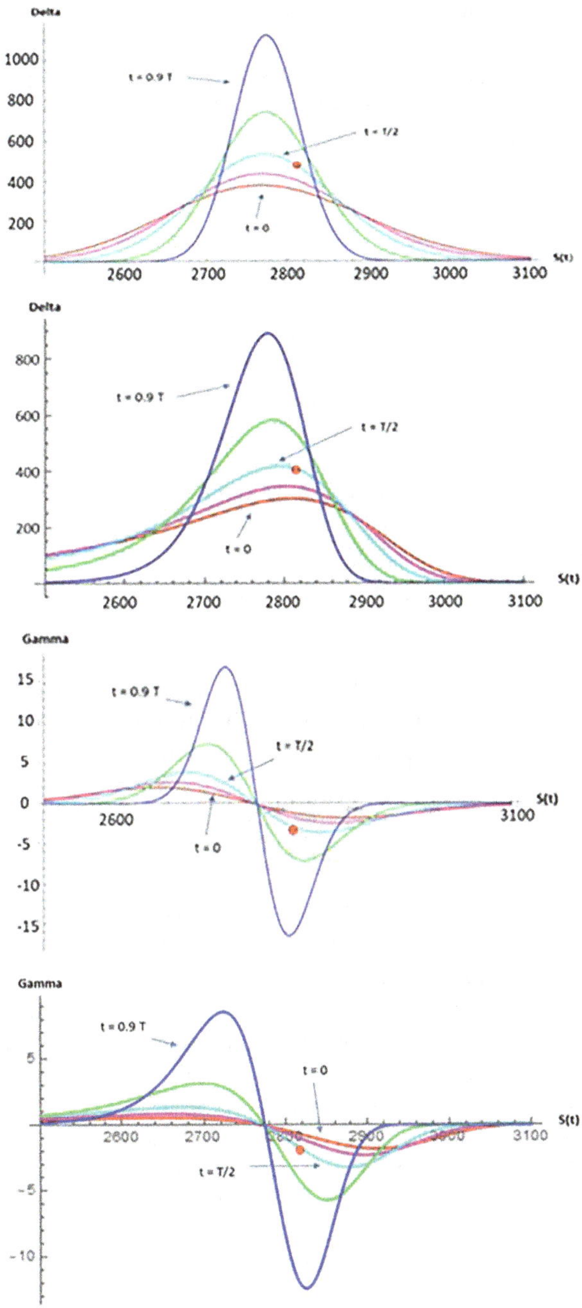

Fig. 4.184 Delta (top) and gamma (bottom) of the bull put spread for times $t = 0, \frac{T}{4}, \frac{T}{2}, 3\frac{T}{4}$ und $0.9T$ as a function of $S(t)$, assuming constant volatility (first and third graph) and assuming volatility that is negatively correlated with the underlying asset's price (second and fourth graph)

Some comments on the delta and gamma values of the bull put spread:

- The difference between the results for constant versus correlated volatility is not big; it consists mainly in a certain "left skew" of the delta curve in the case of correlated volatility.
- The delta values are consistently positive. However, this fact is clear given the form of this strategy: An increase in the price of the underlying asset will always have a positive effect on the price of this option strategy.
- Let's look at a specific example: The red dots in Fig. 4.184 denote delta and gamma values after passage of half of the strategy's life $\left(t = \frac{T}{2}\right)$, at 2820 points of the S&P500 (i.e. 20 points above the strike of the short position, thus in a range where a loss at expiration is not necessarily to be expected, yet where the strategy is beginning to enter a critical phase).

  In the case of constant volatility:
  The delta value at this point is 454.77 and the gamma value at this point is -2.91.
  In the case of negatively correlated volatility:
  The delta value at this point is 402.43 and the gamma value at this point is -1.88.

  Thus, if, after passage of half of the option strategy's life, the price of the S&P500 is approximately 2820 points and increases by one point, the value of the option strategy increases by approximately 455 dollars (at constant volatility). (And: If the S&P500 falls by another point, the (price) losses in the option strategy increase by another 455 dollars approximately.)
  If (at midpoint of the option strategy's life) the S&P500 has risen by one point (to 2821) and rises by another point, then an increase in the value of the option strategy of approximately 452 dollars is to be expected.
  This is because due to the increase in the S&P500, the delta has changed by about -2.91 (= gamma value), from about 455 to about 452.
- The delta value $\Delta(t, s)$ tells us how many units of the underlying asset need to be held for a perfect hedging strategy at time t and an underlying asset value s. So, at two different but closely spaced points in time t and $t + dt$ (dt representing a very small time span), we need to first hold $\Delta(t, S(t))$ and then $\Delta(t + dt, S(t + dt))$ units of the underlying asset. If no other rebalancing occurs between these time points, it will be necessary to reallocate (buy or sell) $|\Delta(t + dt, S(t + dt)) - \Delta(t, S(t))|$ units of the underlying asset at time $t + dt$. If we subdivide the time interval $[0, T]$ into N parts, each of the length dt, and if we rebalance our strategy only at times $i \cdot dt$, we would have to buy or sell, as the case may be, a total of

$$\sum_{i=0}^{N-2} |\Delta((i+1) \cdot dt, S((i+1) \cdot dt)) - \Delta(i \cdot dt, S(i \cdot dt))|$$

units of the underlying asset over the strategy's life. (There is no rebalancing at time 0 and at time $N \cdot dt$.) To get an approximate idea of how many rebalancing trades would be necessary to achieve (near) perfect hedging in the currently contemplated strategy, we are going to simulate and illustrate five possible price curves of the S&P500 in the following (see Fig. 4.185), using the Wiener model (for the trend $\mu = 0$). For each of these price curves, we will determine and illustrate the delta values for our strategy (23 bull put spread contracts) for each point in time and each associated value of the S&P500. Finally, we are going to calculate

$$\sum_{i=0}^{N-2} |\Delta((i+1) \cdot dt, S((i+1) \cdot dt)) - \Delta(i \cdot dt, S(i \cdot dt))|$$

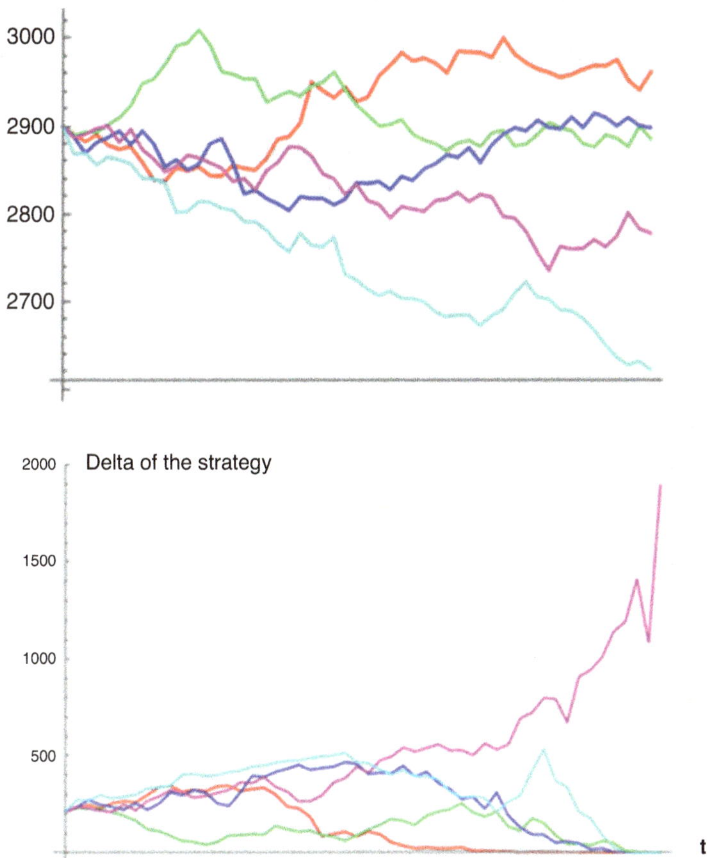

Fig. 4.185 Five simulated price curves for the S&P500 (above) and associated changes in the delta $\delta(t, S(t))$ (below) of the bull put spread strategy (23 contracts)

as an approximation of the number of underlying units that need to be traded for rebalancing purposes. We assume two trades per day.
The number of units that need to be traded are:

Red price: 597 units
Green price: 1828 units
Blue price: 1688 units
Magenta price: 2093 units
Turquoise price: 489 units

The value $\sum_{i=0}^{N-2} |\Delta((i+1)\cdot dt, S((i+1)\cdot dt)) - \Delta(i\cdot dt, S(i\cdot dt))|$ is the approximate variation of the function $\Delta(t, S(t))$, i.e. the approximate total variation of this function on the time interval $[0, T]$. (Take a look at the green curve in the lower picture: This curve starts at a value of approximately 200, then it drops to almost 0, moves back up to about 200, and finally closes at approximately 0. The total variation is therefore approximately $200 + 200 + 200 = 600$. This supports the plausibility of the above value of 597 units that would need to be traded of the underlying asset.)

- If, for example, the delta has to be limited to 400 points due to risk management regulations, then it would make sense to start the strategy at an initial S&P500 value of 2900 points. The delta value would then be approximately 214 points (for both volatility cases). If the S&P500 price remains unchanged, the delta would then obviously decline over time (see Fig. 4.186).
  But if the S&P500 were to drop, even if only to 2840 points (i.e. still well above the critical short strike at 2800), this may cause the delta limit of 400 points to be exceeded. The formula (see corresponding program on this book's website) for the delta of the option strategy over time at an S&P500 value of 2840 points

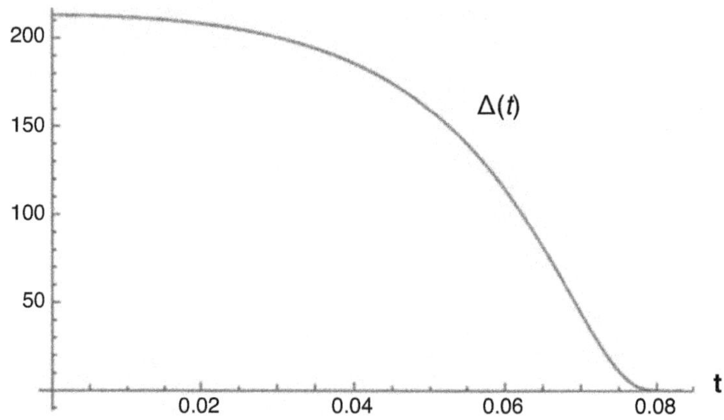

Fig. 4.186 Delta of bull put strategy for constant S&P500 value of 2900 points over time t

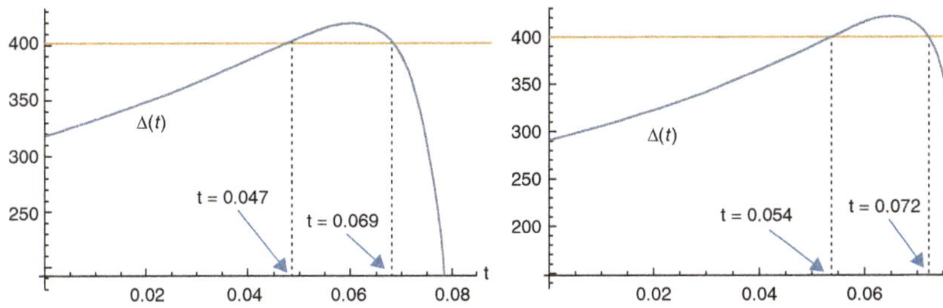

Fig. 4.187 Delta of a bull put spread combination over time with constant S&P500 price of 2840 points (constant volatility left, correlated volatility right)

yields the following Fig. 4.187 (with constant volatility on the left and correlated volatility on the right):

In the range for t between 0.047 and 0.069 (corresponding to the time range between day 17 and day 25)—assuming an S&P500 price of 2840 points—the delta stays above the limit of 400 (at constant volatility). (In the case of negatively correlated volatility, the threshold value is exceeded in the time range from approximately day 19 to day 26.)

- The maximum delta value (in both versions) is obviously (see Fig. 4.184) just to the left of the short strike K_1. In other words: The strongest price decline of the option strategy can be assumed to occur if the price of the underlying asset continues to fall from a value just below 2800 points.
- This maximum delta value (in both versions) increases sharply as time progresses toward expiration (from approximately 400 at the start of the strategy to more than 1000 3 days before expiration).
- The critical value of $S(t)$, i.e. the value for which maximum delta is assumed, remains relatively constant in the case of constant volatility, while it shifts more and more to the left in the case of volatility that is negatively correlated with the price over time. (See Fig. 4.184.)
- In our example here, the values $S(t)$ for which maximum delta is attained, are shown in the following Table 4.11:

Table 4.11 Values of $S(t)$ with maximum delta

	max. delta **const.** vola	at S&P500 value of	max. delta **correl.** vola	at S&P500 value of
$t = 0$	379	2767	301	2809
$t = \frac{T}{4}$	437	2769	344	2802
$t = \frac{T}{2}$	532	2771	418	2794
$t = 3\frac{T}{4}$	742	2773	581	2785
$t = 0.9T$	1125	2774	892	2779

4.41 Test Simulations of Exit Strategies for Bull Put Spread Combinations

As announced above, we are now going to perform some theoretical (!) tests of various possible exit strategies for bull put spread combinations. These are indeed just **theoretical tests** in that we:

1. Assume simulated and not real price movements of the S&P500.
2. Assume that the exit strategies will work precisely, i.e. that we are actually going to be able to execute trades at the specified limits.
3. Assume fair option prices calculated using the Black-Scholes formula.

For this we will proceed as follows:

We will again start with the following basic parameters:
$S_0 = 2900$
$r = 0.02$
$K_1 = 2800$ (strike of the short put position)
$K_2 = 2750$ (strike of the long put position)
$T = \frac{1}{12}$ (1-month expiration for the strategy)

For volatility σ and trend μ (which we need for simulation purposes, not for valuation), we choose different values and combine all possibilities for $\sigma = 0.1, 0.2$ and 0.3 and $\mu = -0.1, 0, 0.1$. All of these choices are just examples. Again, on the website we offer the environment for experimenting with a wide choice of parameters as well as trading strategies and exit strategies. For each of these 9 parameter pairs, we will perform 5000 simulations in which we simulate possible price movements in the Wiener model using the respective parameters. For each of these simulations, the gain or loss is calculated based on the bull put spread strategy. This means: The profit function will be evaluated at time T, unless the strategy was closed prematurely due to specific exit strategy regulations, in which case the losses resulting from closing will be used for valuation purposes.

As we are presenting examples only, we will test only the following variants:

Variant 1: No exit strategy.

Variant 2: Close all contracts as soon as the S&P500 falls below 2800. (The Black-Scholes formula is then used to calculate the costs of closing the option contracts. The loss then consists of these costs minus the initially received option premiums.)

Variant 3: Close all contracts if and when the S&P500 has dropped to such an extent that the loss incurred from the strategy (closing costs minus option premium received) would exceed 10%. (In this case, the loss is calculated as being 10%.)

Variant 4: Same as Variant 3 but with 20% instead of 10%

Table 4.12 Average result of 5000 simulations for each combination of parameter choice and exit strategy

	No exit	Exit variant 1	Exit variant 2	Exit variant 3
$\boldsymbol{\sigma = 0.1/\mu = -0.1}$	+4.170	+6.192	+4.426	+5.765
$\boldsymbol{\sigma = 0.1/\mu = 0}$	+9.707	+9.898	+7.280	+9.404
$\boldsymbol{\sigma = 0.1/\mu = 0.1}$	+12.818	+12.010	+9.011	+11.396
$\boldsymbol{\sigma = 0.2/\mu = -0.1}$	−13.922	−5.833	−2.055	−4.844
$\boldsymbol{\sigma = 0.2/\mu = 0}$	−7.806	−3.373	−1.022	−2.823
$\boldsymbol{\sigma = 0.2/\mu = 0.1}$	−3.376	−1.225	**20**	−1.007
$\boldsymbol{\sigma = 0.3/\mu = -0.1}$	−21.972	−11.779	−4.367	−8.953
$\boldsymbol{\sigma = 0.3/\mu = 0}$	−16.851	−9.951	−3.670	−7.659
$\boldsymbol{\sigma = 0.3/\mu = 0.1}$	−15.616	−9.349	−3.100	−6.965

For each choice of parameters and for each exit strategy, we determine the average result from 5000 simulations and present the results in the following Table 4.12.

So, relative to our investment of 100,000\$, a result of, for example, 4170\$ translates into an average profit of 4.17% per month (based on 5000 test simulations). These results don't appear especially attractive. According to this purely theoretical (!) analysis, it is only at a volatility of 10% that we can expect clearly positive results. These results were all generated assuming constant volatility. For volatility that is negatively correlated with the underlying asset's price, we need to expect even poorer results (for this kind of analysis). So this strategy is generally recommendable only in phases of low volatility.

In Table 4.12, the bold entries represent the best exit strategy for the respective choice of parameters. In six of the nine cases, exit variant 2 (exit as soon as a loss of 10% is reached) proved to be best. This was especially the case when losses were to be expected in the strategy. If we calculate the average result for a fixed exit strategy, taking the average over all of our parameter pairings, the only exit strategy that we get a positive average result for, of approximately 6.5%, is exit variant 2.

For illustration, let's pick the parameter pairing $\sigma = 0.1$ and $\mu = 0$ and look at the results and the simulation process for this case in much greater detail. The first Fig. 4.188 shows the convergence of our simulation results toward the estimated profit/loss in the strategy.

The charts show the average profit of the strategy calculated on the basis of the first n simulated prices for n from 1 to 5000. You can see a strikingly fast convergence toward a relatively stable average value (shown in red). Put simply, and purely informally, this means that the estimates obtained through the simulations are quite reliable. In the future, we will often encounter such graphs, in frequently used applications of so-called Monte Carlo methods, to illustrate the rate of convergence.

For now, let us not look at the average over the results, however; instead we are going to take a closer look at these individual results in detail, starting with illustrative histograms (see Fig. 4.189).

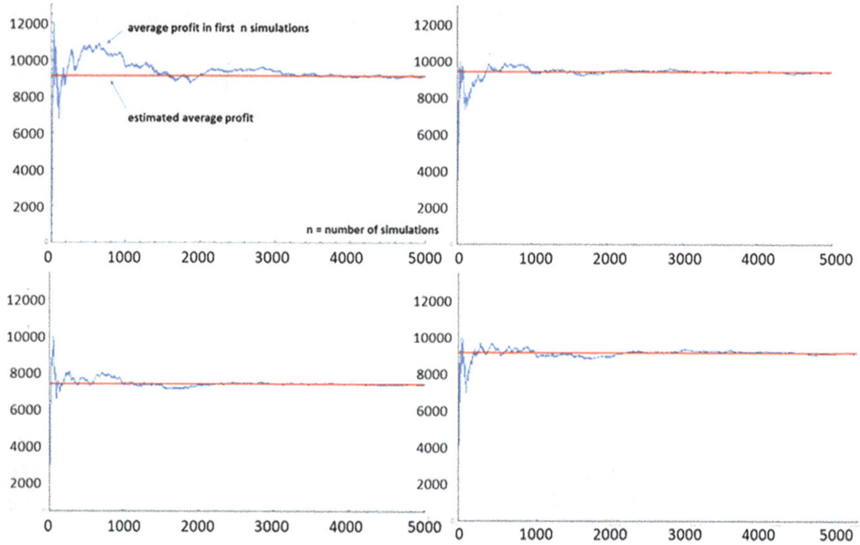

Fig. 4.188 Convergence speed of simulations (no exit (top left), exit variant 1 (top right), exit variant 2 (bottom left), exit variant 3 (bottom right))

In their basic form, the histograms look quite similar: Relatively many individual results in the potentially maximum profit range of close to 20,000$ and a significant number of loss results heavily in the negative range. What is essential here is where most of these losses occur for the individual variants. In this regard, we immediately recognize a fatal fact for the no-exit variant: This variant shows a number of losses close to 100,000$, i.e. losses that essentially mean we lose the entire investment. In exit variant 1, a significant number of losses occur in the range of about −25,000 to −20,000 dollars, and in exit variants 2 and 3, the loss-recording months are essentially found, not surprisingly, at 10,000$ and 20,000$, respectively. Let us illustrate these facts once more in a different way: Instead of performing 5000 simulations in parallel, we assume that this strategy is run 60 months in a row (reinvesting profits or assuming a lower base sum after sustaining losses, as the case may be). Simulating the price movements and the strategy results for these 60 months, we typically get results as shown in Fig. 4.190.

While all three strategies with strict exit variants show a clearly (exponentially) positive development, an (almost) total loss has completely destroyed the variant with no exit strategy. Since a (near) total loss event practically erodes the entire investment, a realistic chance of recovery is no longer possible. Seen from this perspective, it becomes much clearer, much more so than could be seen from the computations of average returns, that implementing a bull put spread strategy in the setting we chose (using the available investment to the full) proves fatal without a strict exit strategy.

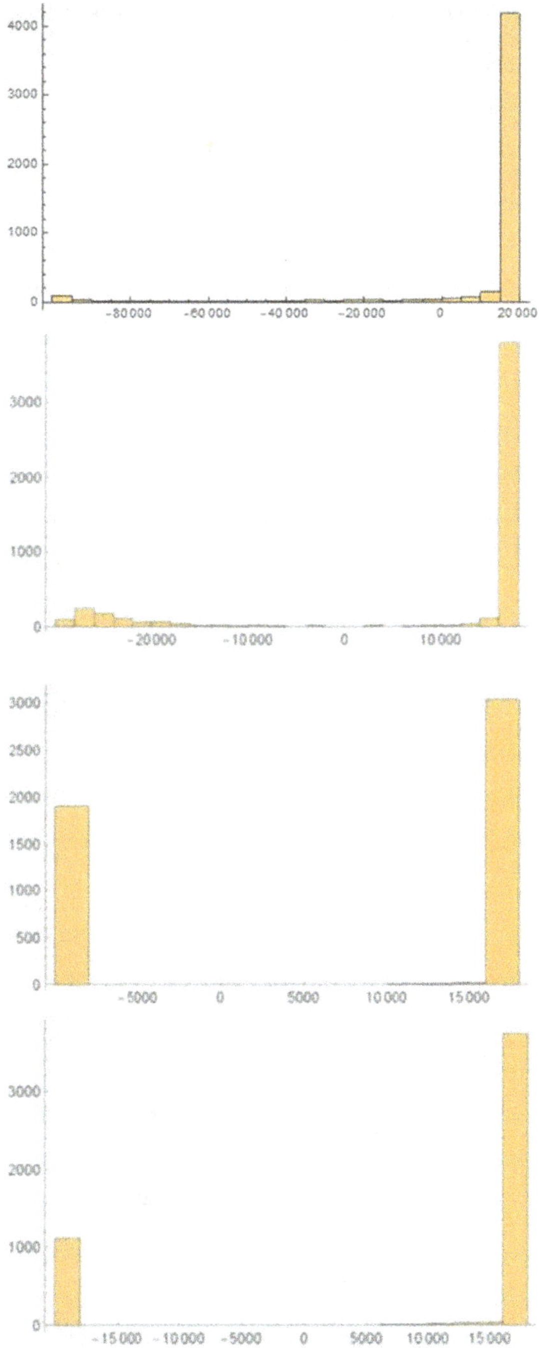

Fig. 4.189 Histograms of the individual strategy results per simulation (no exit (top), exit variant 1 (second from above), exit variant 2 (second from below), exit variant 3 (bottom))

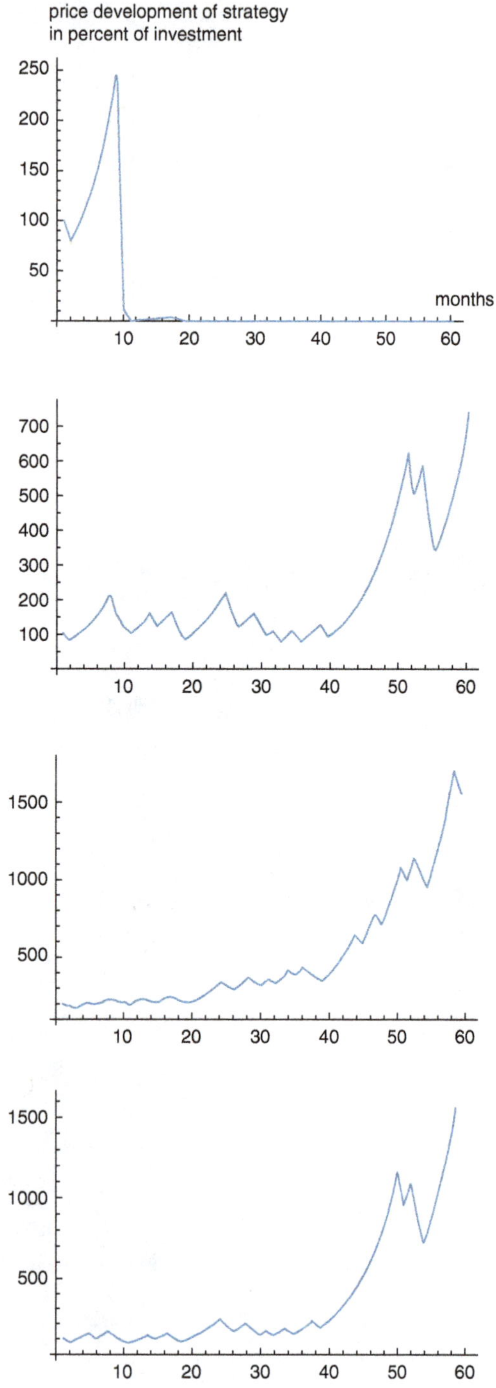

Fig. 4.190 Simulation of the bull put spread strategy over 60 months as a percentage of the initial investment (no exit (top), exit variant 1 (second from above), exit variant 2 (second from below), exit variant 3 (bottom))

Later, we will carry out analog investigations, in the form of back tests, using real S&P500 data and real option data. These results will then be much more instructive in terms of conclusions for real options trading. In many cases, however, such as when there aren't enough reliable real data available to perform an analysis, theoretical simulation-based analyses will provide basic insights to use as starting points.

4.42 Delta/Gamma Hedging

We already know how to perfectly hedge a derivative trading strategy in a time range $[0, T]$ over a Wiener model:

- You calculate the fair price $F(0)$ of the strategy at time 0. (This is the sum of the fair prices of all products included in the strategy at time 0.)
- For each time $t \in [0, T]$, you calculate the fair price $F(t, s)$ of the strategy at time t subject to the condition that the underlying asset has the value s at that point.
- For each time $t \in [0, T]$, you calculate the derivative of $F(t, s)$ with respect to s, i.e. the delta $\Delta(t, s)$ of the strategy.
- You start with a hedging investment of $F(0)$ euros and invest that in $\Delta(0, S(0))$ units of the underlying asset (where $S(t)$ denotes the price of the underlying asset at time t). The remainder of the hedging investment is invested at the risk-free spot interest rate r.
- This hedging portfolio of underlying and cash is (theoretically) rebalanced at each point in time $t \in [0, T]$ such that the portfolio always holds $\Delta(t, S(t))$ units of the underlying asset. All remaining cash assets (which can also be negative of course) are invested at the risk-free interest rate.

We have previously demonstrated that in this approach, the value of the hedging strategy at time T is exactly equal to the payoff of the derivative strategy. We have also pointed out that the hedging strategy can, of course, only be approximated in discrete form, by rebalancing the portfolio at a finite number of different points in time, and we have presented some experiments in this regard.

We are now going to present a simpler and at the same time more generalized approach to "hedging a derivative strategy":

Again, we assume a derivative trading strategy—which we are going to denote by D (for "derivative")—on an underlying asset S with price $S(t)$ over the time range $[0, T]$, and we also assume that we know the fair price $F(t, s)$ of the strategy D at any time t and for any price s. Occasionally, for the sake of simplicity, we will state this fair price only as a function of time t, i.e. in the form $F(t)$.

Our **first step** is to introduce another derivative product A (of any kind) with the same underlying S. In fact, A can simply be the underlying S itself! This is because the underlying S can itself be understood to be a derivative on S with the payoff function $\phi(S(T)) = S(T)$. The fair price of derivative A at time t is then of course simply $S(t)$, as otherwise, the underlying and the derivative A could immediately be exploited for arbitrage. In any case, we assume that A is a highly liquid derivative on S that can be traded without any limitations. We further assume that we know the fair price $A(t, s)$ of A at any time t and for any price s of the underlying. Occasionally, for the sake of simplicity, we will state this fair price only as a function of time t, i.e. in the form $A(t)$.

We now assume that we have bought the derivative (strategy) D at the price $F(0)$ and that we want to substantially reduce the risk associated with holding the derivative D, that is, the risk of sharp swings in the derivative price as a result of changes in the underlying price. This is a risk that we want to mitigate by adding derivative A to derivative D. More precisely:

At any time $t \in [0, T]$, we want to hold exactly $x(t, s)$ units of the underlying asset A, if s exactly denotes the underlying price $S(t)$ at time t. This means:

At any time $t \in [0, T]$, our portfolio will consist of 1 unit of D and $x = x(t, s)$ units of A. The aggregate price G of this portfolio at time t and at the price of S the underlying will therefore be:

$$G(t, s) = F(t, s) + x \cdot A(t, s) \tag{4.10}$$

Now, the idea is to choose $x = x(t, s)$ such that the aggregate portfolio's sensitivity to changes in the underlying price is minimized as much as possible. This goal is certainly best achieved if the derivative of G with respect to s, i.e. the delta of G, equals 0 as often as possible, ideally at all times. So, we differentiate (4.10) on both sides and get:

$$0 = \Delta_G(t, s) = \Delta_F(t, s) + x \cdot \Delta_A(t, s), \text{ i.e. } x = x(t, s) = -\frac{\Delta_F(t, s)}{\Delta_A(t, s)}$$

Hedging the strategy D by holding $x(t, s) = -\frac{\Delta_F(t,s)}{\Delta_A(t,s)}$ units of derivative A at any point in time t is called **delta hedging** the strategy D through A.

In the special case where $A = S$, we get $\Delta_A(t, s) = \frac{dA(t,s)}{ds} = \frac{ds}{ds} = 1$.

This gives us $x = -\Delta_F(t, s)$, and we run the perfect hedging strategy that we are already familiar with.

What is the advantage of using a different derivative than the underlying S itself to hedge a portfolio? Two frequently relevant reasons are:

1. It often becomes necessary during a hedging operation to also short the hedging product A. In these cases it is often much easier to short a derivative than the underlying asset.
2. During the hedging process, the hedging portfolio has to be continuously rebalanced, meaning that derivative A needs to be continuously traded. Another argument in support of using derivatives for hedging purposes is that the prices of derivatives on an underlying S as well as the transaction costs for

derivatives trades are significantly lower (relatively speaking) than for trades in the underlying itself.

Again, only discrete hedging is possible in this process, i.e. trades at a finite number of different points in time. This again means that the aggregate portfolio's continuous **delta neutrality** which we created above applies only locally at the respective trading times. If for a short period of time, the portfolio is not adequately rebalanced, the delta of the portfolio moves away from the optimum value of 0 during that period. The hope is, of course, that the delta does not move too far away from 0 during such non-rebalancing periods. We also want to minimize the local change behavior of delta.

This can be achieved by using another liquid derivative B (with the fair price $B(t,s)$) on the underlying S. At each time t and for each price s of the underlying, we then additionally hold $y = y(t,s)$ units of B. The aggregate portfolio then has the price

$$G(t,s) = F(t,s) + x \cdot A(t,s) + y \cdot B(t,s) \tag{4.11}$$

Our requirements to this aggregate portfolio are as follows: For all t and s, both the delta and the gamma of the aggregate portfolio should be equal to 0. Since the gamma is the derivative of the delta with respect to s, there will be little change in the delta (at least locally), which means that the hedging portfolio will not need much rebalancing. By differentiating (4.11) once or twice, we then obtain the conditions

$$0 = \Delta_G(t,s) = \Delta_F(t,s) + x \cdot \Delta_A(t,s) + y \cdot \Delta_B(t,s)$$

$$0 = \Gamma_G(t,s) = \Gamma_F(t,s) + x \cdot \Gamma_A(t,s) + y \cdot \Gamma_B(t,s)$$

and solving this equation system yields

$$x = x(t,s) = \frac{\Gamma_F \Delta_B - \Delta_F \Gamma_B}{\Gamma_B \Delta_A - \Gamma_A \Delta_B} \text{ and } y = y(t,s) = \frac{\Gamma_F \Delta_A - \Delta_F \Gamma_A}{\Gamma_A \Delta_B - \Gamma_B \Delta_A}.$$

In order for these values x and y to exist, the denominators of the two above expressions must not be equal to 0, of course. This means that $\Gamma_B \Delta_A - \Gamma_A \Delta_B \neq 0$ must always hold. Not wanting to discuss this denominator in too much detail, let us say just as much about this requirement:

$\Gamma_B \Delta_A - \Gamma_A \Delta_B = 0$ simply means that the two vectors $\binom{\Delta_A}{\Gamma_A}$ and $\binom{\Delta_B}{\Gamma_B}$ are parallel. Intuitively, this means that the two derivatives used for hedging are very similarly dependent on the price movements of the underlying asset S. For hedging purposes, it makes good sense therefore to select two derivatives that respond very differently to changes in the underlying price.

If A is again the underlying S itself, then of course $\Delta_A = 1$ and therefore $\Gamma_A = 0$ and so in this special case, we get

$$x(t, s) = \Gamma_F \cdot \frac{\Delta_B}{\Gamma_B} - \Delta_F \text{ and } y(t, s) = -\frac{\Gamma_F}{\Gamma_B}.$$

The condition for the existence of this hedging portfolio is then simply $\Gamma_B \neq 0$. We refer to this type of hedging as **delta gamma hedging**.

Upon rebalancing the hedging portfolio at a finite number of points in time, the aggregate portfolio will become **delta and gamma neutral** locally. The hope is that the aggregate portfolio's delta neutrality will be somewhat longer-term compared to plain delta hedging. We illustrate the process and effect of delta hedging and delta/gamma hedging on a numerical example in the next section.

4.43 Delta/Gamma Hedging: A Realistic Example

We will again start with the S&P500 index and the following parameters:

$S_0 = 2900$
$r = 0.02$
$\sigma = 0.15$
$\mu = 0.10$

And we will consider the specific example of a derivative D on the S&P500 index with expiration $T = 1$ year

and the following payoff function $\Psi(s) := \begin{cases} 100 & \text{if } 2900 < s < 3100 \\ 0 & \text{otherwise} \end{cases}$.

The option therefore pays a profit of 100 dollars precisely when the price $S(T)$ of the underlying asset at time T (in 1 year) is between 2900 and 3100 points. Otherwise, the derivative expires worthless.

We start by determining the fair price $F(t, s)$ and its derivative with respect to s, i.e. $\Delta_F(t, s)$. The general Black-Scholes formula gives us (with $a = 2900$ and $b = 3100$):

$$\boldsymbol{F(t, s)} = 100 \cdot e^{-r(T-t)} \cdot \mathbb{E}\left(\mathbb{1}_{[a,b]}\left(s \cdot e^{\left(r-\frac{\sigma^2}{2}\right)(T-t)+\sigma\sqrt{T-t}w}\right)\right) =$$

(Here $\mathbb{E}$ denotes the expected value, $\mathbb{1}_{[a,b]}$ the characteristic function of the interval $[a, b]$, thus $\mathbb{1}_{[a,b]}(x) := \begin{cases} 1 & \text{if } x \in [a, b] \\ 0 & \text{otherwise} \end{cases}$

and w, as usual, a standard normally distributed random variable.)

$$= 100 \cdot e^{-r(T-t)} \cdot W\left(a < s \cdot e^{\left(r-\frac{\sigma^2}{2}\right)(T-t)+\sigma\sqrt{T-t}w} < b\right) =$$

(Here, W denotes the probability of the subsequent event.)

$$= 100 \cdot e^{-r(T-t)} \cdot W \times \left(\frac{\log\frac{a}{s} - \left(r-\frac{\sigma^2}{2}\right)(T-t)}{\sigma\sqrt{T-t}} < w < \frac{\log\frac{b}{s} - \left(r-\frac{\sigma^2}{2}\right)(T-t)}{\sigma\sqrt{T-t}}\right) =$$

$$= 100 \cdot \boldsymbol{e^{-r(T-t)}} \cdot (\mathcal{N}(\boldsymbol{L_2}) - \mathcal{N}(\boldsymbol{L_1})),$$

where $\mathcal{N}$ denotes the distribution function of the standard normal distribution, and

$$L_2 := \frac{\log\frac{b}{s} - \left(r-\frac{\sigma^2}{2}\right)(T-t)}{\sigma\sqrt{T-t}},$$

and

$$L_1 := \frac{\log\frac{a}{s} - \left(r-\frac{\sigma^2}{2}\right)(T-t)}{\sigma\sqrt{T-t}}.$$

Differentiating $F(t,s)$ with respect to s using the chain rule, and as $(\mathcal{N}(x))' = \phi(x) = \frac{1}{\sqrt{2\pi}}e^{-\frac{x^2}{2}}$, we get

$$\boldsymbol{\Delta_F(t,s)} = 100 \cdot e^{-r(T-t)} \cdot \left(\phi(L_2) \cdot \frac{dL_2}{ds} - \phi(L_1) \cdot \frac{dL_1}{ds}\right) =$$

$$= \boldsymbol{-100 \cdot \frac{e^{-r(T-t)}}{s \cdot \sigma\sqrt{T-t}} (\phi(L_2) - \phi(L_1))}.$$

We are going to test three hedging variants below:
Variant 1:
Conventional hedging (delta hedging) using underlying A
Variant 2:
Delta hedging using call option B
Variant 3:
Delta gamma hedging with underlying A and call option B

We will therefore also need Δ_B, Γ_B, and Γ_F:

$$\begin{aligned}
\boldsymbol{\Gamma_F(t,s)} &= 100 \cdot \left(\frac{e^{-r(T-t)}}{s^2 \cdot \sigma\sqrt{T-t}} \left(\phi\left(L_2\right) - \phi\left(L_1\right)\right) - \right. \\
&\left. - \frac{e^{-r(T-t)}}{s \cdot \sigma\sqrt{T-t}} \left(\phi'\left(L_2\right) \cdot \frac{dL_2}{ds} - \phi'\left(L_1\right) \cdot \frac{dL_1}{ds} \right) \right) \\
&= 100 \cdot \left(\frac{e^{-r(T-t)}}{s^2 \cdot \sigma\sqrt{T-t}} \left(\phi\left(L_2\right) - \phi\left(L_1\right)\right) - \right. \\
&\left. - \frac{e^{-r(T-t)}}{s \cdot \sigma\sqrt{T-t}} \left(L_2\phi\left(L_2\right) - L_1\phi\left(L_1\right)\right) \frac{1}{s\sigma\sqrt{T-t}} \right) \\
&= \boldsymbol{-100 \cdot \frac{e^{-r(T-t)}}{s^2 \cdot \sigma\sqrt{T-t}} \left(\phi\left(L_2\right) \cdot \left(\frac{L_2}{\sigma\sqrt{T-t}} - 1 \right) - \right.} \\
&\boldsymbol{\left. - \phi\left(L_1\right) \cdot \left(\frac{L_1}{2\sqrt{T-t}} - 1 \right) \right)}
\end{aligned}$$

We illustrate $F(t,s)$, $\Delta_F(t,s)$, and $\Gamma_F(t,s)$ in Figs. 4.191 and 4.192 for $t = 0, \frac{T}{4}, \frac{T}{2}, 3\frac{T}{4}$ and $0.9T$.

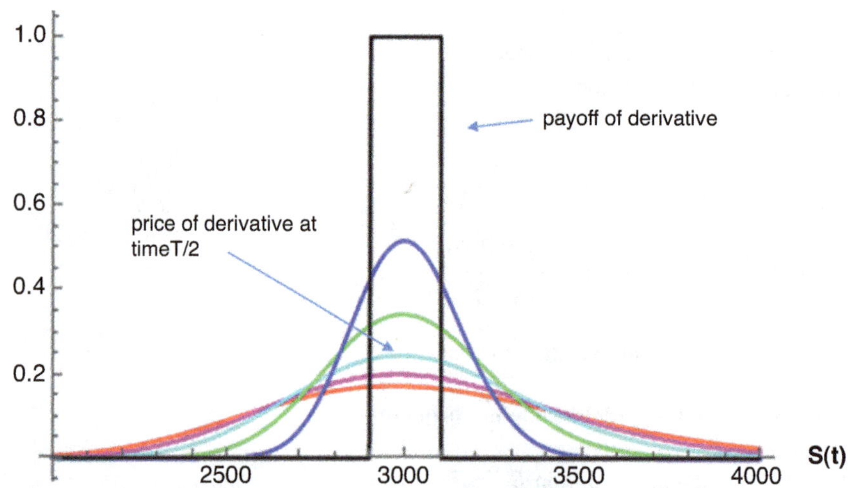

Fig. 4.191 Price of derivate D at times $t = 0$ (red), $t = \frac{T}{4}$ (magenta), $t = \frac{T}{2}$ (turquoise), $t = 3\frac{T}{4}$ (green), and $t = 0.9T$ (blue) and payoff function (black) as a function of $S(t)$, in multiples of 100

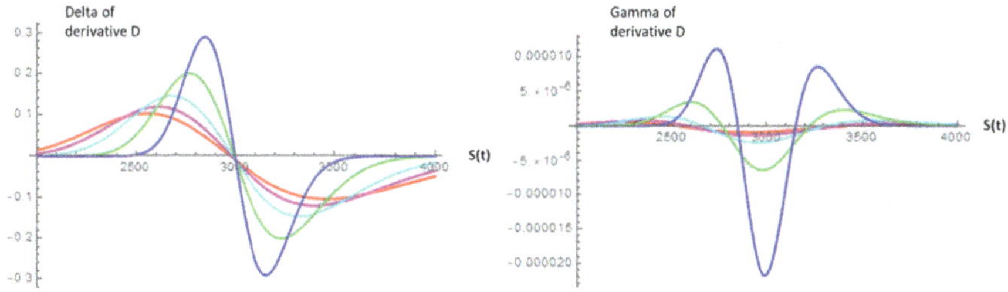

Fig. 4.192 Delta (left) and gamma (right) of derivative D at times $t = 0$ (red), $t = \frac{T}{4}$ (magenta), $t = \frac{T}{2}$ (turquoise), $t = 3\frac{T}{4}$ (green), and $t = 0.9T$ (blue) as a function of $S(t)$, in multiples of 100

We already know Δ_B and Γ_B as delta and gamma of a call option:

$$\Delta_B(s,t) = \mathcal{N}(d_1) \text{ and } \Gamma_B(s,t) = \phi(d_1) \cdot \frac{1}{s \cdot \sigma\sqrt{T-t}} \text{ with } d_1 = \frac{\log\left(\frac{s}{K}\right) + \left(r + \frac{\sigma^2}{2}\right)(T-t)}{\sigma\sqrt{T-t}}.$$

Now the hedging variants 1, 2, and 3 can be tested in detail (using the delta gamma hedging program on our website).

See: https://app.lsqf.org/book/delta-gamma-hedging

The Wiener model is then used to simulate various potential price paths for the S&P500 over the next year. The next step is to select the number of N equidistantly distributed adaptation times $(0, dt, 2 \cdot dt, \ldots, (N-1) \cdot dt)$ and, consequently, to calculate the trajectory of each selected hedging strategy, which should replicate the derivative as closely as possible. As we need to take into account transaction costs too, we also specify the number of hedging instruments we intend to trade.

Each hedging strategy starts with cash equal to the price of the derivative. For the above parameters, the price of the derivative is $F(0) = 17.02$ dollars. Then, at each of the points in time $0, dt, 2 \cdot dt, \ldots, (N-1) \cdot dt$, the portfolio is rebalanced so that the requisite number of hedging instruments is then held in the portfolio. The cash balance bears interest at the risk-free interest rate. The final value of this strategy at time T should be as close as possible to the payoff of the derivative. We also specify, for each individual test run, how many units of the hedging instruments are traded in total. This number is also relevant because of the transaction costs incurred. As already mentioned, using the simulation program on our website, readers can easily carry out a multitude of further tests with any parameters they choose. Here we will present only a few sample test results for our above example and a few comments.

One important **caveat**:

The derivative that we are dealing with in this example is extremely tricky to hedge, of course! This is because the payoff is not continuous. If the S&P500 is close to 2900 or 3100 just before the derivative expires, even small changes in the S&P500

can cause massive changes in the payoff. So the absolute delta is enormously high in these ranges. In such a situation it is practically impossible to achieve any reliable hedging, of course.

In the following, we will run all three hedging variants once for adjustments every 2 months (i.e. $N = 6$) and once for approximately daily adjustment ($N = 250$) and present four typical scenarios of each.

Test examples for **adjustments every 2 months**:

Scenario 1: Strong increase in the S&P500 price (compare Figs. 4.193 and 4.194):

Derivative payoff = 0 dollars

Variant 1:

Final value for delta-only hedging strategy with underlying = 1.12$
traded underlying units = 0.067 units

Variant 2:

Final value for delta-only hedging strategy with call option = 0.58$
traded call option units = 0.079 units

Variant 3:

Final value for delta-gamma hedging strategy $= -0.28\$$
traded underlying units = 2152.96 units
traded call option units = 2153.04 units

Scenario 2: S&P500 in critical range close to expiry (see Figs. 4.195 and 4.196):

Derivative payoff = 0 dollars

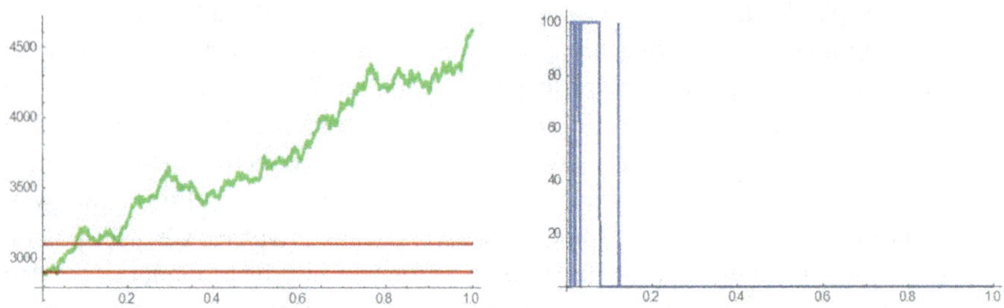

Fig. 4.193 Simulated price movements of the S&P500 (green), with barriers for profit range (red, left) and payoff if exercised immediately on given path (blue, right)

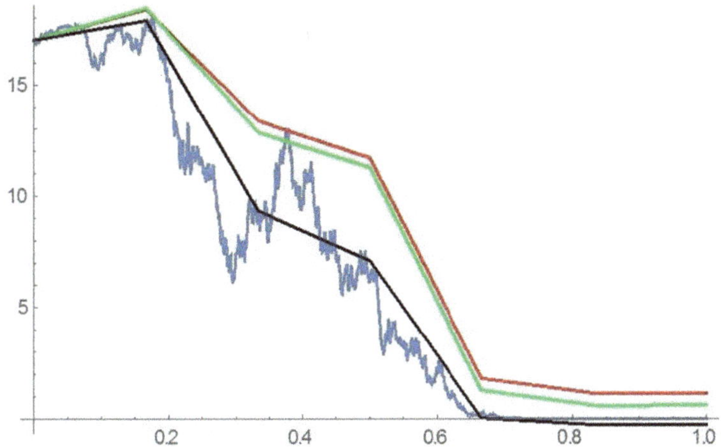

Fig. 4.194 Price movements of the derivative (blue), hedging strategy variant 1 (red), hedging strategy variant 2 (green), and hedging strategy variant 3 (black) with the above price movements of the S&P500

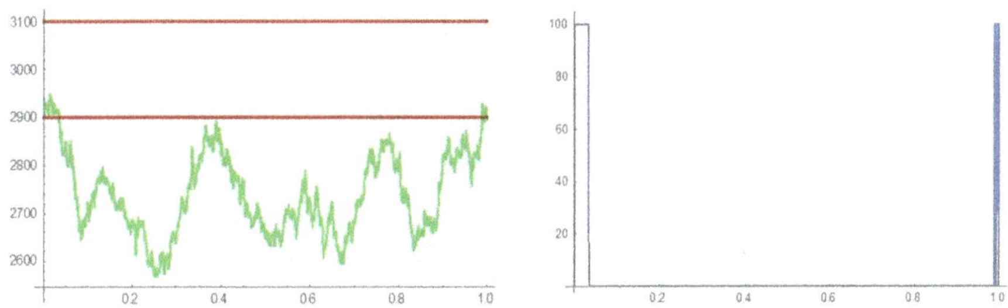

Fig. 4.195 Simulated price path of the S&P500 (green), with barriers for profit range (red, left) and payoff if exercised immediately on given path (blue, right)

Variant 1:

Final value for delta-only hedging strategy with underlying = 29.48$
traded underlying units = 0.069 units

Variant 2:

End value for delta-only hedging strategy with call option = −1.18$
traded call option units = 0.99 units

Variant 3:

Final value for delta-gamma hedging strategy = 11.73$
traded underlying units = 0.155 units

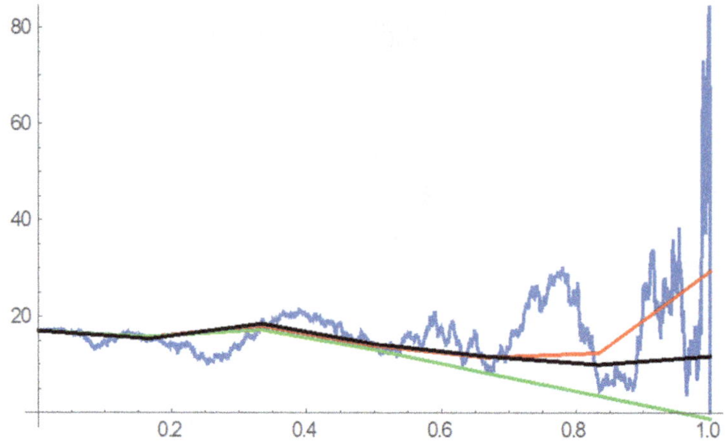

Fig. 4.196 Price movements of the derivative (blue), hedging strategy variant 1 (red), hedging strategy variant 2 (green), and hedging strategy variant 3 (black) with the above price movements of the S&P500

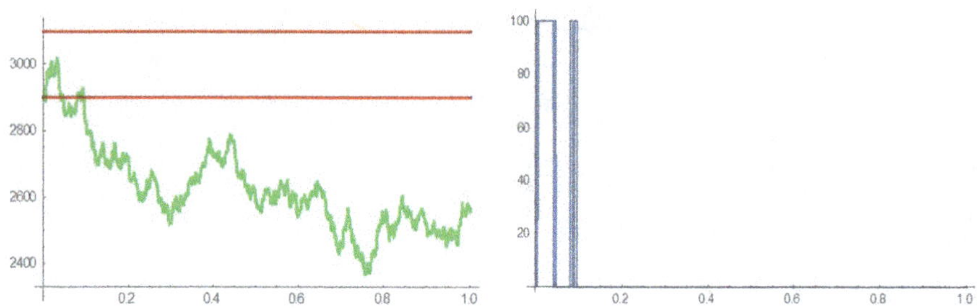

Fig. 4.197 Simulated price path of the S&P500 (green), with barriers for profit range (red, left) and payoff if exercised immediately on given path (blue, right)

traded call option units = 0.962 units

Scenario 3: Sharp decline in S&P500 price (see Figs. 4.197 and 4.198):

Derivative payoff = 0 dollars

Variant 1:

Final value for delta-only hedging strategy with underlying = 8.75$
traded underlying units = 0.072 units

Variant 2:

Final value for delta-only hedging strategy with call option = −0.17
traded call option units = 1.31 units

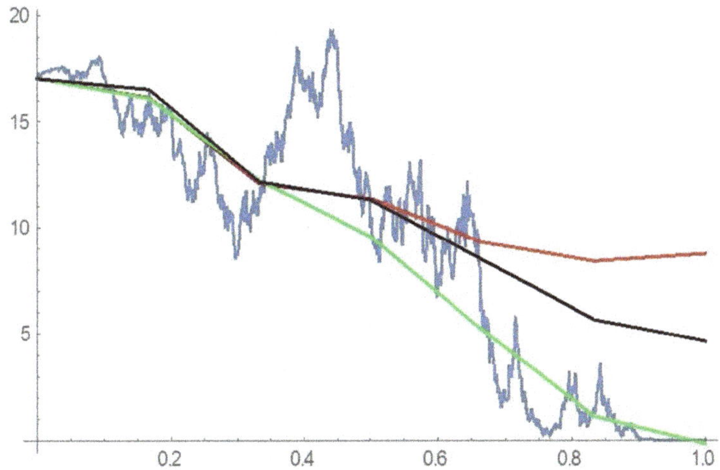

Fig. 4.198 Price movements of the derivative (blue), hedging strategy variant 1 (red), hedging strategy variant 2 (green), and hedging strategy variant 3 (black) with the above price movements of the S&P500

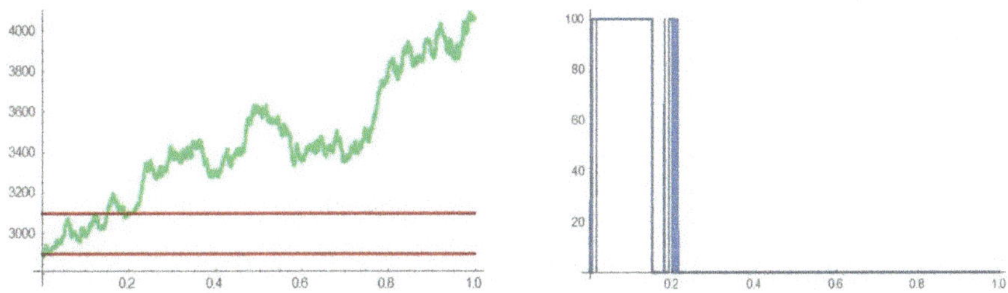

Fig. 4.199 Simulated price path of the S&P500 (green), with barriers for profit range (red, left) and payoff if exercised immediately on given path (blue, right)

Variant 3:

Final value for delta-gamma hedging strategy = 4.65$
traded underlying units = 0.123 units
traded call option units = 1.284 units

Test examples for **daily adjustment**:

Scenario 1: Strong increase in the S&P500 price (see Figs. 4.199 and 4.200):
Derivative payoff = 0 dollars

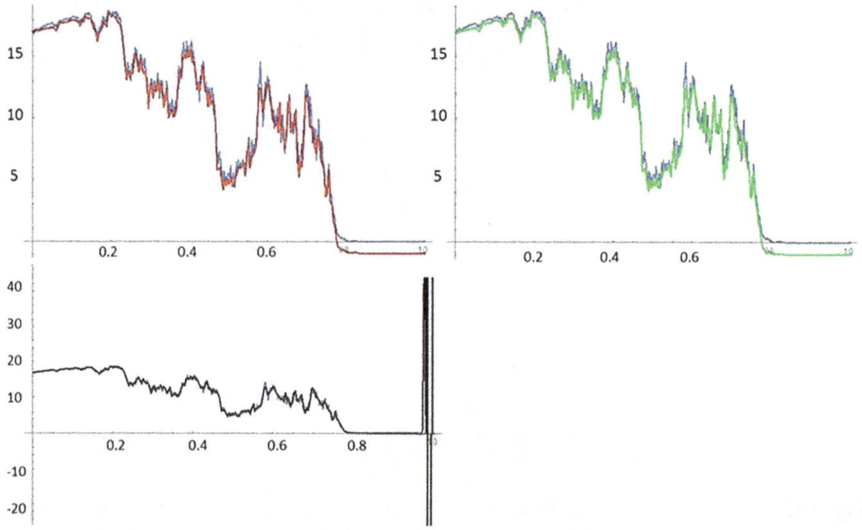

Fig. 4.200 Price movements of the derivative (blue), hedging strategy variant 1 (red), hedging strategy variant 2 (green), and hedging strategy variant 3 (black) with the above price movements of the S&P500

Variant 1:

Final value for delta-only hedging strategy with underlying $= -0.94\$$
traded underlying units = 0.419 units

Variant 2:

Final value for delta-only hedging strategy with call option $= -0.99$
traded call option units = 0.480 units

Variant 3:

Final value for delta-gamma hedging strategy $= 1.59 \times 10^{88}\$$
traded underlying units $= 2.93 \times 10^{100}$ units
traded call option units $= 2.93 \times 10^{100}$ units

Scenario 2: S&P500 in critical range close to expiry (see Figs. 4.201 and 4.202):
Derivative payoff = 100 dollars
Variant 1:

Final value for delta-only hedging strategy with underlying = 115.35$
traded underlying units = 3.913 units

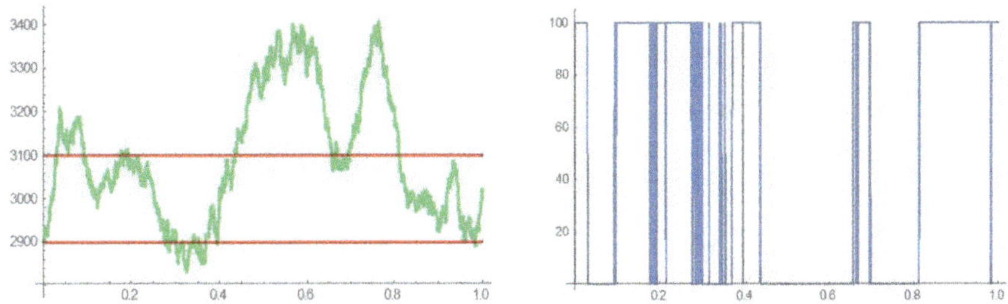

Fig. 4.201 Simulated price path of the S&P500 (green), with barriers for profit range (red, left) and payoff if exercised immediately on given path (blue, right)

Variant 2:

Final value for delta-only hedging strategy with call option = 133.82$
traded call option units = 7.21 units

Variant 3:

Final value for delta-gamma hedging strategy = 108.78$
traded underlying units = 26.01 units
traded call option units = 30.19 units

Scenario 3: Sharp decline in S&P500 price (see Figs. 4.203 and 4.204):
Derivative payoff = 0 dollars
Variant 1:

Final value for delta-only hedging strategy with underlying = 0.64$
traded underlying units = 0.514 units

Variant 2:

Final value for delta-only hedging strategy with call option = -0.94$
traded call option units = 59.34 units

Variant 3:

Final value for delta-gamma hedging strategy = 0.01$
traded underlying units = 0.497 units
traded call option units = 60.085 units

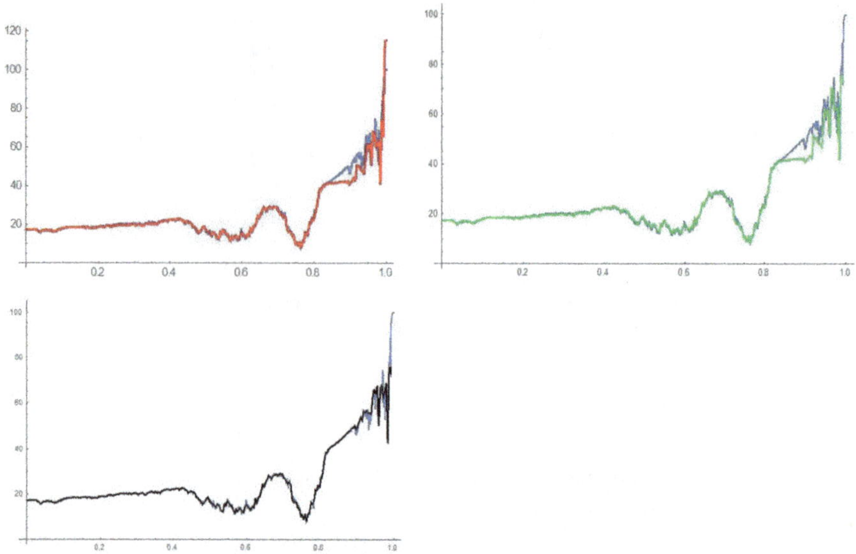

Fig. 4.202 Price movements of the derivative (blue), hedging strategy variant 1 (red), hedging strategy variant 2 (green), and hedging strategy variant 3 (black) with the above price movements of the S&P500

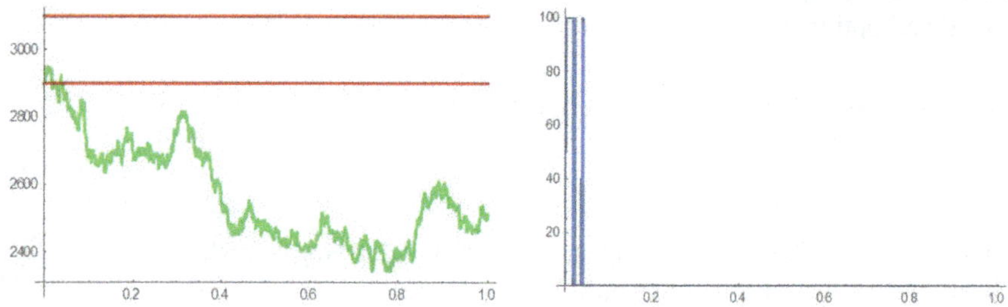

Fig. 4.203 Simulated price path of the S&P500 (green), with barriers for profit range (red, left) and payoff if exercised immediately on given path (blue, right)

Comments

- As noted earlier, the above is purposely not a comprehensive analysis of delta (gamma) hedging techniques. We think that even the few examples listed above show some of the possibilities and limitations of these hedging methods. Especially in the case of delta-gamma hedging (when proceeding strictly "by the rules"), enormous problems can arise with regard to the number of hedging instruments to be traded and with regard to price fluctuations (see e.g. daily adjustment, scenario 1).
- The techniques that we have presented here as examples are basic techniques. The examples show that in some cases, adaptations could certainly be helpful.

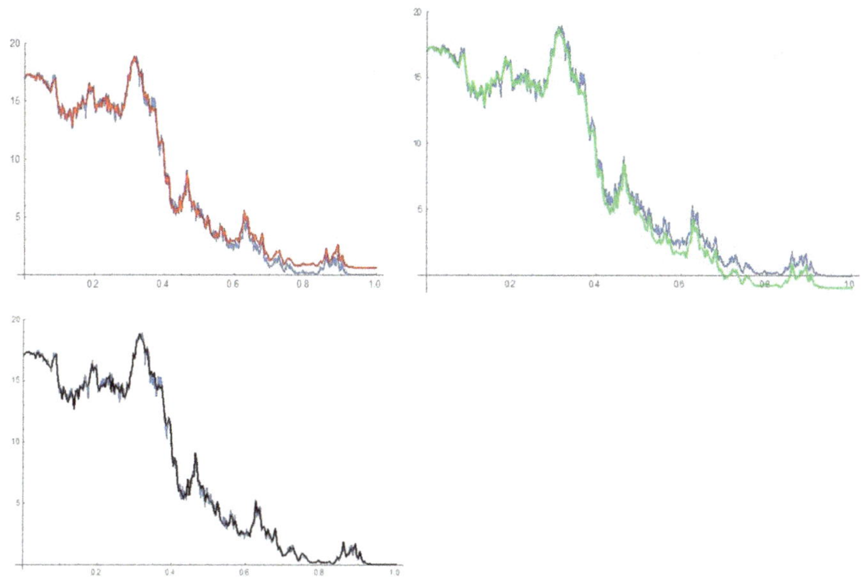

Fig. 4.204 Price movements of the derivative (blue), hedging strategy variant 1 (red), hedging strategy variant 2 (green) and hedging strategy variant 3 (black) with the above price movements of the S&P500

These include, for example, using shorter rebalancing intervals in critical situations, changing the strike of the call option used for hedging, and switching from delta-gamma hedging to delta-only hedging in certain situations.

- If the S&P500 price moves such that it is at a significant distance from the critical limits 2900 and 3100 shortly before the derivative expires, then the hedging results (especially in delta-only hedging strategies with underlying asset or call option) will be quite good even if adjustments are made only every 2 months. In addition, delta hedging requires only relatively small trades.
- If, toward the end of the derivative's life, the S&P500 is close to the limits at which a payoff becomes payable, we see (again especially in pure delta hedging with underlying or call option) that a higher rebalancing frequency generally leads to significantly better results.
- In some cases, delta-gamma hedging requires trades of exorbitantly high volumes of hedging instruments. This is especially the case (in our example) when the underlying asset experiences a strong price increase. The reason for this is obvious: As an example, the number of underlying units that need to be traded is given by $x(t, s) = \Gamma_F \cdot \frac{\Delta_B}{\Gamma_B}$. In this case, Γ_F is the gamma of the derivative D, Δ_B is the delta of the call option, and Γ_B is the gamma of the call option. If the underlying price rises sharply and t approaches expiration date T, then Δ_B will tend toward 1, the (positive) Γ_B will tend toward 0, and thus $\frac{\Delta_B}{\Gamma_B}$ will tend toward

plus infinity. Although the value Γ_F will tend toward 0, the value $\Gamma_F \cdot \frac{\Delta_B}{\Gamma_B}$ and thus the value $x(t, s)$ can still fluctuate very strongly.

References

1. Friedrich Schmid and Markus Trede. *Finanzmarktstatistik*. Springer, 2006.
2. Fisher Black. How we came up with the option formula. *Journal of Portfolio Management*, 15(2):4–8, 1989.
3. Darrell Duffie. Black, Merton and Scholes—their central contributions to economics. *The Scandinavian Journal of Economics*, 100:411–423, 1998.
4. Fischer Black and Myron Scholes. The pricing of options and corporate liabilities. *Journal of Political Economy*, 81:637–654, 1973.

www.ingramcontent.com/pod-product-compliance
Lightning Source LLC
Chambersburg PA
CBHW060948280126
38801CB00003B/115
9783031238727